edition

Discovering GIS and ArcGIS Pro

Bradley A. Shellito

Youngstown State University

Austin • Boston • New York • Plymouth

DEDICATION

This book is dedicated to my parents, David and Elizabeth, whose encouragement showed me how to roll the dice and make them come up sevens.

VICE PRESIDENT, STEM: Daryl Fox
EDITORIAL PROGRAM DIRECTOR: Andrew Dunaway
SENIOR PROGRAM MANAGER: Jennifer Edwards
DEVELOPMENTAL EDITOR: Andy Newton
EDITORIAL ASSISTANT: Nathan Livingston
MARKETING MANAGER: Leah Christians
MARKETING ASSISTANT: Madeleine Inskeep
MEDIA PROJECT MANAGER: Dan Comstock
DIRECTOR, CONTENT MANAGEMENT ENHANCEMENT: Tracey Kuehn
SENIOR MANAGING EDITOR: Lisa Kinne
SENIOR CONTENT PROJECT MANAGER: Harold Chester
SENIOR WORKFLOW PROJECT MANAGER: Paul Rohloff
PRODUCTION SUPERVISOR: Robert Cherry
SENIOR PROJECT MANAGER: Misbah Ansari, Lumina Datamatics, Inc.
PERMISSIONS MANAGER: Jennifer MacMillan
SENIOR PHOTO EDITOR: Sheena Goldstein
RIGHTS AND BILLING ASSOCIATE: Alexis Gargin
DIRECTOR OF DESIGN, CONTENT MANAGEMENT: Diana Blume
PHOTO RESEARCHER: Krystyna Borgen, Lumina Datamatics, Inc.
DESIGN SERVICES MANAGER: Natasha Wolfe
COVER DESIGN: John Callahan
INTERIOR DESIGN: Lumina Datamatics, Inc.
ART MANAGER: Matthew McAdams
COMPOSITION: Lumina Datamatics, Inc.
PRINTING AND BINDING: LSC Communications
COVER PHOTO: imaginima/E+/Getty Images
TITLE PAGE AND CHAPTER OPENER PHOTO CREDITS: kloromanam/Shutterstock

ISBN-13: 978-1-319-23075-3
ISBN-10: 1-319-23075-X

Library of Congress Preassigned Control Number: 2019953635

© 2021, 2017, 2015 by W. H. Freeman and Company

All rights reserved.

Printed in the United States of America

1 2 3 4 5 6   25 24 23 22 21 20

W. H. Freeman and Company
One New York Plaza
Suite 4600
New York, NY 10004-1562
www.macmillanlearning.com

This book was not prepared, approved, or endorsed by the owners or creators of any of the software products discussed herein. The graphical user interfaces, emblems, trademarks and associated materials discussed in this book remain the intellectual property of their respective owners.

ArcGIS for Desktop, ArcGIS Pro, ArcGIS Online, Collector for ArcGIS, and The National Map.

In 1946, William Freeman founded W. H. Freeman and Company and published Linus Pauling's *General Chemistry*, which revolutionized the chemistry curriculum and established the prototype for a Freeman text. W. H. Freeman quickly became a publishing house where leading researchers can make significant contributions to mathematics and science. In 1996, W. H. Freeman joined Macmillan and we have since proudly continued the legacy of providing revolutionary, quality educational tools for teaching and learning in STEM.

About the Author

Bradley A. Shellito is a geographer whose work focuses on the application of geospatial technologies. Dr. Shellito has been a professor at Youngstown State University (YSU) since 2004 and was previously a faculty member at Old Dominion University. He teaches classes on GIS, remote sensing, GPS, and 3D visualization, and his research interests involve applying these concepts to a variety of real-world issues. His first book, *Introduction to Geospatial Technologies* was also published by Macmillan Learning. He also serves as YSU's Principal Investigator in OhioView, a statewide geospatial consortium. A native of the Youngstown area, Dr. Shellito received his bachelor's degree from YSU, his master's degree from the Ohio State University, and his doctorate from Michigan State University.

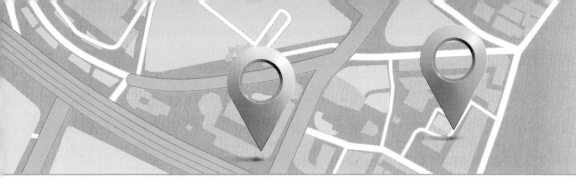

Brief Contents

Preface		xi
Module 1	How to Use Geospatial Data with ArcGIS Pro	1
Module 2	How to Use Tables, Attributes, and Queries in ArcGIS Pro	34
Module 3	How to Create a Layout with ArcGIS Pro	61
Module 4	How to Publish Layers from ArcGIS Pro and Build Web Apps with ArcGIS Online	92
Module 5	How to Obtain Online GIS Data and Use Them in ArcGIS Pro	119
Module 6	How to Create Geospatial Data with ArcGIS Pro	143
Module 7	How to Edit Data with ArcGIS Pro	168
Module 8	How to Perform Spatial Analysis in ArcGIS Pro	190
Module 9	How to Perform Geoprocessing in ArcGIS Pro	215
Module 10	How to Perform Geocoding in ArcGIS Pro	239
Module 11	How to Perform Network Analysis in ArcGIS Pro	261
Module 12	How to Use Raster Data in ArcGIS Pro	284
Module 13	How to Use Remotely Sensed Imagery in ArcGIS Pro	303
Module 14	How to Perform Spatial Interpolation with ArcGIS Pro	328
Module 15	How to Work with Digital Elevation Models in ArcGIS Pro	349
Module 16	How to Work with Contours, TINs, and 3D Imagery in ArcGIS Pro	375
Module 17	How to Work with Lidar Data in ArcGIS Pro	394
Module 18	How to Represent Geospatial Data in 3D with ArcGIS Pro	412
Module 19	How to Use Distance Calculations and Cost Distance in ArcGIS Pro	435
Module 20	How to Perform Map Algebra in ArcGIS Pro	463

Module 21	How to Build a Model in ArcGIS Pro	487
Module 22	How to Use Hydrologic Modeling Tools in ArcGIS Pro	513

Appendix A: Transitioning from ArcMap to ArcGIS Pro — 537

Appendix B: Using Coordinate Systems in ArcGIS
(online only at **http://www.macmillanlearning.com**)

Glossary — 539

Index — 548

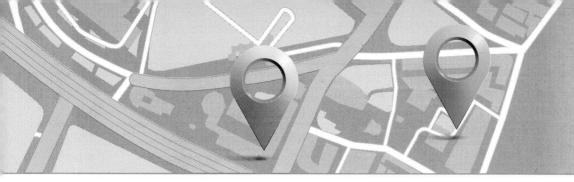

Contents

Preface · xi

Module 1
How to Use Geospatial Data with ArcGIS Pro · 1

Smartbox 1.1	What is an Esri account, and why do I need one?	5
Smartbox 1.2	What are projects and templates in ArcGIS Pro?	6
Smartbox 1.3	How can I access other folders and data through the Catalog pane?	11
Smartbox 1.4	What are the types of vector data, and how are they referenced in ArcGIS Pro?	12
Smartbox 1.5	What is a map, and how does it work in ArcGIS Pro?	13
Smartbox 1.6	How do I display data layers in the Contents pane in ArcGIS Pro?	16
Smartbox 1.7	What are datums, coordinate systems, and projections in ArcGIS?	17
Smartbox 1.8	How do I change projections in ArcGIS Pro?	20
Smartbox 1.9	How do I define a projection in ArcGIS Pro?	23
Smartbox 1.10	How do I save my work as a package in ArcGIS Pro?	29
Troublebox 1.1	What happened to the panes and views I was using in ArcGIS Pro?	9
Troublebox 1.2	Why don't I have access to my data in ArcGIS Pro?	10
Troublebox 1.3	What happened to the map I was using in ArcGIS Pro?	14
Troublebox 1.4	How can I retrieve the symbol options?	28

RELATED CONCEPTS FOR MODULE 1 ArcGIS Pro, Python, and Arcade · 32

Module 2
How to Use Tables, Attributes, and Queries in ArcGIS Pro · 34

Smartbox 2.1	What are the components of an attribute table in ArcGIS Pro?	39
Smartbox 2.2	What are the different types of GIS attribute data in ArcGIS Pro?	41
Smartbox 2.3	How are tables joined in ArcGIS Pro?	43
Smartbox 2.4	What does *selection* mean in ArcGIS Pro?	46
Smartbox 2.5	What is the structure of a simple query in ArcGIS Pro?	48
Smartbox 2.6	What is the structure of a compound query in ArcGIS Pro?	51
Smartbox 2.7	What is an attachment, and how is it used in ArcGIS Pro?	57

RELATED CONCEPTS FOR MODULE 2 Using U.S. Census Data and Attributes in ArcGIS Pro · 59

Module 3
How to Create a Layout with ArcGIS Pro · 61

Smartbox 3.1	What are normalized data, and how are they used in ArcGIS Pro?	65
Smartbox 3.2	What are the differences among the data classification methods in ArcGIS Pro?	67
Smartbox 3.3	Why is color choice important when designing a layout in ArcGIS Pro?	72
Smartbox 3.4	What is a layout, and what are map elements in ArcGIS Pro?	73
Smartbox 3.5	How is scale represented on a layout in ArcGIS Pro?	78
Smartbox 3.6	Why is the choice of fonts important when designing a layout in ArcGIS Pro?	87
Smartbox 3.7	What are some strategies for designing an effective map in ArcGIS Pro?	88
Troublebox 3.1	What happened to the layout I was using in ArcGIS Pro?	75
Troublebox 3.2	Why can't I view or select the map element that I want in the ArcGIS Pro layout?	82

RELATED CONCEPTS FOR MODULE 3 Map Labels and Annotation · 90

Module 4
How to Publish Layers from ArcGIS Pro and Build Web Apps with ArcGIS Online · 92

Smartbox 4.1	How are my data and maps shared to ArcGIS Online?	99

Smartbox 4.2	What are web maps, and how are they built in ArcGIS Online?	103
Smartbox 4.3	What kinds of basemaps are available in ArcGIS Online?	106
Smartbox 4.4	What are web apps, and how are they built in ArcGIS Online?	109
Troublebox 4.1	How do I set up my Esri account to access ArcGIS Online organization-level features?	98
Troublebox 4.2	How can I successfully publish my web layers?	102
Troublebox 4.3	How can I return to configuring my web app?	113
RELATED CONCEPTS FOR MODULE 4 ArcGIS StoryMaps		116

Module 5
How to Obtain Online GIS Data and Use Them in ArcGIS Pro 119

Smartbox 5.1	What are The National Map and the National Geospatial Program?	122
Smartbox 5.2	How does the Locate tool operate in ArcGIS Pro?	130
Smartbox 5.3	What kinds of data are available through the ArcGIS Online portal?	132
Smartbox 5.4	What is VGI, and how can it be used in ArcGIS Pro?	136
Smartbox 5.5	What other GIS data are available online?	138
RELATED CONCEPTS FOR MODULE 5 How Metadata Are Used in GIS		141

Module 6
How to Create Geospatial Data with ArcGIS Pro 143

Smartbox 6.1	Why is scale important when digitizing?	148
Smartbox 6.2	What is a geodatabase, and how does it store geospatial data?	149
Smartbox 6.3	What other data formats can be used in ArcGIS Pro?	150
Smartbox 6.4	How is digitizing performed in ArcGIS Pro?	157
Troublebox 6.1	Is there any better-quality imagery available to use as a basemap for digitizing?	147
Troublebox 6.2	How can ArcGIS Pro properly work with Microsoft Excel tables?	151
Troublebox 6.3	What if some templates are missing from the Create Features pane?	156
RELATED CONCEPTS FOR MODULE 6 Using GNSS and Mobile GIS Apps for Data Creation		165

Module 7
How to Edit Data with ArcGIS Pro 168

Smartbox 7.1	What is temporal accuracy?	172
Smartbox 7.2	What is positional accuracy?	176
Smartbox 7.3	How do the snapping options function in ArcGIS Pro?	178
Smartbox 7.4	What is attribute accuracy?	182
Smartbox 7.5	What is completeness?	185
Smartbox 7.6	What is logical consistency?	185
RELATED CONCEPTS FOR MODULE 7 Topology and Topological Editing in ArcGIS Pro		188

Module 8
How to Perform Spatial Analysis in ArcGIS Pro 190

Smartbox 8.1	How are Summarize tools used in ArcGIS Pro?	198
Smartbox 8.2	How are spatial joins used in ArcGIS Pro?	200
Smartbox 8.3	How does Select Layer By Location work in ArcGIS Pro?	206
Troublebox 8.1	Why doesn't the Measure tool snap to a point?	197
RELATED CONCEPTS FOR MODULE 8 Analysis of Patterns and Clusters in ArcGIS Pro		213

Module 9
How to Perform Geoprocessing in ArcGIS Pro 215

Smartbox 9.1	How can multiple layers be merged in ArcGIS Pro?	220
Smartbox 9.2	How does Dissolve operate in ArcGIS Pro?	223
Smartbox 9.3	How does Clip operate in ArcGIS Pro?	225
Smartbox 9.4	How do buffers operate in ArcGIS Pro?	226
Smartbox 9.5	What are the different types of polygon overlays in ArcGIS Pro?	229
Smartbox 9.6	How is a point-in-polygon overlay performed?	233
RELATED CONCEPTS FOR MODULE 9 Workflows and Tasks in ArcGIS Pro		236

Module 10
How to Perform Geocoding in ArcGIS Pro 239

Smartbox 10.1	How is a streets layer used as a reference database for geocoding in ArcGIS Pro?	243
Smartbox 10.2	What is an address locator and how does it work in ArcGIS Pro?	244
Smartbox 10.3	How does the geocoding process work in ArcGIS Pro?	248
Smartbox 10.4	Why are geocoded results sometimes inaccurate in ArcGIS Pro?	254

Troublebox 10.1	How can I return to the Rematch Addresses pane in ArcGIS Pro?	251

RELATED CONCEPTS FOR MODULE 10 Geocoding Services and ArcGIS Pro 259

Module 11
How to Perform Network Analysis in ArcGIS Pro — 261

Smartbox 11.1	What is an extension in ArcGIS Pro and how is it used?	265
Smartbox 11.2	What are the elements of a network dataset in ArcGIS Pro?	266
Smartbox 11.3	How is a shortest route calculated in ArcGIS Pro?	269
Smartbox 11.4	How does reordering stops affect the shortest routes in ArcGIS Pro?	273
Smartbox 11.5	What are service areas and how are they used in ArcGIS Pro?	277
Smartbox 11.6	What is the closest facility function, and how does it work in ArcGIS Pro?	279
Troublebox 11.1	How can I activate an extension in ArcGIS Pro?	265

RELATED CONCEPTS FOR MODULE 11 Utility Networks in ArcGIS Pro 282

Module 12
How to Use Raster Data in ArcGIS Pro — 284

Smartbox 12.1	How do environment settings affect analysis and usage of raster data in ArcGIS Pro?	287
Smartbox 12.2	What does a raster dataset consist of?	289
Smartbox 12.3	What are zones of raster data in ArcGIS Pro?	290
Smartbox 12.4	What is the National Land Cover Dataset?	291
Smartbox 12.5	How is a raster's attribute table set up in ArcGIS Pro?	293
Smartbox 12.6	What is a region in raster data?	294
Smartbox 12.7	How do raster data compare with vector data?	296
Smartbox 12.8	How does resolution affect the raster dataset?	299

RELATED CONCEPTS FOR MODULE 12 Using Subsets and Mosaics of Raster Data 300

Module 13
How to Use Remotely Sensed Imagery in ArcGIS Pro — 303

Smartbox 13.1	What is orthoimagery?	306
Smartbox 13.2	How is visual image interpretation performed using remotely sensed imagery?	309
Smartbox 13.3	What is NAIP imagery?	311
Smartbox 13.4	What is spatial resolution?	312
Smartbox 13.5	What is a remotely sensed image actually showing?	313
Smartbox 13.6	What is the Landsat program, and what are its capabilities?	319
Smartbox 13.7	What satellites have high-resolution capabilities?	324
Troublebox 13.1	Why is the imagery in grayscale (black and white) instead of color?	307

RELATED CONCEPTS FOR MODULE 13 Georeferencing an Image 326

Module 14
How to Perform Spatial Interpolation with ArcGIS Pro — 328

Smartbox 14.1	How does inverse distance weighted (IDW) interpolation function?	335
Smartbox 14.2	How do the zonal tools operate in ArcGIS Pro?	340
Smartbox 14.3	How do Spline functions operate in spatial interpolation?	342

RELATED CONCEPTS FOR MODULE 14 Geostatistical Methods for Spatial Interpolation 346

Module 15
How to Work with Digital Elevation Models in ArcGIS Pro — 349

Smartbox 15.1	What is a scene in ArcGIS Pro and how does it operate with 3D visualization?	354
Smartbox 15.2	How can I navigate in the scene environment in ArcGIS Pro?	356
Smartbox 15.3	What is the 3D Elevation Program?	357
Smartbox 15.4	What is a hillshade?	359
Smartbox 15.5	What do slope and aspect measure?	360
Smartbox 15.6	What is visibility analysis?	367

RELATED CONCEPTS FOR MODULE 15 Interactive 3D Visibility Analysis 373

Module 16
How to Work with Contours, TINs, and 3D Imagery in ArcGIS Pro — 375

Smartbox 16.1	How can contour lines be used in ArcGIS Pro?	379
Smartbox 16.2	How is a TIN created in ArcGIS Pro?	381
Smartbox 16.3	What is vertical exaggeration?	386
Smartbox 16.4	How can animations be used in ArcGIS Pro?	387

RELATED CONCEPTS FOR MODULE 16 USGS Topographic Maps and US Topos 391

Module 17
How to Work with Lidar Data in ArcGIS Pro 394

- **Smartbox 17.1** What is a LAS file, and how is it used in ArcGIS Pro? 397
- **Smartbox 17.2** How are LAS classification values used in ArcGIS Pro? 401
- **Smartbox 17.3** What do lidar measurements represent? 403
- **Troublebox 17.1** What happened to the reclassified lidar points? 403
- **Troublebox 17.2** How do I restore my lidar measurements to points? 408
- RELATED CONCEPTS FOR MODULE 17 Terrain Datasets in ArcGIS Pro 410

Module 18
How to Represent Geospatial Data in 3D with ArcGIS Pro 412

- **Smartbox 18.1** What is extrusion, and how does it work in ArcGIS Pro? 416
- **Smartbox 18.2** What is a multipatch and how is it used in ArcGIS Pro? 418
- **Smartbox 18.3** What is a web scene in ArcGIS Pro? 419
- **Smartbox 18.4** How do I work with a web scene in ArcGIS Online? 422
- **Smartbox 18.5** What is COLLADA and how is it used with 3D visualization in ArcGIS Pro? 424
- **Smartbox 18.6** How does an offset function in ArcGIS Pro? 427
- **Smartbox 18.7** How can 3D objects be represented in ArcGIS Pro? 430
- **Troublebox 18.1** How can I successfully publish a web scene? 421
- RELATED CONCEPTS FOR MODULE 18 Working with More Detailed 3D Models in ArcGIS Pro 434

Module 19
How to Use Distance Calculations and Cost Distance in ArcGIS Pro 435

- **Smartbox 19.1** How is Euclidean distance calculated and used in ArcGIS Pro? 440
- **Smartbox 19.2** What is allocation, and how is it used in ArcGIS Pro? 442
- **Smartbox 19.3** How is cost distance used in ArcGIS Pro? 444
- **Smartbox 19.4** How is a cost path calculated in ArcGIS Pro? 449
- **Smartbox 19.5** How does Slice work with raster layers in ArcGIS Pro? 453
- **Smartbox 19.6** How are raster layers reclassified in ArcGIS Pro? 456

- RELATED CONCEPTS FOR MODULE 19 Using Path Distance in ArcGIS Pro 462

Module 20
How to Perform Map Algebra in ArcGIS Pro 463

- **Smartbox 20.1** What is ModelBuilder and how is it used in ArcGIS Pro? 400
- **Smartbox 20.2** What is a relational operator and how is it used in map algebra? 472
- **Smartbox 20.3** How are more complex expressions built using map algebra? 475
- **Smartbox 20.4** How are rasters overlaid with Boolean operators using map algebra? 480
- **Troublebox 20.1** Why can't I reopen my saved model? 469
- RELATED CONCEPTS FOR MODULE 20 Using Weighted Raster Overlay 484

Module 21
How to Build a Model in ArcGIS Pro 487

- **Smartbox 21.1** What are model elements and how are they represented in ModelBuilder? 491
- **Smartbox 21.2** What are model parameters and how are they used in ArcGIS Pro? 492
- **Smartbox 21.3** What is a suitability index? 505
- RELATED CONCEPTS FOR MODULE 21 Advanced ModelBuilder Options 511

Module 22
How to Use Hydrologic Modeling Tools in ArcGIS Pro 513

- **Smartbox 22.1** How does flow direction operate in ArcGIS Pro? 519
- **Smartbox 22.2** What are sinks in hydrologic modeling? 521
- **Smartbox 22.3** How does flow accumulation work in ArcGIS Pro? 525
- **Smartbox 22.4** How can a stream network be extracted in ArcGIS Pro? 526
- **Smartbox 22.5** How does stream ordering work in ArcGIS Pro? 529
- **Troublebox 22.1** How do I build a raster attribute table for a raster layer? 522
- RELATED CONCEPTS FOR MODULE 22 Using Pour Points in ArcGIS Pro 536

Appendix A: Transitioning from ArcMap to ArcGIS Pro 537
Appendix B: Using Coordinate Systems in ArcGIS (online only at **http://www.macmillanlearning.com**)
Glossary 539
Index 548

Preface

Why I Wrote *Discovering GIS and ArcGIS Pro*

My approach to teaching GIS courses involves two goals for the students. First, they should be able to do hands-on work with GIS software. Second, and more importantly, they should know the "how" and "why" behind what they're doing. Software changes quickly; the theory has a longer shelf life. The goal of *Discovering GIS and ArcGIS Pro* is to teach students how to combine GIS concepts with ArcGIS Pro software skills. Students learn to use the software, apply it to real-world tasks, and discover why they are doing things. They learn background and theory subjects, when appropriate, as they work through the hands-on application of ArcGIS Pro in each module.

For instance, when students are constructing a map and choosing a data classification method, an accompanying theory-based "Smartbox" provides background on the different classification methods, how each method classifies data, and which method should be used based on the distribution of the data. Similarly, when students are preparing to publish their data to ArcGIS Online as a web layer, an accompanying Smartbox provides background and information on publishing; the difference between feature layers, tile layers, and vector tile layers; and how these items are set up and used. In this way, students encounter the theory as they work with the software. In short: Rather than read pages of background before getting to the hands-on application, students encounter the underlying concepts and explanations while working with the software. Students learn while doing, gaining an understanding of GIS as they are implementing GIS concepts with ArcGIS Pro.

When using a workbook type of text, students all too often click all the right buttons, input the correct values, and end up with the correct answers without having a clear idea of why they were clicking those buttons or using those particular settings. For instance, students can turn a set of two-dimensional building footprints into a 3D representation of the structures by clicking the mouse a few times and entering values in an ArcGIS dialog. Similarly, they can open the Create TIN geoprocessing tool, select some inputs, click Run, and get a TIN (triangulated irregular network) to work in ArcGIS Pro without getting an understanding of what a TIN is or how to create one properly. The goal of *Discovering GIS and ArcGIS Pro* is to provide step-by-step instructions that allow students to use ArcGIS for real-world tasks and applications while also helping them understand the "hows" and "whys" behind their actions.

The target audience for *Discovering GIS and ArcGIS Pro* include introductory and advanced GIS courses or applied GIS courses. Colleges and universities often offer more than one GIS course—such as one at an introductory level and another at an advanced level. *Discovering GIS and ArcGIS Pro* would be appropriate for one or both courses. Within the text, instructors will be able to find the topics they want to present whether they are teaching a single intro-level course, a single advanced or applied course, or a two-course sequence. For instance, at Youngstown State University, I teach a two-course upper-division GIS sequence of classes and teach Modules 1 through 11 in the first class (focusing on GIS basics, working with GIS data, and spatial analysis methods) and Modules 12 through 22 in the second class (focusing on raster data, surfaces and visualization, and building models).

Each part of the book is called a *module* because the book is designed to be as modular as possible. From these 22 modules of material, instructors are able to pick and choose the topics they wish to present in a class, particularly when using *Discovering GIS and ArcGIS Pro* for a single course. This buffet-style approach allows instructors to select the topics they wish to cover in one class. The modules are largely designed to each stand alone, so that if instructors teach working with raster data in Module 12 and then teach working with DEMs in Module 15, the students won't be lost in methods or techniques if the intervening modules are not presented to them.

Through its generous software promotion program, Esri makes it easy for students to access ArcGIS Pro for educational purposes. Instructors at schools with an Esri site license can request a free version of ArcGIS Pro to distribute to their students. (The online request form is at **https://www.esri.com/en-us/landing-page/lp/education-promo**.) This is a full version of ArcGIS Pro with all the extensions and functions used in this book (and it also comes with a full version of ArcMap/ArcGIS Desktop) that students can install on their own desktop or laptop computers. Instructors can request either a DVD to distribute or a code for students to use in downloading the software. The methods used in this book are compatible with ArcGIS Pro 2.4 or higher.

This book also exclusively uses ArcGIS Pro and ArcGIS Online. For nearly 20 years, ArcGIS Desktop has been the standard software product from Esri for working with GIS data and analysis. For a long time, ArcMap was the standard component of ArcGIS Desktop for this type of work, but ArcGIS Pro is the new standard in desktop GIS software available from Esri. ArcGIS Pro is a 64-bit application and a reinvention of the ArcGIS software for doing GIS work. ArcGIS Pro contains new and updated functionality unavailable in ArcMap, and many features work differently in the two packages. In case you're more familiar working with ArcMap (and its related components, such as ArcCatalog and ArcScene), Appendix A in this book provides a guide to transitioning from tasks and functions in ArcMap to ArcGIS Pro.

Organization

Discovering GIS and ArcGIS Pro is organized as follows:

Module 1: How to Use Geospatial Data with ArcGIS Pro. This module explains how items in the real world are represented with vector objects, shows that each object has a spatial reference (in terms of real-world geographic or projected coordinates), and explains how those real-world items can be used and viewed in ArcGIS.

Module 2: How to Use Tables, Attributes, and Queries in ArcGIS Pro. This module explains how each vector object has a series of attributes that relate to it and how those attributes can be accessed and queried for analysis.

Module 3: How to Create a Layout with ArcGIS Pro. This module explains how to use GIS data layers to make a professional-looking map in ArcGIS Pro.

Module 4: How to Publish Layers from ArcGIS Pro and Build Web Apps with ArcGIS Online. This module explains how to use GIS data layers in ArcGIS Pro to create an interactive web app using ArcGIS Online.

Module 5: How to Obtain Online GIS Data and Use Them in ArcGIS Pro. This module explains how to obtain GIS data (for free) via the Internet and then use those data in ArcGIS Pro, with a focus on obtaining data from The National Map and ArcGIS Online.

Module 6: How to Create Geospatial Data with ArcGIS Pro. This module explains how users can create their own GIS vector data to use in ArcGIS Pro.

Module 7: How to Edit Data with ArcGIS Pro. This module explains how users can edit their data (whether created or obtained) and assess the quality of those data.

Module 8: How to Perform Spatial Analysis in ArcGIS Pro. This module explains the basics of spatial analysis using ArcGIS Pro, including working with different types of spatial queries and spatial joins.

Module 9: How to Perform Geoprocessing in ArcGIS Pro. This module describes how geoprocessing and map overlay concepts are used in ArcGIS Pro and how multiple layers can be combined for analysis.

Module 10: How to Perform Geocoding in ArcGIS Pro. This module explains how to turn a spreadsheet of addresses into a point layer and use it for spatial analysis.

Module 11: How to Perform Network Analysis in ArcGIS Pro. This module explains how network analysis (as it is used in transportation studies) is performed using ArcGIS Pro.

Module 12: How to Use Raster Data in ArcGIS Pro. Whereas the previous 11 modules work with vector data, this module explains what raster data are and how these data are used in ArcGIS Pro.

Module 13: How to Use Remotely Sensed Imagery in ArcGIS Pro. This module explains what remotely sensed data (both satellite and aerial imagery) are and how they can be used in ArcGIS Pro as a raster data layer.

Module 14: How to Perform Spatial Interpolation with ArcGIS Pro. This module explains how spatial interpolation methods can create a continuous surface from a set of points.

Module 15: How to Work with Digital Elevation Models in ArcGIS Pro. This module introduces digital elevation models, which allow you to work with GIS data that have a third dimension. It examines how to use data such as slopes, hillshades, and visibility analysis.

Module 16: How to Work with Contours, TINs, and 3D Imagery in ArcGIS Pro. This module explains how TINs are built and used in ArcGIS Pro with other data sources such as contours and also how high-resolution imagery can be used with TINs and in constructing animations in ArcGIS Pro.

Module 17: How to Work with Lidar Data in ArcGIS Pro. This module introduces lidar data and how lidar points can be used to measure heights and elevations of the landscape and objects on the surface.

Module 18: How to Represent Geospatial Data in 3D with ArcGIS Pro. This module explains how structures (and other non-landscape features) can be designed and visualized in 3D in ArcGIS Pro and also published to ArcGIS Online.

Module 19: How to Use Distance Calculations and Cost Distance in ArcGIS Pro. This module introduces how distance measurements are made in ArcGIS Pro and how distance layers are treated, as well as how to use cost distances and least-cost paths.

Module 20: How to Perform Map Algebra in ArcGIS Pro. This module explains how to overlay two or more raster layers using map algebra techniques, and it also introduces how to set up workflows with ModelBuilder.

Module 21: How to Build a Model in ArcGIS Pro. This module explains how to create a separate model (using ModelBuilder) that can be shared or used as a tool in the context of developing a site suitability model.

Module 22: How to Use Hydrologic Modeling Tools in ArcGIS Pro. This module explains how hydrologic modeling tools are used in ArcGIS Pro by examining water flow, stream extraction, and watershed delineations (using ModelBuilder).

Appendix A: Transitioning from ArcMap to ArcGIS Pro. This appendix provides an overview of information for individuals familiar with the operations of ArcMap to transition to using operations in ArcGIS Pro. It also contains a guide to common ArcMap functions and how to access them in ArcGIS Pro.

Appendix B: Using Coordinate Systems in ArcGIS Pro. This appendix provides more technical and computational details on the geographic coordinate systems (GCS) and projected coordinate systems, specifically Universal Transverse Mercator (UTM) measurements and State Plane Coordinate System (SPCS) measurements. It supplements Module 1 and provides a resource for using these spatial references in other modules.

Features of Each Module

Each module presents the material using the following structure and special features:

An application-based approach. Each module focuses on using a variety of ArcGIS Pro tools in a real-world context. At the start of each module, a scenario puts the student in a particular role with a number of tasks to accomplish. These scenarios include the role of a park ranger trying to find the best overland route to rescue a group of stranded hikers (Module 19), a delivery person finding the shortest driving route between a set of libraries (Module 11), and an urban planner trying to determine the minimum height necessary for constructing an observation platform (Module 15). Because the early modules teach basic GIS and ArcGIS Pro skills using data that can be easily replicated for a local area, the scenarios used are simple. The scenarios used in later modules become more specific as the book assumes that students have mastered the basic skills and can work with more difficult or involved tasks. In addition, each module describes professions and applications in the real world that make use of the module's theory and skills. For example, a module may discuss how archeologists, geologists, biologists, civil engineers, city planners, realtors, water resource managers, or law enforcement officers can make use of GIS.

The ability to localize each module with freely available data. Some students may respond better to working with GIS applications if they get to work with local data. For instance, being able to geocode the locations of nearby libraries or determine the sites for an ecological preserve within their own county may better connect students to these tasks. Although each module's data are taken from various locations within Ohio, the data being used are freely available; more importantly, the same kind of data can be obtained for other local areas within the United States. Each module provides brief instructions on how to obtain the same kind of local data for performing the module's tasks. For instance, Module 1 uses five data layers for Mahoning County, Ohio. These layers came from a free download via The National Map and the U.S. Census, which have those same five layers available for all U.S. counties. If you want to perform the module's actions in Ingham County, Michigan, you can download the same five layers for Ingham via The National Map and the Census Bureau. All of the modules are designed such that activities can be easily replicated for other locations using free data.

Smartboxes, Troubleboxes, and Questions. The bulk of each module consists of hands-on use of ArcGIS Pro to complete the tasks involved with the module's scenario. Important key terms are highlighted in blue. Modules are divided into subheadings such as Step x.1, Step x.2, and so forth, to break up specific tasks and skills. Each module also contains "Smartboxes" that present theory (and background information) at an appropriate time to help readers understand what their actions are really doing. For instance, one of the steps in Module 2 walks the reader through the process of joining two tables in ArcGIS Pro. A nearby Smartbox succinctly describes the theory behind joining tables, what is necessary to join tables, and the rules governing an effective join. Similarly, a step in Module 1 shows the reader how to project a dataset. A nearby Smartbox provides the theory and information on what projection is, how it works, and what the end result will be.

Some modules also contain "Troubleboxes," which contain troubleshooting tips for common problems that may arise. For instance, when first working with ArcGIS Pro, students frequently close necessary windows or panes and then don't know how to get them back. A Troublebox placed in an appropriate location in Module 1 describes how to easily restore the missing windows. Similarly, in Module 6, students use an Excel table for the first time in ArcGIS Pro. A Troublebox is placed at that spot to give students the information they need to properly utilize an Excel file in ArcGIS Pro and some alternate means of accessing the data in that file if they encounter errors. I often find myself answering the same questions over and over again to get students back on track; the Troubleboxes were all created based on these student questions that commonly crop up in my classes.

Each module contains a set of questions interspersed throughout the module. There are two goals for these questions. First, they are intended to keep students on track as they move through the module, connecting the buttons and dialogs they're using to the goals of the project they are working on. (Sometimes the questions are simple questions related to the number of selected records or the size of a grid cell.) Second, many questions ask students to think about what they're doing, such as evaluating the output of a shortest path calculation or figuring out why an address didn't geocode properly. By pausing to answer these questions, students will be more connected to the activities they're working on. In addition, the end of each module asks students to present their results either in the form of a map layout, a shared web map or app on ArcGIS Online, or (in some cases) an interactive web scene or a video file.

Related Concepts for each module. There are many topics in GIS—and for each one there is more information than can be covered in a single module without making that module seem like a very contrived cookbook. The Related Concepts section at the end of each module presents important information about related topics that weren't covered in the module. For example, in Module 4, students create a web map using ArcGIS Online. The Related Concepts section for Module 4 describes how to create ArcGIS StoryMaps, which are more in-depth web apps related to the ones students created in the module. Similarly, in Module 15, students work with visibility analysis using viewsheds and lines of sight. The Related Concepts section for Module 15 describes new techniques of working with interactive visibility analysis techniques such as view domes. The Related Concepts section can be presented to the class as part of a lecture; alternatively, students may choose to read these sections on their own for further expanded information about the module's topics.

New for this edition. Several new updates, both small and large, have been made since the second edition, and many new features have been added to this third edition of *Discovering GIS and ArcGIS Pro*. First and foremost, the book has been redone from the ground up using ArcGIS Pro rather than ArcMap, which was used in the first two editions. As noted previously, ArcGIS Pro is the new standard in desktop GIS from Esri, and the book has been updated to utilize it in each module. However, there is not a simple one-to-one correspondence between ArcMap and ArcGIS Pro in terms of features and functions; in several cases, new methods and approaches were utilized in these modules to take advantage of new concepts and techniques. For instance, Modules 3 and 4 (on making a map layout and publishing a web application, respectively) were remade to contain different types of approaches using ArcGIS Pro techniques. Notably, Modules 13, 16, and 18 received rewrites to utilize the new ArcGIS Pro approaches to remotely sensed imagery, animations, and 3D visualization. In addition, many new and revised Smartboxes and Related Concepts sections are included throughout the text to address these concepts. New and revised Troubleboxes throughout the book address common questions related to ArcGIS Pro and ArcGIS Online.

Online and Supplementary Materials

Instructors can download supplements to accompany *Discovering GIS and ArcGIS Pro* from the Macmillan catalog at **https://www.macmillanlearning.com/college/us/product/Discovering-GIS-and-ArcGIS-Pro/p/131923075X**. Instructors or students can download the following resources from the catalog page:

ArcGIS Pro Installation Guide: A PDF of step-by-step directions walks through installing and licensing the student version of ArcGIS Pro so that students in a GIS class can utilize ArcGIS Pro on their home desktop or laptop computers.

Instructors' Guide to ArcGIS Online: A PDF guide for instructors unfamiliar with ArcGIS Online helps navigate working with the cloud-based environment, properly setting up software licenses, and getting student Esri accounts connected to their organization.

Module datasets: The folders containing the GIS datasets used in each module can be downloaded as compressed zip files and unzipped to access the data.

Acknowledgments and Thanks

Books like this don't just spring out of thin air. I owe a great deal to several people who have provided inspiration, help, and support for what would eventually become this book:

- Jennifer Edwards and Andy Newton of Macmillan Learning for invaluable help, advice, and guidance throughout this entire project. I would also very much like to thank Andy Dunaway, Sheena Goldstein, Harold Chester, Paul Rohloff, Lisa Kinne, and Diana Blume at Macmillan Learning and Misbah Ansari, my photo researcher, Krystyna Borgen, and my copy editor, Kitty Wilson, at Lumina Datamatics for their extensive "behind the curtain" work that shaped this book into a finished product.

- Sherry Rogelberg, for great representation and advice.

- Sean Young, for an extensive expert review of the text and ArcGIS Pro methods used in this book.

- The students who contributed to the development of the YSU 3D campus model: Rob Carter, Ginger Cartright, Paul Crabtree, Jason Delisio, Sherif El Seuofi, Nicole Eve, Paul Gromen, Wook Rak Jung, Colin LaForme, Sam Mancino, Jeremy Mickler, Eric Ondrasik, Craig Strahler, Jaime Webber, Sean Welton, and Nate Wood. The 3D data utilized and presented in Module 18 come directly from their hard work.

Also, a very special thanks to all of my professors, instructors, colleagues, and mentors past and present (who are too numerous to list) from Youngstown State University, the Ohio State University, Michigan State University, Old Dominion University, OhioView, and everywhere else for the help, knowledge, notes, information, skills, and tools they've given me over the years. I am also deeply indebted to the work of Tom Allen and Nicole Eve for some of the concepts and methods used in various modules.

I would also like to thank the following colleagues, who reviewed this book:

- Shivanand Balram, Simon Fraser University
- W. B. Clapham, Jr., Cleveland State University
- Alison E. Feeney, Shippensburg University
- Matthew Fockler, Augustana College
- Arlene Guest, Naval Postgraduate School
- Reuben Heine, Augustana College
- Shelley Judge, Wooster

- John McGee, Virginia Tech
- Rafael Moreno, University of Colorado Denver
- Thomas Mueller, California University of Pennsylvania
- Nathan Niemi, University of Michigan
- Reed Perkins, Queens University of Charlotte
- Milda Vaitkus, University of Nebraska–Lincoln

I'd very much like to hear from you regarding any thoughts or suggestions you might have for the book. You can follow me on Twitter @GeoBradShellito or reach me via email at **bashellito@ysu.edu**.

Bradley A. Shellito
Youngstown, Ohio

Accessing Folders and Datasets for Each Module

Each module in this book requires you to download a folder for use within Arc-GIS Pro. To download these free folders, visit **https://www.macmillanlearning.com/college/us/product/Discovering-GIS-and-ArcGIS-Pro/p/131923075X**.

Online Updates for Each Module

ArcGIS Pro is constantly changing, and Esri makes new versions and updates available to users on an ongoing basis. This book was put together using ArcGIS Pro 2.4, but the software never stands still. If there is a significant difference between a method used in this book and changes in a newer version of ArcGIS Pro, please visit **https://www.macmillanlearning.com/college/us/product/Discovering-GIS-and-ArcGIS-Pro/p/131923075X** for downloadable updates for this book.

How to Use Geospatial Data with ArcGIS Pro

ArcGIS Pro Skills

In this module, you will learn how to do the following in ArcGIS Pro:
- Create a project using a template.
- Use the ribbon, tabs, and various panes.
- Access the Catalog pane to examine available data.
- Add point, line, and polygon GIS layers to a map.
- Change the appearance and symbology of GIS layers.
- Access the geospatial characteristics of data layers, including their coordinate system, datum, and projection information.
- Change a data layer from one projected coordinate system to another.
- Define a projection for data layers that are missing this information.
- Export a map to PDF format.
- Save a project.

Learning Outcomes

After studying this module, you should be able to:
- Demonstrate how to begin using ArcGIS Pro with a project template.
- Explain how a map is used in ArcGIS Pro.
- Explain how the Catalog pane can be used to connect to folders and move GIS data from one place to another.
- Describe the three components of the vector data model.
- Describe what a datum is.
- Explain the difference between a geographic coordinate system and a projected coordinate system.
- Explain what is saved in an ArcGIS Pro project.
- Describe what a package is and what the advantage of using a project package is.
- Describe one scripting language used to customize ArcGIS Pro.

Introduction

Here's something to think about: When a wastewater treatment plant takes in raw sewage, its goal is to separate out water from the rest of the material. The treated water flows back to local streams or rivers, and the remainder is pressed into sludge. This sewage sludge is treated and then referred to as *biosolids*, which have the look and feel of dirt, or potting soil. Often, the final destination of biosolids is a landfill or an incinerator, but under certain conditions, landowners use this treated sludge as a fertilizer (Figure 1.1). Because biosolids sometimes contain pathogens, the Environmental Protection Agency regulates their application as fertilizer.

Still, the practice of applying biosolids to the land should raise many questions, including the following:

- What are the present and past locations of the agricultural fields that have been fertilized with biosolids?
- How close is your home to an agricultural field where application is taking place?

- How close to these agricultural fields are the wells from which you get your water?
- If land application of biosolids took place in the past at a particular site, how is that land being used today?

All these questions involve examining locations (such as the past and present land-applied fields), qualities or attributes of those areas (such as the type of land use at the sites), and how one location relates to another (such as the distance from your home or water well to a land-applied field). Answering these questions involves working with large-scale location-based processes, thousands of location data points, and multiple types of data. For this reason, you're going to need a powerful tool to help you conduct your analysis.

FIGURE 1.1 Land application of biosolids on an agricultural field.

geographic information systems (GIS) Hardware and software that allow for computer-based analysis, manipulation, visualization, and retrieval of location-related data.

geospatial or **spatial** Referring to items that are tied to a specific real-world location.

geospatial technologies A term that encompasses many types of methods and techniques for the collection, analysis, modeling, and visualization of geospatial data.

Esri Environmental Systems Research Institute—the market leader in GIS software and the developer of ArcGIS.

ArcGIS A software platform developed by Esri.

ArcGIS Desktop The version of ArcGIS that runs on a personal computer.

ArcGIS Pro A 64-bit standalone GIS program that comes with ArcGIS Desktop.

Geographic information systems (GIS) are hardware and software that allow for computer-based analysis, manipulation, visualization, and retrieval of location-related data. In GIS terms, we refer to the data associated with real-world locations as **geospatial** data (or often simply as **spatial** data). GIS provides a means of solving problems and answering questions related to geospatial data and information; it is used for countless different applications (like the issues surrounding the biosolids scenario). GIS is part of a larger field of technologies, referred to as **geospatial technologies**, which encompasses many types of methods and techniques for the collection, analysis, modeling, and visualization of geospatial data.

Several types of GIS software are available today, ranging from expensive commercial products to freely available open-source alternatives. The industry leader in GIS is **Esri** (Environmental Systems Research Institute), a company based in Redlands, California. Esri was founded in 1969 and released its first version of its Arc/Info GIS software in 1982. Arc/Info has evolved into Esri's current ArcGIS software. A recent salary survey conducted by Geospatial Training Services showed that Esri's GIS products are among the most popular software packages in the industry.

ArcGIS is a software platform with utilities for working with GIS in the cloud, on mobile devices, and on personal computers. **ArcGIS Desktop**, the version of Esri's GIS software for use on personal computers, comes in three different licensing levels: Basic, Standard, and Advanced. Although all three are the same software, the Basic level contains fewer features than Standard, which in turn contains fewer features than Advanced. This book uses the Advanced version, which contains all available features.

When you work with ArcGIS Desktop, there are several subprograms you can work with, including the traditional ArcGIS components ArcMap, ArcCatalog, ArcScene, and ArcGlobe. However, this book focuses on a program installed as part of ArcGIS Desktop called **ArcGIS Pro**, the newest standalone ArcGIS program that contains the functions of all these other programs rolled into one. ArcGIS Pro is not merely an extension or add-on but is a 64-bit program that is separate from the other ArcGIS Desktop programs. If you have experience with the previous components, such as ArcMap and ArcCatalog, see Appendix A for some information that can help you transition from familiar functions in those programs to their counterparts in ArcGIS Pro.

The goal of this first module is to introduce you to some key GIS and geospatial concepts that will be used throughout the book as important building blocks for using GIS. You'll be using several functions of ArcGIS Pro to see how GIS data represent real-world items such as roads, airport locations, and water bodies. You'll also examine the geospatial characteristics of these data (such as the coordinate system in which the data are set up) and how ArcGIS Pro handles these data. Finally, you'll learn how to present these data effectively by using various symbols and colors.

Throughout the text (and in many other GIS applications you may work with), you will use many of the ArcGIS Pro skills and tools you learn in this module. "Smartboxes" within this module explain many of these primary concepts. The goal of the Smartboxes is to provide you with background on and theory related to the concepts you're working within ArcGIS Pro as you encounter them. For example, in this module, you'll be instructed to use the Define Projection tool to complete a task, and then you'll be referred to a nearby Smartbox that explains why you would use that, when you would use it, and the theoretical concepts behind defining projections.

Also throughout the text are "Troubleboxes," which help you troubleshoot common problems that come up in ArcGIS Pro. You may not need them, but they're here to help you overcome issues that may turn up in ArcGIS Pro. In this first module, there are several Troubleboxes to help keep you on track with your initial foray into using ArcGIS Pro.

Module Scenario and Applications

This module puts you in the role of a GIS analyst for a county engineer's office. Your first task will be to assemble a basic set of data for the county's many features (including airports, roads, bodies of water, and locations of county structures) within ArcGIS Pro for a presentation to the general public. You will compile this dataset with an eye to creating future presentations of this material (such as making a map of the data or creating a basic web mapping application).

The following are examples of other real-world applications of this module's theory and skills:

- An EMT operator has been assigned to give a presentation regarding the impact of current road closures on a county's hospitals. She will use ArcGIS Pro to show the locations of home injuries, the locations of hospitals, the county's road network, and the locations of road closures. She will then display these data for the presentation.

- To prepare for a presentation to the board of trustees, a manager of a local outdoor recreation area wants to examine the property boundaries between the lands he manages and the neighboring properties. He'll use ArcGIS Pro to present the rec area's property boundaries, the local road network, walking paths into and out of the area, and the locations of "No Trespassing" signs posted on the property.

- A local government planner needs to present information to a group of volunteers about the new routes the Labor Day parade will take. Using ArcGIS Pro, he'll be able to show the route boundaries, roads, and locations of emergency facilities to the volunteers.

Study Area

- For this module, you will be working with data from Mahoning County, Ohio.

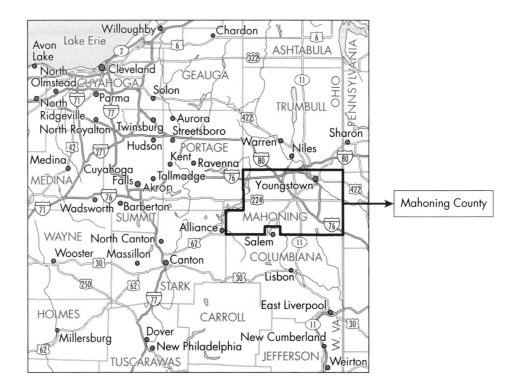

Data Sources and Localizing This Module

The data in this module focus on features and locations within Mahoning County, Ohio. However, this module can easily be modified to use data from your own county or local area instead.

The water bodies data for this chapter were downloaded from the TIGER/Line files of the Census Bureau website at **https://www2.census.gov/geo/tiger/TIGER2018/**. (We'll do more with TIGER files from the Census in Modules 2, 3, and 4.) If you want to use data for a different county, from the AREAWATER folder, download the zip file (containing the water polygons as a shapefile) for the appropriate county. The files are sorted by county FIPS code (for instance, Mahoning County, Ohio, is FIPS code 39099). For example, if you wish to use data from Ingham County, Michigan, you would download the AREAWATER file for FIPS code 26065. A full list of FIPS codes for each state and county is available from the National Weather Service at **https://www.weather.gov/pimar/FIPSCodes**. For this module, after downloading the AREAWATER shapefile, delete the .prj file and then import the shapefile into a geodatabase; doing so will remove the projection information that you will create later in this module, in Step 1.7.

The remaining data layers were downloaded (for free) from The National Map website at **https://viewer.nationalmap.gov/basic/**. Step-by-step details on downloading and extracting GIS data from The National Map are provided in Module 5.

When downloading data for another county (for instance, Ingham County, Michigan), the feature classes to use for this module are struct_point (the structures layer), trans_roadsegment (the roads layer), trans_airportpoint (the airports layer), and GU_countyorequivalent (the boundary layer).

Step 1.1 Getting Started with ArcGIS Pro

- This "Getting Started" step of each module instructs you on which data you'll be using and preliminary things you'll do during the hands-on portion of the module. Because this is your first time through this step, this module provides more details than other modules.

- This module's hands-on applications use the data folder called Module1. Your instructor will be able to supply you with this data, or you can download it directly from this book's website at **https://www.macmillanlearning.com/college/us/product/Discovering-GIS-and-ArcGIS-Pro/p/131923075X**. The text in this module assumes that you have this Module1 folder in a computer location referenced as C:\GIS; if you have it somewhere else (for instance, in a flash drive referenced as G:\GISClass), substitute that location and path to the Module1 folder throughout this module.

- Before you can get started with ArcGIS Pro, you need to obtain an Esri account. For more about Esri accounts and their importance to ArcGIS Pro, see **Smartbox 1.1**.

SMARTBOX 1.1

What is an Esri account, and why do I need one?

An Esri account is a free online account with Esri that you'll use for a variety of things, including publishing web maps through ArcGIS Online (see Module 4) and, more importantly, licensing ArcGIS Pro so you can run it properly. Before beginning work with ArcGIS Pro, you need to have a license to work with it. Licenses for running ArcGIS Pro are managed through your organization's ArcGIS Online setup.

In order to get a license, you need to acquire a free Esri account and then log in with your Esri account to allow ArcGIS Pro to verify with your organization that you have been assigned a license to run the software. If you don't have an Esri account or don't have it set up so that you can run ArcGIS Pro properly, follow these steps:

1. To get your account, point your web browser to **https://accounts.esri.com** and select the option Create a Public Account. Follow the onscreen steps to set up your Esri account.

2. Next, submit your account information to your instructor or the administrator in charge of your organization's ArcGIS Online setup. That individual will work with you to get your account established with your organization's ArcGIS Online setup and then activate your license to use ArcGIS Pro. The good news is you'll only have to do all this once. After the instructor or

6 CHAPTER 1 How to Use Geospatial Data with ArcGIS Pro

administrator has assigned a license to you, ArcGIS Pro will work just fine until the administrator chooses to take that license away from you. Note, however, that the license is tied to your Esri account.

extension An add-on set of functions for ArcGIS Pro.

In addition to needing to have a license to run ArcGIS Pro connected to your Esri account, you will want to use **extensions** in ArcGIS Pro. Extensions are add-ons that contain additional tools and features, and your ability to use them is also tied to having licenses for them connected to your Esri account. In this book, you will be using the extensions *Network Analyst*, *Spatial Analyst*, and *3D Analyst*. For ease of use, when your organization sets up your Esri account with a license to run ArcGIS Pro, licenses for these three extensions should also be activated for you as well.

ArcGIS Pro

- Start ArcGIS Pro by clicking its icon, which is a blue-and-white version of the globe.
- When ArcGIS Pro prompts you to do so, enter your Esri account username and password. If ArcGIS Pro does not prompt you, click on the **Sign In** option on the upper-right side of the ArcGIS Pro screen and then enter your username and password.
- Once you have logged in, ArcGIS Pro starts prompting you with options related to *projects*, such as asking you to open a recent project or to create a new project. If you are creating a new project, you'll be using a project *template*. For more information about projects and templates, see **Smartbox 1.2**.

SMARTBOX 1.2

What are projects and templates in ArcGIS Pro?

ArcGIS Pro treats all the maps, functions, tools, and so on that you're working with in one session as a *project*, which is a file that ends with the extension .aprx. When you start ArcGIS Pro, you can either open a previously saved **project (.aprx)** or start with a new blank project. Think of a blank project as a blank workspace to which you can add data in order to start creating maps or doing data analysis.

project (.aprx) A file that contains information about where all the data layers used in a session are located, as well as their appearance and settings.

template The ArcGIS Pro project function that establishes what will be available when a new project is created.

ArcGIS Pro allows you to choose a **template** to start with for a project—whether you want to begin with a map, the catalog, a local scene, or a global scene. (We'll explain what the map and catalog are in this module, and we'll talk about the scene options in Modules 15 and 18.) A template establishes how ArcGIS Pro will open and what options will be available to you as you begin working with the software. ArcGIS Pro also allows you to start working without a template if you don't want to work with a full-blown project but just want to do some quick analysis; however, you can save your work as a project later if you want to.

When you create a project based on a template, the folder containing that project has two additional things created for you: a file geodatabase (see

Module 6) for that project and a toolbox (see Module 20). These items can both be used to hold data layers, models, or tools that you create over the course of a project, and they enable you to keep all these items centralized with the project.

When you save your work to a project (.aprx) file, note that you are *not* saving all your work into a single file. The .aprx file contains reference information about where your data are, what format they are in, the colors used in your map, and so on. It does *not* contain the actual data you are using but just references to where your data can be found. This means that if you save your work on a GIS project, copy the .aprx file to a flash drive, go to a different computer, and then try to open the map again, it will not open properly, and you will not have your data. The layers and files you used will appear in ArcGIS Pro's Contents pane but with a red "!" next to them, indicating that the pathway to that particular layer cannot be found (or that the path to these files is broken). The project holds information about the path to the files that are located on your computer, but it does *not* hold the files themselves.

For instance, say that you save a project, and it is using all the files on your flash drive (which is mapped on the computer as a G: drive); then if you move to a different computer (which maps your flash drive as an H: drive) and attempt to reopen the files, ArcGIS Pro will be unable to load all the data. It will search for files with a pathway to a particular folder on a G: drive, but those files are now inside a folder on an H: drive. In this case, you see your files in the project's Contents pane with the red "!" symbol next to them. In essence, ArcGIS Pro knows that those files should be in that project, but it can't find the correct pathway to get to them.

When the project was saved, it understood that those files were on a G: drive. However, when the project was opened again, ArcGIS Pro does not see them on a G: drive (because the flash drive is now mapped as the H: drive on the new computer) and thus cannot find the path to those files. When you move your projects from machine to machine, you have to take all your files with you and place them on the new computer in the same path that the project is looking for; alternatively, you can rebuild the links and references stored in the .aprx file on the new machine.

- When prompted, select the **Map** option and click **OK**.
- The Create a New Project dialog box opens. Under Name, type **Module1** as the name for this project file. Under Location, tell ArcGIS Pro where you want to store this project file; in this case, click the **yellow folder button** and navigate to the **C:\GIS\Module1** folder. Do not put a checkmark next to the **Create a new folder for this project** box. When all the settings are correct, click **OK**. ArcGIS Pro now creates a project file called Module1.aprx and saves it to the C:\GIS\Module1 folder. It also creates some other things in the C:\GIS\Module1 folder, such as a geodatabase called Module1 and a toolbox called Module1. (We'll get to those items later.)

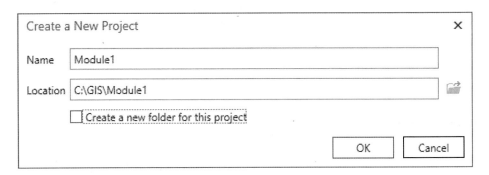

- Finally, ArcGIS Pro opens (Figure 1.2).

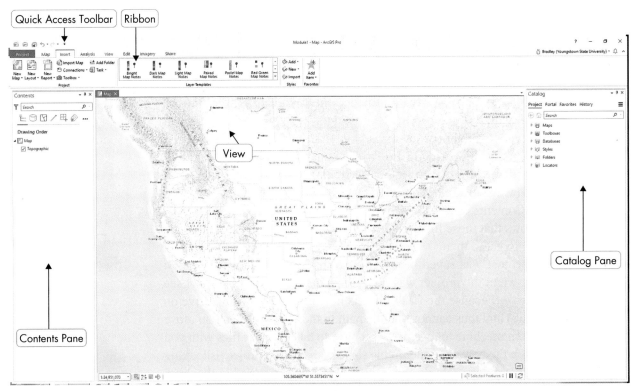

FIGURE 1.2 The basic ArcGIS Pro layout.

view The window serving as the central work area in ArcGIS Pro.

Contents pane A component of ArcGIS Pro that shows all items or layers that are being used in a project.

pane A movable and dockable window that contains items or commands used in ArcGIS Pro.

- The large area in the center of the screen is the **view**, which is where you can visualize your data and create map layouts; the view will be the central point of your work with ArcGIS Pro. When ArcGIS Pro opens, the default view is a map with a blue tab at the top of the view that says "Map." The blue of this tab indicates that the Map tab is the active view. As you add more items to the view, you'll be able to move back and forth between them by using the tabs.

- The large rectangular box on the left side of ArcGIS Pro is the **Contents pane**. It lists all layers and content available for use in the active view. As the name indicates, this box is a **pane**, or a movable window that can be resized. Right now, the Contents pane is showing items related to the Map tab as that is the currently active view (as indicated by the blue tab). If another view were active, the Contents pane would show information about that view instead.

- The large rectangular box on the right side of ArcGIS Pro is the **Catalog pane**, which you use to manage your data, available folders, tasks, and other items. Like the other panes, it's movable and can be resized as needed.

- The **ribbon** contains numerous items that you will use in ArcGIS Pro. Each tab has different groups to organize similar items together, and each group features different buttons; some buttons activate items directly, and others allow you to access pull-down menus. (See Figure 1.3 for the organization of the ribbon.)

Catalog pane A dockable window component of ArcGIS Pro that you use to manage your data.

ribbon The section of ArcGIS Pro that contains tabs, groups, and buttons for various functions.

Quick Access Toolbar The section of ArcGIS Pro that is available to perform basic functions such as opening or saving projects.

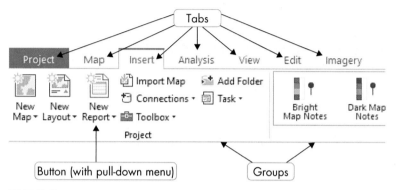

FIGURE 1.3 The basic layout of the ribbon.

- Finally, the **Quick Access Toolbar**, available above the ribbon, provides several basic functions (Figure 1.4).

- When you start to work with ArcGIS Pro, be sure to have the Catalog pane and the Contents pane available to use. See Troublebox 1.1 for information on accessing panes and views when they're not already available.

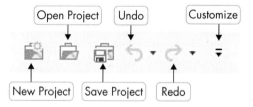

FIGURE 1.4 The functions of the Quick Access Toolbar.

TROUBLEBOX 1.1

What happened to the panes and views I was using in ArcGIS Pro?

If you close a pane or view in ArcGIS Pro, don't panic: Getting them back again is easy. From the **View** tab, within the **Windows** group, click **Catalog pane** to reopen the Catalog pane or **Contents** to reopen the Contents pane. Other windows can be accessed and reopened from this menu as well, such as the Python window or the Tasks pane (see Module 9).

- To access the data you'll be using in this module, in the Catalog pane, expand the **Folders** item, and you see that it contains a single folder called Module1, which is the folder of data you added on page 7. Expand the **Module1** folder, and you see that it contains several items:
  - A red box icon (called a *toolbox*) named Module1.tbx
  - A gray barrel icon (called a *file geodatabase*) named Module1.gdb

10 **CHAPTER 1** How to Use Geospatial Data with ArcGIS Pro

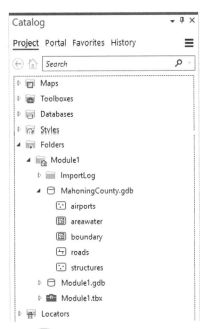

- Another file geodatabase named MahoningCounty.gdb, which you can expand to see the following GIS layers, each with its own gray icon:
  - airports: the locations of airports in the county
  - areawater: the locations of lakes and rivers in the county
  - boundary: the outline of Mahoning County
  - roads: the roads within the county
  - structures: the locations of many different kinds of buildings (schools, colleges, fire stations, bridges, and so on) in the county
- If you do not have access to the MahoningCounty.gdb file or the other items described here, see **Troublebox 1.2** for help in getting things set up correctly before proceeding.

TROUBLEBOX 1.2

Why don't I have access to my data in ArcGIS Pro?

It's important to correctly set up your project location when beginning to use ArcGIS Pro to make sure that your data and your project are together in the same place for ease of use. If you don't do so, you may begin using ArcGIS Pro with your project in one location on your computer and then become unable to see your preexisting data without taking further steps.

In this book's modules, you will begin with a folder full of data, such as Module1. On the computer's hard drive, you'll place this folder in its own dedicated location, such as C:\GIS\. Then when you create a project, you'll place it within that folder. Figure 1.5 shows an example in which a new project called Module1 is being created in the C:\GIS\Module1 folder. When the project is created, it will be in the same folder as the preexisting data being used in the project, and the project file and the datasets will all be in the same folder.

FIGURE 1.5 The Create a New Project dialog being used to keep a project called Module1 in a folder also called Module1 that contains preexisting data.

When you create a project, make sure the box next to **Create a new folder for this project** is not checked. When that box is checked, ArcGIS Pro creates a separate subfolder at that location. Your data ends up in C:\GIS\Module1, but your project has a separate C:\GIS\Module1\Module1 folder created, which throws everything off.

At this point, if you can't see both the Module1 project you've made within the Module1 folder of data, go to the **Project** tab, choose **New**, and then choose the **Map** template and begin this module again but make sure to use the proper settings in the Create a New Project dialog.

- When you created a new project, you chose the Module1 folder to place your project in. The Module1 folder already contained the MahoningCounty geodatabase (the data you'll be working with), and creating the new project also created a new empty geodatabase called Module1 and a new empty toolbox called Module1 in which you can store any new data. Right now, this is the only data you have access to for your project. For more information about how to access other folders and data to use when working with ArcGIS Pro, see **Smartbox 1.3**.

SMARTBOX 1.3

How can I access other folders and data through the Catalog pane?

In ArcGIS Pro, the Catalog pane can be used to examine all the available folders (whether local on the computer's hard drive, on a USB drive, or on a network drive) and the data they have available. You can use the **Folder Connection** function to select new folders (such as your computer's C: or D: drives, a mapped network drive, a data server, or your USB drive); simply right-click on the **Folders** option in the Catalog pane, and the Add Folder Connection button appears. By clicking this button, you can select a new folder, which will be added to your available folder options.

By expanding the folder options, you can see the data contained in the folders. You can then copy, paste, and remove items as in Windows. In this way, you can move geographic data from one location to another. Moving files in this fashion instead of using utilities like Windows Explorer ensures that any extra or ancillary files associated with your data are properly moved or deleted.

Important Note: Although you are working with the Catalog pane in this module, the same functions can also be used in the **catalog view** as well. The catalog view contains many more data management functions than simply being able to browse your data in the Catalog pane. If you had a project focused on managing your data rather than on creating maps, you would use the catalog view instead.

catalog view A view used in ArcGIS Pro to manage data.

- You probably saw each of your layers represented in the Catalog pane by an icon that either showed a set of dots, a set of lines, or two-dimensional shapes. These icons are a shorthand for how your data will be represented in GIS by using *vector data*. See **Smartbox 1.4** for more about what GIS vector data represent, as well as how each type is referenced in ArcGIS Pro.

SMARTBOX 1.4

What are the types of vector data, and how are they referenced in ArcGIS Pro?

Items from the real world are modeled or represented in GIS. For example, the Mahoning County dataset used in this module contains representations of airport locations, roads that pass through the county, and the locations and sizes of county lakes. In the **discrete object view** of the world, each of these things can be considered a different entity with a specific location, length, size, or area. In this way, the world can be observed as a series of objects. For instance, each lake in the county has a fixed boundary, just as each road has definite starting and stopping points. If there are multiple lakes in the county, then each one is considered its own object. The same is true with each leg of a county road.

Each object can be represented in ArcGIS Pro in one of three ways (Figure 1.6): as a **point** (such as an airport location), a **line** (such as a road), or a **polygon** (such as a lake). A point is a zero-dimensional object that represents just a simple set of coordinate locations. A line is a one-dimensional object created by connecting the starting and stopping points (and any points in between that give the line its shape). A polygon is a two-dimensional object that forms an area from a set of lines (or that defines an area from a line that creates a boundary).

discrete object view A conceptualization of the world in which all items can be represented by objects.

point A zero-dimensional vector object.

line A one-dimensional vector object.

polygon A two-dimensional vector object.

vector objects Points, lines, and polygons that are used to model real-world phenomena using the vector data model.

vector data model A model that represents geospatial data with a series of vector objects.

shapefile A series of files that make up one vector data layer.

file geodatabase A single folder that can contain multiple datasets, each as its own feature class.

FIGURE 1.6 Representations of the three vector objects: points, lines, and polygons.

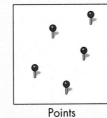 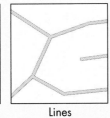 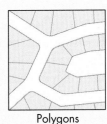

Points   Lines   Polygons

These points, lines, and polygons are referred to as **vector objects** and make up the basis of the GIS **vector data model**. By using these three vector objects, you can represent any real-world phenomena. For instance, the locations of stop signs, water wells, or groundwater testing sites can be modeled as points; streams, railroads, or power lines can be modeled as lines; and residential neighborhoods, congressional districts, or watershed boundaries can be modeled as polygons.

In ArcGIS Pro, there are different data storage formats for the vector data model (and we'll go into more depth about them in Module 6). For now, however, you need to know about only two common ones that we'll use in this book:

- A **shapefile** is a simple data structure composed of several computer files that can store points, lines (referred to as *polylines*), or polygons. A shapefile can hold only one type of data. Thus, a line shapefile would hold roads data, while a polygon shapefile would hold county boundary data.

- A **file geodatabase** is a single folder that can hold one or more feature classes, and each feature class can hold a different type of data. For instance, a line feature class could hold walking path data, while a point feature class could be used to mark a trailhead.

In this module, the MahoningCounty.gdb file that you are using is a file geodatabase, and each of the layers you're using is a feature class. The airports and structures layers are point feature classes, the roads layer is a line feature class, and the boundary and areawater layers are polygon feature classes.

You'll be working exclusively with vector data in Modules 1 through 11 of this book. There is a second type of data storage, called *raster data*, that is completely different from vector data. You'll begin incorporating raster data into ArcGIS Pro starting in Module 12.

Step 1.2 Working with Maps in a Project

- It's time to add data layers to ArcGIS Pro to start making a map and doing GIS analysis. In order to do so, you'll have to use a map in the view to hold your data. When you began the Module1 project, you told ArcGIS Pro to start you with the Map template so you have one available.
  - Note that if you hadn't done this or if you wanted another map, you could easily access one; on the **Insert** tab, within the **Project** group, click the **New Map** button and choose **New Map**. For more information about what this map is in ArcGIS Pro terminology, see **Smartbox 1.5**.

SMARTBOX 1.5

What is a map, and how does it work in ArcGIS Pro?

In ArcGIS Pro, **map** is the term used for the view that contains all the data and layers being used in a single project. There is also a corresponding tab called Map that is added to a view. You can change the name of the map view to something more distinctive by right-clicking the **Map** icon in the Contents pane, choosing **Properties**, and then, in the Layer Properties dialog box that opens, choosing the **General** tab and typing a new name for the map to replace Map.

When you inserted a new map into your project, not only did ArcGIS Pro give you a new view to work with, but it also added a backdrop image to the map. A corresponding item called Topographic also was added to the Contents pane. This image is intended to help give your data layers context and display them using a particular real-world coordinate system. Whenever you add data to ArcGIS Pro, the available layers in the project are listed in the Contents pane and are then displayed in the map. If you change the color of a layer in the Contents pane, those changes will then be displayed in the map. Whatever is listed in the Contents pane is then available to be displayed in the map.

One advantage to working with ArcGIS Pro is that you can have multiple maps available in a project. If you inserted a second map, and then a third, you would have three different maps available to you that you could access by choosing a different tab in the view. For example, Figure 1.7 shows three maps: one called Mahoning County, a second just called Map, and a third called Map1. Only one of these maps can be active at a time; as mentioned earlier in this module, the active map's tab is blue to distinguish it from the other map tabs. In Figure 1.7, the active map is the Mahoning County map. The Contents pane lists the layers that are part of the active map tab.

map (view) The area used for displaying GIS data in two dimensions.

CHAPTER 1 How to Use Geospatial Data with ArcGIS Pro

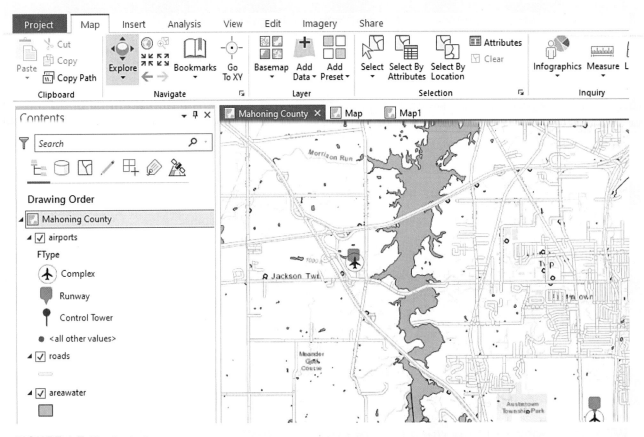

FIGURE 1.7 The Contents pane shows the layers in the active map.

Note that ArcGIS Pro considers a map to be a 2D view. This is because it shows your layers flat on the map in two dimensions. In Module 15, you'll create other views, called scenes capable of showing your data in three dimensions.

- By default the Map tab contains an image of a map of the United States. This image serves as a backdrop on which you can place your own GIS data layers. You'll also see that the Contents pane shows what is in the map view. If you ever close the map (by clicking on the X in its tab) and want to get it back, see **Troublebox 1.3**.

TROUBLEBOX 1.3

What happened to the map I was using in ArcGIS Pro?

If you close a map tab in ArcGIS Pro, getting it back again is easy. In the Catalog pane, you can see an expandable Maps option. Expand it, and you see a listing of all the maps in this project, including the one you closed. Double-click the map listing in the Catalog pane, and the corresponding map tab reopens.

- To examine the properties of the map, in the Contents pane, right-click **Map** and select **Properties**.
- In the Map Properties dialog box, click the **General** tab and type the new name **Mahoning County** for the map.

- Also under the General tab, look at the information about the units of the map and the units of display. Map units are the units of measurement in which the data were created, such as decimal degrees, feet, meters, or miles. Display units represent how you are measuring things on the map. Answer Question 1.1.

> **Question 1.1** In what measurement units was the initial map made?

- You can't change the map units (how the map was originally made), but you can change the display units to measure things differently. Select **US Feet** for Display Units and click **OK**. Now when you make measurements on the map, you will receive distances in feet. Note that the only thing that has changed is the way ArcGIS Pro reports results to you. It does not change what units the data were created in.
- Return to the map and move the cursor around the screen. In the lower-right portion of the screen, you can see distances and coordinates reported in US Feet.
- Go back to Map Properties and change Display Units back to **Decimal Degrees**. Now, in the map, you see measurements in decimal degrees.

Step 1.3 Adding Data to a Map

- Now that you have your map set up, it's time to add your GIS data to it. To begin, on the **Map** tab on the ribbon, within the **Layer** group, click the **Add Data** button and choose **Data**.
- In the Add Data dialog box, navigate to the C:\GIS\Module1\ folder and double-click the **MahoningCounty.gdb** geodatabase to access the feature classes it contains.
- Select the **boundary** feature class and click **OK** to place it on the map.

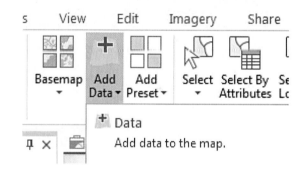

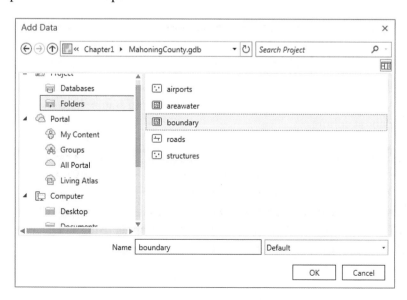

- The layer corresponding to the boundary of Mahoning County, Ohio, appears in the Mahoning County map on top of its proper location on the map of the United States. Also, the layer appears in the Contents pane. Note that there is a checkmark in the box next to the layer's name in the Contents pane. By clicking in this box, you can check and uncheck it. When the box has a check in it, the data layer is displayed. When there is no checkmark, the data are not displayed.

- Note that another way of adding data to the map is to drag and drop it from the Catalog pane into either the Contents pane or the map itself. Either by using this method or by clicking the Add Data button, add the other four feature classes (airports, areawater, roads, and structures) to the map. You then see the layers added to the Contents pane.

- Data layers are displayed in ArcGIS Pro by their placement in the Contents pane. (See **Smartbox 1.6** for more information.)

SMARTBOX 1.6

How do I display data layers in the Contents pane in ArcGIS Pro?

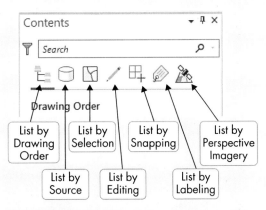

FIGURE 1.8 The different ways of displaying data layers in the Contents pane.

There are seven ways of presenting your data in the Contents pane (Figure 1.8):

- List by drawing order: This is the default setting. The layer at the top of the stack will be drawn on top of all other layers, then the layer underneath it is drawn, and so on. By using this option, you can drag layers from all sources to different positions in the Contents pane.

- List by source: Layers are listed according to their source. For instance, all the layers from one folder will be listed, then all of the layers from a second folder, and so on.

- List by selection: Layers will be listed by how many objects are selected in each layer and which layers are unavailable for selection. (See Modules 2, 8, and 9 for more about selection.)

- List by editing: Layers will be listed by their editing status. (See Module 7 for more about editing.)

- List by snapping: Layers will be listed based on their snapping status. (See Module 7 for more about snapping in relation to editing.)

- List by labeling: Layers will be listed based on their labels.

- List by perspective imagery: Layers will be listed according to whether they contain oblique imagery. (See Module 13 for more about using imagery in ArcGIS Pro.)

- Select the **List by Drawing Order** option. The layer at the top of the Contents pane is displayed first. The layer displayed second from the top is displayed next, and so on. The top layer is displayed "on top of" the other layers. Thus, if you have two layers that cover the same area, the one at the top of the Contents pane list will be displayed on top of the others.

- You can move a layer by clicking on the name of the layer and dragging it to another place in the Contents pane.

- Rearrange the layers on your map so that water bodies are displayed at the top, then the structures next, then the airports, then the roads, and finally the county boundary at the bottom.

Step 1.4 Examining GIS Data and Working with Coordinate Systems, Datums, and Projections

- When using geospatial data in ArcGIS Pro, it's a good idea to keep all the data you are using in the same projection, coordinate system, and datum to avoid any computational or analysis problems. For more information about working with these topics, see **Smartbox 1.7**; for more in-depth technical details, see Appendix B.

SMARTBOX 1.7

What are datums, coordinate systems, and projections in ArcGIS?

Each layer of data used in ArcGIS is a piece of geospatial data—that is, it's tied to a specific location on Earth's surface. Earth is not perfectly round; it's more of an ellipsoid (or spheroid) shape, meaning it's slightly larger at its center than at the poles. To create geospatial data, you need to have a **datum**, which is a reference for coordinate systems used to determine locations around the ellipsoid. The problem is that there's no one universal datum used for all measurements. Literally hundreds of different datums exist; some of them reference the entire globe, and others are used for more local references. Because a datum is the starting point for determining coordinates, measurements made in one datum won't necessarily line up with measurements made in another datum. If you're determining the coordinates of a manhole cover using a system derived from one datum, and a friend is measuring the same location but using a different datum, the two sets of coordinates (for the same object) will be different.

Some common datums that you'll encounter with geospatial data in GIS are:

- **NAD27**: the North American Datum of 1927. This datum was used for large-scale mapping of the United States and North America. Its center is at Meades Ranch in Kansas.

- **NAD83**: the North American Datum of 1983. This datum is used for much of the current data measured in the United States and across North America today. Measurements from NAD83 are very different from measurements from NAD27 (differing by up to a couple hundred meters).

datum A reference surface of Earth.

NAD27 The North American Datum of 1927.

NAD83 The North American Datum of 1983.

NATRF2022 The North American Terrestrial Reference Frame of 2022.

WGS84 The World Geodetic System datum of 1984.

geographic coordinate system (GCS) A set of global latitude and longitude measurements used as a reference system for finding locations.

projected coordinate system (PCS) A set of measurements made on a flat grid system, initially derived from a GCS.

UTM The Universal Transverse Mercator projected coordinate system, which divides the world into 60 reference zones with measurements in meters.

SPCS The State Plane Coordinate System projected coordinate system, which divides states into zones with measurements made in feet or meters.

SPCS2022 The State Plane Coordinate System of 2022, a revised and updated version of SPCS.

Transverse Mercator A projected coordinate system used for north–south data.

- **NATRF2022**: the North American Terrestrial Reference Frame of 2022. This new datum is being established to eventually replace the more common NAD83 data by the year 2022.

- **WGS84**: the World Geodetic System of 1984. The Global Positioning System (GPS) uses this datum for its measurements. Its coordinates are very similar, but not exactly the same, as those you'd get using NAD83.

Having a reference model allows you to determine the coordinates of locations. ArcGIS Pro references coordinate systems in two different ways (see Appendix B for further information):

- **Geographic coordinate system (GCS)**: This is the system of measurements made in latitude and longitude and measured in degrees. In ArcGIS, this reference model is commonly listed as GCS with the datum that was used when the data were created.

- **Projected coordinate system (PCS)**: Often, GCS data are transformed (or projected) into a flat two-dimensional grid system that allows you to make measurements in units such as feet or meters (rather than degrees) with consistent angles or lengths. Because a PCS is a flat grid system, coordinates are measured as an x (such as an east and west measurement) and y (such as a north and south measurement). As is the case with datums, there are a multitude of different projections, and measurements of distance, area, or perimeter made in one projection will not necessarily match up with measurements made of the same locations in another projection.

Some common projected coordinate systems you'll likely encounter with geospatial data and GIS are:

- **UTM**: the Universal Transverse Mercator. This system divides the world into 60 zones, with each zone being 6 degrees of longitude wide. It covers all of Earth except for the poles. Measurements made in each zone are independent of measurements in other zones, so (for example) coordinates for zone 17 cannot be compared with coordinates from zone 16. UTM uses meters for its units of measurement.

- **SPCS**: the State Plane Coordinate System. This system is used only for the United States. It divides a state into different zones that correspond to county boundaries or similar lines. For instance, Ohio is divided into a north zone and a south zone, while Arizona is divided into an east zone, a central zone, and a west zone. Measurements made in each zone are independent of those made in another (so Ohio north zone measurements won't match up with Ohio south zone measurements). There are two types of SPCS coordinates—an old one based on NAD27 that uses feet and a modern one based on NAD83 that uses either feet or meters.

- **SPCS2022**: the State Plane Coordinate System of 2022. This updated version of the SPCS 83 version of the State Plane Coordinate System is set to debut in 2022. It is currently being designed so that each state will consist of just a single zone for measurements.

- **Transverse Mercator**: This projection is commonly used for data with a north–south orientation, and UTM uses this projection as its basis. Units are usually measured in meters.

Step 1.4 Examining GIS Data and Working with Coordinate Systems, Datums, and Projections

- **Lambert Conformal Conic**: This projection is commonly used for data that have an east–west orientation at middle latitudes (such as the United States). It keeps the shapes of locations the same as they are in the real world, and its units are usually meters.

- **Albers Equal Area Conic**: Like Lambert, this projection is used for east–west data at mid-latitudes, so it's commonly used with U.S. data. It keeps the areas (sizes) of locations the same as they are in the real world, and its units are usually meters.

- **WGS84 Web Mercator**: This projection is a standard projection used for web maps, and in particular it is the standard used for ArcGIS Online. Its units are meters.

Lambert Conformal Conic A projected coordinate system used for east–west data.

Albers Equal Area Conic A projected coordinate system used for east–west data at mid-latitudes.

WGS84 Web Mercator A projected coordinate system often used with web maps.

- Each data layer you're using in this module comes with information about the geographic coordinate system and the datum it was created in, as well as the projected coordinate system that each layer was projected to (and the units of measurement that this projected coordinate system uses). To examine this information about your data, do the following:
  - In the Contents pane, right-click on the layer and select **Properties**.
  - Click on the **Source** tab and then scroll down and expand the **Spatial Reference** portion.
- Look at the coordinate system information for each of your five feature classes and answer Question 1.2. (*Hint:* Think carefully about the information you're getting about the datum and the coordinate system.)

> **Question 1.2**
> Fill out the following chart with the information on the datum, projected coordinate system, and units of measurement of each of the five layers.

Layer	GCS Datum	Projected Coordinate System	Units of Measurement
airports			
areawater			
boundary			
roads			
structures			

- As you go through the layers, you'll find that one of them (airports) is in a different projected coordinate system than the rest of your data, and another layer has no defined coordinate system at all.
- To make sure all of your data properly align, you'll want to make these two data layers match the rest of your data in terms of their datum and projected coordinate system. This involves two different tasks: changing the projection of the airports layer (in Step 1.5) to match the others and establishing a coordinate system for the layer that's missing that information (in Step 1.6).

Step 1.5 Changing a Data Layer to a Different Projection

- To get the airports data to match up with the other layers, you have to change its projection to match your other data sources. See **Smartbox 1.8** for more information about changing projections.

SMARTBOX 1.8

How do I change projections in ArcGIS Pro?

When you project a layer, you're creating a new layer that has a new datum, coordinate system, or projection. For instance, if you want to change a roads layer set up in GCS NAD83 measured in degrees and turn it into a UTM Zone 20 layer measured in meters, you have to project the data. **Projection** is a mathematical process that sets up your data in a new projected system (Figure 1.9). You would project your data to change them from their existing coordinate system (whether it's a geographic coordinate system or a projected coordinate system) to a different one. When changing from one datum to another, ArcGIS Pro may ask you to specify or select a particular datum transformation method so that the newly projected layer is set up correctly.

projection A process that changes a layer from one coordinate system into another.

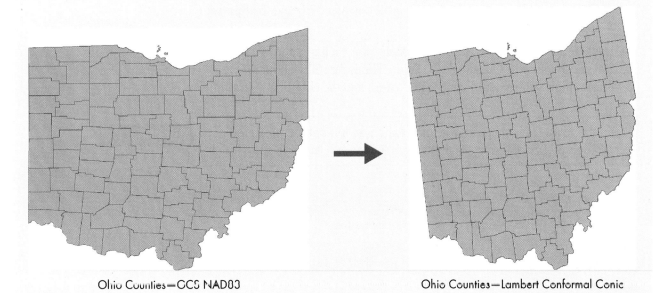

FIGURE 1.9 Ohio county boundary files in the GCS NAD 83 projection and then projected into the Lambert Conformal Conic projection.

By default, a new map in ArcGIS Pro has the WGS 84 Web Mercator projection. The projection that is displayed in ArcGIS Pro is controlled by the coordinate system that the map is set to. You can set the coordinate system by right-clicking on the map in the Contents pane, selecting the map's **Properties**, and selecting the **Coordinate Systems** tab.

In addition, the map has the coordinate system properties of the first layer that is added to it. By default, the map has the coordinate system of the initial topographic backdrop, but if the first layer you add has a different coordinate system or projection, ArcGIS Pro changes the map to have those measurements or that appearance. Subsequent layers that you add to the map are then projected on-the-fly to match the projected appearance of the initial layer. ArcGIS Pro does this to try to properly line up the two datasets instead of showing how they don't fit with each other. Note that ArcGIS Pro does this for display purposes only. ArcGIS Pro doesn't actually change the properties of the layer; it simply alters the appearance to match the other data in the map.

- ArcGIS Pro contains numerous different functions called **geoprocessing tools** (also referred to simply as *tools*) that let you analyze and manipulate GIS data, including a tool to change projections. To access these tools, go to the **Analysis** tab, within the **Geoprocessing** group, and click the **Tools** button.

geoprocessing tools
Specific functions in ArcGIS Pro that are used to analyze and manipulate GIS data.

- A new **Geoprocessing pane** opens on the right-hand side of the screen; it is the pane in which tools are run. In the Geoprocessing pane you can type the name of a tool to have ArcGIS Pro search for it, and you can also go directly to a tool by clicking on **Toolboxes** in this pane.

Geoprocessing pane
A window in ArcGIS Pro where tools are run.

- You see a new arrangement of red "box" icons called *toolboxes*. In ArcGIS Pro, toolboxes hold tools and also toolsets full of similar tools that are grouped together. Expand the **Data Management Tools** toolbox (the red icon that looks like a box) and select the **Projections and Transformations** toolset (the red and gray icon that looks like a hammer lying on top of the box). From there, select the **Project** tool (the hammer icon itself).

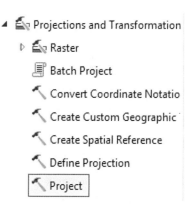

- The Geoprocessing pane switches to become the **Project** tool's dialog.

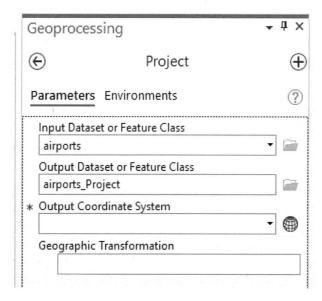

- For **Input Dataset or Feature Class**, click the **browse button** in the dialog box to navigate to your C:\GIS\Module1\ folder and select **airports** from inside the MahoningCounty file geodatabase.

- For your output dataset, you will create a new feature class called airports_Project, which ArcGIS Pro will automatically name for you. By default this will be placed in your C:\GIS\Module1 folder and in the Module1 file geodatabase that was created for this project.

- You now have to select what coordinate system you want to reproject your data into so that it will match up with the NAD83 UTM Zone 17 coordinate system of the rest of your data. Click the button to the right of the Output Coordinate System option (its icon looks like a gray sphere) to select a coordinate system from the Coordinate System dialog.

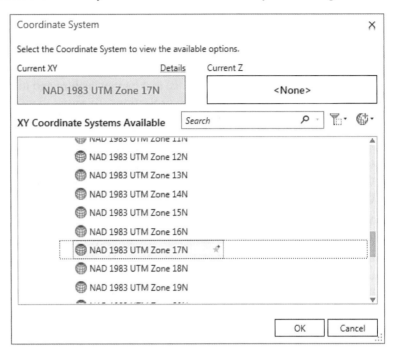

- Under the XY Coordinate Systems Available heading, you see new options. From them, expand the following:
  - **Projected Coordinate Systems**
  - **UTM**
  - **NAD 1983**
- Finally, choose **NAD 1983 UTM Zone 17N**. The Current XY box changes to reflect this newly chosen projected coordinate system. Click **OK** to return to the Project dialog.

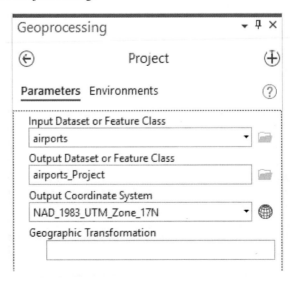

- When all options are properly set, at the bottom of the Geoprocessing pane click **Run** to run the tool. ArcGIS Pro reprojects your data and creates a new feature class called airports_Project in your Module1 file geodatabase. After a few seconds, you see scrolling text across the bottom of ArcGIS Pro indicating that the Project tool is in progress. A green pop-up box appears at the bottom of the screen to let you know when the tool has completed the process.

- You can see that an airports_Project layer is added to the Contents pane. You can delete the old airports layer at this point by right-clicking it and choosing **Remove**.

Step 1.6 Defining a Projection for a Layer That Is Missing This Information

- The next step is to take care of the areawater feature class that has no coordinate system information attached to it. Unfortunately, you cannot simply project this layer as you did with the airports layer. Answer Question 1.3.

> **Question 1.3**
> Why can't you just use the Project option to fix the coordinate information for the areawater layer, as you did for the airports layer? (*Hint*: Think carefully about what Project is really doing.)

- Because the areawater layer is missing its coordinate system and projection information, you need to define a projection for it. See **Smartbox 1.9** for more background information about defining projections.

SMARTBOX 1.9

How do I define a projection in ArcGIS Pro?

Sometimes, you'll receive geospatial data to work with (or create it yourself), and its coordinate system will be listed as "undefined." The word *undefined* often indicates that the creator of a dataset did not explicitly tell ArcGIS Pro what coordinate system was used in making the layer. In the shapefile data format, a separate .prj file is required to hold this information, and other GIS data formats often have something similar—related information that goes with the layer and holds the coordinate information. If this file or information is missing (or not created), you have to supply ArcGIS Pro with the information yourself. The Define Projection tool in ArcGIS Pro allows you to do this. Think of it as a person showing up at a party but forgetting to put on a name tag; he knows what his name is, but nobody else does. Using Define Projection is like putting a name tag on a person to properly identify him to everybody else—in this case, making sure that ArcGIS Pro understands which coordinate system the layer has to try to match with other datasets. In short, Define Projection allows you to tell a layer that is missing the coordinate information exactly what its coordinate system is.

Be very careful to provide the correct projection information when using this tool. If you define the wrong projection, you will encounter a lot of problems. Think of name tags again: If Bob arrives at the party but wears Dave's name tag, everybody at the party will assume that Bob is Dave, has Dave's job, drives Dave's car, and has Dave's characteristics. The same concept holds true for geospatial data. If a data layer should be defined as UTM but you incorrectly define it as State Plane, ArcGIS Pro will try to make measurements of the UTM layer by using the State Plane system and will try to project the layer on-the-fly by using its information about State Plane. The layer will not match the others because ArcGIS Pro is working with incorrect information. Often, you'll see all your regular data shown in one place on the map, with the incorrectly identified layer shown in a completely different location.

Note: If you receive a layer that has no geospatial reference whatsoever (such as a scanned photograph, a scanned map, or an architectural diagram not drawn with real-world grounding), Define Projection won't solve the problem (because the dataset isn't just missing its coordinate system but had no coordinate system to begin with). In this case, a new solution—georeferencing—will have to be used. (See Module 13 for more on this topic.)

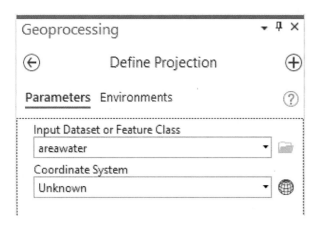

- In the Geoprocessing pane, use the Toolboxes option to see the full range of available tools and toolboxes and expand the **Data Management Tools** toolbox and select the **Projections and Transformations** toolset. From there, select the **Define Projection** tool.

- Select the **areawater** layer in Input Dataset or Feature Class.

- Right now, the areawater layer has an unknown coordinate system assigned to it. To explicitly select a coordinate system to assign to this layer, click the gray sphere button next to the Coordinate System pull-down menu to open the Coordinate System dialog box.

- Under the XY Coordinate Systems Available heading, you see new options. From these, expand the following:
  - **Projected Coordinate Systems**
  - **UTM**
  - **NAD 1983**

- Finally, choose **NAD 1983 UTM Zone 17N**. The Current XY box changes to reflect this newly chosen projected coordinate system. Click **OK** to return to the Define Projection dialog.

- In the Geoprocessing pane, click **Run** again and wait as ArcGIS Pro defines the projection. The green pop-up box at the bottom of the Geoprocessing pane lets you know when it's done.

- From the Contents pane, bring up **Properties** for the areawater layer and look at the **Source** tab under the **Spatial Reference** heading. You should see the proper projection information in place.

- *Important Note:* Luckily, the coordinate system that the areawater layer was missing happened to be the same as for one as the other layers. If it had been different (for instance, if areawater had been missing State Plane coordinate information), you would have had to use the Project tool to change it to match the coordinate and projection information of the other layers.

Step 1.7 Changing the Appearance of a Layer

- Now that all your data are properly set with their spatial reference, it's time to make them look good. Since you're compiling these data for future presentations, the first thing you want to do is change the color and appearance of the data. When layers are added to ArcGIS Pro, they're assigned a random color, so you could, for example, end up with orange lakes or green roads. To alter the appearance of a layer, you need to change its **symbology** through the Symbology pane. You can bring up this pane for a layer in two different ways in ArcGIS Pro:

  1. Right-click layer and choose **Symbology**.
  2. Click once on a layer in the Contents pane to choose it and then go to the **Appearance** tab, within the **Drawing** group, and click the **Symbology** button.

- Whichever method you choose, do this with the areawater layer.

- The Symbology pane opens on the right side of the screen. For the areawater layer, you see that it is being shown as Single Symbol (which means that all the objects in this layer, no matter what they represent, will be drawn with the same color, outline, or symbol). To change this to a different color, click on the **colored box** next to the word Symbol.

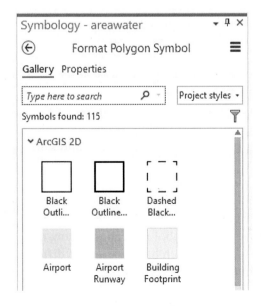

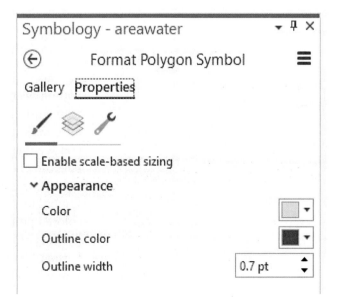

- In the Symbology pane, a new set of options appear. Under the **Gallery** option, you can choose from a predefined set of symbols; for instance, the predefined color for Building Footprint is a shade of tan, and Cemetery is a shade of light green. You can choose one of these presets as a color for your areawater later, but more options are available under the **Properties** option. There you can directly select from many different colors, choose a different outline color for the polygons, or change the width of the outline border around the polygons.

- Using these options, choose a new appropriate color and look for Mahoning County's water bodies. If you chose one of the predefined choices under Gallery, the polygons automatically changes to that symbology, but if you made custom choices under the Properties option, you need to click **Apply** at the bottom of the Symbology pane for ArcGIS Pro to accept the changes.

- At the top of the Symbology pane, click on the **left-pointing arrow** to return to the main part of the Symbology pane. You again see that your areawater polygons are being displayed as a Single Symbol, but now with the color scheme you've assigned in the colored box.

- Next, change the symbology of the roads layer to select a different color and appearance for the roads, using either the presets in the Gallery option of the Symbology pane or the Properties option. You might want to adjust the width of the roads as well as use a different type of symbol instead of the default thin line.

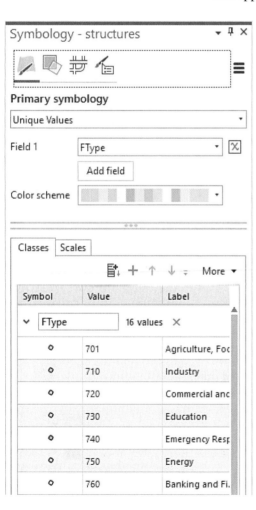

- Note that the areawater and roads layers contain several different types of water bodies or road types. ArcGIS Pro allows you to display these layers according to these different types instead of displaying all objects with the same symbol—for instance, showing interstates with a different symbol than residential roads. The structures layer contains several different types of buildings, such as schools, bridges, and hospitals, and you can show the layer in ArcGIS Pro by giving each building type its own symbol. To do this, bring up the Symbology pane for the structures layer. In the Symbology pane, instead of working with Single Symbol, select **Unique Values**.

- For the Value Field, choose **FType**. In the table at the bottom of the Symbology pane, you see each category with a different colored point. These points all represent different types of structures. By scrolling across the table, you see that the point associated with value 830 represents government structures, while the point associated with 730 represents education structures. You can choose a better symbol to visually represent your data than a randomly colored point. In the Symbology pane, click on one of the **point symbols** to change the Symbology pane to show the Format Point Symbol dialog, which has a variety of different symbols to use for points.

Step 1.7 Changing the Appearance of a Layer   27

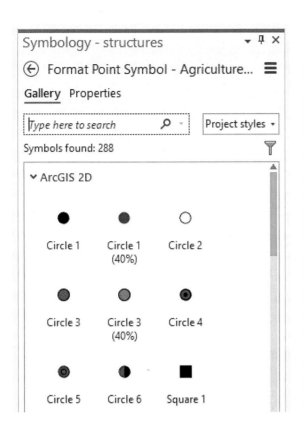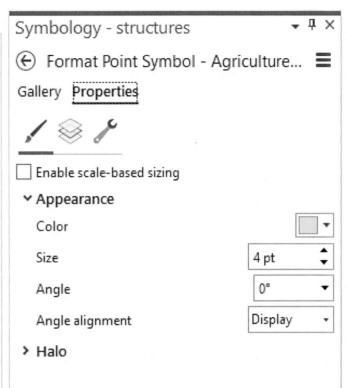

- Using these options, choose a new appropriate symbol for Mahoning County's various structures. Not all of the classes of structures listed in the Contents pane are actually in the layer. The ones you want to change are Education (Ftype 730), Emergency Response (Ftype 740), Health and Medical (Ftype 800), Industry (Ftype 710), Information and Communication (Ftype 880), Transportation Facilities (Ftype 810), and Water Supply and Treatment (Ftype 850). If you chose one of the predefined choices under Gallery, the points automatically change to that symbol, but if you made custom choices under the Properties option (such as changing the size or outline), you need to click **Apply** at the bottom of the Symbology pane for ArcGIS Pro to accept these changes. (If you happen to close out of one of the symbol options, close all the preset choices in the Gallery, or have no options available in the Gallery, see **Troublebox 1.4** for help getting your choices back.) When choosing your symbols, keep two things in mind:
  - You can use the **Search** box under the Gallery option to find different symbols. For instance, you can type "hospitals" to select a variety of different hospital symbols from several of the different symbols available.
  - Additional symbol options can be added. From the **Insert** tab, in the **Styles** group, click the **Add** button and choose the option **Add System Style**. This allows you to add more symbol choices to the Symbology pane.

TROUBLEBOX 1.4

How can I retrieve the symbol options?

If you find that you don't have any preset options for symbols in the Gallery section of the Symbology pane (for instance, if you accidentally closed them), it's easy to get them back. From the **Insert** tab, in the **Styles** group, click the **Add** button and select **Add System Style**. This allows you to select whichever types of symbology (such as styles) you want to use in your project (Figure 1.10). This includes various preset color choices or color schemes (such as Esri preset colors for parks or recreational land uses), 2D symbols (to have distinctive points for objects such as hospitals or fire stations), and 3D symbols (see Modules 15 and 18).

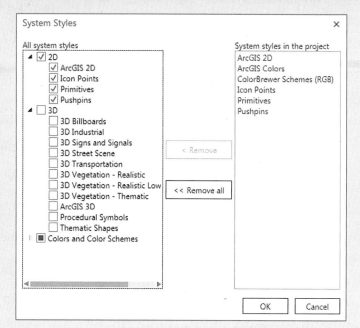

FIGURE 1.10 Available system styles in ArcGIS Pro.

- Next, change the symbology of the airports the same way, using all of the Unique Values with the Ftype value in the airports layer, and display a new, more representative symbol for the airport types. As with the structures layer, not all of the airport types from The National Map are in this layer. You only need to change the symbols for airport complexes (Ftype 200) and airport runways (Ftype 201).

- Next, change the symbology of the boundary feature class to be transparent but still have a solid outline. You can do this either be choosing **Black Outline** from the Gallery on the Symbology pane or by choosing **No Color** under the Fill Color options for the Properties part of the Symbology pane.

- Finally, just so there's no confusion about what a layer represents, give the layers better names in the Contents pane than "areawater" or "airports_Project." Bring up the **Properties** of the areawater layer and select the **General** tab. In the **Name** box, type a better, more descriptive name for this layer (such as **County Water Bodies**) and click **OK**. You see the name changed in the Contents pane. Give each of your other four layers better names than the defaults.

- When you have your new symbols changed, answer **Question 1.4**.

Question 1.4 Which symbology did you choose for each of the following for your new map of Mahoning County?

Airports (complexes)	
Airports (runways)	
Areawater	
Roads	
Structures (Education)	
Structures (Emergency Response)	
Structures (Health)	
Structures (Industry)	
Structures (Information)	
Structures (Transportation)	
Structures (Water Supply)	

Step 1.8 Saving Your Work

- All the work that you've done in creating new projected layers and altering symbology is part of your project, and you'll want to save your project. (Note that it will be saved as Module1.aprx, the name you gave it back on page 7 when you got started.) There are two ways to save your project in ArcGIS Pro:
  - On the **Quick Access Toolbar**, click on the **Save** icon.
  - Click on the **Project** tab and then click on **Save** (or, if you want to save the project under a different name, click **Save As**).
- As discussed in **Smartbox 1.1**, all you've saved in the project is a file that points to where your data layers are stored and how they should appear when this project is reopened. In order to place all your work, data layers, tools, scripts, and results in a single compact file that you can easily move around (to take home, take to a conference, or take to another computer lab to work on later), you would want to also create a package for your project. See **Smartbox 1.10** for more details.

SMARTBOX 1.10

How do I save my work as a package in ArcGIS Pro?

A **package** is the term used in ArcGIS Pro for a single compressed file containing everything that you're working with. A **project package** collects your project plus all the data layers that you've used or created in the project (as well as their assigned symbology and appearance) into a single compressed file. The project package also contains other elements that can be part of projects, such as maps, tools, scripts (see *Related Concepts for Module 1* on page 32 of this module), layouts (see Module 3), scenes (see Module 15), models (see Module 20), or tasks (see Module 9). A project package is much more portable than a project because it contains all the related information in a single file.

When you create a project package, you are effectively "zipping" up all your maps, the layers that are in them, and any other related tools, toolboxes,

package A single file that contains work or data used in ArcGIS Pro.

project package A single file that contains all the tools, scripts, tasks, and data layers used in a project, as well as their appearance.

and data you're working with into one single file with a .ppkx extension. You can then either save that project package to disk or upload it to your ArcGIS Online account (which you established at the start of this module).

An advantage of a project package is that it can be reopened on another computer that does not have the same path names or folder structure as the original. For instance, if you saved your Module1 project in your D:\GIS\Module1 folder and took the project to another computer, you would have to place it on the computer's D: drive in a folder named GIS to properly reopen it and have all the links work correctly. With a project package, you could reopen the project from a different drive (for instance, a flash drive that your computer identifies as a J: drive), and all links would reopen correctly. A project package can also be uploaded to the cloud via ArcGIS Online and then downloaded to a different computer and opened there.

To create a project package, from the **Share** tab, in the **Package** group, click on the **Project** button to open the Package Project pane. In this pane, you can either choose to save the package to your own computer (as a .ppkx file) or share the project package via ArcGIS Online (see Module 4 for more about this option). ArcGIS Pro requires you to write a short summary of the project package and also include tags, which can be used for searching for a particular package on ArcGIS Online (Figure 1.11).

At the bottom of the Package Project pane, click **Analyze**. ArcGIS Pro checks your work for any outstanding issues and shows you the errors with a red symbol indicating what the error is or a yellow symbol indicating a warning to be mindful of. You have to fix or address these issues before ArcGIS Pro allows the project package to be created. Finally, at the bottom of the pane, click on **Package** to complete the project package.

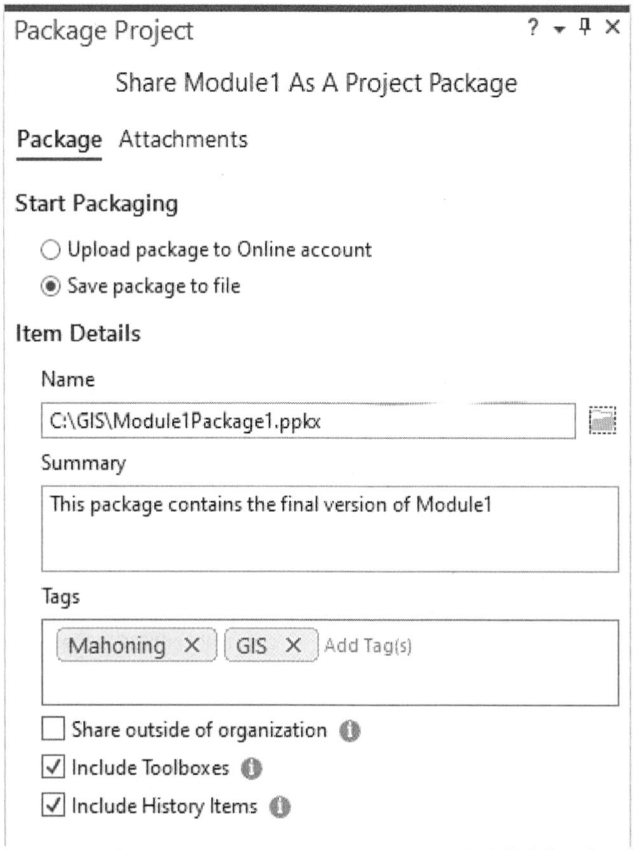

You can reopen a project package the same way you would open a project: either by clicking the **Open** icon on the Quick Access Toolbar or clicking on the **Project** tab and choosing **Open**. When you reopen a project package, files and items from the package load into ArcGIS Pro as a project.

Also note that on the Share tab within the Package group, you have several other packaging options available as well. You can use them if you don't want to package up the entire contents of a project but instead only want to package and share specific parts of what you're working with. These options include the following:

- Geoprocessing Package: This option is used to package and share items such as models. (See Module 20 for more about creating geoprocessing packages.)

FIGURE 1.11 The Package Project pane in ArcGIS Pro.

- Layer Package: This option packages and shares a single layer (and its appearance). A **layer package** has the extension .lpkx.
- Map Package: This option produces a package and shares the layers and contents of a single map in ArcGIS Pro. A **map package** has the extension .mpkx.
- Mobile Map Package: This option packages and shares the layers and contents of a single map but in a format that can be used with ArcGIS mobile apps. (See Module 6 for more about using these apps.)

layer package A single file that contains one data layer.

map package A single file that contains the contents of one map in ArcGIS Pro.

Step 1.9 Finishing Up and Exporting Your Work

- When you've finished displaying the layers the way that you want them, your task is complete; you've compiled your set of data layers in ArcGIS Pro for your presentation, set them all up in the proper spatial reference, and set up their appearance to match your audience's expectations. If you have zoomed in or out of any of your layers, you should restore the view to the dimensions of your layers. To do so, right-click on a layer (in this case, the **boundary layer**) and select **Zoom to Layer**. This resets the view to the extent of the chosen layer (in this case, the extent of the boundaries of Mahoning County).

- In future modules, you will usually either create a professional-looking map layout to print (see Module 3) or digitally share your finished product with others via ArcGIS Online (see Module 4). For now, however, you can save your completed display as a PDF file that you can print or share with others. From the **Share** tab, in the **Export** group, click the **Map** button. Give it the filename **Module1** and for Save as Type, select **PDF**. Place this PDF file in your C:\GIS\Module1\ folder. When you're done, answer Question 1.5.

> **Question 1.5**
> Submit your final PDF file to your instructor, who will check your work for completeness and accuracy to make sure you get credit for this question.

Closing Time

This module introduces the basic principles of working with geospatial data in ArcGIS Pro. You have seen that things in the real world can be represented by one of three objects (points, lines, or polygons), these objects must have a defined spatial reference and coordinate system, and they can be visualized in different ways in ArcGIS Pro. In the next module, you'll start looking beyond just the locations and geospatial coordinates of these objects and see how descriptive information can be tied to these locations as well. For instance, each of the points in the structures layer has information about the type of building the point represents (education, industry, health, and so on). You'll begin working with these types of attributes in Module 2.

ArcGIS Pro is more than a bunch of predefined buttons and tools. You can do a lot of varied things with ArcGIS Pro and diversify your GIS skill set by writing computer scripts with the Python and Arcade programming languages. See *Related Concepts for Module 1* for information about these types of programming resources available for use with ArcGIS Pro.

RELATED CONCEPTS FOR MODULE 1

ArcGIS Pro, Python, and Arcade

You may often want to automate processes in ArcGIS Pro, customize activities, or create more in-depth operations. For instance, if you need to define the projection of 20 different layers, it would be easier to find a way to automate that task rather than run the Define Projection tool 20 times by hand. In ArcGIS Pro, you can accomplish automation by writing a **script**, or a section of computer code, and running that script in the software. Within ArcGIS Pro you can write scripts by using two different programming languages.

By writing scripts in the **Python** scripting language, you can execute command-line operations rather than use the tools; you can also create more in-depth operations. Python (created by Guido van Rossum and named for the Monty Python comedy team) is an object-oriented, open-source programming language that is very versatile and used in several different fields. In ArcGIS, the **ArcPy** add-in allows you to gain access to the tools of ArcGIS and use them when writing Python scripts or executing code. When using Python in ArcGIS, the first command you should execute is **import arcpy**, which gives you access to the ArcPy add-in (and, thus, the ArcGIS Pro commands and geoprocessing tools).

There are two ways to work with Python in ArcGIS Pro. The first of them is by clicking the **Python** button in the Python window, which is accessible from the **View** tab, in the **Windows** group. The Python window is a command prompt that allows you to type and execute Python code directly; it's best used for short statements, such as running tools. See Figure 1.12 for an example of using the Python window to run the Project tool that was used in this module. The Python code is directly typed in the Python window.

Two commands are shown in the Python window in Figure 1.12. The first is the **import arcpy** command, which enables the various ArcPy functions and allows you to run ArcGIS Pro tools from within the Python window. The second command, **arcpy.Project_management**, is the Project tool and its three required parameters: the input feature class, the output feature class, and the spatial reference information. This example shows 26917 as the spatial reference information, which is a shorthand code for UTM NAD 83 Zone 17.

script A short section of computer code.

Python An object-oriented, open-source programming language that is used for developing scripts for ArcGIS Pro.

ArcPy The Python add-in that enables ArcGIS Pro tools and settings to be used in Python scripts.

```
Python
import arcpy
arcpy.Project_management(r"D:/GIS/Chapter1/mahoningcounty.gdb/airports", r"D:/GIS/Chapter1/mahoningcounty.gdb/airports_project", arcpy.SpatialReference(26917))
```

FIGURE 1.12 A sample Python script used for running the Project tool in ArcGIS Pro.

A second way of working with Python in ArcGIS is to use a separate program to write lengthier or more complex scripts. Keep in mind that Python scripts are simply text files, so you could use a text-editing utility like Notepad to write a script. However, a more effective means of preparing scripts would be to use a program specifically designed to aid you in writing Python scripts, called an IDE (Integrated Development Environment) note that you should use Python version 3.x because that's the version supported by ArcGIS Pro, not the older version 2.7). Using a program like an IDE as an environment for writing scripts will help you keep your coding, commenting, and syntax clearer when you're writing lengthy scripts. A Python script is saved with the extension .py and can be run through a program like an IDE. For more about transitioning older ArcMap Python 2.7 scripts to use in ArcGIS Pro, see Appendix A.

Python is commonly used with ArcGIS for writing scripts related to geoprocessing or running tools. In the online help for ArcGIS Pro (located at **https://www.esri.com/en-us/arcgis/products/arcgis-pro/resources**), the web page for each tool also lists examples of Python code that can be used in the Python window and also as standalone scripts. For instance, if you want to use Python to run the Define Projection tool, look up that tool in the online help, and you see examples of Python code showing how Define Projection can be implemented in Python scripting.

The second scripting language available in ArcGIS Pro is **Arcade**, which is referred to as an *expression language*. Short scripts written in Arcade are used in ArcGIS Pro to change the appearance of data layers you're working with, such as altering labels, computing unit conversions, grouping multiple features together, or converting raw numbers into percentages. An advantage to writing scripts in Arcade is that they are compatible across the ArcGIS platform, which means they will work on mobile devices as well as in ArcGIS Pro on the desktop.

Arcade A scripting language for writing expressions in ArcGIS.

Key Terms

geographic information systems (GIS) (p. 2)
geospatial or spatial (p. 2)
geospatial technologies (p. 2)
Esri (p. 2)
ArcGIS (p. 2)
ArcGIS Desktop (p. 2)
ArcGIS Pro (p. 2)
extension (p. 6)
project (.aprx) (p. 6)
template (p. 6)
view (p. 8)
Contents pane (p. 8)
pane (p. 8)
Catalog pane (p. 9)
ribbon (p. 9)
Quick Access Toolbar (p. 9)
catalog view (p. 11)
discrete object view (p. 12)
point (p. 12)
line (p. 12)
polygon (p. 12)
vector objects (p. 12)
vector data model (p. 12)
shapefile (p. 12)
file geodatabase (p. 12)
map (view) (p. 13)
datum (p. 17)
NAD27 (p. 17)
NAD83 (p. 17)
NATRF2022 (p. 18)
WGS84 (p. 18)
geographic coordinate system (GCS) (p. 18)
projected coordinate system (PCS) (p. 18)
UTM (p. 18)
SPCS (p. 18)
SPCS2022 (p. 18)
Transverse Mercator (p. 18)
Lambert Conformal Conic (p. 19)
Albers Equal Area Conic (p. 19)
WGS84 Web Mercator (p. 19)
projection (p. 20)
geoprocessing tools (p. 21)
Geoprocessing pane (p. 21)
package (p. 29)
project package (p. 29)
layer package (p. 31)
map package (p. 31)
script (p. 32)
Python (p. 32)
ArcPy (p. 32)
Arcade (p. 33)

2

How to Use Tables, Attributes, and Queries in ArcGIS Pro

Learning Outcomes

After studying this module, you should be able to:

- Indicate how attributes for objects are stored in an attribute table.
- Define four different types of data classifications.
- Explain how to join two tables together using a key.
- Describe the format of an SQL query.
- Explain how to construct a compound query.
- Describe what a datum is.
- Explain how to obtain both spatial and attribute data from the U.S. Census Bureau.

ArcGIS Pro Skills

In this module, you will learn how to do the following in ArcGIS Pro:
- Open a layer's attribute table.
- Add a table (such as an Excel spreadsheet) to ArcGIS Pro.
- Join a spreadsheet table to a GIS layer's attribute table.
- Sort an attribute table.
- Perform both simple and compound Select By Attributes queries.
- Export the results of a selection to a new feature class.
- Choose an attribute field to label and apply labels to features.
- Add a new field to an attribute table and add data to the new field.
- Add an attachment of a photo to the pop-up window for an object.

Introduction

You can represent all types of geospatial data using GIS. For instance, a set of lines can represent the length of roads, and several polygons can represent the area and dimensions of parcel boundaries. However, there's a lot more to capturing these features in GIS than just specifying their spatial properties such as lengths or coordinates. For example, a transportation engineer is going to need information about the road's name, speed limit, covering type, and condition, and a county auditor would need information about the owner, zoning, property valuation, and taxes on the land parcels. All of these kinds of data are **non-spatial data**, which are data that are descriptive and not location based. For instance, a spreadsheet could contain information such as the address of a house, the owner's name, and the house's tax-assessed value without having any ties to geospatial data.

Fortunately, GIS easily allows you not only to handle non-spatial data but also to connect non-spatial data to geospatial locations. (After all, having accurate geospatial representations of land parcels separate from tabular data about those parcels wouldn't be very helpful for analysis.) In ArcGIS Pro, all the information (spatial and non-spatial) about a layer is stored in that layer's **attribute table**. This table consists of a series of rows (referred to as **records**) and columns (referred to as **fields**). Each of the table's records represents a separate object.

non-spatial data Descriptive information that does not have location-based qualities.

attribute table A spreadsheet-style form in which the rows consist of individual objects and the columns are the attributes associated with these objects.

record A row of an attribute table.

field A column of an attribute table.

For instance, if a layer contains 100 polygons representing the boundaries of lots in a new subdivision, then its attribute table would contain 100 records, one for each polygon/land lot. If you have a point layer representing the locations of 150 different water wells, then its attribute table will have 150 records.

The table's fields contain all the **attributes** (or related information) about each record, such as any non-spatial data. For instance, each record in the lots layer would have a field indicating the name of the lot's owner, the zoning code for the lot, the monetary value of that lot, a parcel ID, and so on. Figure 2.1 shows an example of an ArcGIS Pro attribute table of polygons representing the boundaries of counties in Ohio.

attribute Non-spatial data that can be associated with a geospatial location.

Each record is a separate county polygon, and the fields represent the attributes of each county (county name, the state that county is in, the FIPS code for that county, and so on). This module introduces you to working with attribute data in ArcGIS Pro, building queries of GIS data based on attributes, and joining non-spatial tabular data to geospatial feature class data.

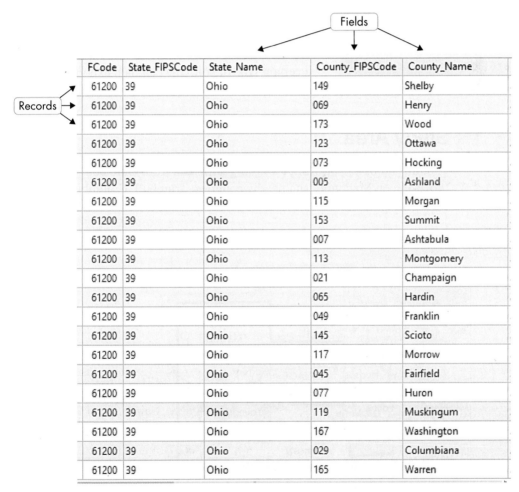

FIGURE 2.1 An ArcGIS Pro attribute table.

 ## Module Scenario and Applications

This module's scenario puts you in the role of an education coordinator for the state of Ohio. As part of a study related to higher education in the state, your job is to assess the state's population distribution and then examine the locations of the colleges and universities in a set of counties in northeastern Ohio. Eventually, you'll be presenting your results in an interactive format that shows which colleges are located in which cities and includes graphics of the colleges.

Aside from this module's scenario, the following are examples of other real-world applications of this module's theory and skills:

- A county auditor is updating the GIS database following countywide property tax assessment. Rather than retyping the database, she would join the new assessment spreadsheet to the geospatial data of the land parcel locations.

- A marketing analyst for a department store is working on a focused advertising campaign. To determine where shoppers are traveling from, he can pose a query to the attribute table to find certain zip codes and examine their locations and boundaries on a map.

- An environmental engineer is performing a soil analysis in GIS. By posing a query to the attribute table, she can quickly select all areas that contain a certain type of soil. She can then export those features to their own layer for further analysis.

Study Area

- For this module, you will be working with data from Ohio counties, focusing on Mahoning and nearby counties: Trumbull, Columbiana, Stark, and Portage.

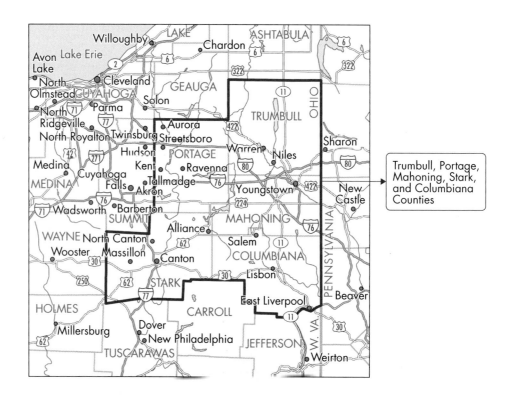

Data Sources and Localizing This Module

The data in this module focus on features and locations within Ohio's counties. However, this module can easily be modified to use U.S. Census Bureau data on your own county or local area instead. (See *Related Concepts for Module 2* for more about using Census data in ArcGIS Pro.) This module used county boundaries from the TIGER/Line files from the Census, but the way Census data are served and formatted has been updated since then. Comparable Census data for your own county are available from the U.S. Census for free. You need to download a county boundary dataset and attribute tables from the Census and then join them together.

The county boundaries can be obtained as Cartographic Boundary files from **https://www.census.gov/geographies/mapping-files/time-series/geo/carto-boundary-file.html**. The counties layer you download contains all the counties in the United States. You can then query the shapefile and extract the appropriate counties into a separate feature class for your state, based on the STATEFP attribute. This attribute references each state with a two-digit FIPS code. For instance, Ohio is referenced with state FIPS code 39. If you were using this module in Arkansas, for instance, you would use FIPS code 05. A full list of FIPS codes for each state and county is available from the National Weather Service at **https://www.weather.gov/pimar/FIPSCodes**. Note that some of the U.S. Census Bureau's county boundaries extend to the national boundaries when they border water features (such as in the Great Lakes area between the United States and Canada).

Attribute data must be downloaded separately and joined to the counties shapefile. The attribute data (the population statistics) for each county in the state were downloaded from the Census Bureau's American FactFinder website, but these data are now available through the Census data website at **https://data.census.gov/cedsci/**. The table used in this module is "G001: Geographic Identifiers (2010 SF1 100% data)." On the U.S. Census data website, these can be found by querying for Population and Housing. Census attribute data were downloaded as a CSV file, edited and cleaned up, and converted to an Excel file for use in this module. The tabular data from the Census data website can be joined to the Cartographic Boundary layer by using the AFFGEOID attribute in the Cartographic Boundary file and the GEO ID attribute in the Census data.

The structures layer was downloaded (for free) from The National Map website at **https://viewer.nationalmap.gov/basic/**. The colleges were subsequently extracted from this layer by using the FCode field and querying for the value 73006, which is the code assigned to colleges and universities. If you were working with Arkansas data, for instance, you could download the structures layer for the entire state and extract only the colleges for your county and its immediate neighbors. Step-by-step details on downloading and extracting GIS data from The National Map are provided in Module 5. You can get the same types of data for your own county from The National Map and supplement it with some local digital camera photos for the attachments section in Step 2.10.

Step 2.1 Getting Started

- This module's hands-on applications use the data folder called Module2. Your instructor will be able to supply you with this data, or you can download it directly from this book's website at **https://www.macmillanlearning.com/college/us/product/Discovering-GIS-and-ArcGIS-Pro/p/131923075X**. The text in this module assumes that you have this Module2 folder in a computer location referenced as C:\GIS; if you have it somewhere else (for instance, in a flash drive referenced as G:\GISClass), substitute that location and path to the Module2 folder throughout this module.

- The Module2 folder contains three digital camera pictures (YSU.jpg, ITT.jpg, and MCCTC.jpg) that you will use in Step 2.10. The Module2 folder also contains a file geodatabase called Ohiocounties, which contains the following items:
  - CensuscountiesOhio: a polygon feature class of Ohio county borders
  - DEC_10_SF1_G001: a geodatabase table created from an Excel spreadsheet containing information about Census 2010 population and housing values
  - Colleges: a point feature class of colleges and universities within a five-county region of Ohio

- Start ArcGIS Pro.
- Sign in with your Esri account username and password.
- Create a new project using the **Map** template. Call this project **Module2** and place it in your **C:\GIS\Module2** folder. Ensure that there is not a checkmark in the **Create a new folder for this project** box.
- When ArcGIS Pro opens, change the map's name to **Ohio**.

Step 2.2 Interactively Obtaining Attribute Information

- Add the **CensuscountiesOhio** feature class to the Ohio map.
- Each of the polygons shown represents a county in Ohio, but various attributes are assigned to each one. ArcGIS Pro makes it easy to obtain the attribute data associated with a layer of geospatial data by using the Explore tool. Click on the **Map** tab, and within the **Navigate** group, click the **Explore** button. The Explore tool is then highlighted, and you can use it to zoom in and out of the map view, or you can click on an object (such as a point, line, or polygon) in the view and retrieve its attribute information. See Figure 2.2 for how to use the various functions of the Explore tool.

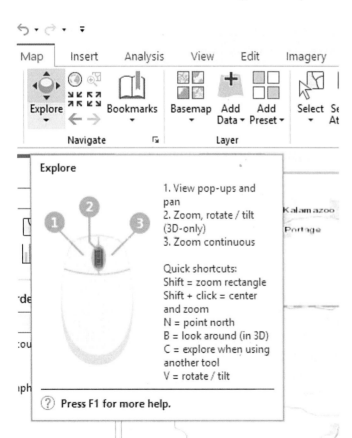

FIGURE 2.2 The functions of the Explore tool in ArcGIS Pro and how they relate to a mouse.

- For instance, click on the county in the northeastern corner of Ohio, and a pop-up box appears, showing the available attribute data connected to that polygon. Answer Question 2.1 and then close the pop-up window when you're done.

> **Question 2.1** What is the name of this county?

- Click on the county in the northwestern corner of Ohio. Answer Question 2.2 and then close the pop-up window when you're done.

> **Question 2.2** What is the name of this county?

Step 2.3 Examining the Records and Fields of a Layer's Attribute Table

- The attribute data connected to each of the objects in the Censuscounties layer is stored in that layer's attribute table. There are two different ways to open it in ArcGIS Pro:

  1. In the Contents pane, right-click on the layer whose attribute table you want to open (in this case, the **CensuscountiesOhio** layer) and select **Attribute Table**.

  2. In the Contents pane, click once on the layer whose attribute table you want to open (in this case, the **CensuscountiesOhio** layer) and then, on the ribbon, on the **Data** tab, within the **Table** group, click the **Attribute Table** button.

A new window appears at the bottom of ArcGIS Pro, and you see a tab with the CensuscountiesOhio name attached to the window; this is the attribute table for the layer. For further information about what you're examining in this attribute table, see **Smartbox 2.1**.

SMARTBOX 2.1

What are the components of an attribute table in ArcGIS Pro?

When an attribute table (or any other type of table, for that matter) is opened in ArcGIS Pro, a single movable and resizable window is added to ArcGIS Pro. You see a tab at the top of the table window indicating the name of the attribute table (in this case, CensuscountiesOhio). As you open more attribute tables, you see them added to the tabs at the top. No matter how many attribute tables you have open, ArcGIS Pro has just one window, and each attribute table is accessible as a different tab; this is similar to how you would use multiple tabs for different maps in a view.

As shown in Figure 2.3, there are several buttons and controls available for use in the table window.

FIGURE 2.3 The components of the ArcGIS Pro table window.

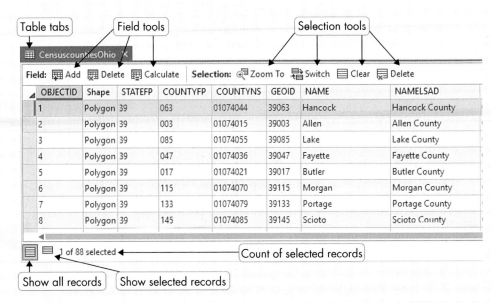

- The table tabs show which of the opened attribute tables are being displayed in the table window.

- The field tools are used when you want to add a new empty field to an attribute table, delete a field, or calculate a new value for a field.

- The selection tools are used when querying or selecting records (and also when clearing the results of a query or selection), as well as for zooming the map to the selected results of a query or deleting selected records. See Step 2.5 for more information.

- The Show All Records and Show Selected Records buttons are used to switch what is being displayed in the table window. When records are selected (see Step 2.5), clicking Show Selected Records shows only those records in the table. Clicking Show All Records displays all the records, whether or not they are selected.

- The count of selected records shows the total number of records in the attribute table, as well as how many have been selected.

- Each polygon in the feature class is shown as a different record. ArcGIS Pro uses the ObjectID field internally to identify object number 1, object number 2, and so on. By examining the count of selected records at the bottom of the table window, you can see that there are a total of 88 records in this attribute table. Each record corresponds with a specific object in the layer; in this case, each record represents an individual county, and because there are 88 counties in Ohio, there are 88 records in the table. By examining the fields in the CensuscountiesOhio attribute table, you can see that there are several different types of data available, including different numerical values and text strings. Each field of attributes can be placed into one of four types of data classifications. See **Smartbox 2.2** for more information about these different data types.

SMARTBOX 2.2

What are the different types of GIS attribute data in ArcGIS Pro?

Fields of non-spatial data values in attribute tables represent one of four types of data: nominal, ordinal, interval, or ratio data (Figure 2.4).

Nominal data represent some type of unique identifier, such as your phone number, bank account number, or Social Security number. Mathematical operations with nominal data that are numerical just don't make sense; for instance, you can't add two phone numbers together in order to get a mutual friend's phone number. Names or descriptive information associated with a location or phenomenon are also nominal data. Even if the name isn't unique (for example, several points on a map may represent different franchises of a national coffee shop, such as Starbucks, which all have the same name), the fact that it is descriptive or name data classifies it as nominal.

Ordinal data are used to represent a ranking system of some kind with respect to the data. If a field represents a phenomenon where something is in first place, something else is in second place, and another item is in third place, then you're dealing with ordinal data. In the Olympic Games, the first-place winner of the event gets the gold, the second-place winner gets the silver, and the third-place winner gets the bronze; this ranking represents ordinal data.

nominal data A type of data that is a unique identifier of some kind. If the data are numerical, the differences between numbers are not significant.

ordinal data A type of data that refers solely to a ranking of some kind.

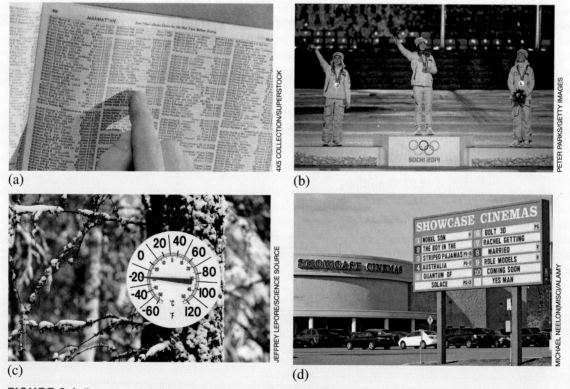

FIGURE 2.4 Examples of the four data types: (a) nominal data (names and numbers in a phone book), (b) ordinal data (gold, silver, and bronze medal winners at the Olympics), (c) interval data (temperatures Fahrenheit and Celsius), and (d) ratio data (the box office grosses from Hollywood movies).

Here's another example: If you have a map of the addresses of the winners of a horse-racing event and the data indicate who placed first, who placed second, and so on, these would be ordinal data. Ordinal data deal solely with the rankings, and not with anything else associated with these values (such as each horse's finishing time, or the differences in finishing times between competitors). Ordinal data don't tell us anything about how much faster the first-place horse was than the second-place horse—only that one is first and one is second.

Interval data are used when the difference between numbers is significant but there is no fixed zero point. With interval data, the value of zero is just another number on the scale and does not represent the bottom of the scale. Celsius temperature is an example of interval data; in the Celsius scale, zero degrees represents the freezing point of water, but Celsius measures temperatures below zero (and thus zero is not the bottom of the Celsius scale). Because there is no fixed zero point, it's fine to subtract numbers, but dividing numbers makes no sense. For instance, if the temperature is 15 degrees yesterday and 30 degrees today, we can subtract the two values and say that today is 15 degrees warmer than yesterday. However, we cannot divide the numbers and say that today is twice as warm as yesterday (because 30 degrees is not twice as warm as 15 degrees).

Ratio data are used when the difference between numbers is significant, and there exists a fixed and non-arbitrary zero value as the bottom of the measurement scale. For instance, your age and weight are both ratio data because you can't be less than zero years old or weigh less than zero pounds. In terms of temperature, the Kelvin scale would be considered ratio data because zero degrees Kelvin is considered the bottom of the scale (such that zero degrees Kelvin is the coldest that something can be measured). With ratio data, we can subtract values, but we can also multiply and divide them to get meaningful information. For instance, the gross box office receipts of one movie may be $100 million, while another film's gross box office receipts may be $300 million. We can say that the second film took in $200 million more than the first film, but we can also divide the numbers and say that the second film took in three times the amount of money that the first film did (since $100 million is one-third of $300 million).

> **interval data** A type of numerical data in which the difference between numbers is significant but there is no fixed non-arbitrary zero point associated with the data.
>
> **ratio data** A type of numerical data in which the difference between numbers is significant, and there exists a fixed, non-arbitrary zero point associated with the data.

- Examine the different fields of data and answer Questions 2.3 and 2.4.

Question 2.3 What type of data classification (nominal, ordinal, interval, or ratio) does the Shape_Area field represent, and why?

Question 2.4 What type of data classification (nominal, ordinal, interval, or ratio) does the COUNTYFP field represent, and why?

Step 2.4 Joining a Table to a Layer's Attribute Table

- There are several attributes for each record in the counties attribute table, but as part of your project, you're to look into population distributions related to colleges. In addition, there are no population attributes for the counties in the attribute table. While you could manually look up the population attributes and type them in, there's a better approach available: You can take another spreadsheet and join its data and attributes to their corresponding records in the counties attribute table. See **Smartbox 2.3** for more information about joining tables in ArcGIS Pro.

SMARTBOX 2.3

How are tables joined in ArcGIS Pro?

ArcGIS Pro gives you the capability to link two or more data tables together. Often, non-spatial data (such as a tabular dataset or a spreadsheet) are linked to geospatial data (using a layer's attribute table). For instance, a geospatial layer might represent the polygon boundaries of each building on a college campus. A separate spreadsheet of non-spatial data might provide information such as the building's name, number of classrooms, number of computers, and so on. Linking these two tables together would enable you to examine the geospatial layer, find a record for a particular building, and then access all of the building's non-spatial data as well.

In ArcGIS Pro, tables are connected by a **join**, or the merging of two (or more) tables into one. A join can be performed only if the two tables have a field in common (referred to as a **key**). It's on the basis of this key that data from one set of records can be linked to another set of records. Figure 2.5 shows an example of a join, where one of the tables is a layer's attribute table and the second is a tabular spreadsheet. Their common field (the key) is the COUNTYFP attribute in the first table and the County* attribute in the second table. These two tables can be joined, using these two fields as their key, and the records of the attribute table can thus be linked to their corresponding records in the tabular spreadsheet.

Keep in mind that while the common field can have a different name or title in each table, it must be of the same type (that is, both must be numbers, or both must be text strings). If the fields are represented differently (for instance, if the values in the key are represented with numerical values in one table and represented with text strings in the second table), then the field cannot be used as a key, and the join will not work. Table values that are left-justified represent text strings (even if they are shown as numbers). Table values that are right-justified represent actual numerical values (not strings of text). In addition, the field used as the key should not have any missing or null values because records from the two tables will not link together properly without the value in the key field.

A join is considered a **one-to-one join** if there is only a single record from each table being joined to a single record in another table. For instance,

join A method of linking two or more tables together.

key The field that two tables have to have in common with each other in order for the tables to be joined.

one-to-one join A join in which a single record is linked to another single record.

Key

OBJECTID *	Shape *	STATEFP	COUNTYFP	COUNTYNS	GEOID	NAME	NAMELSAD	LSAD
1	Polygon	39	063	01074044	39063	Hancock	Hancock County	06
2	Polygon	39	003	01074015	39003	Allen	Allen County	06
3	Polygon	39	085	01074055	39085	Lake	Lake County	06
4	Polygon	39	047	01074036	39047	Fayette	Fayette County	06
5	Polygon	39	017	01074021	39017	Butler	Butler County	06
6	Polygon	39	115	01074070	39115	Morgan	Morgan County	06
7	Polygon	39	133	01074079	39133	Portage	Portage County	06
8	Polygon	39	145	01074085	39145	Scioto	Scioto County	06
9	Polygon	39	163	01074093	39163	Vinton	Vinton County	06
10	Polygon	39	125	01074075	39125	Paulding	Paulding County	06
11	Polygon	39	069	01074047	39069	Henry	Henry County	06
12	Polygon	39	051	01074038	39051	Fulton	Fulton County	06
13	Polygon	39	015	01074020	39015	Brown	Brown County	06
14	Polygon	39	091	01074058	39091	Logan	Logan County	06
15	Polygon	39	155	01074089	39155	Trumbull	Trumbull County	06
16	Polygon	39	131	01074078	39131	Pike	Pike County	06
17	Polygon	39	111	01074068	39111	Monroe	Monroe County	06
18	Polygon	39	087	01074056	39087	Lawrence	Lawrence County	06
19	Polygon	39	001	01074014	39001	Adams	Adams County	06

+

OBJECTID *	Id	Id2	State	County *	Population	Housing
1	0500000US39001	3900	39	001	28550	12978
2	0500000US39003	3900	39	003	106331	44999
3	0500000US39005	3900	39	005	53139	22141
4	0500000US39007	3900	39	007	101497	46099
5	0500000US39009	3900	39	009	64757	26385
6	0500000US39011	3901	39	011	45949	19585
7	0500000US39013	3901	39	013	70400	32452
8	0500000US39015	3901	39	015	44846	19301
9	0500000US39017	3901	39	017	368130	148273
10	0500000US39019	3901	39	019	28836	13698
11	0500000US39021	3902	39	021	40097	16755
12	0500000US39023	3902	39	023	138333	61419
13	0500000US39025	3902	39	025	197363	80656
14	0500000US39027	3902	39	027	42040	18133
15	0500000US39029	3902	39	029	107841	47088
16	0500000US39031	3903	39	031	36901	16545
17	0500000US39033	3903	39	033	43784	20167
18	0500000US39035	3903	39	035	1280122	621763
19	0500000US39037	3903	39	037	52959	22730

=

OBJECTID *	Shape *	STATEFP	COUNTYFP	COUNTYNS	GEOID	NAME	NAMELSAD	LSAD	OBJECTID *	Id	Id2	State	County *	Population	Housing
1	Polygon	39	063	01074044	39063	Hancock	Hancock County	06	32	050000	39063	39	063	74782	33174
2	Polygon	39	003	01074015	39003	Allen	Allen County	06	2	050000	39003	39	003	106331	44999
3	Polygon	39	085	01074055	39085	Lake	Lake County	06	43	050000	39085	39	085	230041	101202
4	Polygon	39	047	01074036	39047	Fayette	Fayette County	06	24	050000	39047	39	047	29030	12693
5	Polygon	39	017	01074021	39017	Butler	Butler County	06	9	050000	39017	39	017	368130	148273
6	Polygon	39	115	01074070	39115	Morgan	Morgan County	06	58	050000	39115	39	115	15054	7892
7	Polygon	39	133	01074079	39133	Portage	Portage County	06	67	050000	39133	39	133	161419	67472
8	Polygon	39	145	01074085	39145	Scioto	Scioto County	06	73	050000	39145	39	145	79499	34142
9	Polygon	39	163	01074093	39163	Vinton	Vinton County	06	82	050000	39163	39	163	13435	6291
10	Polygon	39	125	01074075	39125	Paulding	Paulding County	06	63	050000	39125	39	125	19614	8749
11	Polygon	39	069	01074047	39069	Henry	Henry County	06	35	050000	39069	39	069	28215	11963
12	Polygon	39	051	01074038	39051	Fulton	Fulton County	06	26	050000	39051	39	051	42698	17407
13	Polygon	39	015	01074020	39015	Brown	Brown County	06	8	050000	39015	39	015	44846	19301
14	Polygon	39	091	01074058	39091	Logan	Logan County	06	46	050000	39091	39	091	45858	23181
15	Polygon	39	155	01074089	39155	Trumbull	Trumbull County	06	78	050000	39155	39	155	210312	98163
16	Polygon	39	131	01074078	39131	Pike	Pike County	06	66	050000	39131	39	131	28709	12481
17	Polygon	39	111	01074068	39111	Monroe	Monroe County	06	56	050000	39111	39	111	14642	7567
18	Polygon	39	087	01074056	39087	Lawrence	Lawrence County	06	44	050000	39087	39	087	62450	27603
19	Polygon	39	001	01074014	39001	Adams	Adams County	06	1	050000	39001	39	001	28550	12978

FIGURE 2.5 Joining two tables together on the basis of their common field (key).

in our campus building example, the geospatial layer has a single record for each building, and the campus spreadsheet also has a single entry for each building.

many-to-one join A join in which many records in an attribute table are linked to one record from another table.

relate A join operation that establishes a connection between tables but does not append fields from one table to another table.

A different type of join is a **many-to-one join**, in which many of the records in an attribute table are joined to the same one record in the other table. An example of this type of join would be a geospatial layer of land cover for a county. Each polygon in the land-cover layer would be tagged with a particular code indicating whether that piece of land represented water, forests, agriculture, urban, barren, or wetland land cover. However, the county is covered with tens of thousands of polygons, each with one of six land-cover codes. A separate spreadsheet with more detailed information about each of the six land covers (such as a description and attribution) will be joined to the geospatial layer. In this case, each of the thousands of records in the geospatial layer will be linked to only one of the six possible records in the spreadsheet, resulting in a many-to-one join. As in a one-to-one join, the result is a single table.

When other types of linkages (such as a one-to-many join, in which several different records can be linked to a single record in another table—this is the reverse of a many-to-one join) need to be performed, a **relate** operation is used. A relate links the information in the two tables rather than merging them together into a single table.

- Add the **DEC_10_SF1_G001** table to ArcGIS Pro; you can add a table the same way you add other data. You then see a new item in the Contents pane with the table name appear under the heading "Standalone Tables."
  - *Note:* This geodatabase table was derived from an Excel spreadsheet and thus is only one table. If you had the Excel table with multiple sheets, you would have to select the sheet to add.
- In the Contents pane, open the **DEC_10_SF1_G001** table (which you do the same way you would open a layer's attribute table). In the table window, you see a second tab (corresponding to this spreadsheet) added next to the CensuscountiesOhio attribute table. Select the **DEC_10_SF1_G001** tab to examine this table.
- The DEC_10_SF1_G001 table has several more attributes. You can use one of them (the attribute called **COUNTYFP** in the CensuscountiesOhio attribute table and the attribute called **County** in the DEC_10_SF1_G001 table) for the join. Answer Question 2.5.

> **Question 2.5**
> Each of the two tables has another field with the same information (in the CensuscountiesOhio table, it is called STATEFP, and in the DEC_10_SF1_G001 table, it is called State). Why can these fields not be used as the key when joining the population values to the county feature-class records?

- When joining two tables, you begin with the table you want to join attributes to. To join the non-spatial attributes in DEC_10_SF1_G001 to the geospatial features in CensuscountiesOhio, you begin with the geospatial layer. There are two ways of initiating this join procedure in ArcGIS Pro:
  1. In the Contents pane, right-click on the layer you want to join attributes to (in this case, the **CensuscountiesOhio** layer) and then select **Joins and Relates** and then **Add Join**.

  OR:

  2. In the Contents pane, click once on the layer you want to join attributes to (in this case, the **CensuscountiesOhio** layer), and then, on the ribbon, in the **Data** tab, within the **Relationships** group, click the **Joins** button and choose **Add Join**.
- The Geoprocessing pane opens on the right-hand side of the screen where the Catalog pane was, and it contains the Add Join tool dialog box.
- In the Add Join tool, under Layer Name or Table View, select **CensuscountiesOhio** (which is the layer to which you want to join the table).
- Under Input Join Field, choose **COUNTYFP** (which is the field in the CensuscountiesOhio layer you will use as the key for the join).
- Under Join Table, choose **DEC_10_SF1_G001** (which is the table with the non-spatial information you want to join).

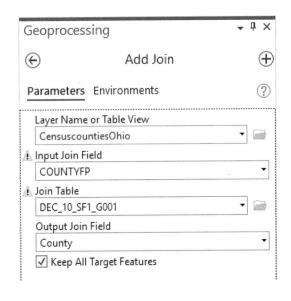

- Under Output Join Field, choose **County** (which is the field in DEC_10_SF1_G001 you want to use for the join).
- Leave the other options alone and click **Run** in the lower-right corner of the Geoprocessing pane.
- Back in the table window, select the **CensuscountiesOhio** tab (the one you joined the table to) and scroll across it. You see that the fields from the DEC_10_SF1_G001 table have been added via the key. Each record in CensuscountiesOhio now also has its corresponding data from DEC_10_SF1_G001 properly in place.

Step 2.5 Selecting Records in an Attribute Table

- Now that the tables are joined, you can find out which county in Ohio has the largest population. Scroll over to the Population attribute in CensuscountiesOhio. Right-click in the header of the field (the gray Population name), and a new menu of possible actions appears.
- Select **Sort Descending** to sort the entire table using the values in the Population field.
- With the table sorted, the largest population value is now at the top of the table. Scroll across to find the NAME field to see which county has the largest population and answer Question 2.6.

Question 2.6 Which Ohio county has the largest population?

- Look at the far left-hand side of the attribute table and click the mouse in the small box to the left of the first record. You see the record highlighted in a cyan color; this indicates that the first record is now selected in ArcGIS Pro. Minimize the table, and you also see the corresponding county polygon outlined in cyan. For more information about selection in ArcGIS Pro, see **Smartbox 2.4**.

SMARTBOX 2.4

What does *selection* mean in ArcGIS Pro?

selection The process of choosing certain records or features and setting them aside from the remainder of the records or features.

In ArcGIS Pro, **selection** refers to the process of choosing records from an attribute table or choosing their corresponding features from a layer and holding them aside to work with those records or features separately. By default, ArcGIS Pro uses a cyan color to denote selected features or records. When records are selected in an attribute table, they are highlighted in cyan, and their corresponding features in the view are also highlighted in cyan.

For instance, the records that correspond to the results of a query (see Step 2.6) are considered selected records. The table window allows you to examine all records or only selected records; this is often useful for viewing only those records that were the result of a query. The table window also gives you a count of how many records are selected at any given time. Often, actions that are taken (such as computing statistics or exporting data to another layer) are performed only on the selected features. Thus, it's advisable to check which items (if any) are selected before performing an action. If necessary, you can clear all the selected features before proceeding with another task.

There are several ways of selecting features and records in ArcGIS Pro. A Select By Attributes query (see Step 2.6) is commonly used to select records through a query. However, features can also be selected through spatial queries (see Module 8) or interactively by the user (see Module 9). When performing a selection, you can choose from the following options:

- **New selection**: The query results wipe any existing selection and specify a new set of selected features or records.

- **Add to current selection**: The query results are selected, and any existing selection is retained as well.

- **Remove from current selection**: The pool of selected results is retained except for the results of the query.

- **Select subset from current selection**: The query results are drawn only from the pool of selected records, not any unselected records.

- **Switch the current selection**: Any selected records become deselected, and any previously unselected records become selected. In essence, this reverses which records are selected and which are not selected.

- You'll do much more with selection in the next steps, but for now, clear the one selected record by clicking the **Clear** button at the top of the ArcGIS Pro table window. You can also clear selected records by going to the **Map** tab on the ribbon, and in the **Selection** group, clicking the **Clear** button.

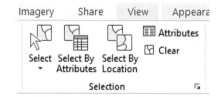

Step 2.6 Performing a Simple Query (Select By Attributes)

- Often, you'll find you want to obtain more information from attribute tables besides simply sorting fields and digging through the attributes. For instance, in our joined table, there's information about population and housing characteristics. If you wanted to find the counties with a certain level of population *and* a minimum number of housing units, you would need to do a lot of cross-checking of columns, numbers, and names.
  To obtain this kind of information from a table in GIS, you can perform a **query** of the table. For more information about building simple queries in ArcGIS Pro, see **Smartbox 2.5**.

query The conditions used to retrieve data from a database or table.

SMARTBOX 2.5

What is the structure of a simple query in ArcGIS Pro?

SQL The Structured Query Language—a formal language for building queries.

simple query A query that contains only one operator.

relational operator One of six connectors used to build a query.

A query in ArcGIS Pro is done in **SQL** (Structured Query Language) format and works like a mathematical function. A **simple query** involves choosing one field (or attribute), choosing an operator that will affect that attribute, and then choosing a value. For instance, as a result of the join, your Ohio counties layer now contains an attribute field called Population. If you wanted to find all counties that have a population of more than 100,000 persons, you would first select the proper field (in this case, **Population**), select an operator (in this case a Greater Than, >, operator), and then choose a value to use in the query (in this case, the number **100000**). Your SQL query expression would therefore be **Population > 100000**. After ArcGIS Pro evaluates this simple query, all records that have a value for the Population attribute that is greater than a value of 100000 will be selected.

In ArcGIS Pro, you have a choice of **relational operators** to use. There are six main relational operators:

- Equal (=)
- Not Equal (<>)
- Greater Than (>)
- Greater Than or Equal To (>=)
- Less Than (<)
- Less Than or Equal To (<=)

These operators work like their mathematical equivalents and are useful when selecting population values greater than or less than a certain amount. The Equal and Not Equal operators are useful when selecting text-based attributes as well. For instance, your Ohio counties layer also has an attribute field called NAME that contains the names of each county as a string of text. If you wanted to select the record corresponding with a particular county (such as Lake County), you could build a query using the NAME field, the Equal operator, and a string of text (for example, 'Lake') that you are looking for. In this case, your query expression would read **NAME is Equal to 'Lake'**. When ArcGIS Pro evaluates this query, the record whose NAME field contains the string 'Lake' is selected. Note that single quotes are necessary when building a query for a text string.

To reverse this operation and find all counties that are *not* named Lake, you could use the Not Equal operator and build this query expression: **NAME is Not Equal to 'Lake'**. When this query is evaluated, all counties except for the record whose NAME field contains the string 'Lake' are selected.

In ArcGIS Pro, the Select Layer By Attribute tool is used to build and run SQL queries. Inside this tool is a function called the Query Builder that you can use to construct a query. You can either manually choose the field, operator, and value from pull-down menus (what ArcGIS Pro refers to as "Clause mode"), or if you are more experienced with SQL, you can use the SQL Edit tools to type your own SQL expressions by hand.

- To build a simple query in ArcGIS Pro, select the **Map** tab, and within the **Selection** group, click on the **Select By Attributes** icon.
- The Geoprocessing pane opens on the right-hand side of ArcGIS Pro, showing the Select Layer By Attribute tool.
- To choose which layer to query, under Layer Name or Table View, choose **CensuscountiesOhio**.
- Under Selection type, choose **New selection**.
- Next, build the SQL query by clicking on **New expression** to use the interactive query building features of ArcGIS Pro. From the options that appear, choose **Create new expression**. A new pop-up window that allows you to build the query opens.

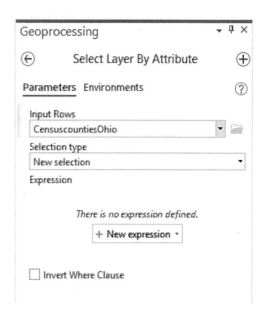

- From the first pull-down menu, choose the field you want to use. In this case, choose **Population**.
- From the second pull-down menu, choose the operator you want to use. In this case, choose **is greater than**.
- From the last pull-down menu, choose the value you want to use. For this query, you won't be using any values from the POPULATION field in the attribute table, so you can just type **127500**.
- Your final query should read: **Population is Greater Than 127500**. Click on the **green checkmark** icon to verify whether the expression is valid. If it is (and it should be), ArcGIS Pro gives you a message saying that the SQL expression is valid. If you've completed the query incorrectly, click the **red X** icon to wipe the boxes clean and build the query again. Back in the Select Layer By Attribute dialog box, click **Run** to execute the SQL query. Open the CensuscountiesOhio layer's attribute table. The number of selected records (that is, those that met the results of your query) is shown in the attribute table's window, and you can also see that counties that met the results of your query are selected in the map. Both selected records in the attribute table and features in the map are shown in a cyan color.
- Click the **Show Selected Records** button in the table to show only the selected records. Answer Questions 2.7, 2.8, and 2.9.

Question 2.7 How many Ohio counties have a population of more than 127,500 persons?

Question 2.8 Where (in general terms) are these higher-population counties located in Ohio?

Question 2.9 Do any of the following counties have more than 127,500 persons (and if so, which ones): Stark, Portage, Trumbull, Mahoning, or Columbiana?

- In the CensuscountiesOhio attribute table click the **Clear** button at the top of the table to clear the results of the selection and restore the table to not have any records or features selected.
- Return to the Geoprocessing pane and the Select Layer By Attributes tool to build a fresh new query. Because you're starting with a brand-new query, you'll want to clear the existing expression by clicking on the **Remove** icon (the red X) in the tool.

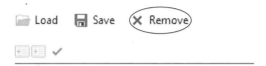

- Build a new query to find which counties in Ohio have the lowest overall population. To do so, perform a Select By Attributes query to find all counties with a population of fewer than 40,000 persons. Answer Questions 2.10, 2.11, and 2.12.

Question 2.10 How many Ohio counties have a population of fewer than 40,000 persons?

Question 2.11 Where (in general terms) are these lower-population counties located in Ohio?

Question 2.12 Do any of the following counties have fewer than 40,000 persons (and if so, which ones): Stark, Portage, Trumbull, Mahoning, or Columbiana?

- Clear the selection again.

Step 2.7 Performing a Compound Query (Select By Attributes)

- Often, when using queries for analysis or to examine subsets of your data, a simple query does not provide enough flexibility. For instance, if you wanted to know something about both population and housing characteristics, a simple query wouldn't be useful because you could build an expression of only one attribute at a time. The same holds true if you wanted to simultaneously determine the counties with both the largest and smallest populations in a single query; you could not accomplish this with one simple query. To perform these more complicated actions, you need to build a compound query. (See **Smartbox 2.6** for more about writing compound queries.)

SMARTBOX 2.6

What is the structure of a compound query in ArcGIS Pro?

A simple query allows you to evaluate only *one* attribute using *one* relational operator (such as **NAME = 'Lake'**). However, there will be plenty of times when you're going to want to perform a query that has multiple criteria (such as selecting both **NAME = 'Lake'** and **NAME = 'Mahoning'** simultaneously). A **compound query** can combine two (or more) simple queries. For instance, your Ohio counties layer has attribute fields for both population (the number of persons in the county) and housing (the number of homes in the county). If you wanted to find all counties with both a population of more than 100,000 persons and also more than 100,000 homes, you would need to create the first simple query (**Population > 100000**) and connect it with the second simple query (**Housing > 100000**) so that you could effectively query both attribute fields at the same time.

A compound query relies on a special kind of connector called a **Boolean operator** that explains how the multiple queries are tied together. In an SQL query, the Boolean operators AND and OR are used to build compound queries. The **AND** operator is used when finding the commonalities (or the intersection) of both parts of the query. For instance, in the population and housing example above, using the AND operator to make the query read **(Population > 100000) AND (Housing > 100000)** would select all records whose value for Population is above 100,000 *and* whose value for Housing is above 100,000. If only one of these two criteria is met (for instance, a record with a Population value of 250,000 but a Housing value of only 99,000), that record would not be selected. To satisfy AND as the Boolean operator, *both* parts must be valid.

The **OR** operator is used when either one part or the other (or both) of a query is valid. To continue with the same example, suppose the query were constructed as follows: **(Population > 100000) OR (Housing > 100000)**. Any record whose value for Population is above 100,000 would be selected, in addition to any record whose value for Housing is above 100,000. If a record had a Population value of 250,000 and a Housing value of 99,000, it would still be selected because one part of the query is met (the Population value). To satisfy OR as the Boolean operator, at least one part of the expression must be valid, and then that record is selected.

Another type of Boolean function available in a query is the **NOT** operator, which is used to effectively reverse the results that would normally be selected. For instance, a query that reads **NOT (NAME = 'Lake')** would select all records that do not have the text string 'Lake' in their NAME field (likely all records except for the one representing 'Lake' county).

Compound queries can be used to create more complex types of query expressions. For instance, if you wanted to determine all records with either a population of greater than 100,000 or less than 50,000 and to exclude Lake County from your analysis, your query expression would read **((Population > 100000) OR (Population < 50000)) AND NOT (NAME = 'Lake')**.

compound query A query that contains more than one operator.

Boolean operator (query) One of the connectors used in building a compound query.

AND (query) The Boolean operator that chooses features that meet both criteria in a query.

OR (query) The Boolean operator that chooses features that meet one or the other (or both) of the criteria in a query.

NOT (query) A Boolean operator used to negate a function or query.

- You now need to make a compound query that determines which Ohio counties have a population of more than 127,500 persons and also contain more than 100,000 housing units. Return to the Select Layer By Attribute tool.
- Build the first part of the expression as you did before, so that it reads **Population is greater than 127500**, and click **Add** (but don't click Run just yet as this is just one part of the compound query).
- Click **Add Clause** to add the second part of the compound query.

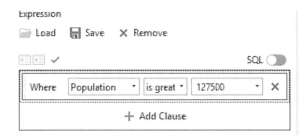

- In the new dialog box, from the first pull-down menu, choose **And**. This is the Boolean connector that will join the first part of the query with the second part of the query.
- From the other pull-down boxes complete the query so that it reads **Housing is greater than or equal to 100000**.

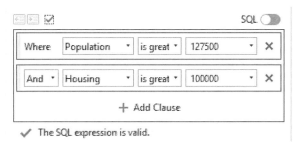

- Back in the Select Layer By Attribute tool, click on the **green checkmark**, and ArcGIS Pro verifies that everything is okay with your compound query.
- Click **Run** to execute your compound query. Again, you see the selected records and polygons in cyan. Answer Question 2.13.

Question 2.13 How many Ohio counties have a population of more than 127,500 persons and also have more than 100,000 housing units? Which counties are they?

- Clear the selection.
- As noted earlier in this module, part of your project is to examine the characteristics of a small set of counties in northeastern Ohio: Mahoning, Columbiana, Stark, Portage, and Trumbull. Rather than work with all 88 Ohio counties, you can select these five counties to work with as a subset of the county data. You can build a new compound query to select only these five counties.
- Return to the Select Layer By Attribute tool and remove any existing queries that are in it. Your compound query will be looking for five

different counties, and thus you'll be making the equivalent of five simple queries, joined by Boolean operators. For this compound query, you'll be using the NAME field, as it contains text strings of county names.

- Build the first part of the compound query by finding the NAME equal to Columbiana. First, choose **NAME** for the field and then choose **is equal to** as the operator, and then in the third pull-down menu, choose **Columbiana** from the list of values.

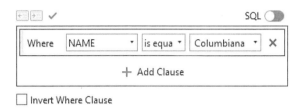

- To add the second part of the SQL query using a Boolean operator, click **Add Clause**.

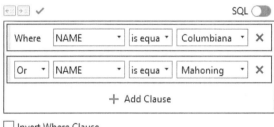

- Select **Or** from the initial pull-down menu and then build the second part of the compound query so that it reads **NAME is equal to Mahoning** (by choosing **Mahoning** from the Values pull-down menu).
- Click the **Add Clause** button and add the third part of the compound query so that it reads **Or NAME is equal to Portage**.
- Click the **Add Clause** button again and add the fourth part of the compound query so that it reads **Or NAME is equal to Stark**.
- One more time, click the **Add Clause** button and add the fifth part of the compound query so that it reads **Or NAME is equal to Trumbull**.
- Your compound query should be choosing all five county records based on their NAME attribute. Before proceeding, click on the **green checkmark** icon to verify that everything is valid. If it isn't, clear the query and rebuild it again. When everything is valid, answer Question 2.14. Finally, click **Run** to run the Select Layer by Attribute tool and execute the query.

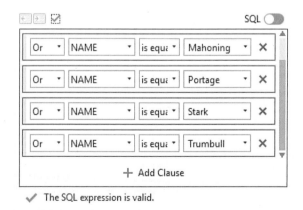

> **Question 2.14**
> Why are you using OR as the Boolean connector in all four places in the compound query? What would be the result of using AND in place of OR all four times?

Step 2.8 Exporting the Results of a Selection to a New Layer

- Now that the five counties are selected, you can create a new feature class that contains only these counties. Because all further analysis is limited to these five counties, you don't need the other 83 Ohio counties in the layers you're working with. You can take selected features (such as your selected five county polygons) and create a new dataset from them in one of two ways:

  1. Right-click on the layer you want to export from in the Contents pane (in this case, **CensuscountiesOhio**) and choose **Data** and then choose **Export Features**.

     OR:

  2. Click once on the layer you want to export from in the Contents pane (in this case, **CensuscountiesOhio**) and then, on the **Data** tab of the ribbon, within the **Export** group, click the **Export Features** button.

- The Geoprocessing pane appears, this time with the Feature Class to Feature Class tool.

- Under Input Features, choose **CensuscountiesOhio** as the layer from which you want to take features.

- Under Output Location, the default should be **Module2.gdb**. This is the default geodatabase created with your Module2 project.

  - For purposes of this module, just use this as the place where you want to put the new exported feature class. If at a later time you wanted to export the data to a different location, you would click the **browse** button next to Output Location and navigate to a different file geodatabase and choose that instead.

- Under Output Feature Class, type **Fivecounties**. This is what your new feature class will be called.

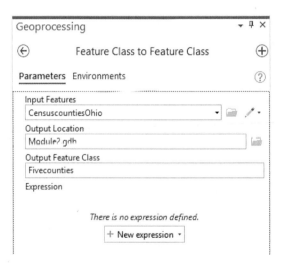

- Leave the other settings alone and then click **Run**.

- A new polygon layer of only the five selected counties is placed in the Contents pane. You will not be using the old CensuscountiesOhio layer, so right-click on it in the Contents pane and select **Remove** to delete it from the Contents pane.

- Right-click on the **Fivecounties** layer and choose **Zoom to Layer** to adjust the view to only the five counties.

Step 2.9 Using Queries for Analysis

- The next part of your project involves examining the locations of colleges and universities within this five-county region. Add the **colleges** feature class to the Contents pane. This layer contains points showing the locations of those places tagged as a college or university by the USGS. Place this layer on top of the counties so that the points are visible. Change the symbology of the layer to a single symbol using something more prominent than a simple point. (See page 25 in Module 1 for directions on how to change a layer's symbology.)

- Open the colleges attribute table and answer Question 2.15.

> **Question 2.15**
> How many points in this five-county region are considered colleges or universities?

- For ease of use, you can now place labels next to each of the points to help identify them. In the Contents pane, click on the **colleges** layer to tell ArcGIS Pro that this is the layer you want to label features on, and then click on the **Labeling** tab on the ribbon.

- On the **Labeling** tab, in the **Label Class** group, place a checkmark in the **Label Features In This Class** box, and from the **Field** pull-down menu, choose **Name**. This allows you to specify which attribute field ArcGIS Pro will be drawing from to label the points—in this case, you'll use the field called Name that contains the name of the college or university.

- To actually place labels next to the points, on the **Labeling** tab, in the **Layer** group, click the **Label** button. You see the name of each college or university appear next to its respective point. Note that you can click the **Label** button to turn the labels off and then back on at any time.

- The Labeling tab gives you a multitude of options for customizing the labels of your features, including the range at which the labels will be visible; the symbology, color, and font used for the labels; and where the labels will be shown in relation to their features. Try out some of the options to change the appearance of the labels to whatever you find gives the best-looking labels for this layer.

- Next, you can examine some major cities within this five-county region and see how many colleges and universities are located within each one. You can do this by performing an SQL query using the colleges layer and its City attribute.

- Bring up the Select Layer By Attribute tool and build a query to find all records where the **City** attribute is equal to **Canton**. Answer Questions 2.16 and 2.17.

Question 2.16 How many colleges or universities are in the city of Canton?

Question 2.17 In which county is Canton located? (*Hint:* Use the Explore tool to examine the attributes of the Fivecounties layer to answer this question.)

- Clear the selected features and build a new query to find all records where the **City** attribute is equal to **Youngstown**. Answer Question 2.18.

Question 2.18 How many colleges or universities are in the city of Youngstown?

- Clear the selected features.

Step 2.10 Configuring Attribute Pop-ups and Working with Attachments

- Return to the **Map** tab, and in the **Navigate** group, click the **Explore** button to activate the **Explore** tool. Next, click on the point representing Youngstown State University. The pop-up window that opens shows all the values from the attribute table that connect with that point; you can see that there are a lot of attributes. If you were going to be using this information in a presentation, you wouldn't need to display all those attributes; rather, you would just want to get across pertinent information about a particular point. ArcGIS Pro gives you the ability to adjust the attributes shown in the pop-up. In this case, right-click on the **colleges** layer and select **Configure Pop-ups**.

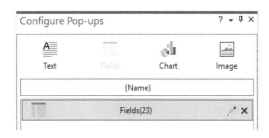

- The Configure Pop-ups pane opens on the right-hand side of the screen. To change which fields of data are displayed, next to the Fields heading, click on the **edit** icon (the pencil).

- Under Fields Options, you see a list of the 23 different fields in the colleges attribute table. If a checkmark is in the Display box, that field's attribute will appear in the pop-up you get when clicking with the Explore tool; if there is no checkmark, that attribute will not appear. You should set checkmarks for only the following seven fields: FType, FCode, Name, Address, City, State, and Zipcode.

- Click the **left-facing arrow** in the upper-left corner of the Pop-ups pane to save your edits.

- Back in the main Configure Pop-ups pane, you see that the Fields indicator has the number 7 next to it, indicating that only seven fields will be shown when you click on a college point. To see this in action, use the **Explore** tool to click on one of the colleges, and you see that only those seven attributes are displayed in the pop-up.

- Pop-ups can contain more than just information from a feature's attribute table; items called *attachments* can be used as well (see **Smartbox 2.7**). To use attachments, they must first be enabled for a layer, using a tool. If the Geoprocessing pane is closed, go to the **Analysis** tab, and in the **Geoprocessing** group, click on the **Tools** button to reopen it. The tool to use is the Enable Attachments tool; either search for it in the Find Tools box or, in the **Toolboxes** option, open the **Data Management** toolbox, within the **Attachments** toolset, and choose the **Enable Attachments** tool.

> **attachment** A function that allows a user to connect non-spatial media, such as documents, files, photos, or charts, to a geospatial feature and make them part of the triggered pop-up.

SMARTBOX 2.7

What is an attachment, and how is it used in ArcGIS Pro?

ArcGIS Pro makes it possible to have attachments or other non-spatial items, such as photos, reports, graphs, or HTML documents, also appear in the pop-up when an object is clicked on. For instance, a utility inspector with a point layer of fire hydrants in an area could go to the site of a hydrant, take a photograph of it, and then set it up as an attachment for a point in a layer representing that particular hydrant. Then, when that point in the layer with the attachment is clicked on in ArcGIS Pro, the photo of the hydrant appears in the pop-up along with its attributes.

Similarly, if you have a point layer representing points of interest along a hiking trail, you could set up attachments so that, whenever you click on one of the points, the pop-up contains a photo, or when you click on the polygon representing the park boundary, a chart of recreational usage is part of the pop-up. Graphs, photos, and reports are all examples of non-spatial data, and attachments allow you to **geotag** them, or link them to spatial locations.

> **geotag** A process whereby non-spatial media are linked to geospatial features.

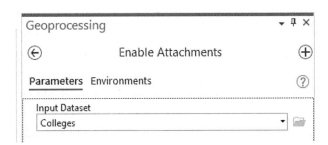

- In the Enable Attachments tool, under Input Dataset, choose **Colleges**. Then click **Run**. This allows the colleges layer to have attachments added to it.

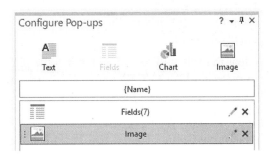

- Return to the Configure Pop-ups pane. Click the button that says **Image** to add another option, called Image, to the pop-up. Click on the **pencil** icon next to the **Image** option.

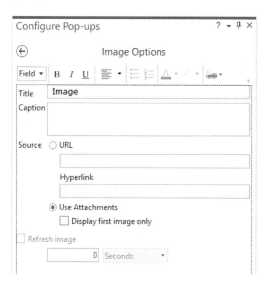

- In the Image Options dialog, you need to tell ArcGIS Pro the source of the image it should display. Click on the radio button next to **Use Attachments** to tell ArcGIS Pro to display an image when that image is attached to a particular point. Then click the left-facing arrow at the top of the dialog box.

- For attachments in the colleges layer, you need to choose a photo and connect it to a particular point. On the **Edit** tab, within the **Selection** group, click on the **Attributes** button. The Attributes pane opens on the right-hand side of the screen.

- The Attributes pane contains a message asking you to select one or more features. You can attach photos one feature at a time. In the colleges layer's attribute table, select the record for Youngstown State University. The record is highlighted in cyan, and the contents of the Attributes pane change to show the various attributes associated with the Youngstown State University record.

- To attach the photo to this selected record, in the Attributes pane, click on the **Attachments** option.

- Under the Attachments option, click the **Add** button (which has the green plus symbol next to it). From there, navigate to your **C:\GIS\Module2** folder and choose the **YSU.JPG** file and click **Open**. You see that YSU.JPG (which is the name of the photo of YSU) is added as the attachment for this particular point/record.

- On the map, you see the Youngstown State University point selected in cyan. Click on it with the Explore tool, and you see the pop-up appear with the seven displayed fields as well as the attached photo.

- Clear the selected record and attach the photo called **ITT.JPG** to the record for ITT Technical Institute—Youngstown.

- After you've attached that photo, clear the selected record and attach the photo called **MCCTC.JPG** to the record for Mahoning County Career and Technical Center.

- Answer Question 2.19.

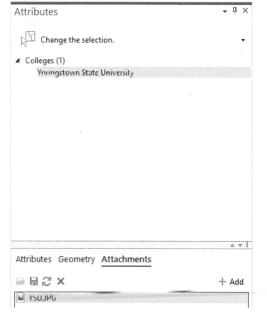

> **Question 2.19**
> Demonstrate your three working attachments for your instructor, who will check over your work for completeness and accuracy to make sure you get credit for this question.

- Save your Module2 project and, if necessary, export it to a project package. (See Module 1 for more about saving projects and project packages.)

Closing Time

The ability to link non-spatial data to geospatial locations makes GIS different from other types of information systems. Linking attribute data to geospatial features is a common process in GIS. A good source for large amounts of different types of data is the U.S. Census. The Census Bureau doesn't provide data such as population demographics or housing characteristics in a geospatial format but rather as a series of tables that you can use in ArcGIS Pro by joining them to features. See *Related Concepts for Module 2* for more information about how to access and use these kinds of data in ArcGIS Pro.

Handling attribute information, querying attribute tables, and joining tables together are all common tasks in GIS, and throughout this book, you'll often find yourself working with attributes tied to geospatial data. As noted in Module 1, a different kind of GIS data (called *raster data*) doesn't have the same format as the vector data you've been working with, and thus it handles its attributes a bit differently, but you'll get to that in Module 12. Starting in the next module, you'll be applying your knowledge of geospatial data and attributes to creating a professional-looking map of your results.

RELATED CONCEPTS FOR MODULE 2

Using U.S. Census Data and Attributes in ArcGIS Pro

The U.S. Census Bureau conducts a census every 10 years, collecting a vast amount of data about the U.S. population. This rich dataset includes information on population demographics, as well as the country's social, economic, and housing characteristics. Different characteristics are available depending on the unit of analysis being examined (for example, the state, county, Census Tract, or Block Group level). The Census Bureau makes all these datasets available to the public (for free) via the Census website at **https://data.census.gov** (Figure 2.6).

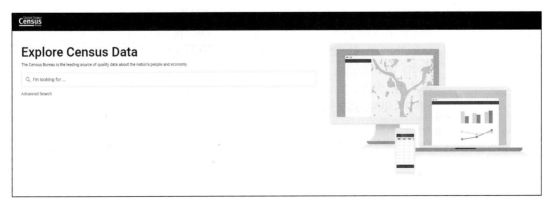

FIGURE 2.6 The U.S. Census Bureau's data website.

Prior to 2020, these data were served out through the American FactFinder (AFF) program, but these data and their functionality have been rolled into the main Census data website.

Using the Census data website, you can specify the characteristics you wish to examine and the geographic level, and then you can access all this information in tabular format. These tables of data can also be downloaded to use as attributes in GIS, as each data table is similar in structure to an attribute table. Each unit being examined (such as each state or each county) is a record, and the various Census values (such as total number of houses or total population) are stored as fields. However, even though this attribute data is organized geographically, it is still non-spatial in nature. While you could examine characteristics for a particular county, you could not perform spatial analysis with that information. To do so, you would have to join these tables to geospatial data layers within ArcGIS Pro.

The U.S. Census Bureau also provides separate Cartographic Boundary files and TIGER/Line files (in shapefile format) for a variety of different areal units (such as state, county, congressional district, Census Tract, Block Group, or county subdivision) through its Geography website at **https://www.census.gov /programs-surveys/geography.html**. You can download the Cartographic Boundary files for the areas you want (such as your state's counties or your state's Census Tracts) from the U.S. Census TIGER/Line resources and then compile and download the tabular data you wish to examine through Census data website. Then, in ArcGIS Pro, you can join the non-spatial Census data to the geospatial layers of the Cartographic Boundary files and examine U.S. Census data in ArcGIS Pro. This is how you used the data in this module; the county boundaries file was downloaded from TIGER/Line resources, and the non-spatial data on population and housing came from the Census data and American FactFinder.

Another source for Census data to use in ArcGIS Pro is the National Historical Geographic Information System (NHGIS), available online at **https://www .nhgis.org**. NHGIS allows to download (for free) current and historical Census data in tabular format, as well as current and historical GIS layers as shapefiles. The data from NHGIS has already brought together the Cartographic Boundary files with selected Census attributes for use in ArcGIS.

Key Terms

non-spatial data (p. 34)
attribute table (p. 34)
record (p. 34)
field (p. 34)
attribute (p. 35)
nominal data (p. 41)
ordinal data (p. 41)
interval data (p. 42)
ratio data (p. 42)

join (p. 43)
key (p. 43)
one-to-one join (p. 43)
many-to-one join (p. 44)
relate (p. 44)
selection (p. 46)
query (p. 47)
SQL (p. 48)
simple query (p. 48)

relational operator (p. 48)
compound query (p. 51)
Boolean operator (p. 51)
AND (p. 51)
OR (p. 51)
NOT (p. 51)
attachment (p. 57)
geotag (p. 57)

How to Create a Layout with ArcGIS Pro

ArcGIS Pro Skills

In this module, you will learn how to do the following in ArcGIS Pro:
- Symbolize data by multiple values using graduated colors.
- Create a choropleth map and examine different types of data classifications.
- Create a layout and use the layout navigation tools.
- Add a map frame to a layout and modify it.
- Activate a map frame.
- Add various map elements (scale bar, north arrow, and legend) to a layout and modify them.
- Add text to a layout and modify it.
- Design a layout that has artistic and aesthetic merit.
- Print a completed layout or export it to another file format.

Introduction

One of the key uses of GIS is to make a map of your data or the results of your analysis. A **map** (not to be confused with the map view in ArcGIS Pro that you've used for two modules now) is a visual representation of geospatial data that conveys a message about location-based concepts. For instance, GIS can be used to create a map to illustrate property boundaries, hot spots of criminal activity, or the extent of a watershed. This is a unique feature of geospatial data: It can be visualized using a map rather than a chart or a graph (Figure 3.1).

However, there's more to creating an effective map than adding a title and legend to your data and printing it out. **Cartography** is the art and science of mapmaking. Learning good map-design (cartographic) skills will help you create maps that are more useful for your target audience. ArcGIS Pro allows you to quickly make professional-looking maps from your geospatial data and also allows you the freedom to design a map the way you want to.

Several types of maps can be created using GIS. **Reference maps** convey location information or highlight various features of an area. All of the following are examples of reference maps: a map of the roads and highways between Cleveland

Learning Outcomes

After studying this module, you should be able to:
- Explain what cartography is.
- Explain the difference between normalized data and regular count data.
- Describe the differences between four types of data classification methods in choropleth maps.
- Explain how scale is represented on a map.
- Describe a minimum of three different map elements used in a layout.
- Explain the concepts of good cartographic design.
- Demonstrate how to compose a layout in ArcGIS Pro.

map A visual representation of geospatial data that conveys some sort of message to its reader.

cartography The art and science of making maps.

reference map A map that shows the location of features rather than thematic information.

Solar Radiation for Buildings in Central Wellington

Average Annual Values for Rooftops in Kilowatt Hours per Square Meter

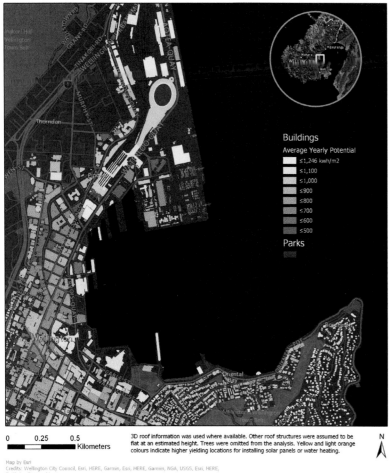

FIGURE 3.1 A map of Wellington, New Zealand, showing average yearly solar radiation.

topographic map A map created by the USGS to show landscape and terrain as well as the location of features on the land.

thematic map A map that displays a particular theme or feature.

choropleth map A type of thematic map in which data are displayed according to one of several different classifications.

and Pittsburgh, a map of tourist locations in San Diego, a map showing the location of abandoned housing in a city, and a map of the location of water wells in a rural area. **Topographic maps** are specific kinds of reference maps that show factors such as landforms, land-cover types, and other built or natural features. (See Module 16 for more about topographic maps.)

A **thematic map** illustrates a specific theme. The following are examples of thematic maps: a map of the United States showing whether or not a state has a smoking ban in public buildings, a map that shows whether a state's electoral votes went to Donald Trump or Hillary Clinton in the 2016 presidential election, and a map showing which countries are members of the European Union and which are not.

Choropleth maps are thematic maps that show multiple values (rather than, say, the two choices of "red state" or "blue state"). One example of a choropleth map is a map showing the percentage of the popular vote in each state for a single Presidential candidate. Another example is a map showing a state's federal income tax rate, or a map that shows each county's population per square mile.

In this module, you will take a value from a layer's attribute table and create a choropleth map from it. Then, from the choropleth map you'll create a professionally designed map layout to print out. Specifically, you'll be making design choices to create an effective map, choosing color and scale, and adding map elements such as a scale bar, legend, north arrow, title, and descriptive text. Unlike most of the other modules in this book, this module does not provide a series of questions for you to answer but rather includes a checklist of items to aid you in making sure your map is complete.

In many later modules you will compose a map layout at the end of your analysis in order to present your results. While you won't be specifically using a choropleth map in other modules, the skills and knowledge you gain from this module will enable you to create professional-quality maps from other types of geospatial data.

 ## Module Scenario and Applications

In this module, you will take the role of a real estate agent who is involved with the sales of vacation homes in Ohio. Your goal is to create an effective map that shows seasonal homes in each Ohio county (as the percentage of the total housing stock for that county). You can then distribute your map to colleagues within your agency and to potential home buyers to show them which counties have the highest percentages of seasonal homes.

Here are some other real-world applications of this module's theory and skills:

- A researcher at the Centers for Disease Control and Prevention needs to prepare a map showing the rates of bird flu in each township in the state of Georgia. The map needs to be colorful and well designed so that it can be distributed to the general public.

- An analyst at a police department is preparing a map showing the number of 911 calls in a city per household. She is designing the map at the Census block level for distribution to other police analysts.

- An economist is examining population and income demographics for several states and is mapping them at the county level. He is creating several maps of these factors using Census data prior to performing more detailed analysis.

Data Sources and Localizing This Module

The data in this module focus on features and locations within Ohio's counties. However, this module can easily be modified to use U.S. Census Bureau data on your own county or local area instead. (See *Related Concepts for Module 2* for more about using Census data in ArcGIS Pro.) This module used county boundaries from the TIGER/Line files from the Census, but the way Census data is served and formatted has been updated since then. Comparable Census data for your own county are available from the U.S. Census for free. You need to download a county boundary dataset and attribute tables from the Census and then join them together.

The county boundaries can be obtained as Cartographic Boundary files from **https://www.census.gov/geographies/mapping-files/time-series/geo/carto-boundary-file.html**. The counties layer you download contains all the counties in the United States. You can then query the shapefile and extract the appropriate counties into a separate feature class for your state, based on the STATEFP attribute. This attribute references each state with a two-digit FIPS code. For instance, Ohio is referenced with state FIPS code 39. If you were using this module in Florida, for instance, you would use FIPS code 12. A full list of FIPS codes for each state and county is available from the National Weather Service at **https://www.weather.gov/pimar/FIPSCodes**. Note that some of the U.S. Census Bureau's county boundaries extend to the national boundaries when they border water features (such as in the Great Lakes area between the United States and Canada). Also note that the county boundary file was projected into the Web Mercator (auxiliary sphere) projection for purposes of this module. There might be another projection for your own state (such as the Lambert Conformal Conic projection) that you want to use instead. (See Module 1 for more about projecting data.)

Attribute data must be downloaded separately and joined to the counties shapefile. The attribute data (the population statistics) for each county in the state were downloaded from the Census Bureau's American FactFinder website, but these data are now available through the Census data website at **https://data.census.gov/cedsci/**. The tables used in this module are Vacancy Status (and the "for seasonal, recreational, and occasional use" attribute) and Occupancy Status (for the total number of houses) for each county. Census attribute data were downloaded as a CSV file, edited and cleaned up, and converted to an Excel file for use in this module. The tabular data from the Census data website can be joined to the Cartographic Boundary layer by using the AFFGEOID attribute in the Cartographic Boundary file and the GEO ID attribute in the Census data.

Step 3.1 Getting Started

- This module's hands-on applications use the data folder called Module3. Your instructor will be able to supply you with this data, or you can download it directly from this book's website at **https://www.macmillanlearning.com/college/us/product/Discovering-GIS-and-ArcGIS-Pro/p/131923075X**. The text in this module assumes that you have this Module3 folder in a computer location referenced as C:\GIS; if you have it somewhere else (for instance, in a flash drive referenced as G:\GISClass), substitute that location and path to the Module3 folder throughout this module.

- The Module3 folder contains a file geodatabase called OhioInfo, which contains Ohiocountyhousing, a polygon feature class of Ohio county borders. (Housing data from the U.S. Census Vacancy and Occupancy information have been joined to this feature class.)

- Start ArcGIS Pro.

- Sign in with your Esri account username and password.

- Create a new project using the **Map** template. Call this project **Module3** and place it in your **C:\GIS\Module3** folder. Ensure that there is not a checkmark in the **Create a new folder for this project** box.

- When ArcGIS Pro opens, change the map's name to **Ohio**.

Step 3.2 Setting Up a Choropleth Map

- Add the **Ohiocountyhousing** feature class to the Contents pane.

- Right-click the **Ohiocountyhousing** layer and select **Properties**. Under the **General** tab, change the name of the layer to **Ohio Seasonal Homes**.

- Right now, all the Ohio counties are shown with the same color, and no attribute values are displayed in the Contents pane. Each county contains different attributes, and what you want to do is display the one that shows the percentage of the county's housing stock that is seasonal homes. Open the layer's attribute table and scroll across it. There are several fields, but two in particular will be useful: (1) a field called *seasonal* that represents the number of seasonal homes in each county and (2) a field called *PerSeas* that represents the percentage of the total number of houses that are seasonal homes. To create a choropleth map, you'll be using the *PerSeas* attribute instead of the *seasonal* attribute because the *PerSeas* attribute has been normalized. For more about using normalized data on a map, close the attribute table and see **Smartbox 3.1**.

SMARTBOX 3.1

What are normalized data, and how are they used in ArcGIS Pro?

When you're dealing with data values to be presented on a choropleth map, you must first normalize those data values before displaying them on the map. To **normalize** values, you must convert them to a consistent level of data representation, such as a percentage. Normalized values can be compared to each other independently of the size of the polygons (or areal units) that contain those values. A choropleth map that shows count values (the actual numbers constituting whatever is being mapped) will be different from another choropleth map showing the normalized version of those values.

For instance, if you were creating a choropleth map of unemployment for each state in the United States, chances are that larger states with higher populations, such as California, would also have very high numbers of employment; smaller or less populated states, such as Rhode Island, would have lower numbers. In fact, according to U.S. Bureau of Labor Statistics data for unemployment in 2018 (found at **https://www.bls.gov/news.release/srgune.t01.htm**), California had an estimated 920,000 unemployed persons, while Rhode Island had 23,000 unemployed persons. Thus, your choropleth map using these count values would show a massive amount of unemployment in California and very little unemployment in Rhode Island. However, California's size and population are much larger than those of Rhode Island, and a greater number of people will likely equate with a similarly higher number of unemployed persons. Thus, your choropleth map would be showing very skewed results.

A better mapping strategy would be to normalize the data and, instead of showing the number of unemployed persons, show the percentage of those who are employed—by dividing the number of unemployed persons by the total civilian labor force. By mapping these normalized data, you can show that the two states' unemployment rates are about the same: California's 2018 unemployment rate was 4.2%, while Rhode Island's was 4.1%. By normalizing your data, you ensure that the values on your map are comparable.

In this module, you're working with the percentage of the Ohio county housing stock that is seasonal, which is normalized data. Figure 3.2 shows the seasonal home data in two ways. One map contains the raw count values, and the other contains the normalized version of the data (dividing the number of seasonal homes by the total number of houses). The maps show two very different versions of seasonal home distribution in Ohio. In mapping count values, more highly populated counties (with more houses), such as Lucas, Hamilton, or Cuyahoga Counties, have correspondingly more seasonal homes. The choropleth map with the normalized data, however, shows each county mapped as a percentage of the total housing stock and thus keeps the results on the same level (and is the map that should be used).

normalize To alter count data values so that they are at the same level of representing the data (such as using them as percentages).

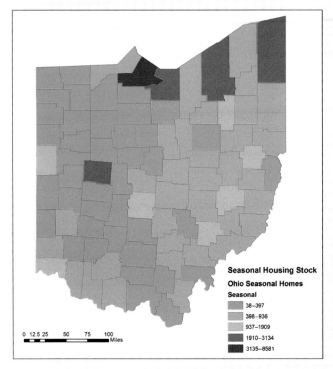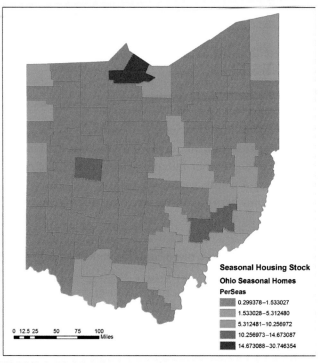

FIGURE 3.2 Count values of Ohio counties' seasonal homes versus normalized percentage values for the same seasonal homes.

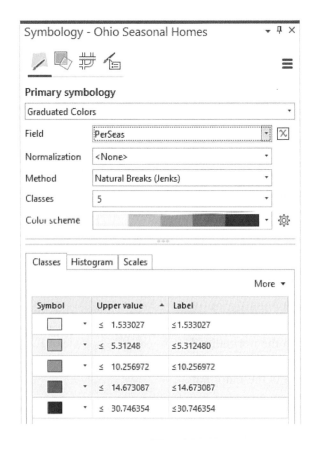

- Back in Module 1, you changed the colors and symbology for individual items, but that won't work with a choropleth map. You could display the unique value for each county, but you'd then end up with 88 different values (or different colors) in the Contents pane, which would be really difficult to work with or effectively display on a map. Instead, you can group several data values together into one category and display a single color for the entire category. To display the PerSeas attribute for each state, bring up the Symbology pane for the **Ohio Seasonal Homes** layer. (See page 25 in Module 1 for how to activate the Symbology pane to change a layer's symbology.)
- From the top pull-down menu, choose **Graduated Colors**.
- In the **Field** pull-down menu, choose **PerSeas**.
- In the **Normalization** pull-down menu, choose **<None>**. (Since the PerSeas attribute is already normalized, you don't need to normalize it a second time.)

- In the **Method** pull-down menu, choose **Natural Breaks (Jenks)**.
- In the **Classes** pull-down menu, choose **5**.
- A default color ramp will be provided, and you can use that for now. You'll see that the display in the Ohio map view will be automatically changed to reflect your settings. You'll see that each county is now displaying the PerSeas attribute, but the values have been grouped into five categories, with each category given its own color. You're using the Natural Breaks method of separating the data into the categories, but there are several different methods of classifying data, and the one you choose depends on the nature of the data being separated. See **Smartbox 3.2** for more information.

SMARTBOX 3.2

What are the differences among the data classification methods in ArcGIS Pro?

When setting up a choropleth map, you divide multiple values into smaller groupings. For instance, the 88 seasonal home values you're working with in this module can get grouped into three categories (highest, lowest, and average) or five categories (highest, high, average, low, and lowest). However, there are several different, more scientific, ways of doing this kind of **data classification** when setting up a choropleth map. ArcGIS Pro has many different types of data classification methods available, and each of them produces a different map, depending on the method used. The way the values are distributed helps determine which of the methods you should apply when making a map. For instance, are all of the seasonal home values evenly separated (that is, are there the same number of high values as low values), or are they mostly clustered around an average value with only a few outlying high and low values?

There are four main data classification methods available, with options for setting up your own customized classification. The first of these is the **Natural Breaks** method (also called the Jenks Optimization method). This method looks for naturally occurring gaps between values and uses these gaps to establish the start of each category into which the values will be placed. As Figure 3.3 shows, there are a lot of Ohio counties with low seasonal home percentages, so they get placed into one category, and the small handful of counties with high values get placed into a different category. Natural Breaks is most useful when there are nicely defined break points in the data values.

The second classification is the **Equal Interval** method (Figure 3.4), which involves creating category ranges of equal sizes. The category size is dependent on the range of values in the dataset. In this case, seasonal home percentages for Ohio counties are between 0.299% (in Fayette County) and 30.746% (in Ottawa County). The difference between these values is 30.447, and since we're using five categories, each category will contain

data classification Various methods used for grouping together (and displaying) values on a choropleth map.

Natural Breaks A data classification method that involves selecting class break levels by searching for spaces in the data values.

Equal Interval A data classification method that involves selecting class break levels by dividing the total span of values (from highest to lowest) by the number of desired classes.

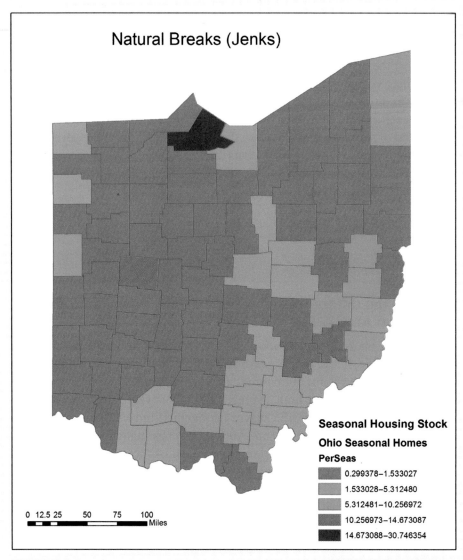

FIGURE 3.3 A choropleth map of Ohio counties' seasonal home percentages, as shown with the Natural Breaks data classification method.

6.089%. Thus, all values between 0.299% and 6.388% will be placed in the first category. All values between 6.388% and 12.478% will be placed into the second category, and so on. As Figure 3.4 shows, because there are so many Ohio counties with low percentages of seasonal homes, most of them fall into the first category (less than 6.388%), and only a handful fall into the other defined categories. Equal Interval is most useful when the data values are continuous, without a large number of very high or very low values.

The third data classification method is **Quantiles** (Figure 3.5), in which the categories each contain an equal (or nearly equal) number of entries, regardless of their actual values. Because there are 88 counties in Ohio and we are mapping five different categories, each category will contain 17 or 18 different values. Thus, the lowest 17 values get placed into the first category,

Quantiles A data classification method that involves placing an equal number of values in each class.

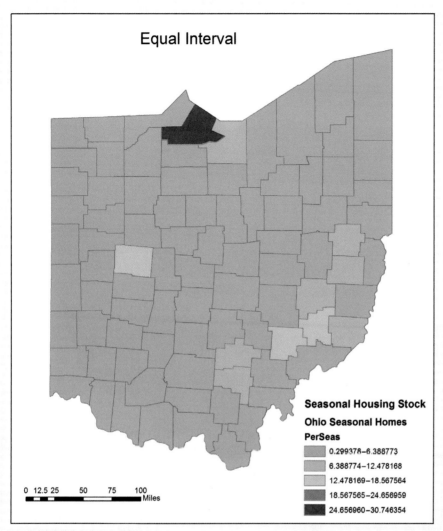

FIGURE 3.4 A choropleth map of Ohio counties' seasonal home percentages, as shown with the Equal Interval data classification method.

and the highest 17 values get placed into the last category. In Figure 3.5, the counties are very evenly distributed; that is, there are nearly the same number of the highest seasonal home percentages as there are of the lowest. However, look at the legend for the map. The highest category has values of 3.4% to 30.74%, while the lowest category shows values between 0.299% and 0.50%. The Quantiles method is best used when the data values are evenly distributed (which is not the case in this dataset).

The fourth method, the **Standard Deviation** classification (Figure 3.6), involves calculating the mean and standard deviation of the values in the dataset. A standard deviation is the average distance that a single data value is away from the mean (or average) of all the data values. The categories are then defined by these values. The mean of the seasonal home percentages is 2.58%, and the standard deviation is 4.31%. These figures are used in establishing the break points between categories (see Figure 3.6). For instance, the highest

Standard Deviation A data classification method that involves computing break values by using the mean of the data values and the average distance of a value from the mean.

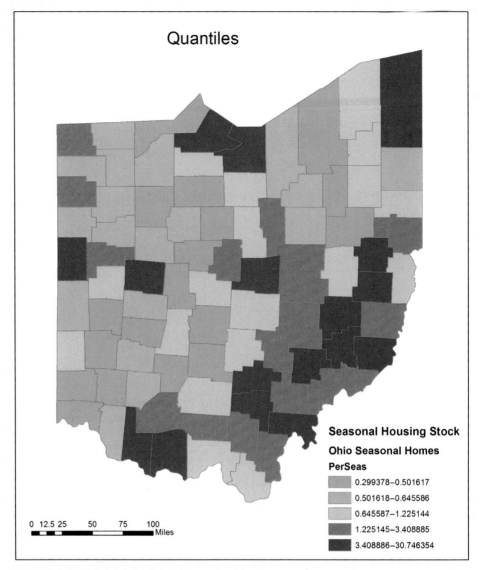

FIGURE 3.5 A choropleth map of Ohio counties' seasonal home percentages, as shown with the Quantiles data classification method.

category has values of greater than 2.5 times the standard deviation greater than the mean ((2.5 × 4.31) + 2.58), or 13.355%. Similarly, the lowest category has values of less than 0.50 times the standard deviation less than the mean ((20.50 × 4.31) + 2.58), or 0.425%. The Standard Deviation method produces the best results when the data values follow a normal distribution (where the values are well distributed around the mean).

A few other data classification methods available in ArcGIS Pro, including the following, allow a greater degree of customization:

- Manual Interval: This method allows the user to define break points for categories.

- Defined Interval: This method works like Equal Interval, except the user can specify how large each of the categories should be.

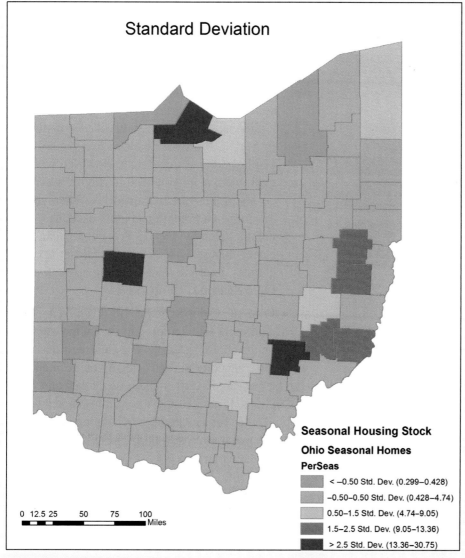

FIGURE 3.6 A choropleth map of Ohio counties' seasonal home percentages, as shown with the Standard Deviation data classification method.

- Geometrical Interval: This method attempts to have the same number of values in each category but is useful for continuous data. (It's a blend of the Natural Breaks, Equal Interval, and Quantiles methods.)

In the Symbolization pane of the layer for which you are setting up the data classification, you see a set of buttons above the color ramp. By default, your data are shown on the Symbology pane with the first button, the label view. This lets you see the ranges of values as they appear with their labels. The button next to this is the histogram view button. Clicking it switches the appearance of the data so you can see a chart of the distribution of the data, view the values used to define the starting and ending points of the categories, and change both the method used and the number of categories (Figure 3.7).

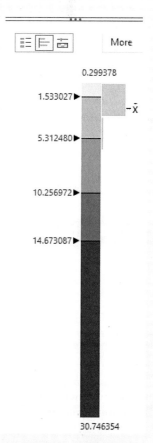

FIGURE 3.7 The histogram view of data distribution in the Symbology pane of ArcGIS Pro, containing information for data classification.

CHAPTER 3 How to Create a Layout with ArcGIS Pro

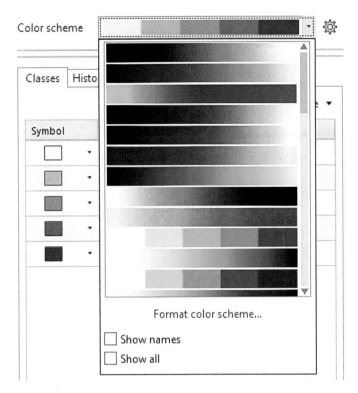

Step 3.3 Choosing Colors for the Map

- In Module 1, you displayed data layers (for example, airports, water bodies, roads, and different types of structures) using different sizes, symbols, and colors, such as an airplane symbol to denote airports. If you were creating a reference map, those symbols you displayed in ArcGIS Pro would be the same symbols displayed on a printed map. However, with a choropleth map, you need to choose different colors for the various categories. When you set up the choropleth map, you used the default given by ArcGIS Pro, but you can easily change those colors. To adjust the color choices, return to the Symbology pane for the **Ohio Seasonal Homes** layer.

- From the **Color Scheme** pull-down menu, select a different set of colors for each of the categories on the choropleth map. The **color ramp** assigns a different color to each category on the map, often with a light shade as the first one and a darker shade as the last one, but there are many different options. If none of the color ramps appeal to you, you can select an individual color for a category by changing its display color (as you did in Module 1). However, several factors influence the colors you should select. See **Smartbox 3.3** for more information about color choices in map design.

color ramp A range of colors that are applied to the thematic data of a map.

SMARTBOX 3.3

Why is color choice important when designing a layout in ArcGIS Pro?

The colors used on the map are important to the overall map design, especially when it comes to presentation. However, colors may appear one way on a computer monitor, a different way when printed out, and an entirely different way when projected onto a screen from an LCD projector. Shades of green and blue that look great on your monitor will not necessarily look the same when printed on an inkjet printer, or they may appear "washed out" when projected on a screen.

ArcGIS Pro allows you to select from a variety of graduated colors when selecting a color ramp. **Graduated colors** are often different shades of a single color (such as several shades of blue, including light blue, medium blue, dark blue, and an even darker shade of blue); they may also take on multiple hues (for example, from shades of green through shades of blue).

graduated colors The use of various hues in representing ranges of values on a map.

graduated symbols The use of different sized symbols to convey thematic information on a map.

Another option for symbology is the use of **graduated symbols**, in which different sizes of the same symbol are used (but the color for all of them remains the same). Figure 3.8 provides examples of graduated colors and graduated symbols.

FIGURE 3.8 The use of graduated colors and graduated symbols in ArcGIS Pro.

Step 3.4 Creating an Initial Layout in ArcGIS Pro

- To begin laying out the print-quality version of the map, you need to work with a **layout** instead of the map. This mode of ArcGIS Pro works like a blank canvas, allowing you to construct a map using various elements. For more information about working with a layout and map elements, see **Smartbox 3.4**.

layout The ArcGIS Pro interface used for composing a printed map.

SMARTBOX 3.4

What is a layout, and what are map elements in ArcGIS Pro?

A layout in ArcGIS Pro is like a blank canvas on which you can build a professional-quality map. When you are working with a layout in the view, a new Layout tab appears on ribbon, giving you a new set of tools for navigating and zooming in and out of the layout (Figure 3.9).

In the Navigate group, the tools and their uses are as follows:

- The Navigate tool is used to pan around the layout and move the layout as a whole.
- The magnifying glass icon is used to zoom the view to the extent of any selected features.
- The icon with the four triangles pointing outward is used to zoom the map to its full extent.
- The 1:1 icon is used to zoom the layout to 100 percentage.
- The white pages with the two black arrows are used for zooming in to the extent of the entire width of the page.
- The white pages with the blue arrows are used for switching between extents that you previously zoomed to.

A layout is constructed from the numerous **map elements** that can be added. A map element is an individual item placed on the layout, such as a scale bar,

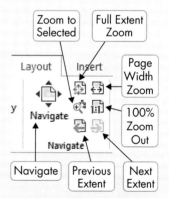

FIGURE 3.9 The controls used for navigating and interacting with a layout in ArcGIS Pro.

map elements The various items used in composing a layout.

map frame The map element that allows for the display of the contents of a map on a layout.

FIGURE 3.10 The tools for selecting map elements in a layout.

a north arrow, or a legend. Another element used in a layout is a **map frame**, which is a box filled with the contents of a specific map in ArcGIS Pro. When you insert a map frame element, what is displayed in a map can be displayed in a box within the layout.

Also, each element has properties (such as creating borders, filling colors, or fixing size and position) that you can access individually after you have selected a particular element. Once you have selected map elements, you can resize them by dragging at their corners, and you can also drag them to different locations. You can delete map elements by selecting them with the cursor and pressing the Delete key on the keyboard. On the ribbon, in the Layout tab, within the Elements group, there are tools and buttons that let you interact with the map elements as follows (Figure 3.10):

- The Select tool allows you to click on an element to work with it or to select multiple elements at once. The pull-down menu under Select allows you to specify how you want to select elements, such as by drawing a rectangle or a polygon and then selecting all elements that touch or are within the boundaries of the shape you drew on the layout.

- The Select All button allows you to choose all elements on the layout.

- The Clear Selection button allows you to deselect all selected elements.

- The Delete button removes all selected elements.

In this module, you will be adding map elements for two map frames, a north arrow, a scale bar, a legend, and different types of text, but many additional elements can be used in a layout. Although they're not used in this module, the following are some of the other map elements that can be added from the Insert tab:

- In the Map Frames group, Grid allows you to add a grid to your layout to showcase things such as latitude/longitude lines on a map.

- In the Text group, Symbol allows you to add specific symbols such as road signs or callouts to the layout.

- Also in the Text group, Dynamic Text allows you to insert dynamic text; this refers to a variety of different types of text, including the map's representative fraction information, the current date and time, the current user of ArcGIS Pro, the coordinates of the map frame, the coordinate system of the layers, and so on. With dynamic text, as you change these items in the map itself (as you can the scale), their corresponding text on the layout changes as well.

- In the Graphics group, the Rectangle, Line, and Point options allow you to draw things on the map that are not GIS layers (such as drawing a circle around a point of interest).

- Also in the Graphics group, Picture allows you to add an image to your map, including downloaded pictures, images from your camera or phone, or similar graphics.

- A layout in ArcGIS Pro appears as a separate tab so that you can switch back and forth between the map and the layout as needed. To create a layout, select the **Insert** tab, and within the **Project** group, click on **New Layout**.

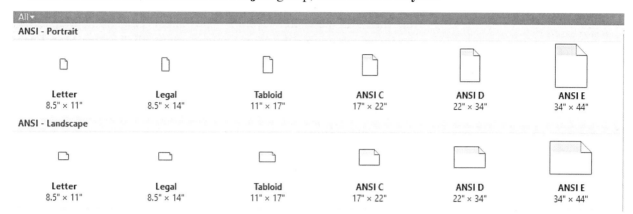

- ArcGIS Pro gives you a wide variety of orientation and size choices for your layout. Before you can go any further with your map, you need to choose whether the map should be oriented in portrait or landscape mode. **Portrait** aligns the layout vertically (so that it's longer than it is wider), and **landscape** aligns the layout horizontally (so that it's wider than it is longer). If you were mapping the United States, you'd likely choose landscape to place the entire country to fill the page. However, to map Ohio, you'll likely use portrait mode as it better fits the dimension of the state. For this module, choose the option **ANSI - Portrait – Letter 8.5" × 11"**.

  - *Important Note:* If you find that later on during map design you want to change the orientation from portrait to landscape (or vice versa), you don't have to start from scratch with a new blank layout. Instead, on the ribbon choose the **Layout** tab, and within the **Page Setup** group, click on the **Orientation** icon. You can then choose either **Portrait** or **Landscape** as the orientation for your layout.

portrait The vertical orientation of a layout.

landscape The horizontal orientation of a layout.

- A new Layout tab opens next to your Map tab. The Layout tab works like a blank canvas and allows you to construct a map using various elements. You can switch back and forth between the map and the layout by clicking on the appropriate tab. If you close a layout (by clicking the X on its tab in the view) before you want to, you can get it back as described in **Troublebox 3.1**.

TROUBLEBOX 3.1

What happened to the layout I was using in ArcGIS Pro?

If you close a Layout tab in ArcGIS Pro, getting it back again is easy. In the Catalog pane, you'll see an expandable option for Layouts. Expand it and you see a list of all the layouts in this project, including the one you closed. Double-click the layout listing in the Catalog pane, and the corresponding layout tab reopens.

Step 3.5 Inserting a Map Frame into the Layout in ArcGIS Pro

- The first element to add to the layout is a map frame, which is an element that shows all the visible layers in the Contents pane of the Map tab. From the **Insert** tab, within the **Map Frames** group, click on the **Map Frame** icon and then select the **Ohio** option for the map with the scale values after it. This is a reference to the open Ohio Map tab that contains your Ohiocountyhousing layer; by choosing this option, you're letting ArcGIS Pro know that you want to add the visible layers from that map to this layout's map frame.

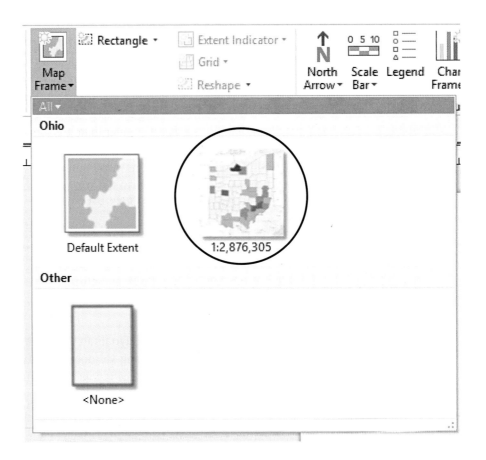

- Next, on the layout, draw a box with the cursor that will be the boundary of the map frame. The map frame element is added and filled with the visible layers of the Ohio map view (that is, the Ohiocountyhousing layer and the topographic map background). The map frame element in the layout contains only the layers being displayed from the corresponding map's Contents pane.
  - ***Important Note:*** The actual boundary of the printed page is *not* the map frame. The outer boundary of the layout shows the extent of the printed page, and the inner boundary shows the outline of the map frame.

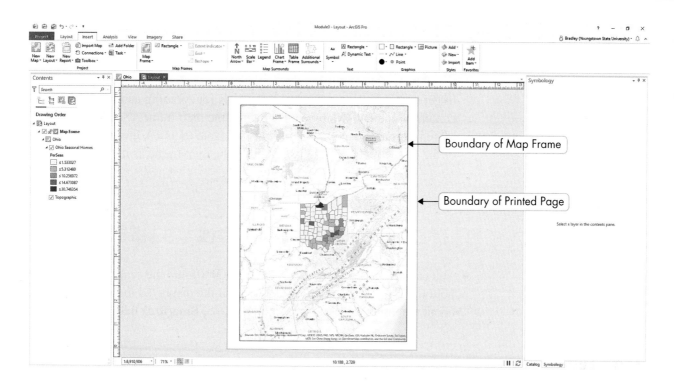

Step 3.6 Working with a Map Frame in the Layout in ArcGIS Pro

- The map frame is the map element that contains a link to a particular map in your project, and it shows all the visible layers in that map's Contents pane. Like all other map elements, the map frame can be resized and moved around in the layout. On the ribbon, from the **Layout** tab, within the **Elements** group, click on the **Select** button and then click and drag the map frame around the layout or resize it as you see fit (by using the eight white boxes at its corners and midpoints). When you're placing and resizing the map frame, keep in mind that you'll have several other map elements on the final layout, so be mindful of the position and size of the frame, as well as the available blank whitespace.

- When you resize the map frame, what is being displayed in it will change. For instance, by reducing the size of the map frame, you may also cut off part of the state boundaries. In that case, you need to adjust the appearance of what is being shown inside the map frame or change the scale of the layers displayed in the map frame. In order to adjust the position or scale of the contents of the map frame, on the ribbon, select the **Layout** tab, and within the **Map** group, click on **Activate**.

- With the map frame activated, you can adjust the appearance of what is being shown in the map frame by zooming, panning, or changing the scale. A new contextual tab, **Activated Map Frame**, appears at the top of the ribbon, and from there you can click on the **Layout** tab on the ribbon to get access to the regular Navigate tools for zooming and panning the content of the OhioMap being shown in the map frame. (See Module 1 for more information on how to use the Navigate tools.) Adjust the appearance of the Ohiocountyhousing layer within the map frame as necessary.

- You can also type a scale (such as 1:6000000) into the scale box at the bottom of the layout to adjust the scale of the displayed map. For more about the importance of scale in layouts, see **Smartbox 3.5**.

SMARTBOX 3.5

How is scale represented on a layout in ArcGIS Pro?

In GIS, there are a couple of different ways of thinking about scale. **Geographic scale** refers to the real-world size or area of something. If you're making a map of the entire United States, that map would cover a very large geographic scale, while a map of your property boundaries would reflect a very small geographic scale.

Map scale reflects how many units of measurement on the map are equal to a number of units in the real world. For instance, 1 inch measured on a map might be equivalent to 5000 inches in the real world. Map scale is usually measured as a **representative fraction (RF)**, such as 1:150000. This ratio indicates that 1 unit of measurement on the map (such as 1 inch or 1 centimeter) represents 150,000 of the same units in the real world.

Depending on the representative fraction and map scale, maps are considered **small-scale maps** or **large-scale maps**. Small-scale maps often show a larger geographic area but have a smaller RF value (such as 1:250000). Large scale maps show a smaller geographic area and have a larger RF value (such as 1:4000). The largest-scale map you could make would be 1:1, in which one unit of measurement on the map would reflect one unit of real-world measurement; in other words, a 1:1 map would be the same size as the area that it represents.

The choice of map scale is crucial when designing a map, as this choice affects how much area can be displayed, as well as the types of symbology that can be used. For instance, on a small-scale map (such as a 1:1000000 map), cities would be represented as points. Individual features of cities (such as park boundaries) could not be properly represented at that scale. Major roads could be represented, but individual streets could not. However, on a larger-scale map (such as a 1:24000 map), individual park boundaries

geographic scale The real-world size or extent of an area.

map scale A metric used to determine the relationship between measurements made on a map and their real-world equivalents.

representative fraction (RF) A value indicating how many units of measurement on the map are the equivalent of a number of units of measurement in the real world.

small-scale map A map with a lower value for its representative fraction. Such maps usually show a large geographic area.

large-scale map A map with a larger value for its representative fraction. Such maps usually show a smaller geographic area.

could be shown as polygons, and many residential roads could be represented as lines.

For instance, setting the map scale of the layout to 1:3000000 would produce a very small-scale map. At this scale, a distance of 1 inch measured on the map of Ohio would equal to 3,000,000 inches (or 250,000 feet) in the real-world Ohio. Zooming in or out of the map causes the scale to change accordingly. You can also type a new RF into the scale box to manually change the display scale for the layout.

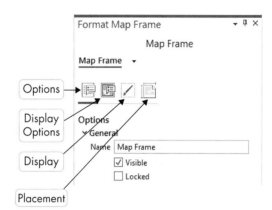

- When you have everything the way you want it and are ready to return to working with the content of the layout, on the ribbon select the **Layout** tab, and within the **Map** group, click on the **Close Activation** button, and you are returned to the regular layout controls. (In addition, the contextual Activated Map Frame tab at the top of the screen disappears to remind you that you are no longer working with an activated map frame.)

- To make further adjustments to the map frame, go to the layout's Contents pane, right-click on the **Map Frame** element, and then choose **Properties**. A new pane called Format Map Frame appears on the right-hand side of the screen.

- Four icons at the top of the Format Map Frame pane give you new choices for changing the map frame to appear as you want it:
  - Options: This lets you choose whether the map frame is visible or not and what name the map frame will contain, and it also allows you to switch the contents of the map frame to have those of another map in your project.
  - Display Options: This allows you to change things such as the scale of the map frame, the location of the map frame's center, the rotation of the contents of the map frame, or the units of measurement being used.
  - Display: This allows you to change the border, background, and shadow of the map frame. By default, the map frame has a black line border, but you can make it invisible by choosing No Color for this option.
  - Placement: This allows you to change the size and position of the map frame or to rotate the frame itself.

Step 3.7 Placing an Inset Map into the Layout in ArcGIS Pro

- One of the elements you will add to your layout is a small inset map showing Ohio's location in the greater context of the entire United States. To create this inset map, you'll first create it as a map and then put a second map frame into the layout to show the contents of this second map. To get started, you need to add a second map to the project. On the ribbon, on the **Insert** tab, within the **Project** tab, click on the **New Map** button and then choose **New Map**.

- A new Map tab appears in the view. Change its name to **OhioContext**.

- Next, add the Ohiocountyhousing layer to the OhioContext map. By default, it should be displayed as a single symbol (that is, each county polygon should have the same color).
- Zoom out so that you can see the boundary of Ohio displayed in the context of the whole of the United States.
- From the **Map** tab, within the **Layer** group, click the **Basemap** button. Choose the option **Light Gray Canvas**. The basemap changes from being a topographic map to being a white-and-gray outline of the geography of North America.
- Make sure that Ohiocountyhousing is displayed as a single symbol and use the Symbology options (again, on the **Appearance** tab, within the **Drawing** group, click the **Symbology** button) to change the color of Ohio so that it stands out from the rest of United States, as shown here.

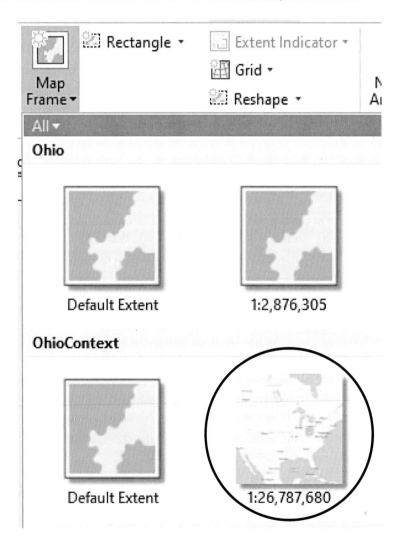

- Return to the Layout tab in the view. You need to now insert a second map frame linked to the OhioContext map. On the ribbon, on the **Insert** tab, within the **Map Frames** group, click the **Map Frame** button. From the options that appear, choose **OhioContext**. Click in the layout and draw a box with the cursor to insert the second map frame.

- Change the size and appearance of the second map frame showing this inset map so that you can fit it properly on your layout and then activate this map frame and adjust its contents so that you can see Ohio in the context of the entire United States on the map frame.

Step 3.8 Placing a Scale Bar into the Layout in ArcGIS Pro

- Even though you've set the map scale in ArcGIS Pro, you need to communicate this information to the map reader. A graphical device for displaying the map scale is a **scale bar**. To add a scale bar to the map, first click on the Ohio Seasonal Homes layer, then choose the **Insert** tab, and within the **Map Surrounds** group, click on the **Scale Bar** button.
- A number of options for scale bars appear, as shown here.

scale bar A graphical device used to represent scale on a map.

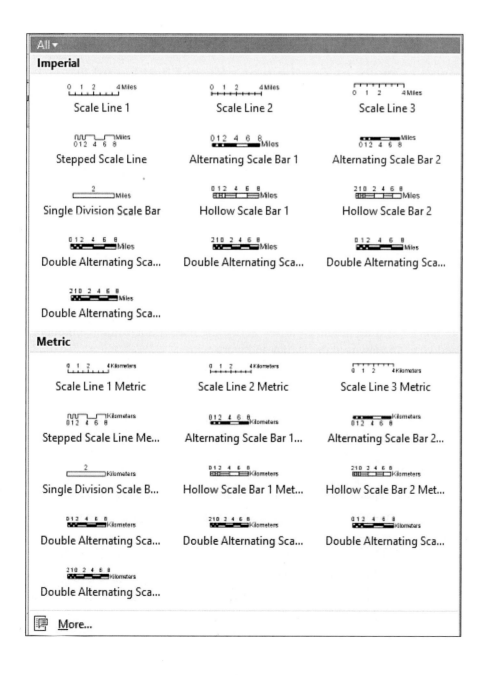

- Additional scale bars are available by clicking the **More...** option.
- Select an appropriate scale bar by clicking on it.
- The cursor turns into a crosshairs; click on the Layout where you want to place the scale bar.
- The scale bar is added to the layout as a map element that you can select and move around, as previously described. In the layout's Contents pane, the scale bar is placed in the drawing order above the *two* map frames. Just as with adding layers to the map's Contents pane, elements are placed on top of or underneath other map elements, depending on where they are ranked in the layout's Contents pane. If, after adding map elements, you find that you're having difficulty finding a particular map element in the Contents Pane for the layout you're working with, see **Troublebox 3.2**.

TROUBLEBOX 3.2

Why can't I view or select the map element that I want in the ArcGIS Pro layout?

When elements are added to the layout, you see the corresponding item appear in the Contents pane when the layout is the active tab in the view. For instance, the layout's Contents pane for this module contains items such as the map frame, the north arrow, the scale bar, and the percentage of seasonal homes (the legend). The order in which these items are drawn on the map is similar to how the GIS layers from a map are shown in the Contents pane: Elements at the bottom of the Contents pane are drawn first, and then elements are drawn on top of these items as you move up the Contents pane.

If you have added a north arrow and a map frame, but if the north arrow is at the bottom of the pane, then the map frame may be drawn over top of it, causing the north arrow to be covered. When you have many elements in a map, the priority for how they are drawn is affected by their position in the Contents pane. In this case, drag the north arrow above the map frame in the Contents pane, and you can view it in the layout and work with it.

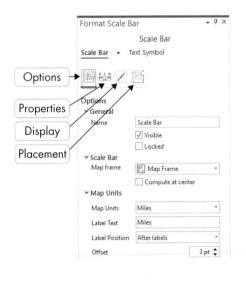

- To change the appearance of the scale bar, double-click on it as a map element on the layout, and a new Format Scale Bar pane opens on the right-hand side of the screen.
- Four icons available at the top of the Format Scale Bar pane give you new choices for changing the scale bar to appear as you want it:
  - Options: This lets you choose whether or not the scale bar is visible and name the scale bar, and it also allows you to choose which map frame in the layout the scale bar is for (if you have multiple map frames).
  - Properties: This allows you to change things such as the position of the numbers on the scale bar, the number of decimal places on the scale bar, where the marks on the scale bar are shown, and similar visual properties of the scale bar.

- Display: This allows you to change the border, background, and shadow of the scale bar.
- Placement: This allows you to change the size and position of the scale bar or to rotate the scale bar.

Step 3.9 Placing a North Arrow into the Layout in ArcGIS Pro

- A **north arrow** is a graphical indicator of what direction north is facing on the layout. Sometimes when you're making a map (depending on the map projection or the map design), north is not always straight up, and a north arrow helps orient the map reader. To add a north arrow to the map, choose the **Insert** tab, and within the **Map Surrounds** group, click on the **North Arrow** button.

north arrow A graphical device used to show the orientation of the map.

- A number of options for north arrows appear, as shown here.
- Additional north arrows are available by clicking the **More...** option.
- Select an appropriate north arrow by clicking on it.
- The cursor turns into a crosshairs, and you can click on the layout where you want to place the north arrow.
- The north arrow is added to the layout as a map element that you can select and move around, as previously described.
- To change the appearance of the scale bar, double-click on it as a map element on the layout, and a new Format North Arrow pane opens on the right-hand side of the screen.

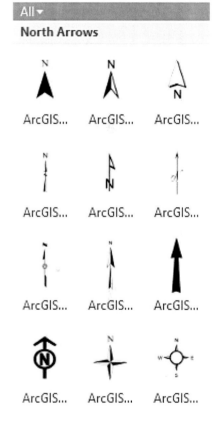

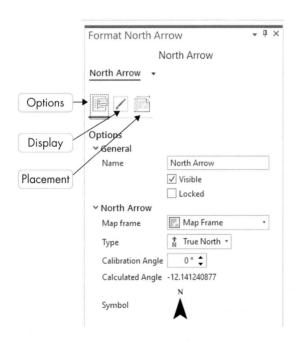

- The three icons at the top of the Format North Arrow pane give you new choices for changing the north arrow to appear as you want it:
  - Options: This lets you choose whether if the north arrow is visible or not and name the north arrow, and it also allows you to choose whether the north arrow represents true north or map north.

- Display: This allows you to change the border, background, and shadow of the north arrow.
- Placement: This allows you to change the size and position of the north arrow.

Step 3.10 Placing a Legend into the Layout in ArcGIS Pro

legend A graphical device used on a map to explain what the various map symbols and colors represent.

- Next, you'll add to your layout a map **legend**, a device that explains to the map reader what the various symbols and colors on the map represent. When you place a legend, ArcGIS Pro wants to know which frame you want to use as the basis for the legend. In the layout, click in the map frame that corresponds with the Ohio map. Then, to add a legend to the map, on the ribbon choose the **Insert** tab, and within the **Map Surrounds** group, click on the **Legend** button.

- If you move your cursor into the layout, you see it change shape to look like a crosshairs. Use this cursor to draw a box on the layout where you want the legend to go. ArcGIS Pro then inserts the legend into the layout. Just as the map frame shows the contents of the map it is linked to, the legend shows the information for the visible layers of the map frame you chose to show it for.

- Double-click on the legend element in the layout, and the Format Legend pane opens on the right-hand side of ArcGIS Pro.

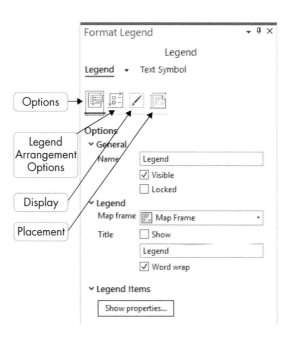

- Four icons at the top of the Format Legend pane give you new choices for changing the legend to appear as you want it:
    - Options: This lets you change the title of the legend from Legend to something else, choose whether or not the title of the legend is visible, choose whether or not the legend is visible, determine how words wrap

in the legend, and choose how it synchronizes with new layers being added to the map. In this case, change your legend's title to **Percentage of Seasonal Homes**. Also, to display this new title, put a checkmark next to the **Show** box.

- Legend Arrangement Options: This allows you to change how things like text are displayed in the legend, the font of the legend, and details such as how words in the legend wrap and how they are spaced out.

- Display: This allows you to change the border, background, and shadow of the legend.

- Placement: This allows you to change the size and position of the legend.

• Return to the **Options** section in the legend. By clicking on the **Show Properties…** button, you can bring up the Format Legend Item pane, which allows you to choose exactly what is shown in the legend and how things are arranged in columns in the legend.

• For instance, your legend should currently read **Percentage of Seasonal Homes** and have a heading under this that reads **Ohio Seasonal Homes** (the name of the feature class) and then another heading underneath that reads **PerSeas** (that is, the name of the field from the layer's attribute table). Having these extra headings on your map probably wouldn't mean much to the casual map user, and besides, the descriptive title you gave the legend will convey to the reader what the map is. You can remove these items from the legend by removing the checkmarks next to **Layer Name** and **Headings**. Your legend should now just simply read **Percentage of Seasonal Homes** and have the appropriate values beneath it.

- Click on the **left arrow** in the upper-left corner of the pane to save your changes to the legend.

Step 3.11 Adding Text and a Title to the Layout in ArcGIS Pro

- To add text to the map (for instance, for the name or the date), on the ribbon choose the **Insert** tab, and within the **Text** group, click on the **Symbol** button.

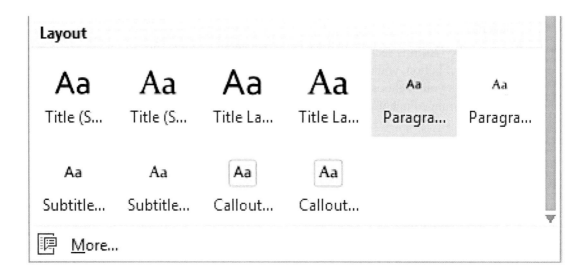

- A number of different types of text symbols appear. Scroll down the list to the options under **Layout** and choose one of the **Paragraph** options.
- Draw a box on the layout where you want to place the text. A small box that reads **Text** is added to the layout.
- To change the appearance of the text you inserted, choose it as a map element on the layout, and a new Format Text pane opens on the right-hand side of the screen.
- Three icons at the top of the Format Text pane give you new choices for changing the text to appear as you want it:
  - Options: This allows you to choose whether the text is visible, as well as what the text says. In the box, type the text that you want to add to the layout—in this case, your name, today's date, and the source of the data being used in the layout. With the second of the light-blue tag options chosen (the blue tag with the </> symbol in it), the text in the layout updates as you type.
  - Display: This allows you to change the border, background, and shadow of the text box.
  - Placement: This allows you to change the size and position of the text box.

Step 3.11 Adding Text and a Title to the Layout in ArcGIS Pro 87

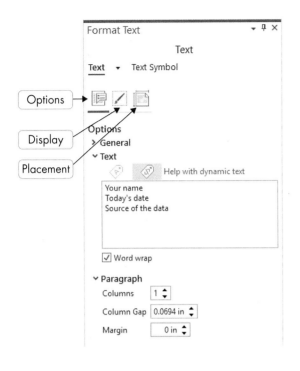
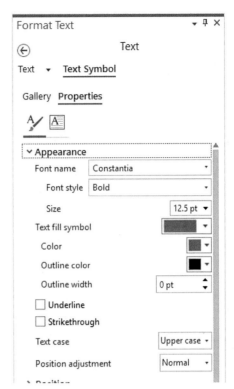

- To change the size and font of the typed text, in the Format Text pane click on **Text Symbol** and then click on **Properties**. Expand the heading for **Appearance**, and new options for the font to use, the font style, the size of the font, the color of the text, and other choices are available to you. For more information about the use of type and fonts in a layout, see **Smartbox 3.6**.

 SMARTBOX 3.6

Why is the choice of fonts important when designing a layout in ArcGIS Pro?

Although ArcGIS Pro gives you a wide variety of choices when it comes to selecting fonts for text, titles, and labels, usually a map should contain only two different fonts, carefully chosen to complement each other (and the map as a whole). Using too many different fonts on a map, or selecting several of the more esoteric fonts is likely to make the map more difficult to read (and thus less helpful to its intended audience). By selecting the **Text Symbol** option, you can choose from several premade fonts that Esri has set up as standard fonts for map items such as historic regions, coastal areas, or oceans (Figure 3.11).

AaBbYyZz	*A a B b Y y Z z*	A a B b Y y Z z	AaBbYyZz	*A a B b Y y Z z*
Coastal Region	Ocean	Physical Region	Historic Region	Sea

FIGURE 3.11 Examples of designated fonts available in ArcGIS Pro.

- Under the other headings—Position, Rotation, Halo, Shadow, and Callout—are more options for changing the appearance of the text.
- After you have the text set up the way you want it, you need to add a new set of text for the title of the map. To add more text, return to the ribbon and choose the **Insert** tab, and within the **Text** group, click on the **Symbol** button. Scroll down the list to the options for **Layout** and choose one of the **Title** options.
- Draw a box on the layout where you want the title to be, and a new text box containing the word **Text** is added. Use the same techniques you used to modify your previous text to set up a title for the map with appropriate size and font. Make sure that your title is descriptive of what your layout is showing. When naming a map, don't use the word "map" in the title, as that's pretty self-evident (it would be like naming your dog "Dog" or your cat "Cat"). Choose a more descriptive title that sums up your map of seasonal homes in Ohio counties.

Step 3.12 Evaluating the Map: Refining and Editing Map Elements

- You can arrange the elements on your map to produce the best designed map possible. For example, you can resize and move map elements by selecting them with the mouse and resizing them like you would the map frame.
- You can delete map elements by selecting them with the cursor and pressing the Delete key on the keyboard.
- Also, each element has a set of properties you can adjust as desired. See **Smartbox 3.7** for further information about map design.

SMARTBOX 3.7

What are some strategies for designing an effective map in ArcGIS Pro?

Designing a good, effective map to communicate its message is crucial when setting up a layout. You don't want to add a data frame, throw on a legend and a north arrow, and then print out the map. The map would look sloppy and unprofessional.

visual hierarchy The way features are displayed on a map to emphasize their level of prominence.

A map layout should follow a **visual hierarchy**, with some items more prominently displayed than others. The purpose of the map will help you determine which map elements should be the most prominent in the map's visual hierarchy. For instance, if you're creating a map of a local park, is the purpose to show the park's location in relation to its surroundings (such as the roads needed to reach the park), or is it to highlight the locations and dimensions of the walking paths? Maps with these different goals should be designed differently and should have different elements prominently displayed. Without a clear hierarchy of its items, a map is likely to look cluttered or be confusing to the reader. A clear visual hierarchy also aids in balancing the empty spaces on a map.

ArcGIS Pro gives you a great deal of flexibility when putting together a layout. Each element (such as the north arrow, the text boxes, or the legend) you place on the map can have its borders prominently highlighted or made transparent. To access these options, right-click an element and select **Properties**, and a new pane opens, allowing you to change the appearance of that element. In the options for the border, for example, you can change the thickness of the lines, or by selecting **No color** you can make the element's border transparent. In the options for the background, you can change how the color behind the element appears; the shadow options allow you to add extra effects (if desired) to the element's appearance. In some cases, you may find that placing (or removing) borders around the boxes of each element enhances the map's design.

 Checklist

Use the following checklist to make sure you have put together your map layout for this module in a complete, visually appealing fashion that conveys information effectively.

_____ The percentage of the seasonal housing stock in all 88 Ohio counties, displayed with an appropriate color scheme

_____ Ohio counties displayed at an appropriate scale (which should be the focus of your layout)

_____ An inset map showing the state of Ohio in relation to the rest of the United States, shown at an appropriate scale

_____ An appropriate scale bar

_____ A map legend with an appropriate name (not "legend"), size, font, number of columns, and so on

_____ A north arrow of appropriate size and position

_____ An appropriate title (without the words "map" or "title") in an appropriate size and font

_____ Your name, the date, and the source of the data in an appropriate size and font

_____ Overall map design, including appropriate borders, color schemes, balance of items and placement, and so on

Step 3.13 Printing or Exporting a Layout in ArcGIS Pro

- When you have constructed the layout the way you want, choose the **Share** tab in the ribbon.
- To print the layout, within the **Print** group, click on the **Layout** button. Printer options appear for you to choose from.

- To export the layout to another digital format (such as JPEG or PDF), within the **Export** group click on the **Layout** button. You can then specify where you want to save the exported version of the layout, the filename you want to give it, and (from the Save As Type pull-down menu) how you want to save the file (for example, in a graphic format, such as PNG, JPG, or TIFF, or as a PDF).
- ***Important Note:*** If some borders don't appear, if borders are cut off, or if portions of the map are cut off from the boundary of the printed page, make sure to go back and adjust those items prior to submitting your final results.

Closing Time

This module explores how to use GIS data to create a professional-looking map. ArcGIS Pro provides a wide variety of options for designing and customizing maps for a wide variety of purposes. While in this module you produced a choropleth map, the layout tools in ArcGIS Pro allow you to create many other types of maps, such as thematic maps or reference maps. Features on choropleth maps are not commonly labeled; for instance, you wouldn't identify individual Ohio cities or interstates on the housing stock map you created in this module. However, labeling features (such as the names of streets or subdivisions) on a reference map would be critical. See *Related Concepts for Module 3* for more about ArcGIS Pro's map labeling properties.

A drawback to creating a layout is that it's a static image; once it's produced, users can't do any of the usual GIS tricks, like zooming in or out of the map, turning layers on or off, or querying a polygon to find out the percentage of housing stock for a particular county. Also, once the map is produced and distributed (as a printout or a PDF or a posted graphic), if the map needs to be updated, new maps need to be created and distributed. It might be useful to design a web-based interactive map that can be easily updated. In Module 4, you'll be doing just that—starting with the data you worked with in this module but creating a web-based map instead of a printed one.

RELATED CONCEPTS FOR MODULE 3

Map Labels and Annotation

Although you didn't do it for the map you created in this module, chances are good that you'll want to label features when you're creating a reference map. For instance, if you're designing a map of your college campus, you probably want to provide labels for the names of the buildings, parking lots, roads, and athletic fields shown on the map. In Module 2, you labeled the features you were working with by selecting a field in the attribute table and then assigning that field to be displayed in conjunction with a feature in the view.

Labels in ArcGIS Pro are all displayed with the same font and style, and ArcGIS Pro automatically places these labels. While this characteristic of ArcGIS Pro does quickly provide names or values, it gives you a limited amount of design flexibility. For instance, you can't adjust the position or size of each individual label, and you are unable to move an individual label if you don't like its location.

In ArcGIS Pro, the default font for labels is Tahoma, the default font size is 10, and the default font style is regular—though you can easily change all of these options. ArcGIS Pro gives you access to the **Maplex Label Engine**, which allows for more flexibility with placing labels on maps. The Maplex Label Engine provides a variety of labeling tools—for instance, fitting labels within polygons, allowing labels to curve, determining the best placement for labels, and emphasizing label placement for features such as land parcels, rivers, or contours. Using Maplex Label Engine is the default method for labeling; if you want to use ArcGIS Pro's Standard Label Engine, go to the **Labeling** tab, and within the **Map** group, click on the **More** button, and remove the checkmark next to **Use Maplex Labeling Engine**.

In addition, ArcGIS Pro provides you with the ability to create **annotation**, or text that appears on a map. Annotation is a special type of feature class in a geodatabase (see Module 6) that allows information (such as the name of a lake or the designation of a transformer box) to be linked with an object in a separate feature class. For instance, if the owner attribute of a land parcel is changed in a feature class, its linked annotation (the text that accompanies it on the map) is updated as well. Unlike labels, annotation properties can be adjusted individually.

Maplex Label Engine Functionality in ArcGIS Pro that allows for a greater degree of flexibility and options for labeling features.

annotation A method of adding text to a map that allows for individual pieces of text to be edited separately.

Key Terms

map (p. 61)
cartography (p. 61)
reference map (p. 61)
topographic map (p. 62)
thematic map (p. 62)
choropleth map (p. 62)
normalize (p. 65)
data classification (p. 67)
Natural Breaks (p. 67)
Equal Interval (p. 67)
Quantiles (p. 68)

Standard Deviation (p. 69)
color ramp (p. 72)
graduated colors (p. 72)
graduated symbols (p. 72)
layout (p. 73)
map elements (p. 73)
map frame (p. 74)
portrait (p. 75)
landscape (p. 75)
geographic scale (p. 78)
map scale (p. 78)

representative fraction (RF) (p. 78)
small-scale map (p. 78)
large-scale map (p. 78)
scale bar (p. 81)
north arrow (p. 83)
legend (p. 84)
visual hierarchy (p. 88)
Maplex Label Engine (p. 91)
annotation (p. 91)

4

How to Publish Layers from ArcGIS Pro and Build Web Apps with ArcGIS Online

Learning Outcomes

After studying this module, you should be able to:

- Define what the cloud is.
- Explain what ArcGIS Online is.
- Explain how ArcGIS Online works as a SaaS structure.
- Explain the differences between three different types of web layers that can be published to ArcGIS Online.
- Explain how data are transferred from ArcGIS Pro to ArcGIS Online.
- Describe how to create a web map in ArcGIS Onlino.
- Explain the difference between a web map and a web app in ArcGIS Online.
- Describe what a story map is.

ArcGIS Online Esri's cloud-based GIS platform, where data and web mapping software can be accessed via the Internet.

ArcGIS Pro Skills

In this module, you will learn how to do the following:

- Publish web layers from ArcGIS Pro to the cloud structure of ArcGIS Online.
- Create a web map in ArcGIS Online from layers published from ArcGIS Pro.
- Configure the attributes and information that will appear in pop-ups when an item is queried in ArcGIS Online.
- Choose a basemap to accompany the features in ArcGIS Online.
- Create a configurable web app from a web map in ArcGIS Online.
- Customize a web app in ArcGIS Online.
- Share a web app from ArcGIS Online across the Internet.

Introduction

Making a map of your data is a fundamental task in GIS. However, a static map (whether a printed map or an exported digital version of the same) isn't always the best method of presenting your data or results. Consider the following situation: A police GIS analyst has put together a set of layers showing the locations of various types of crimes (robbery, assault, and so on) for the county and has mapped them at the census-block level for several different years. To distribute this information, she would have to make maps for each year and perhaps for different combinations of crimes. The result would be a stack of well-designed maps, all at the same scale, showing different information. While this set of maps might convey the necessary themes, there is a more flexible option: creating an interactive map that allows users to change scales, select which layers to display, or quickly compare different years by turning layers on and off. Such an interactive map could allow its creator to correct errors on the fly or easily update the map with fresh information as it becomes available rather than reprint and redistribute maps.

The idea of sharing and distributing GIS maps and data in an easy-to-use and interactive format is at the heart of **ArcGIS Online** (Figure 4.1). The resources

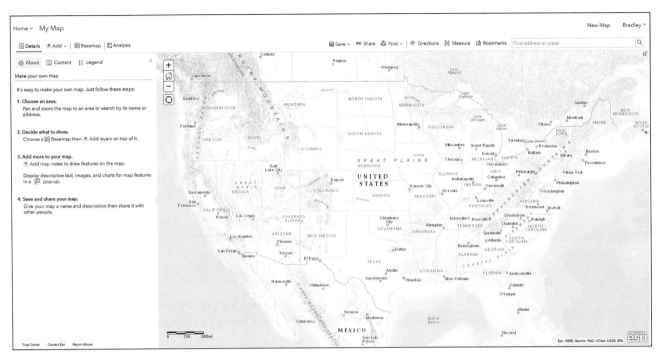

FIGURE 4.1 The ArcGIS Online interface.

of ArcGIS Online are part of a **cloud** structure, in which your data, maps, and applications are stored on Esri's data servers, and users can access these materials via the Internet as needed. For instance, you can save your data as a project package and share it via ArcGIS Online. Other users can then access and download your project package directly into ArcGIS Pro; users therefore have the ability to quickly utilize your data, maps, and project. You can do the same with a map package or layer package as well, making it easy to share and distribute your data across the Internet.

More importantly, however, ArcGIS Pro gives you the ability to transfer your data and layers to ArcGIS Online and create interactive maps and apps online. These ArcGIS Online maps and apps allow users to turn layers on and off, zoom in and out, make measurements, and identify features, among other interactive tasks. The benefit of using ArcGIS Online is that your maps and content are stored on an Esri server somewhere else and can be accessed on-demand as needed through the Internet.

A big advantage of creating web maps and apps from GIS data using ArcGIS Online is that end users don't need ArcGIS Pro installed on their computers to use your map; ArcGIS Online web maps simply run in a web browser. Therefore, ArcGIS Online works as a **SaaS (Software as a Service)** application. With SaaS, all the software you need to build a web map or app is stored on a server and accessed via a web browser; you do not have to download anything to your own computer. Also, with SaaS, things like GIS data, maps, applications, and geospatial content are stored on servers and can be accessed by users via a web browser (Figure 4.2). This means that people who generate the data and maps don't have to serve those items out to others on their own, and they also don't have to be responsible for distributing them on their own servers, and end users only need a web browser to access these materials.

cloud A computing structure in which data, content, and resources are all stored at another location and served to the user via the Internet.

SaaS (Software as a Service) A cloud structure wherein the software being used is stored on a server at another location and accessed on demand via the Internet.

FIGURE 4.2 Using the cloud with GIS.

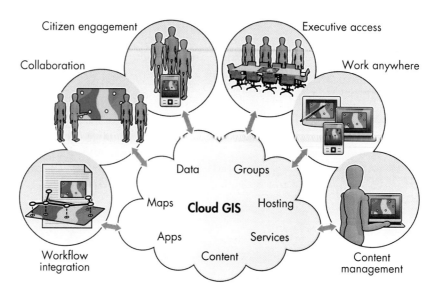

This ease of distribution can be useful for rapidly distributing geospatial information in emergency situations. For example, a rescue team on the ground after a natural disaster can immediately identify areas of concern on the map, upload those data to a web map on ArcGIS Online, and immediately distribute that information over the Internet. This module takes you through the steps involved in sharing your ArcGIS Pro GIS data on the Internet as a web application via ArcGIS Online. You can think of this as a three-step process:

1. Use ArcGIS Pro to assemble the GIS layers, set them up as you want them, set up their symbology, and do the general GIS work described in earlier modules.

2. Use ArcGIS Pro to transfer your data layers and their appearance to a web map in ArcGIS Online, where you configure the web map and layers for proper display and usage online.

3. Turn the web map into an interactive web app that contains additional tools and features.

Note that there are two different versions of ArcGIS Online: a free version and an **organization** version. With the free version (available with the free Esri account that you created in Module 1; see Smartbox 1.1 on page 5 for more information), you can design web maps and applications using data available online. However, with the organization-based version, you can directly publish data layers and maps from ArcGIS Pro to ArcGIS Online and build web maps and applications with those additional data. This module assumes that you have access to the organization version (which is included as part of an Esri site license) and can publish maps with it.

Also keep in mind that there is a cost involved in publishing and sharing map services with your ArcGIS Online organization. Your school or organization receives a certain number of **credits** for its organization. The sharing of hosted services uses up credits from your organization's account. For instance, publishing and sharing a tiled web layer uses up a variable number of credits, depending on factors such as the size of the dataset and the scales at which the tiles can be viewed. Your organization's administrator will have information about the number of credits available to you.

organization The ArcGIS Online version utilized by a group, such as a business or a school.

credits A system used by Esri to control the amount of content that can be served by an organization in the ArcGIS Online subscription model.

Module Scenario and Applications

In this module, you'll again be taking the role of the real estate agent from Module 3 to further your analysis of population and housing trends by county. In this module, you'll be comparing two layers: population density (that is, persons per square kilometer) for each county and the percentage of housing vacancies for each county. However, rather than create a pair of printed layouts, you'll build an interactive web application to compare the two layers and share it through ArcGIS Online. As in Module 3, there are no questions to answer in this module; instead, a checklist of items helps you make sure your web application is complete and well designed.

The following are additional examples of other real-world applications of the theory and skills discussed in this module:

- A park ranger is creating a map of a state park's hiking and driving trails, campsites, and other park features. Due to the change of seasons and landscape conditions, certain trails and areas are not accessible at all times. The ranger will use ArcGIS Online to create a web map of current park features and conditions that can be easily updated based on season and other conditions. The web map will be visited by people who want to use the park, as well as park rangers.

- A city's tourist bureau wants to develop a map of tourist attractions, local restaurants, and features of interest for the city's visitors. Rather than print paper maps, the bureau wants to create an interactive map of the area that shows images of key tourist areas. Visitors will be able to select a location and get information about that tourist site, the hours of operation, and photos of the locale. The bureau plans to use ArcGIS Online to develop a map that can be viewed using only a web browser.

- An archeologist is working at a site in a specific country and is developing a GIS map of the site that can be shared with the local government, as well as other researchers. By sharing this interactive map via the web, he can easily distribute it to interested parties, as well as make updates as new discoveries are found at the site.

Study Area

- In this module, you will be working with data for each county in the state of Ohio.

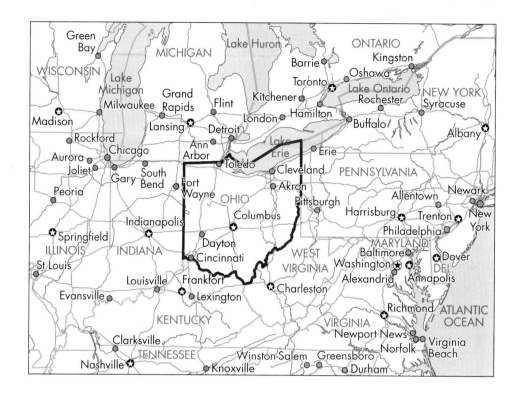

 Data Sources and Localizing This Module

The data in this module focus on features and locations within Ohio's counties. However, this module can easily be modified to use U.S. Census Bureau data on your own county or local area instead. (See *Related Concepts for Module 2* for more about using Census data in ArcGIS Pro.) This module used county boundaries from the TIGER/Line files from the Census, but the way Census data is served and formatted has been updated since then. Comparable Census data for your own county are available from the U.S. Census for free. You need to download a county boundary dataset and attribute tables from the Census and then join them together.

The county boundaries can be obtained as Cartographic Boundary files from **https://www.census.gov/geographies/mapping-files/time-series/geo/carto-boundary-file.html**. The counties layer you download contains all the counties in the United States. You can then query the shapefile and extract the appropriate counties into a separate feature class for your state, based on the STATEFP attribute. This attribute references each state with a two-digit FIPS code. For instance, Ohio is referenced with state FIPS code 39. If you were using this module in Florida, for instance, you would use FIPS code 12. A full list of FIPS codes for each state and county is available from the National Weather Service at **https://www.weather.gov/pimar/FIPSCodes**. Note that some of the U.S. Census Bureau's county boundaries extend to the national boundaries when they border water features (such as in the Great Lakes area between the United States and Canada). Also note that the county boundary file was projected into the *Web Mercator (auxiliary sphere)* projection for purposes of this module. There might be another projection for your own state (such as the *Lambert Conformal Conic* projection) that you want to use instead. (See Module 1 for more about projecting data.)

Attribute data must be downloaded separately and joined to the counties shapefile. The attribute data (the population statistics) for each county in the state were downloaded from the Census Bureau's American FactFinder website, but these data are now available through the Census data website at **https://data.census.gov/cedsci/**. The tables used in this module are Vacancy Status (and the "for seasonal, recreational, and occasional use" attribute), Occupancy Status (for the total number of houses), and Population for each county. Census attribute data were downloaded as a CSV file, edited and cleaned up, and converted to an Excel file for use in this module. The tabular data from the Census data website can be joined to the Cartographic Boundary layer by using the AFFGEOID attribute in the Cartographic Boundary file and the GEO ID attribute in the Census data.

Step 4.1 Getting Started

- This module's hands-on applications use the data folder called Module4. Your instructor will be able to supply you with this data, or you can download it directly from this book's website at **https://www.macmillanlearning.com/college/us/product/Discovering-GIS-and-ArcGIS-Pro/p/131923075X**. The text in this module assumes that you have this Module4 folder in a computer location referenced as C:\GIS; if you have it somewhere else (for instance, in a flash drive referenced as G:\GISClass), substitute that location and path to the Module4 folder throughout this module.
- The Module4 folder contains OhioWeb, a file geodatabase that contains two feature classes:
  - OhioPop: A polygon feature class of Ohio county borders; population data and population density have been calculated for each county.
  - OhioVacant: A polygon feature class of Ohio county borders; housing and vacancy data have been calculated for each county.
- Start ArcGIS Pro.
- Sign in with your Esri account username and password.
- Create a new project using the **Map** template. Call this project **Module4** and place it in your **C:\GIS\Module4** folder. Ensure that there is not a checkmark in the **Create a new folder for this project** box.
- When ArcGIS Pro opens, change the map's name to **CountyPopDens**.

Step 4.2 Setting Up Layers in ArcGIS Pro Before Publishing

- Add the **OhioPop** feature class to the Contents pane of the CountyPopDens map.
- As in Module 3, change the symbology of the Ohio counties to **Graduated Colors**. Use **PopDens** as the value to display, selecting the **Natural Breaks (Jenks)** method and **5** classes. Select an appropriate color ramp for displaying the data. Each Ohio county should now be shown with the population density.

- As in Module 3, create a second map (from the **Insert** tab, within the **Project** group, click the **New Map** button and then choose **New Map**). Change the name of this new map to **CountyVacant**.
- Add the **OhioVacant** feature class to the CountyVacant map.
- Change the symbology of OhioVacant to **Graduated Colors** and display the **PersVacant** attribute (which is the percentage of the housing stock for each county that is considered vacant). Use the **Natural Breaks (Jenks)** method and **5** classes for the symbology. Also choose a color scheme that is distinctively different from the color scheme you used for OhioPop.
- Save your Module4 project. You should now have two maps, each containing one feature class.
- Before proceeding, it's critical that your Esri account has access to the organization-level features of ArcGIS Online. If it doesn't, you won't be able to share your ArcGIS Pro work to ArcGIS Online and complete the rest of this module. See **Troublebox 4.1** for more information.

TROUBLEBOX 4.1

How do I set up my Esri account to access ArcGIS Online organization-level features?

In order to publish your maps directly from ArcGIS Pro, you need to access the organization-level features of ArcGIS Online. To do this, you must have your free Esri account connected to your school's or business's organization-level structure. Just having a free Esri account won't allow you to publish from ArcGIS Pro.

When a business purchases a subscription to ArcGIS Online (or when a college or university acquires a subscription as part of an educational site license), one or more individuals at that business or school will be designated the administrator for ArcGIS Online. The administrator will establish the organization for that place. For instance, if organization-level access to ArcGIS Online is available at your school, an organization should be established by a system administrator. Before you can use all of these ArcGIS Online features, you have to join the organization.

The administrator can allow you to join the organization either as another administrator, as a user, or (as you will be doing in this module) as a publisher. When you set up your Esri account, you identified an e-mail address to be associated with the account. The administrator can have an e-mail message sent to that account to invite you to join the organization at the publisher level. You must click the link in the e-mail invitation and carefully follow the steps and instructions.

At the end of the process, your Esri account will reflect your role as a publisher within the organization, and you'll be able to share your ArcGIS Pro data and maps directly to ArcGIS Online.

Important Note: If you don't know who the administrators are at your school, contact Esri's customer service representatives, who should be able to let you know the contact individual at your school.

Step 4.3 Sharing a Map from ArcGIS Pro to ArcGIS Online

- In this step, you'll **share** each of your maps to ArcGIS Online. In sharing, you'll be uploading your GIS data to ArcGIS Online for use in creating interactive web maps or applications out of that data. For more information about sharing and web layers, see **Smartbox 4.1**.

share To distribute data, content, maps, or applications across the Internet.

SMARTBOX 4.1

How are my data and maps shared to ArcGIS Online?

When you are sharing your data from ArcGIS Pro to the cloud structure of ArcGIS Online, it is set up in a format appropriate for being used on the web instead of the desktop software program. This process is also referred to as **publishing**, and it involves transferring your data to the cloud for use there. Sharing your data means moving to ArcGIS Online the one or more layers being displayed in a map in ArcGIS Pro. Keep in mind that publishing your data consumes credits, and some web-enabled formats (such as tile layers) consume many more credits than others.

When your layers are shared from ArcGIS Pro, you can control who they are shared with (that is, who can access them through ArcGIS Online). ArcGIS Pro gives you the option to share your data with no one (that is, you will be the only one who can use or modify those data in ArcGIS Online), with everyone else in your organization (that is, all the others who also have their Esri accounts connected to the organization you are connected to), or simply with everyone (that is, anyone who is using ArcGIS Online or ArcGIS Pro). Controlling who you are sharing your data with is important, as you can also enable your layers to be edited after they are published. Once your layers are shared and accessible on ArcGIS Online, others with whom you have shared can also access them, and if you have enabled them to be edited, others can then make changes to your data. Sometimes you might want this (for instance, if others in your organization are collaborating together and making edits to datasets as a group), but other times you may not.

However your data is shared, when you are publishing your data, you are setting up your data as a **web layer** in ArcGIS Online; this is the web-enabled format you can work with in the ArcGIS Online environment when constructing a web map or application. If you share an entire map containing multiple data layers in ArcGIS Pro, that map in ArcGIS Online will consist of a web layer. In ArcGIS, these web layers are often referred to as **hosted** layers because someone else (that is, the computers and servers that make up ArcGIS Online) are holding (or *hosting*) the data and enabling the distribution and usage for you.

When your web layers are published to ArcGIS Online, their initial projection (whether geographic or projected coordinate system) is not maintained. ArcGIS Online services use the **Web Mercator auxiliary sphere** map projection (a common setup used for web mapping utilities such as Google Earth and Bing Maps) for their content and basemap, and when you publish your layers, your data are projected on the fly to match this projection.

publish To place data or content onto a cloud server.

web layer A format used when publishing a map of one or more layers to ArcGIS Online.

hosted A term denoting that the layers are stored on a server in the cloud.

Web Mercator auxiliary sphere The projected coordinate system used by ArcGIS Online.

There are several different types of hosted web layers that can be published to ArcGIS Online, including the following:

feature layer A hosted web layer that allows a user to share GIS data layers that can also be displayed, queried, or edited.

tile layer A hosted, cached web layer that sets up GIS data as a series of image tiles that can be displayed but not queried or edited.

vector tile layer A hosted web layer that displays using the vector representation of the data.

- **Feature layer**: This is a web layer that consists of features (that is, points, lines, polygons). With a feature layer in ArcGIS Online, the features can be visualized and have their symbology changed, the layer's attribute table can be opened, and features can be filtered out to control what is displayed. You can also enable editing of a feature layer so that you (or others you have shared your feature layer with) can make changes to it.

- **Tile layer**: This is a web layer that consists of a set of predrawn images (referred to as *tiles*). These layers set up in a format that does not allow you to edit, query, or otherwise access an attribute table. Rather, each tile layer takes the form of an image that can be displayed. Tiled layers are useful for visualizing GIS data, especially large GIS datasets. For instance, a map of population change at the census-block level can be displayed as a tiled layer that can be turned on and off to show other layers. Publishing tile layers requires a considerable number of credits, depending on the level of detail and available scales of the data that are cached (and the size of that cache that is published and stored on ArcGIS Online). Tile layers are commonly used when publishing raster data (see Module 12).

- **Vector tile layer**: This is a web layer that consists of a set of tiles but utilizes the vector representation of the items being published. Vector tile layers have smaller file sizes than tile layers (and so are easier and less costly to store in ArcGIS Online), and they can be quickly rendered at many different scales and levels of detail.

You can also share a map to ArcGIS Online directly as a web map. In this case, you are sharing the layers in your map as web layers, as described above, but you are also sharing your basemap, as well as any web layers you may already be using in the map in ArcGIS Pro.

- You will be publishing both of your maps to ArcGIS Online and then building an app from both of them, but you'll do them one at a time. To start with the CountyPopDens map, choose its tab from the view.
- Select the **Share** tab, and within the **Share As** group, click the **Web Layer** button and then choose **Publish Web Layer**.
- The Share As Web Layer pane opens. By default, the name of the web layer you're sharing is **CountyPopDens** (the same as the map in ArcGIS Pro).
- Write a short summary of what this web layer will be (for example, **This web layer shows the population density for Ohio counties**).
- For Tags, type **Ohio, Population Density**.
- For Layer Type, choose **Feature**.
- The Folder location is your own ArcGIS Online account being used for this module, so choose it from the pull-down menu.

Share As Web Layer

Sharing CountyPopDens As A Web Layer

General Configuration Content

Item Details

Name

CountyPopDens

Summary

This web layer shows the population density for Ohio counties.

Tags

Ohio ×   Population ×

Layer Type ⓘ

- ● Feature
- ○ Tile
- ○ Vector Tile

Location

Folder

bradshellito (root)

Share with

- ☐ Everyone
- ☑ Youngstown State University
  - Groups ▼

Finish Sharing

✓ Analyze   ☁ Publish   ☰ Jobs

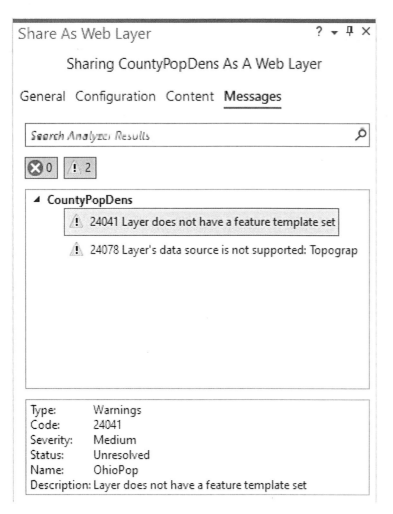

- For the **Share with** options, put a checkmark next to the name of your university or organization (for example, **Youngstown State University**).

- Finally, click **Analyze**. ArcGIS Pro examines your map and layers for any potential errors that need to be fixed prior to publishing.

- ArcGIS Pro flags any errors with a white X in a red circle and any warnings with a black exclamation point in a yellow triangle. You must correct any errors before ArcGIS Pro will allow you to proceed with publishing the map, but you will be able to proceed with only warnings. In this case, you have a set of warnings concerning your layers not having a feature template set; this is okay, as a template will be automatically created after publishing, so you don't have to worry about these warnings. Click on the **Publish** button at the bottom of the Share As Web Layer pane.

- You may encounter an error from ArcGIS Pro saying that a web layer with the same name as the one you are trying to publish already exists. See **Troublebox 4.2** for information on how to deal with that so that you can continue publishing.

TROUBLEBOX 4.2

How can I successfully publish my web layers?

One of the requirements for published web layers is that they be named uniquely from other layers—not only in your own Esri account but also throughout your organization. If someone else in your organization (such as another student or instructor) has already published a layer called CountyPopDens, for example, and you try to publish a layer with that same name, ArcGIS Pro will stop you from doing so. The solution is to just give the layer another name; for instance, you can just add your initials to the end of the CountyPopDens filename, and it should publish fine. Keep in mind that for the remainder of this module, your CountyPopDens layer will have a slightly different name, so be sure to use that new name.

- After a couple of minutes, you should get a message highlighted in green at the bottom of the pane indicating that you have successfully shared your web map.

- Next, you can publish your second map, so in the view, switch over to the CountyVacant map. Follow the same steps to publish the CountyVacant map as a web layer, using the same settings as before, with the following differences: For the name, use **CountyVacant**; for Tags, use **Ohio** and **Vacant**, and for the summary use **This web layer shows the percentage of the housing stock that is vacant for Ohio counties**.

- When you click **Analyze**, you should see the same warnings as before. Again, you can safely ignore them and click **Publish**.

- When the CountyVacant layer is successfully published, you can start working with the online side of things, so open a web browser and go to **https://www.arcgis.com/home/index.html**.

- On the website, click on **Sign In** and enter your Esri account username and password.

Step 4.4 Configuring a Web Map from Published Layers

- After logging in to ArcGIS Online, you should see a banner at the top with your organization's name and a series of tabs above that. Click on **Content** to display all of the various layers and items that are stored with your Esri account on ArcGIS Online.

- In your web browser you see that the two hosted feature layers (CountyPopDens and CountyVacant) have been published to ArcGIS Online. You're now ready to build and configure web maps from them. For more information about working with and designing web maps, see **Smartbox 4.2**.

☐	Title			
☐	CountyVacant	Feature Layer (hosted)	🏠 ☆	⋯
☐	CountyVacant	Service Definition	🔒 ☆	⋯
☐	CountyPopDens	Feature Layer (hosted)	🏠 ☆	⋯
☐	CountyPopDens	Service Definition	🔒 ☆	⋯

SMARTBOX 4.2

What are web maps, and how are they built in ArcGIS Online?

A **web map** is an assemblage of layers used together in an interactive, online environment and is the basic interface for working with web layers and other data in ArcGIS Online. A web map in ArcGIS Online allows you to bring together multiple types of data, change their symbologies, configure pop-ups,

web map An interactive online representation of GIS data that can be accessed via a web browser.

and provide a means of viewing data layers online in a simple web interface. A web map also includes a basemap on which to display your layers. Unlike a static layout, a web map is interactive: Layers can be turned on and off, you can click on objects to see pop-ups about the objects' attributes, you can zoom and pan through the web map, and so on. Also, a web map is assigned its own URL so that it can be easily shared and viewed by others. When someone opens that URL in a web browser, the web map opens even if the user doesn't have an Esri account or ArcGIS Pro.

Web maps can be built with either the free or organization-level versions of ArcGIS Online, but if you have access to the organization-level features, you can take advantage of a variety of analysis tools for use with the layers of the web map, such as spatial joins, summarize operations, and hot spot analysis (see Module 8), as well as merge, dissolve, buffer, and overlay operations (see Module 9). Note that if you want to publish layers to ArcGIS Online from ArcGIS Pro to build a web map, you need to have the full organization permissions.

A web map can be assembled through a variety of data sources. First, some data that are stored locally on a computer can be added directly to the map. Shapefiles (see Modules 1 and 6) containing points, lines, or polygons can be added to the web map if the shapefile is in a zipped format. A simple text file or CSV file containing x/y coordinates can be plotted as a point layer (see Module 6), and a similar file containing addresses can be geocoded and turned into a point layer (see Module 10). Also, GPX files (see Module 6) can be directly added to a web map, allowing things such as points, lines, or polygons recorded by a GPS receiver to become layers in a web map. However, all these are simple representations of data, and while they can easily be used in a web map, ArcGIS Online limits the number of features a shapefile can contain (such as a maximum of 1000 features in a shapefile).

A web layer can be better used to build a web map than one of the locally stored options as it is not limited to the number of features that can be shown on the web map. Publishing layers to ArcGIS Online from ArcGIS Pro enables you to add them to a web map. In addition to your own published layers, you can take advantage of the wealth of other web layers that Esri allows you to add to a web map via the ArcGIS Living Atlas of the World (see Module 5), which is a searchable set of already made web layers that you can pull from for your own web maps.

Also, a web map can have its contents drawn from a variety of web **services**, which are other types of data that have already been published and are available online (possibly outside the ArcGIS Online environment). You can add a web service to a web map by providing the URL of the web service to ArcGIS Online. These services can come in multiple varieties, such as the ArcGIS Server Web Service, WMS (Web Map Service, which allows for map data to be added to a web map), WMTS (Web Map Tile Service, which deals with imagery and cached maps), or WFS (Web Feature Service, which deals with vector data). For example, satellite imagery may be available online at another website as a WMTS, and to use it in a web map, you would only need to provide the URL of that service to add it to the map.

service A format for data that can be accessed as web content.

- Click on the **three dots** in the row of the CountyPopDens feature layer (hosted), and from the pop-up menu that opens, choose **Open in Map Viewer**.
- The basic web map interface opens, and you see a Contents area on the left-hand side. Click on **Details** and then click on **Content**. As in ArcGIS Pro, the layers in the Contents area are those being used in the web map, and if a layer has a checkmark next to it, it is being displayed in the web map. Right now, your web map consists of one layer (CountyPopDens). If you hover your cursor next to or below CountyPopDens, you see a set of icons appear.
- Click on one of the polygons representing an Ohio county on the web map. You see a pop-up filled with all the various attributes of the layer—far more than you need to convey information about population density. The next step is to configure the web map so that only the attributes you want appear in a pop-up if someone clicks on the map. In the Contents area, in the icons that appear when you hover your cursor next to CountyPopDens, click on the icon with the **three blue dots**.
- In the new menu that appears, choose **Configure Pop-up**.
- A new side pane appears, with numerous options for making changes to the pop-up. To change which attributes are displayed, click on **Configure Attributes**.

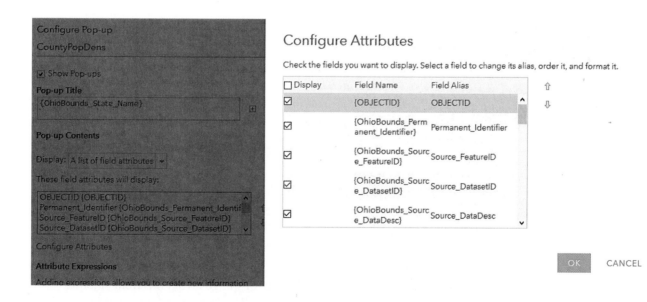

- A new menu appears, showing a list of all the various attributes for each county in the CountyPopDens layer. If a checkmark is next to an attribute, it will be displayed in the pop-up; otherwise, it will not be displayed. Remove the checkmarks next to all the attributes except for **County_Name** and **PopDens** and then click **OK**.

- Back in the web map, you're done configuring the pop-up, so click **OK** in the left-hand side pane.

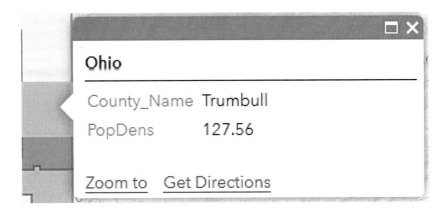

- When configuration is done, click on a county. The pop-up appears, showing only your chosen attributes for one layer.
- Pan the map and zoom in or out of the state of Ohio so that you can clearly see all counties in the entire state.

  Important Note: The way that the state looks in terms of its scale or what is being shown in the view will carry over to the final published form.

- You can now select a new basemap on which to display the CountyPopDens layer. The default layer is topographic, which doesn't fit well with the population theme. For more about the types of basemaps available in ArcGIS Online, see **Smartbox 4.3**. To change basemaps, click the **Basemap** button. For this web map, choose an appropriate basemap from the available options, keeping in mind the nature of the data you're trying to present on the web map.

 SMARTBOX 4.3

What kinds of basemaps are available in ArcGIS Online?

basemap An image layer that serves as a backdrop for the other layers used in ArcGIS Online.

In ArcGIS Online, a **basemap** is a tiled layer on which your other layers are displayed. It serves as a backdrop image on which to place your data and thus can't be queried or edited. Think of a basemap as a georeferenced background image that is used when setting up a web map. ArcGIS Online has several different basemaps available. The choice you make depends on the kind of web map you're designing (Figure 4.3):

- Imagery: This layer contains a variety of images from different satellite sensors (or aerial photography platforms). The imagery that is visible depends on the scale to which you've zoomed or the area of the world you're viewing. (See Module 13 for more about using aerial and satellite imagery in ArcGIS Pro.)

- Imagery with Labels: This is the same layer as the Imagery layer, but it also includes labeled country boundaries, as well as labeled state and county boundaries for the United States.

- Streets: This is a world street map that also labels cities, parks, water features, and some building footprints.

- Topographic: This layer covers the world, showing built features as well as land cover and shaded relief.

- Dark Gray Canvas: This layer is intended to provide geographic reference, but because of its dark gray color, it remains in the background and allows for the appearance of your data to "jump" off the screen. As such, it contains minimal labels and features and is intended for use as a neutral backdrop.

- Light Gray Canvas: This is a version of the Dark Gray Canvas but with a lighter gray background for use with different color schemes on maps.

- National Geographic: This world map shows a variety of roads, city features, water bodies, and landmarks, along with shaded relief and land cover.

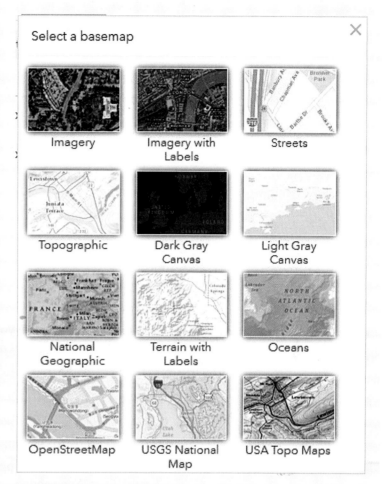

FIGURE 4.3 The available basemaps in ArcGIS Online.

- Terrain with Labels: This layer provides shaded terrain relief, water features, and bathymetry, as well as country, state, and county borders and labels.

- Oceans: This basemap is intended for use in ocean and marine maps. It features water bodies, coastlines, and bathymetry.

- OpenStreetMap: This layer shows road features from the OpenStreetMap project, an open-source online map that can be edited by anyone. (See Module 5 for more information about OpenStreetMap.)

- USGS National Map: This layer shows the features of The National Map. (See Module 5 for more information.)

- USA Topo Maps: This layer shows the historical USGS topographic maps. (See Module 16 for more information.)

- Finally, click the **Save** button at the top of the web map to save all the changes you've made and choose **Save**.

- When you save the web map, a dialog appears, allowing you to give the web map a name, tags, and a short summary. When you're finished, click **Save Map**.

- With your web map saved, you can now set up a second web map. In the upper-left corner of the ArcGIS Online website, click on **Home**, and then from the pull-down set of options that appears, choose **Content** to return to the main Content page, which shows all the things stored in your space on ArcGIS Online.

- Much as you did before, click on the **three dots** next to the CountyVacant feature layer (hosted) but this time choose **Add to New Map**.

- A new web map interface appears, but this time the only layer shown is CountyVacant. You need to create a second web map using this layer, in much the same way as you made the web map of the CountyPopDens layer. Do the following:

  - Configure the pop-up for the CountyVacant layer but set its attributes so that only the **County_Name** and **PersVacant** attributes are displayed.
  - Change the scale of the displayed Ohio to match how you zoomed in on the layer in the CountyPopDens map.
  - Switch the basemap from Topographic to something that better reflects the map's theme.
  - Save the map with the name **Ohio Percentage of Vacant Houses per County**, with appropriate tags and a summary.

Step 4.5 Creating a Web App

- Next, you will use your web map to build an interactive web app. The app that you create will end up including both of the web maps you've made so far. For more information about web apps, see **Smartbox 4.4**. Click on the **Share** button at the top of the Ohio Percentage of Vacant Houses per County web map.

SMARTBOX 4.4

What are web apps, and how are they built in ArcGIS Online?

Creation of web apps lies at the heart of working with GIS data online. A **web app** is a lightweight, standalone, and interactive website that contains more features and widgets than a basic ArcGIS Online web map has to offer. In a web map, you can change the symbology and labels of layers, use different basemaps, and control which attributes are shown in a query, but the design of the map is preset. For instance, you can't move web map elements (such as the text or zoom slider) to another location, and you can't easily add a north arrow or an inset map. In fact, all web maps designed using the basic interface of ArcGIS Online look the same. With a web app, you can customize the appearance of a basic web map by adding numerous additional options, tools, and widgets to the map and creating a fancier or more in-depth application. When a web app is created, it is assigned its own URL, which can be shared with others so they can work with the new app.

Web apps have different layouts than basic web maps, as well as different color schemes, appearances, and usages. For example, Figure 4.4 shows a web app that is different from the one you will create but that allows for comparison between the two web layers used in this module (for housing vacancy and population density). The web app has a very different appearance from the basic web map, as it can present the two layers overlapping one another but with a swipe bar down the middle of the app. By moving the swipe bar, the user can compare one layer on top of another. In addition, the legends of the two layers

web app A standalone website that allows the user to customize its appearance and tools.

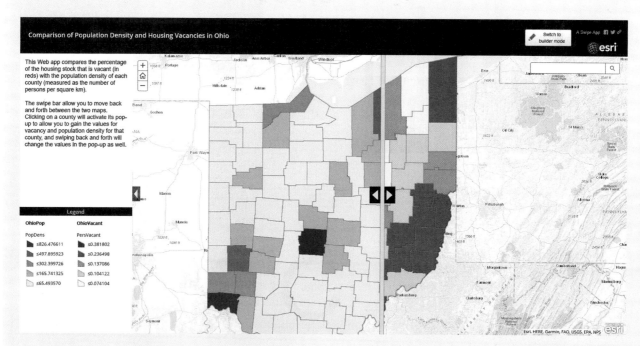

FIGURE 4.4 A web app crafted using the Swipe template.

are presented side by side, a text box with information is available, and a search box and social media links to Facebook and Twitter are positioned in the upper-right corner. This web app has its own unique appearance and has much more versatility to the user than a standard web map.

Web apps are simple and easy to design. In ArcGIS Online, when you share your web map, you have the option of converting it to a web app, using one of several different configurable apps. These configurable apps are like templates that allow you to build apps that compare layers, add directions and routing information, or add tools such as elevation profiles. You can also simply create a basic app that has a more customizable appearance than the generic web map.

Apps can also be built from ArcGIS Online web maps using *Web AppBuilder* for ArcGIS or *Operations Dashboard*, two other web-based Esri programs that put more app design tools and formats at your fingertips. For instance, Figure 4.5 shows a web app called a **dashboard** (created using the Operations Dashboard tools) for the city of Philadelphia. It allows an interactive map to be part of the web app; in addition, data can be displayed in multiple charts or as text, address lists, and other customizable content. With a dashboard, data and analytics can be incorporated together into the app to show things such as counts of crimes by date, the types of crimes, peak times when crimes occurred, the type of crime at an address, and so on.

dashboard A type of web app that can combine maps, data, and analytics together.

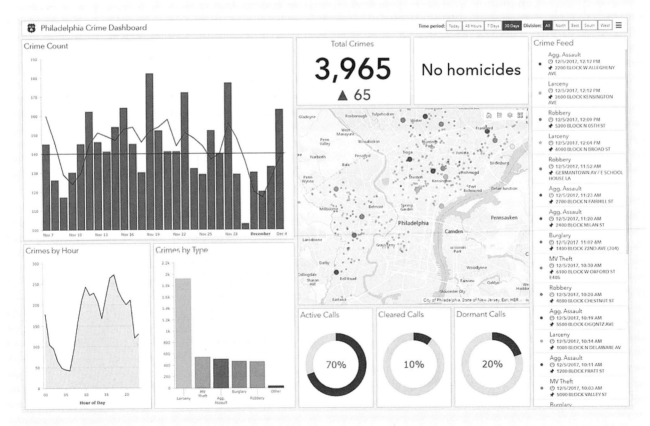

FIGURE 4.5 A web app created using Operations Dashboard.

- By sharing a web map, you enable others to view it, access it, use it, and also potentially download it into ArcGIS Pro. In the Share dialog, you see the shortened URL assigned to the Ohio Percentage of Vacant Houses per County web map. Place a checkmark next to your organization to designate who you want to share this map with. Because the final product you will be creating is a web app, click on **Create a Web App**.

- The Create a New Web App dialog allows you to choose how you want to build your web app—using either a configurable app, Web AppBuilder, or Operations Dashboard. For the app you are building in this module, click on the **Configurable Apps** tab, choose **Compare Maps/Layers**, and click on **Compare**.

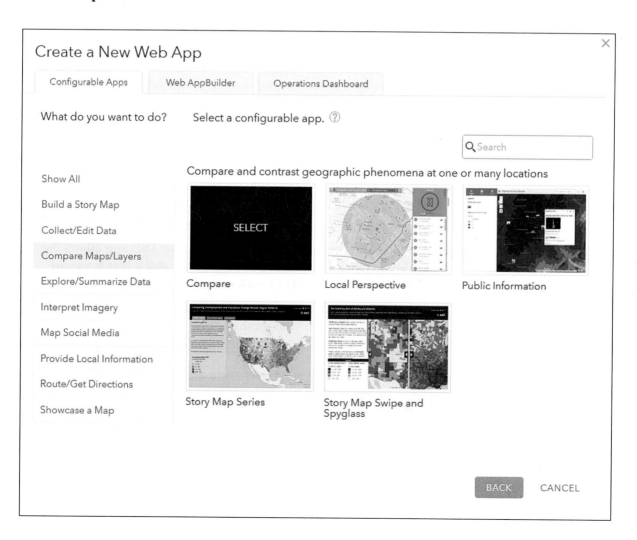

- A sidebar appears, showing the compare app graphic. Click on **Create Web App**.

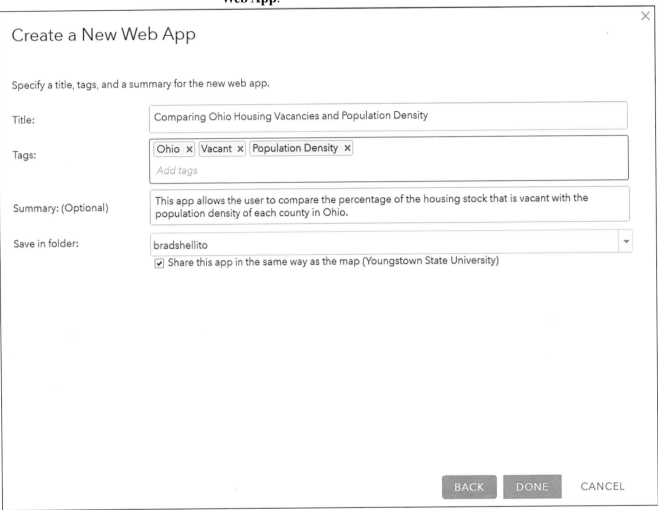

- For the title, type **Comparing Ohio Housing Vacancies and Population Density**.
- For Tags, make sure you specify **Ohio**, **Vacant**, and **Population Density**.
- For the summary, type something appropriate, like **This app allows the user to compare the percentage of the housing stock that is vacant with the population density of each county in Ohio**.
- For the **Save in folder** option, choose your main folder for your Esri account (which should be your username).
- Place a checkmark next to **Share this app in the same way as the map**.
- Click **Done**.

Step 4.6 Configuring a Web App for Online Publishing

- The Configure Web page opens, allowing you to begin configuring the web app as you want it. The Compare app is designed to display two web maps side by side. You see the two panels for the web maps: One currently has the Ohio Percentage of Vacant Houses per County web map, and the other panel contains a gray canvas and does not show the contents of a second web map. There is also a side panel on the left of the web page that shows all the various options for configuring the web app; this is where you will work when building the app.

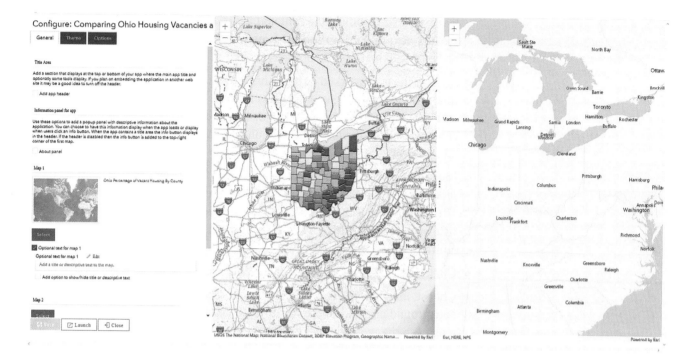

- At any time after making changes, you can click **Save** at the bottom of the side panel to save your changes. Your changes are not auto-saved, so be sure to click on the Save button to store your changes as they will be lost if you click Close or otherwise close the browser tab. If you close the browser and don't know how to get back to it, see **Troublebox 4.3**.

TROUBLEBOX 4.3

How can I return to configuring my web app?

If you should the browser or the tab or otherwise get shut out of your work, you can get back to the Configure page by doing the following:

- From the Content page in ArcGIS Online, locate your web app (in this case, the Comparing Ohio Housing Vacancies and Population Density Web Mapping Application) and click the **three dots** in its row.
- From the pop-up menu that appears, click **View Item Details**.
- On the next screen, click **Configure Web App** to return to the Configure page.

- In the side panel, choose the **General** tab (which should be the default one chosen at the start of the configuration process) and look at the area marked Title. Place a checkmark in the box next to **Add app header**. Give your app an appropriate title that will be displayed when the app opens and use the pull-down menu to choose the location of the title (top or bottom).

- Look at the area marked Map 1. The notes for Map 1 refer to the left-most of the two panels for displaying maps; in this case, Map 1 is the Ohio Percentage of Vacant Houses per County web map. Place a checkmark in the box next to **Optional text for map 1** and type a description of the map there; if desired, adjust the font and text size. Also place a checkmark in the box next to **Add option to show/hide title of descriptive text**. Then place a checkmark in the box next to **Show content when app loads**.

- Look at the area marked Map 2; this corresponds to the right-most panel that currently contains a blank gray canvas. Click the **Select** button, and you see a new window that contains the web maps available in your content. Choose the Ohio Population Density per County web map and click **Select**.

- Next, place a checkmark in the box next to **Optional text for map 2** and type a description of the map there; if desired, adjust the font and text size. Also place a checkmark in the box next to **Add option to show/hide title of descriptive text**. Then place a checkmark in the box next to **Show content when app loads**.

- Click **Save**.

- Choose the **Theme** tab. Using the options here, you can modify the background and text of the app title, the panels, and the buttons. Make whatever changes you feel best suit the app.

- Click **Save**.

- Choose the **Options** tab.

- Under the Compare options, choose **Side by Side** for Compare Layout. (Based on the way the state of Ohio is oriented, placing the two maps side by side works best. For a state that is more horizontally oriented, such as Kentucky, you would likely be better off choosing Stacked for Compare Layout.)

- Place a checkmark in the box next to **Synchronize Maps**. This means when you zoom in on one map or pan it, the other will zoom or pan the same way.

- Place a checkmark in the box next to **Synchronize Popups**. This means when you click on a county in one map to activate its pop-up, the corresponding pop-up in the other map will open as well.

- Under the Tools options, place a checkmark in the box next to each of the following: **Legend**, **Scalebar**, and **Social Sharing** (and then also put a checkmark in the **Include social media options** box). All these options (as well as the ones you did not check) are various icons and widgets that you can add to an app to customize it further. For each of these four options you checked, choose a location for where that control or icon will go on the app.

- Click **Save**. You see a preview of what the final app will look like. Make any changes or adjustments to the app elements now. For instance, you might have too many icons grouped together, the text might not be what you want for the final version, and so on.

- When you are satisfied with the appearance of the app, click **Save** one final time, and you are ready to share the final app.

Step 4.7 Sharing and Viewing a Web App

- When the web app is configured the way you want it, click **Launch** in the side panel. A new tab opens, showing the final version of your web app. Test the app's features, such as the synchronized zoom and pan, the synchronized pop-ups, the legend features, turning text on and off, and so on. If the app is not the way you want it, close this tab and return to the Configure tab to make changes there, save those changes, and then launch the app again.

- When you created the web app, you specified that you wanted it shared the same way that you shared the initial web map (that is, with the organization). If you want to obtain a shortened URL for the final app, click on the **social sharing** icon in the final app (the icon that has three arrows coming out of it), and a pop-up box appears, showing the shortened URL along with icons to share the URL on Facebook, Twitter, LinkedIn, or email. This shortened URL is assigned to the final app, and anyone in your organization who has this shortened URL can access your web app.

Checklist

Use the following checklist to make sure you have put together your web app for this module in a complete fashion:

_____ Two maps of Ohio counties, one showing population density per county and the other showing housing vacancy percentages per county, displayed side by side

_____ The layer showing Ohio population density and the layer showing Ohio vacant housing in distinctive color schemes

_____ A basemap for each of the two maps that is not the default topographic one and that is appropriate for the theme of the maps

_____ The pop-up for the vacant housing layer showing only two attributes (the name of the county and the percentage of the housing stock that is vacant); the pop-up should have a color that is complementary to the layer and an appropriate title

_____ The pop-up for the population density layer showing only two attributes (the name of the county and the population density); the pop-up should have a color that is complementary to the layer and an appropriate title

_____ Appropriate text descriptions for each of the two maps (at least one sentence of descriptive information for each), with a button to hide and show the text on each of the two maps

_____ A legend for each map

_____ A link for social sharing that displays icons for various social media options

_____ A scale bar

Closing Time

In this module you learned how to share your GIS data and content with others via the Internet by setting up a web map and constructing a standalone web app. There is another type of popular web app called a *story map* that can also be created with multiple types of GIS data. See *Related Concepts for Module 4* for more about designing these specific type of web apps.

ArcGIS Online can also be used to share GIS data with others. You didn't learn how to use ArcGIS Online that way in this module, but you could share your ArcGIS Pro work as a map package for others to use via ArcGIS Online. Similarly, rather than set up a web map or web app, you could simply share the population and housing data through ArcGIS Online for others to download. In Module 5, you'll be obtaining data directly through ArcGIS Online for use in ArcGIS Pro (as well as from other free online GIS data sources).

RELATED CONCEPTS FOR MODULE 4

ArcGIS StoryMaps

As you've seen in this module, ArcGIS Online makes it easy to create web applications for all manner of uses. A specific type of web app that is very popular is the ArcGIS StoryMap. As its name implies, a **story map** is a web app that is used to communicate a story (or theme) to its user by combining maps, spatial data layers, descriptive text, attribute pop ups, and geotagged photos and videos. Story maps are highly customizable and allow a designer to tell a story by bringing together all these multiple elements. For much more about designing and publishing ArcGIS StoryMaps, see **https://storymaps.arcgis.com/**.

Figure 4.6 shows an example of a story map that is an interactive guide to the cities that have hosted the Olympics since 1896. On this story map, you can select a particular location on the map where the Olympic games have been hosted, and the story map presents you with a photo and information about the event. By selecting an option from the timeline at the top of the page, you can bring up the photo and information about the Olympic games of that date as well as center the map at that location. Story maps provide a way to bring together geospatial data (such as the locations of cities that have hosted the Olympics) with non-spatial data (photos, attributes, and descriptive text about that city and the event) in an interactive format.

story map A web application designed to convey a specific theme or concept to the user using a special set of web templates.

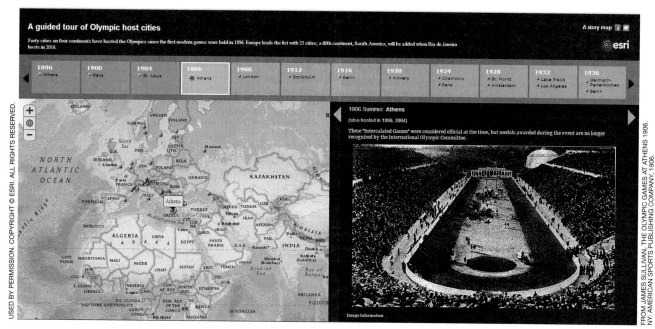

FIGURE 4.6 A story map showcasing the history of the Olympics.

Figure 4.7 shows a different type of story map—this one tracking the voyage of Charles Darwin on the *HMS Beagle* from 1831 to 1836. The story map shows the various locations that Darwin visited, linked to photos and descriptive text about each place, along with excerpts from Darwin's writings about each spot. By choosing one of the flagged locations on the map or one of the photos in the carousel at the bottom of the story map, you can obtain information about that place. By following the locations in order, you get the full story of the *Beagle*'s journey.

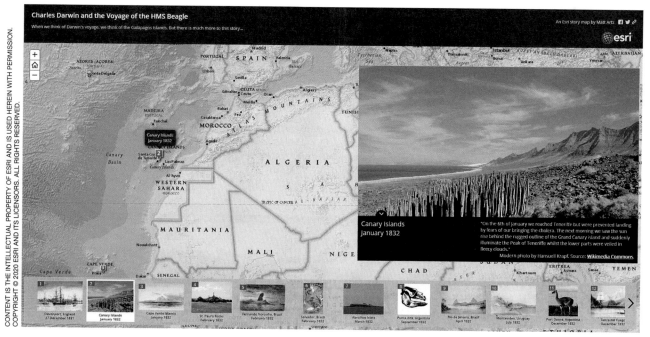

FIGURE 4.7 A story map showing the track of the voyage of Charles Darwin and the *HMS Beagle*.

Key Terms

ArcGIS Online (p. 92)
cloud (p. 93)
SaaS (Software as a Service) (p. 93)
organization (p. 94)
credits (p. 94)
share (p. 99)
publish (p. 99)
web layer (p. 99)
hosted (p. 99)
Web Mercator auxiliary sphere (p. 99)
feature layer (p. 100)
tile layer (p. 100)
vector tile layer (p. 100)
web map (p. 103)
service (p. 104)
basemap (p. 106)
web app (p. 109)
dashboard (p. 110)
story map (p. 116)

5

How to Obtain Online GIS Data and Use Them in ArcGIS Pro

ArcGIS Pro Skills

In this module, you will learn how to do the following:
- Use The National Map viewer to select downloadable geospatial data.
- Download multiple datasets from The National Map.
- Add data from The National Map to ArcGIS Pro.
- Use the Explore and Locate tools in ArcGIS Pro to examine geospatial data.
- Download data layers from the portal of ArcGIS Online into ArcGIS Pro.
- Change basemaps in ArcGIS Pro.

Learning Outcomes

After studying this module, you should be able to:
- Demonstrate how to obtain GIS data from The National Map.
- Describe at least four types of GIS data that are available from The National Map.
- Explain what a portal is in relation to obtaining GIS data.
- Describe the ArcGIS Living Atlas of the World.
- Explain the concept of volunteered geographic information.
- Describe at least two online data sources beyond ArcGIS Online for obtaining GIS data.
- Explain the concept of metadata and its importance.

Introduction

In the modules so far, you've been using data from a variety of sources, and those data were already acquired, created, and/or processed for your use. When you're working with GIS, you sometimes create layers on your own (as you'll do in the next module), but often you use layers created by someone else. For instance, if you need a layer with information on Census Tract boundaries, land parcels, streams, or interstates, there's no need to reinvent the wheel and create it yourself—chances are it's already been made and is available.

There's a lot of GIS data out there. While much of it is created or compiled by private companies that may be willing to sell it to you, there's much more that can be had for free. For instance, U.S. government agencies such as the USGS (United States Geologic Survey) and the U.S. Census Bureau make many GIS datasets available free via the Internet. The goal of this module is to introduce you to some sources of GIS data, show you how to download layers obtained from these sources into ArcGIS Pro, and explain how to do some analysis with these data. You'll be getting data from The National Map (a key source of freely available GIS data provided by the U.S. government) as well as ArcGIS Online. In addition, you'll start using basemaps streamed from ArcGIS Online (which you'll also use in later modules).

Module Scenario and Applications

This module puts you in the role of an urban planner compiling data that you will use to analyze a local smart-growth initiative for the city of Youngstown, Ohio. You need to get your hands on some basic data layers representing city features and examine their geospatial features and attributes. In addition, to assess the quality of the data layers you're working with, you need to obtain additional data from multiple sources and compare them with actual imagery of the city.

Here are some additional examples of other real-world applications of this module's theory and skills:

- A real estate agent is setting up a presentation of new housing developments in the area for prospective buyers. She wants to see the proximity of the houses to the local wastewater treatment plants. Before performing her analysis, she must first determine the locations of the plants. She can accomplish this task by downloading that geospatial information from The National Map.

- A local government is setting up an emergency response plan that will allow police, fire, and EMT units to respond quickly to situations at nearby schools. Before the government can implement the plan, it needs to determine the locations of the emergency response centers as well as all schools in the area. All of these items are available as point layers via The National Map.

- An environmental scientist is going to conduct an assessment of water quality for the county. However, he first needs GIS data layers showing the locations and dimensions of all the county's lakes, rivers, streams, and other water bodies. He can download this dataset from The National Map prior to starting his study.

Study Area

- For this module, you will be working with data from the city of Youngstown, Ohio.

Data Sources and Localizing This Module

The data in this module focus on features and locations within the city of Youngstown, Ohio. However, you can easily modify this module to use data from your own city or local area instead.

In Step 5.2, you will be downloading structures, boundaries, transportation, and hydrography data from The National Map for the city of Youngstown. The National Map will deliver these layers for the entire state or a large region of the state. This same data are available for all other U.S. states and cities. Follow the same steps to obtain the data for your own city.

In Step 5.5, you will use railroads data from the Ohio Department of Transportation that has been uploaded to ArcGIS Online. Similar data may be available

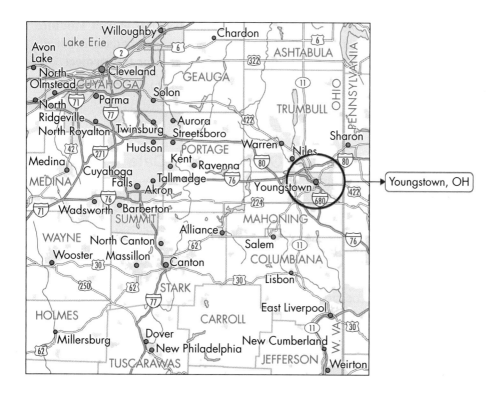

for your own state via ArcGIS Online (search for "rail"); similar datasets may also be available for layers other than railroads (such as more detailed data on water bodies). Another source of national rail lines distinct from The National Map data is ArcGIS Online; this map service, called "USA Railroads," is rail data from the U.S. Census (search for "Census rail" to find it).

In addition, the basemap layers you'll use from ArcGIS Online in Step 5.6 cover the entire United States as well as other countries, ensuring that you'll have access to this type of data for your own local area. (You can then compare the roads layer with the roads around your own local campus as well.)

Step 5.1 Getting Started

- This module's hands-on applications use the data folder called Module5. Your instructor will be able to supply you with this data, or you can download it directly from this book's website at **https://www.macmillanlearning.com/college/us/product/Discovering-GIS-and-ArcGIS-Pro/p/131923075X**. The text in this module assumes that you have this Module5 folder in a computer location referenced as C:\GIS; if you have it somewhere else (for instance, in a flash drive referenced as G:\GISClass), substitute that location and path to the Module5 folder throughout this module.
- The Module5 folder contains the following:
  - TMNData: an empty folder that you will use to hold your data
  - Ytownarea: a file geodatabase containing one feature class:
    - Youngstown: a polygon feature class of the borders of the city of Youngstown
- Start ArcGIS Pro.
- Sign in with your Esri account username and password.

- Create a new project using the **Map** template. Call this project **Module5** and place it in your **C:\GIS\Module5** folder. Ensure that there is not a checkmark in the box next to **Create a new folder for this project**.
- When ArcGIS Pro opens, change the map's name to **National Map Youngstown**.
- For now, minimize ArcGIS Pro.

Step 5.2 Downloading USGS Geospatial Data from The National Map

- Start a web browser.
- Navigate to The National Map data download website: **https://viewer.nationalmap.gov/basic/**. (See **Smartbox 5.1** for more information about The National Map and its place in the U.S. National Geospatial Program.)
- In the Search Location box, type **Youngstown, OH** and click **Go**.
- The map view changes to show Youngstown, Ohio, and its surrounding area.
- The next thing to do is define the area of interest and locate what data are available for that area. Keep in mind, as noted above, that defining the

SMARTBOX 5.1

What are The National Map and the National Geospatial Program?

National Geospatial Program (NGP) A U.S. federal initiative for managing and distributing geospatial resources.

The National Map An online resource that distributes U.S. geospatial data.

The **National Geospatial Program (NGP)** is a federally funded initiative that provides geospatial data (in GIS-ready format) to the public. The NGP uses data from a variety of sources and works to ensure that high-quality data products are available for use in numerous settings. For instance, the Emergency Operations Office component of the NGP is part of an effort to respond rapidly to disasters and emergency situations by quickly deploying necessary geospatial data to emergency services.

The National Map is at the core of the NGP and forms the heart of the geospatial data available through the program. The National Map data are available for free download via The National Map viewer in geodatabase or raster format. Many different types of data are available from The National Map. In this module, you'll be using transportation (roads and railroads), structures, and boundaries layers from the USGS, as well as lakes and rivers layers from National Hydrology Data (Figure 5.1). Beyond these, you can also freely download geographic names, historical topographic maps, and the new interactive US Topos (see Module 16), National Land Cover Data (see Module 12), contour lines (see Module 16), lidar point cloud data (see Module 17), orthoimagery (see Module 13), and 3D Elevation Program files (see Module 15).

The USGS often makes changes and enhancements to The National Map data, the viewer, and the download methods to allow for easier access to its geospatial data. To keep abreast of changes, check out The National Map's website at **https://www.usgs.gov/core-science-systems/national-geospatial-program/national-map**, or follow The National Map on Twitter, at **@USGSTNM**.

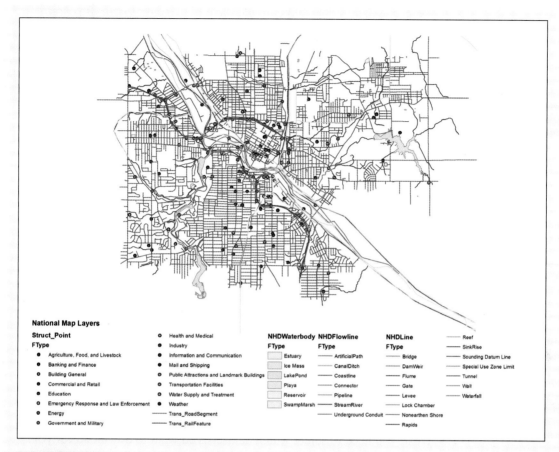

FIGURE 5.1 Various structures, transportation, and hydrography data layers of The National Map.

area of interest does not limit your data download to only that area; it just indicates to The National Map that it needs to identify the data that covers your area of interest. To begin, click the **Draw Rectangle** button.

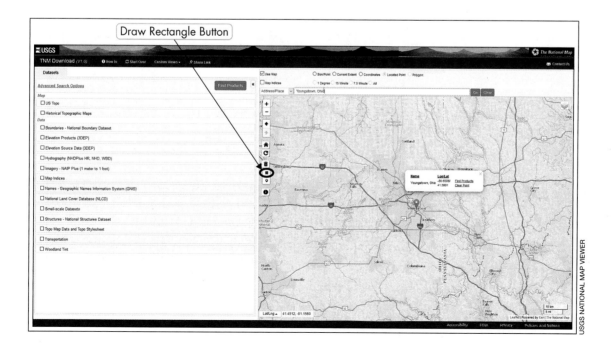

- Next, click and draw a box around the city of Youngstown. Be careful to keep your box only within the confines of the state of Ohio and avoid drawing the box too close to Pennsylvania; if your box goes onto Pennsylvania, The National Map will offer you download choices for datasets for the entire state of Pennsylvania as well as Ohio.

- Next, you need to select which datasets you want to download for your area of interest. For this module, you'll be acquiring four different datasets: boundaries, structures, hydrography, and transportation.

- In the Datasets box on the left-hand side of the screen, put a checkmark in the box next to **Boundaries – National Boundary Dataset**.

- When the Boundaries option expands, choose the option under Data Extent for **State** and the option under File Format for **FileGDB 10.1**.

- In the Datasets box on the left-hand side of the screen, put a checkmark in the box next to **Structures – National Structures Dataset**.
- In the Datasets box on the left-hand side of the screen, put a checkmark in the box next to **Transportation**. Under Transportation, also be sure to select the option **National Transportation Dataset**.
- Finally, in the Datasets box on the left-hand side of the screen, put a checkmark in the box next to **Hydrography (NHDPlus, HR, NHD, WBD)** and also select the **National Hydrography Dataset** option.
- Under the options for Data Extent, choose **HU-4 Subregion**.
- Under the options for File Format, choose **FileGDB 10.1**.

- When you have all four datasets selected, click on the blue **Find Products** button near the top of the website.
- You see the four chosen datasets in the Data panel. Click on the **Results** option for Structures, and you see the entire statewide Structures dataset that you will be downloading. Also click **Results** for Transportation and Boundaries. For these three datasets, you should have only one result: a statewide option for Ohio. If you drew your initial area of interest rectangle into another state, you'll have those data options available as well.
- For the Hydrography (NHDPlus, HR, NHD, WBD) layers, you get more than one result. The Hydrography data are downloaded in smaller chunks than are the statewide layers for transportation, structures, and boundaries. Thus, you'll have to select which of the NHD chunks you want to download.

- By looking at your options for Hydrography (there are two options shown in the graphic above, and you may have additional options, depending on how you drew the initial rectangle around the city), you can click on the word **Thumbnail** for each option to see a box drawn on the screen that shows the extent of that chunk of the Hydrography dataset. In the graphic above, the second option (**USGS National Hydrography Dataset Best Resolution (NHD) for Hydrologic Unit (HU) 4 - 0503**) is the one that covers the selected area of interest, so be sure that is the file you will be using and not any others.

- With your datasets selected, you can begin to download them. To download The National Map data, you can either add the options to a cart and begin a checkout process, or you can directly download each layer. For instance, when that know which Hydrography layer you want, you can click the **Download** option for that layer, and that layer (in a zipped format) begins downloading.

- Once that layer has completed downloading, return to your available options for Boundaries and click **Download** on that layer to directly download the Boundaries layer. Once that has downloaded, download the Structures layer, and finally download the Transportation layer. Given the size of some of these files, the process may take a few minutes, depending on your Internet speed.

- When all four downloads are completed, use Windows to navigate to the download location of your files. You should have four zipped files, one for each of the data layers. Use Windows to copy and paste all four of these zipped files into your **C:\GIS\Module5\TMNData** folder.

Step 5.3 Unpacking The National Map Data for Use

- All of your files have been downloaded in zip format, which is a "zipped," or compressed, file format, and you will have to use Windows to unzip the files before you can get to the actual National Map layers to use in ArcGIS Pro. Use Windows to navigate to the **C:\GIS\Module5\TMNData** folder.

Step 5.3 Unpacking The National Map Data for Use 127

GOVTUNIT_Ohio_State_GDB.zip	7/25/2018 1:13 PM	Compressed (zipp...	17,365 KB
NHD_H_0503_HU4_GDB.zip	7/25/2018 1:13 PM	Compressed (zipp...	85,218 KB
STRUCT_Ohio_State_GDB.zip	7/25/2018 1:12 PM	Compressed (zipp...	1,921 KB
TRAN_Ohio_State_GDB.zip	7/25/2018 1:14 PM	Compressed (zipp...	171,838 KB

- Each of the zip files contains the file geodatabases you downloaded from The National Map. To get to them, you have to unpack the contents of these zip files.

- To start with the Hydrography data, right-click on the hydrography zip file (which is the one that starts with the letters NHD) and select the option **Extract All**.

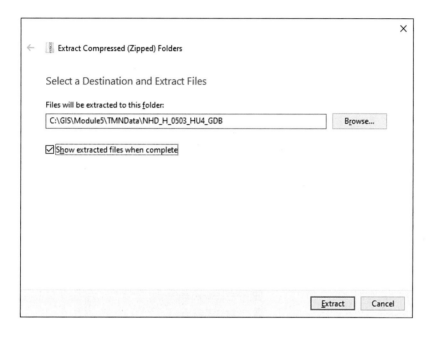

- Set the folder for the files to be extracted to as **C:\GIS\Module5\ TMNData** and click **Extract**. Windows unpacks the contents of the file and places them in the TMNData folder. The new folder labeled as a file folder is the file geodatabase you will use.

- Use the same procedure on the other zip files. You should now have a file geodatabase that corresponds to the downloaded data for each zip file.

- *Important Note:* Before proceeding, make sure that all your files are downloaded and unzipped into the proper folders.

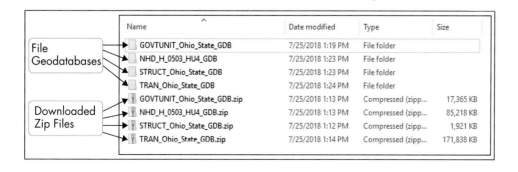

Step 5.4 Examining National Map Datasets

- Return to ArcGIS Pro and use the catalog view to navigate to your **C:\GIS\Module5\TMNData** folder, under which you should see your four folders. Expand each one of them, until you get to the file geodatabase (the gray canister symbol). Expand each of the file geodatabases. Each file geodatabase contains feature datasets (such as GovernmentUnits, Hydrography, and so on) that can also be expanded. Inside these feature datasets is where you'll find the feature classes used in this module.

- Expand the **Government Units, Hydrography, Structures**, and **Transportation** feature datasets inside the geodatabases.

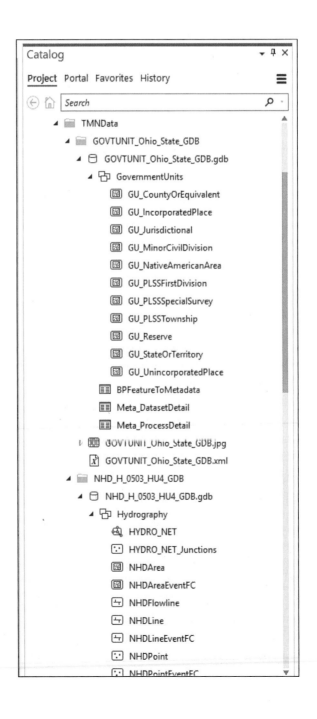

Step 5.4 Examining National Map Datasets   129

- Each geodatabase contains many feature classes and layers. At the scale you're working with, some of them are empty. In this module, you'll be working with only the following feature classes (so drag and drop each of them into ArcGIS Pro's Contents pane from the catalog view):

  - GU_IncorporatedPlace: This is a polygon boundary for the area.
  - Struct_Point: This is a point layer of different structures in the area.
  - Trans_RailFeature: This is a line layer of railroads.
  - Trans_RoadSegment: This is a line layer of roads.
  - NHDLine: This is a line layer of water features (such as levees).
  - NHDFlowline: This is a line layer of streams.
  - NHDWaterbody: This is a polygon layer of water bodies, such as lakes or estuaries.

- Change the symbology of GU_IncorporatedPlace to **Black Outline** fill (so the polygon will be transparent) and use a thick black line as its border (so the study area boundaries stand out). Move it to the bottom of the layers in the Contents pane.

- Turn on all of the layers and answer Question 5.1. (*Hint:* As you did in Module 1, check the Properties of one of the layers and click on its **Source** tab and then **Spatial Reference** to get information needed to help answer this question.)

> **Question 5.1**
> What datum, coordinate system, and units of measurement are used in The National Map data? (Keep in mind that all of the downloaded layers have the same spatial reference.)

- You can see that when you download data from The National Map, datasets such as boundaries, structures, and transportation are delivered to you as the entire statewide coverage. As this module focuses solely on the area around the city of Youngstown, from the **Module 5** folder, in the **Ytownarea** geodatabase, add the **Youngstown** feature class. You see the outline of the city of Youngstown appear on the screen. In the Contents pane, right-click on the **Youngstown** feature class and choose **Zoom to Layer** to center the view on the city.

  - *Important Note:* If at any time during the module, you move outside the boundaries of the city or return to the statewide extent of the datasets, you can always re-center yourself on the city of Youngstown by right-clicking on the **Youngstown** feature class in the Contents pane and choosing **Zoom to Layer**.

- For the remaining questions in this step, you will be looking at the various features in The National Map dataset and how they relate to one another. You can use the Locate and Explore tools on the Map tab to help answer these questions.

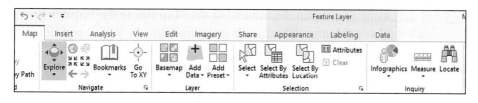

- By choosing the **Explore** tool and clicking on a point, line, or polygon, you can get attribute information about that particular feature (such as its name or other non-spatial data, such as population and length; see Module 2 for more information). However, these datasets you've downloaded contain thousands of different features; you can use the Locate tool to find a particular one so that you can then use Explore to obtain all the attribute information about that one particular feature. See **Smartbox 5.2** for more about how Locate works.

SMARTBOX 5.2

How does the Locate tool operate in ArcGIS Pro?

In ArcGIS Pro, the Locate tool is used to find a particular feature or features out of a single layer or from all of the layers in a project. Some layers contain hundreds or thousands (or more) of different features, and trying to find one visually, by looking at labels or by randomly clicking with the Explore tool, is pointless. Think of Locate as a "GIS search engine" that allows you to find a feature on the map.

When using Locate, the first thing you do is tell it which layers you want to search, by choosing the **Settings** option. There are two defaults under Settings: XY provider and ArcGIS World Geocoding Service. These are searchable items located in the cloud that help determine a location based on coordinates or an address. (See Module 10 for more about address matching.) ArcGIS Pro also allows you to choose a particular layer and the fields in that layer that will be searched by the Locate tool. For instance, in Figure 5.2, the Struct_Point layer has been added under Settings and thus is available to be the layer in which you search for a feature.

FIGURE 5.2 The Settings options for the Locate tool.

When you use Locate, you then indicate to ArcGIS Pro which layer you want to search—either a layer you specify or one of the default items. You can also have Locate search through all the layers added to the map, either by searching for an exact matching string of text or a layer whose attributes contain that text. For instance, if you use the Layer Search (Equals) option, Locate searches all layers in the map for attribute text with the case-sensitive phrase "fire station," and any layer that contains exactly that phrase will be returned to you. Alternatively, if you use the Layer Search (Contain) option, Locate searches for any non-case-sensitive text matching that and returns those layers to you. Regardless of which of these three ways you use Locate, it will find those particular features and show them to you in the Locate pane.

- To start using Locate, on the ribbon, on the **Map** tab, in the **Inquiry** group, click the **Locate** button. The Locate pane opens on the right-hand side of the screen.

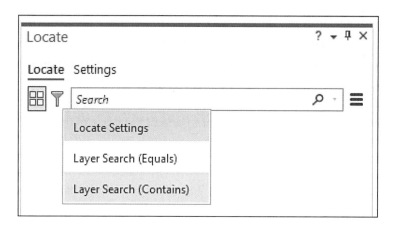

- In the Locate pane, click the **Filter** icon (the gray funnel) and from the pull-down menu, choose **Layer Search (Contains)**.

- In the Locate pane's Search box, type **Acld School and Learning Center** and press the Enter key. This refers to the Acld School and Learning Center school in Youngstown. Because you chose Layer Search (Contains), ArcGIS Pro searches for the non-case-sensitive text "Acld School and Learning Center" through all available layers on the map.

- When Locate finishes its search, the results are shown in the Locate pane. If the search is still going, you can click the **red** button in the Locate pane to stop the search.

- In the Locate pane, right-click on the result **Acld School and Learning Center**, which is in the Struct_Point layer, and choose **Zoom To**. The map changes to show the point in the center of the map. Now use the **Explore** tool (on the **Map** tab, within the **Navigate** group) to click on that Acld School and Learning Center school point. A pop-up appears, showing all the attribute table information about that point.

- Using the Explore tool, click on the road that the Acld School and Learning Center school is located on and answer Question 5.2.

> **Question 5.2** What road is the Acld School and Learning Center school located on?

- Using the Locate tool (using the Layer Search Contains or Layer Search Equals) and the Explore tool with your downloaded National Map data layers, answer **Questions 5.3 to 5.8**. (You might want to change the symbology of features of various layers to make the features stand out better on the map.)

Important Note: After you add your seven layers, you will have structures, boundaries, and transportation layers for the entire state of Ohio as well as National Hydrology Dataset (NHD) layers for a large section of northeastern Ohio. All the questions in this module are focused on only the sections of these data in the city of Youngstown, and your answers should refer only to features and locations within the city boundaries. If you find more than one place that meets your criteria during a Locate operation, you should work only with the location in Youngstown. Remember that you can always return to the boundaries of Youngstown by clicking on the **Youngstown layer** in the Contents pane and choosing **Zoom to Layer**.

- When you've finished answering these questions, turn off all the National Map layers but leave the Youngstown boundary visible.

Question 5.3 What type of water body lies adjacent to one of the parts of the road that make up N Dunlap Ave?

Question 5.4 Youngstown Fire Department Station 7 is at the corner of which two Youngstown roads?

Question 5.5 What water feature flows between Lake Cohasset and Newport Lake? As what type of water body does The National Map data classify this?

Question 5.6 Vittorio Ave is adjacent to what type of water body?

Question 5.7 Coitsville Ditch connects to what body of water?

Question 5.8 Volney Elementary School is closest to what stream/river?

Step 5.5 Using ArcGIS Online Data in ArcGIS Pro

- You can bring a lot of other freely available data directly into ArcGIS Pro; that is, you don't have to download it from websites and then import it. ArcGIS Pro allows you to connect to the available resources of ArcGIS Online (which ArcGIS Pro also refers to as the "portal," or the "Living Atlas") and download those maps and layers directly into ArcGIS Pro. In Module 4, you shared your data through ArcGIS Online, but this time you'll be retrieving and using data shared there by others. These data are delivered to you via the Internet at no cost to you and with no reduction of credits to your organization. (See **Smartbox 5.3** for more information.)

SMARTBOX 5.3

What kinds of data are available through the ArcGIS Online portal?

portal A central repository of available online data.

In GIS terms, a **portal** is a central repository for GIS data and resources, including the layers, packages, web maps, and data available for your organization through ArcGIS Online. ArcGIS Pro allows you to access your organization's content (such as layers you've published to ArcGIS Online) and add them directly into ArcGIS Pro to use alongside your other data. Further, ArcGIS Pro gives you access to data and layers published by others to their own organizations, as long as that data had been shared with everyone. ArcGIS Pro uses the term "All Portal" to refer to the ability to add others' layers into ArcGIS Pro.

Another type of portal that is maintained by Esri is the **ArcGIS Living Atlas of the World**, which contains numerous kinds of data that can be incorporated directly into ArcGIS Pro. For example, the various basemaps that you can stream to ArcGIS Pro are stored as tile layers within the Living Atlas. Many other layers are available through the Living Atlas, such as satellite imagery, aerial photography, soils layers, U.S. interstates, population data, water resources, and much more. For a searchable index of what is available in the Living Atlas, see **https://livingatlas.arcgis.com/en/browse**.

In ArcGIS Pro, you can add data from a portal the same way you would add a layer stored on your own computer—by using the **Add data** dialog box. There are options to search for and add data to ArcGIS Pro from your own layers on ArcGIS Online (the My Content option), layers that have been shared to ArcGIS Online with everyone (the All Portal option), and layers shared by Esri through the ArcGIS Living Atlas of the World (the Living Atlas option).

Many types of packages (such as layer packages; see Smartbox 1.10) and layers are available through the portal. When using data from a portal, keep in mind the distinctions between types of web layers discussed in Smartbox 4.1 in Module 4. When working with feature layers, depending on how they were configured before they were shared to a portal, you will often be able to open their attribute tables, select records or features, and extract them to their own new feature classes or layers. For instance, if you download a feature layer of the state boundaries of the United States, you might be able to select only the boundaries of Arkansas and Oklahoma and then extract those into a separate layer for your own use. However, tile layers serve as a backdrop (like a basemap), and you can't open an attribute table and extract features from these layers. For example, if you download a soils tile layer for the entire United States and need only the soils data from Arkansas and Oklahoma, you will be unable to query the layer, extract the data for only those two states, edit the data, or access the layer's attribute table.

ArcGIS Living Atlas of the World A set of GIS data and resources set up by Esri for access via ArcGIS Online and ArcGIS Pro.

- To directly access data from the portal of ArcGIS Online content, go to the ribbon, on the **Map** tab, within the **Layer** group, click the **Add data** button and then choose **Data (add data to the map)**.

- In the Add Data dialog box, from the options in the left-hand panel, choose **All Portal**. This allows you to search and add data from ArcGIS Online.

- In the Add Data dialog search box (in the upper-right corner), type **ohio rail** and press the Enter key.

- A new set of options appear in the middle panel. Choose the **rail_line** dataset (a layer package), which has been made available from ODOT (Ohio Department of Transportation), and click **OK**.

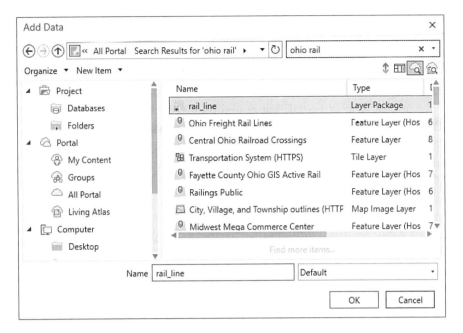

- The rail_line data should appear in the Contents pane, directly added from ArcGIS Online. Because it's a layer package, you can make some changes to the data. Choose that layer in the Contents pane, and from the **Appearance** tab, within the **Drawing** group, click the **Symbology** button and choose **Unique Values**.

- The Symbology pane opens. Under **Primary symbology**, for Field 1, choose **RR_STATUS** to display all the unique data for the RR_STATUS attribute. In this case, a rail line will be classified as either Abandoned or Active. Back in the Contents pane, you see that the lines in the rail_line layer are now being drawn with one line as Abandoned rail lines and another line as Active rail lines. Change the symbology of each one to distinguish their color and line thickness from each other.

- Turn on the Trans_RailFeature layer and change its symbology to be different from the Abandoned and Active rail line layer from ODOT.

- If your map is not at the scale of just the boundary of Youngstown, right-click on the Youngstown layer and choose Zoom to Layer. Next, zoom in to several of the active rail lines (from ODOT via ArcGIS Online) and compare them with the rail features from The National Map.

- Zoom out a bit so that you can see the entire Youngstown area and the rail lines feeding into it from around the region. Run a **Select by Attributes** query (see Module 2) of the rail_line layer to find all rail lines that are considered abandoned (using the RR_STATUS attribute) and that also have the Railroad Name (the RR_NAME attribute) not equal to Unknown. Answer Questions 5.9 and 5.10.

Question 5.9 Which railroads have abandoned lines coming toward Youngstown from the southeast?

Question 5.10 Which railroads have abandoned lines coming toward Youngstown from the northwest?

Step 5.6 Using Basemap Data from ArcGIS Online in ArcGIS Pro

- By examining the different rail lines, you can see that the ODOT dataset and The National Map dataset don't necessarily match up. You therefore need another source to compare these lines so that you can figure out which dataset is the most accurate representation of the county railroads. Seeing how the data layers match up with the actual real-world features will give you a start on assessing the usefulness of the data for future analysis. To begin evaluating the datasets, you can access high-resolution aerial imagery as a **basemap** from the cloud resources of ArcGIS Online. This basemap layer has been spatially referenced so that your other GIS layers align with it in ArcGIS Pro.

> **basemap** A map service available from ArcGIS Online (such as topographic maps or high-resolution imagery) that can be used in ArcGIS Pro.

- There are several basemaps available to use as map services from ArcGIS Online, including topographic maps, street maps, and different types of remotely sensed imagery, all at different levels of detail, depending on the scale at which you're working. Note that the same basemaps available to you for setting up a web map in ArcGIS Online (see Module 4) are also available for use here in ArcGIS Pro. (See Smartbox 4.3 in Module 4 on page 106 for more information.) From the **Map** tab, within the **Layer** group, click the **Basemap** button and choose **Imagery Hybrid**. The Imagery with Labels layer contains high-resolution imagery from Esri, with labels added for things like locations and cities.

Imagery	Imagery Hybrid	Streets
Topographic	Navigation	Streets (Night)
Terrain with Labels	Light Gray Canvas	Dark Gray Canvas
Oceans	National Geographic Style Map	OpenStreetMap

- The basemap is added to the bottom of the Contents pane. You likely saw that the ODOT Active rail data from ArcGIS Online did not match up with rail line data from The National Map. With the layers on top of high-resolution imagery, you can now judge which dataset is the most accurate by examining where they match up with imagery of the landscape. Zoom in on some of the areas for the two rail lines. Answer Question 5.11.

> **Question 5.11** By comparing the Active rail lines from the ODOT data with the rail lines from The National Map and the Imagery with Labels basemap from ArcGIS Online, which of the rail lines is the most accurate? How can you tell?

- Click the **Basemap** button again to switch basemaps, and this time choose the **OpenStreetMap** basemap. OpenStreetMap is an example of volunteered geographic information (VGI) data that can be used in ArcGIS Pro (see **Smartbox 5.4**).

SMARTBOX 5.4

What is VGI, and how can it be used in ArcGIS Pro?

volunteered geographic information (VGI) User-generated geospatial data.

crowdsourcing Leveraging the knowledge and resources of multiple individual users for a larger project.

OpenStreetMap An open-source collaborative mapping project whose data are also available as a basemap in ArcGIS Pro.

Volunteered geographic information (VGI) refers to geospatial data that's supplied by everyday persons, often people without GIS skills. VGI allows people to contribute their geographic knowledge in a way that others can use. For example, when a road is closed in your neighborhood or when construction creates a new traffic pattern on your campus, your firsthand knowledge of the situation can be helpful in quickly updating the geospatial datasets used to make the maps of your area. Online sources such as Google Maps or MapQuest, or a Garmin or Magellan satellite navigation system, provide locations and routing that are only as accurate as the geospatial data that they use. However, it may take the companies that create this data a very long time to make small updates to your local area. This is where VGI comes into play. If you can provide those updates instead, your knowledge of new road patterns, out-of-business restaurants, or new library locations (among many other possible examples) will be invaluable in keeping these datasets up to date. This type of leveraging of user resources toward the completion of larger projects is referred to as **crowdsourcing**, a phenomenon that's becoming increasingly common in building and maintaining geospatial datasets.

There are many VGI resources available on the Web. The editing tools of Google Maps (**https://www.google.com/maps**) allow users to provide updates for Google Maps, and Wikimapia (**https://wikimapia.org**) allows users to identify and label geographic features and locations around the world. **OpenStreetMap** is a VGI collaborative mapping project. Registered users can provide updates or information to the online datasets, and the OpenStreetMap data are freely available (see **https://openstreetmap.org**). The data from OpenStreetMap are available as a basemap to use in ArcGIS Pro (Figure 5.3).

Step 5.6 Using Basemap Data from ArcGIS Online in ArcGIS Pro

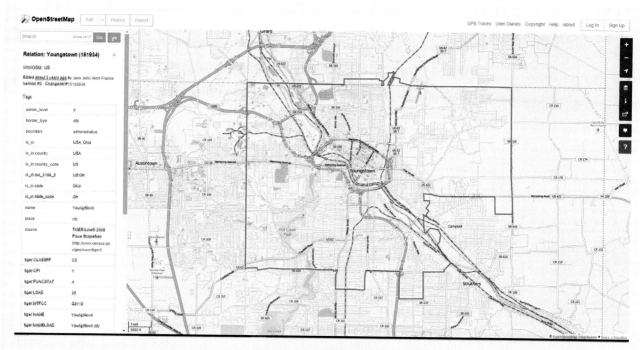

FIGURE 5.3 Youngstown, Ohio, in OpenStreetMap.

- Zoom in to the Youngstown State University (YSU) campus area; to find its general area, locate the point in the structures layer that corresponds with Youngstown State University. Examine the OpenStreetMap imagery with respect to the rail line layer from ArcGIS Online. Answer Question 5.12.

> **Question 5.12** What major road (just south of the Youngstown State University campus) is an abandoned rail line?

- While you're looking at the campus area, turn on The National Map road line layer (Trans_RoadSegment) and examine the OpenStreetMap basemap with respect to the road lines. Answer Question 5.13.

> **Question 5.13** What roads on the YSU campus area (as included in The National Map Trans_RoadSegment layer) don't match up with the OpenStreetMap basemap imagery (and how don't they match)?

- Many sites on the Internet provide freely available GIS data. In this module you have used data from only two of them (The National Map and ArcGIS Online), but there are many more sources of data. See **Smartbox 5.5** for information on other online data sources.

SMARTBOX 5.5

What other GIS data are available online?

Beyond The National Map and ArcGIS Online, there are a lot of GIS data available on the Internet—and a lot of them are free. As GIS usage and the importance of geospatial data have grown, so has the amount of GIS data. Many local county governments have established GIS data and resources online. For instance, Mahoning County, Ohio, maintains an extensive collection of GIS data and makes all of it available for free download. Everything from county culvert location shapefiles to high-resolution orthophotos is available for use in ArcGIS Pro. (For more information, see **http://gis.mahoningcountyoh.gov/161/GIS**.) If you're looking for GIS data for nearby areas, check to see if your county government or county auditor makes GIS data available.

State-level GIS data are often available from state GIS portals or clearinghouses. For example, the state of Pennsylvania has the Pennsylvania Spatial Data Access (Pasda), which serves as a clearinghouse for the state's geospatial data. Pasda is a repository for many types of statewide GIS data, including road layers, watershed data, and elevation surfaces. (For more information and the opportunity to browse the downloadable data, see **http://www.pasda.psu.edu**.) Your state likely has a GIS clearinghouse online with similar GIS data for you to access.

Sometimes, you can find data available for FTP download, and sometimes you can find data by connecting to a GIS server using the catalog functions of ArcGIS Pro that give you direct access to data via the cloud. In addition to these resources, there are numerous other online sources of free GIS data, including the following examples:

Digital Line Graphs (DLGs) The digitized features from USGS topographic maps.

Digital Raster Graphics (DRGs) Scanned and georeferenced versions of USGS topographic maps.

- EarthExplorer: This USGS utility allows you to download GIS land cover and elevation data (including lidar data; see Module 17), as well as **Digital Line Graphs (DLG)** files. DLG files are the digitized versions of features on USGS topographic maps (such as roads, railroads, hydrography, or hypsography contour lines). Because new topographic maps (see Module 16) are no longer being produced, these DLG features extracted from them are considered "legacy" data (that is, they represent GIS data from an older snapshot in time). However, DLGs are still useful sources of data and are available at 1:24000 and 1:100000 scales. Find EarthExplorer online at **https://earthexplorer.usgs.gov** (Figure 5.4).

- Landfire: This website provides access to environmental GIS data related to such attributes as wildlife habitat, vegetation cover, and forest canopy cover. Much of the data from Landfire comes in raster data format (see Module 12). Find Landfire online at **https://www.landfire.gov** (Figure 5.5).

- Libremap Project: This website features different types of GIS data for free download. It is a great source for finding scanned and georeferenced versions of USGS topographic maps (see Module 16), called **Digital Raster Graphics (DRGs)**. Find Libremap online at **http://libremap.org**.

FIGURE 5.4 The USGS EarthExplorer website.

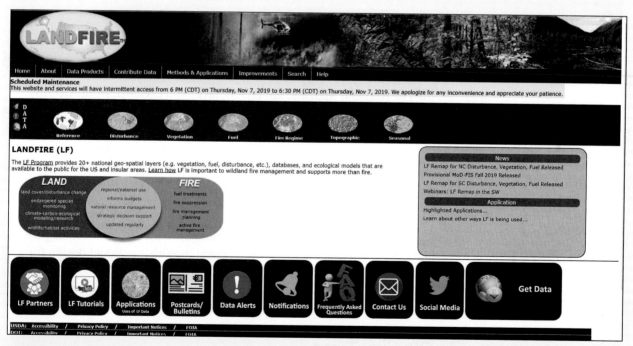

FIGURE 5.5 The Landfire website for acquiring ecological data.

- **U.S. Census Bureau:** The U.S. Census Bureau is a source of free TIGER (Topologically Integrated Geographic Encoded Referencing system) products, including the **TIGER/Line** files of cartographic boundaries such as counties, Census Tracts, or Block Groups (see Module 2). National road data

TIGER/Line Topologically Integrated Geographic Encoded Referencing boundary and road network GIS data created by the U.S. Census Bureau.

Spatial Data Transfer Standard (SDTS) A neutral file format used for the distribution of GIS data.

data interoperability A concept that allows GIS data of many different file types to be imported into ArcGIS Pro or Esri file formats to be converted to other file types.

are also available as TIGER/Line files (see Module 10). However, as noted in Module 2, TIGER/Line boundary files do not contain socioeconomic attributes; for those, you can use the Census Data website to download attribute tables and join them to the geospatial data. You can find TIGER products online at **https://www.census.gov/geographies/mapping-files/time-series/geo/tiger-line-file.html**.

While much of the data you can download is available in raster, shapefile, or geodatabase format, some older data may still be delivered in **Spatial Data Transfer Standard (SDTS)** format. SDTS was set up as a "neutral" file format to allow geospatial data to be transferred to multiple GIS formats, but it is no longer used today. ArcGIS Pro includes tools that allow for the import of SDTS files into Esri file format. In fact, ArcGIS Pro **data interoperability** tools (and an extension) allow the import and translation of different types of GIS data into a class in a file geodatabase.

Note that this Smartbox covers only GIS-specific data. There is plenty of remotely sensed imagery available. For instance, EarthExplorer allows you to download several different types of aerial photography and satellite imagery for use in ArcGIS Pro, and The National Map gives you access to high-resolution orthoimagery. See Module 13 for much more about using remote sensing imagery in ArcGIS Pro.

Step 5.7 Printing or Sharing Your Results

- Change the symbology of your layers so that both the railroad features of The National Map and the ODOT layer you obtained from ArcGIS Online can be clearly distinguished from one another. For now, turn off the other layers from The National Map. To put your two railroad layers into context, turn on the OpenStreetMap basemap and show your rail lines on top of that.
- Zoom to the extent of Youngstown.
- Save your final Module5 project.
- Finally, either print a layout (see Module 3) of the final version of your railroad layers (including all of the usual map elements and design for a layout) or share your results as a web map through ArcGIS Online (see Module 4).

 ## Closing Time

This module examines how to obtain various types of GIS data to use in your projects. While The National Map and ArcGIS Online are easily accessible sources of data, there are plenty more out there. (See **Smartbox 5.5** for more information.) However, no matter what the source of the data, you should have information about the data you're receiving. See *Related Concepts for Module 5* for more information regarding this "data about data."

Starting in the next module, you'll be creating your own GIS layers. While you will often use available data as a starting point for your analysis, many times you will be making your own data to use. For instance, if you're creating a map of the trails of a local metropark, you'll probably find that kind of data unavailable for easy access. Similarly, if you have acquired the coordinates of field locations (such as culverts or water wells), you'll have to convert those numbers into a GIS data layer before you can use them.

RELATED CONCEPTS FOR MODULE 5

How Metadata Are Used in GIS

Almost all GIS data are accompanied by some sort of descriptive information about the data (usually as a separate "readme" text file or an XML document). Useful information about the data might include a statement of how accurate the data are, what geographic or projected coordinate system is being used, and what each of the fields in the attribute table represents. Without this information, you might not be able to use the data properly. For instance, if you have to define the layer's projection, you have to know what projected coordinate system to use. Similarly, if the fields in the attribute table have strange names or if the attributes use coded values, you need something to explain what each of these items means. For instance, The National Map NHDFlowline feature class contains the attribute fields ReachCode, FlowDir, FType, and FCode. When you're using this hydrography data, it would help to know what these things represent. In GIS (and other fields), this descriptive information, or "data about data," is called **metadata**.

As GIS data have proliferated, standards have developed regarding what information should be included with a layer's metadata. The U.S. **Federal Geographic Data Committee (FGDC)** has set up a long-standing series of guidelines called the **Content Standard for Digital Geospatial Metadata (CSDGM)**. These content standards include the following (see **https://www.fgdc.gov/metadata/csdgm-standard**):

- Identification of the dataset (a description of the data)
- Data quality (information about the accuracy of the data)
- Spatial data organization information (how the data are represented, such as in vector format)
- Spatial reference information (the geographic or projected coordinate system used)
- Entity and attribute information (what each of the attributes means)
- Distribution information (how you can obtain the data)
- Metadata reference information (how current the metadata is)
- Citation information (how to cite the data)
- Time period information (what date do the data represent)
- Contact information (whom to contact for further information)

International metadata standards are also being adopted in North America. The **ISO 191xx** (where *xx* refers to a pair of numbers that reflect the most current version of this standard) metadata and associated standards are now becoming a

> **metadata** Descriptive information about data.
>
> **Federal Geographic Data Committee (FGDC)** A committee that is responsible for setting GIS metadata content standards.
>
> **Content Standard for Digital Geospatial Metadata (CSDGM)** A set format created by the FGDC for what items a metadata file should contain.
>
> **ISO 191xx** An international metadata standard adopted in the United States as the North American Profile (NAP) of ISO 191xx.

commonly referenced format and have been endorsed by the FGDC. In the United States, this set of standards is adopted as the North American Profile (NAP) of ISO 191*xx*. This structure contains information similar to that of the CSDGM, but it also contains some new elements, extended elements, and support for new geospatial applications. For more about ISO 191*xx* standards, see **https://www.fgdc.gov/metadata/iso-suite-of-geospatial-metadata-standards**.

In ArcGIS Pro, a layer's metadata are accessible via the Item description, which doesn't adhere to a particular format. However, ArcGIS Pro can be set up to allow the use of specific metadata formats (such as CSDGM and NAP ISO 191*xx*). In ArcGIS Pro's Catalog View, select the **Project** tab, click then the **Options** button, and finally choose the **Metadata** tab, to get several options for structuring how you want to work with metadata in ArcGIS Pro (Figure 5.6).

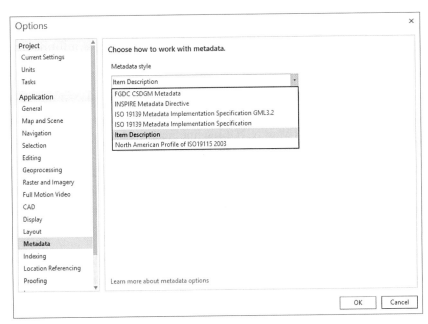

FIGURE 5.6 The available metadata styles in ArcGIS Pro.

Key Terms

National Geospatial Program (NGP) (p. 122)
The National Map (p. 122)
portal (p. 132)
ArcGIS Living Atlas of the World (p. 133)
basemap (p. 135)
volunteered geographic information (VGI) (p. 136)
crowdsourcing (p. 136)
OpenStreetMap (p. 136)
Digital Line Graphs (DLGs) (p. 138)
Digital Raster Graphics (DRGs) (p. 138)
TIGER/Line (p. 139)
Spatial Data Transfer Standard (SDTS) (p. 140)
data interoperability (p. 140)
metadata (p. 141)
Federal Geographic Data Committee (FGDC) (p. 141)
Content Standard for Digital Geospatial Metadata (CSDGM) (p. 141)
ISO 191*xx* (p. 141)

How to Create Geospatial Data with ArcGIS Pro

ArcGIS Pro Skills

In this module, you will learn how to do the following in ArcGIS Pro:
- Create a file geodatabase.
- Convert a table of coordinates into a point layer.
- Create feature classes within a geodatabase.
- Obtain and use a basemap as a data source for digitizing.
- Set a bookmark for a location.
- Digitize points, lines, and polygons and save them as geodatabase feature classes.
- Create a new field in an attribute table and calculate values for it.

Learning Outcomes

After studying this module, you should be able to:
- Define what digitizing is in relation to GIS.
- Demonstrate how to digitize features in GIS.
- Explain what a file geodatabase is and how it stores data.
- Describe what feature classes and feature datasets are.
- Describe what GPS is.
- Explain how GIS data can be collected in the field using a mobile app.

Introduction

In the first five modules, you used geospatial data that had already been created and processed for you to use. While a huge amount of data is available, you sometimes need to create your own datasets to use in GIS. If the data you need is unavailable, out of date, or incomplete, you have only one option: make your own version. Sometimes, you will have to create data from scratch, and there are many ways to create data to use in GIS. For example, measurements taken in the field by hand are often used to build GIS data. GPS devices or surveying can provide you with a set of latitude and longitude coordinates for a location. In turn, you can change these coordinates into a point layer to use in ArcGIS Pro.

However, it is sometimes impossible to obtain data firsthand. In such a case, you have to rely on a secondary source to build GIS data. You can begin with a preexisting map or remotely sensed image and use that as the base to start building your data. For instance, if you're going to create a GIS layer of the walking paths of a local fairgrounds, you could get an aerial photograph that shows an overhead view of the fairgrounds (which would allow you to clearly see the paths) and use that photo to create a set of lines that represent the lengths and dimensions of the walking paths.

The process of creating GIS data (whether points, lines, or polygons) from another secondary source is referred to as **digitizing**. Commonly, **heads-up digitizing** is performed as just described: You add a map or an image to ArcGIS

digitizing The process of creating digital GIS data from a secondary source.

heads-up digitizing Using a map or imagery as a backdrop in the digitizing process.

Pro, and it becomes the basis for creating your own data. In digitizing, you're in essence drawing points, lines, or polygons on the screen, and these sketches are converted to data layers inside the GIS. In the fairgrounds example, you could simply draw lines that follow the walking paths that are visible on the aerial photo, or you could carefully draw polygons by tracing around the dimensions of the buildings.

Module Scenario and Applications

In this module, you'll be taking the role of an event planner for a university campus. Your university is developing a new campus map to be used for planning and hosting conferences, arts festivals, and other community events. The university's marketing department has a lot of ideas about visualizing and presenting this information, but it needs a map. It's your job to create a new map of the campus that shows the roads, sidewalks, parking lots, and university buildings.

The following are some examples of other real-world applications of this module's theory and skills:

- A geologist is drilling wells for groundwater sampling; the locations of these wells are denoted with sets of coordinates (perhaps latitude and longitude obtained through GPS). The geologist can use ArcGIS Pro to plot the coordinates for these wells as points on a map for future analysis.
- An archeologist is recording the locations of artifacts at dig sites. She can use these coordinates to measure these positions and plot them as points using ArcGIS Pro.
- A historian is using old maps of historic buildings in an area to re-create the appearance or layout of these places. He can use digitizing to create new GIS data from these old sources.

Study Area

For this module, you will be examining a portion of the Youngstown State University (YSU) campus in Youngstown, Ohio.

Data Sources and Localizing This Module

The data in this module focus on features and locations within the YSU campus. However, you can easily modify this module to use data from your own campus or local area instead. Because you will be creating the data directly from the Imagery

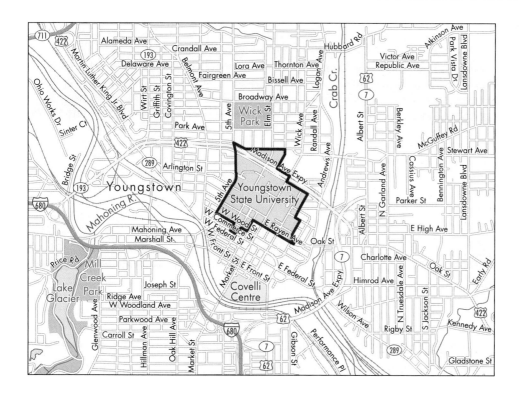

basemap, you can simply find your own campus area that you're familiar with on the basemap and digitize from there. You can quickly create the Campusbound shapefile for your own area as well. You can use a source such as GPS measurements or Google Earth for obtaining coordinates of prominent points in your own local area for Step 6.3.

Step 6.1 Getting Started

- This module's hands-on applications use the data folder called Module6. Your instructor will be able to supply you with this data, or you can download it directly from this book's website at **https://www.macmillanlearning.com/college/us/product/Discovering-GIS-and-ArcGIS-Pro/p/131923075X**. The text in this module assumes that you have this Module6 folder in a computer location referenced as C:\GIS; if you have it somewhere else (for instance, in a flash drive referenced as G:\GISClass), substitute that location and path to the Module6 folder throughout this module.

- The Module6 folder contains the following:
  - Campusbound.shp: A simple polygon shapefile showing a simplified boundary of the YSU campus.
  - Parkingcoords.xlsx: A Microsoft Excel file containing the latitude and longitude of parking lift arms on campus.
  - ParkingMap: A map of the YSU campus in PDF format. You won't be able to see this in ArcGIS Pro because it's just a PDF file. Open this file in Windows and use it as a reference for this module.

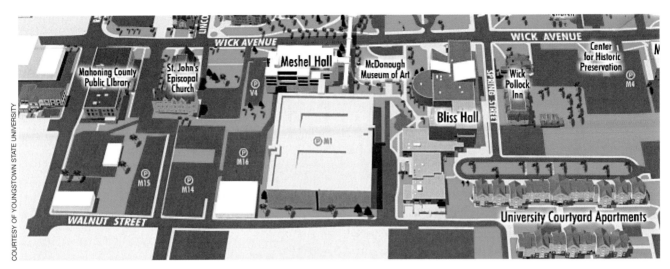

FIGURE 6.1 A section of the YSU campus.

- Start ArcGIS Pro.
- Sign in with your Esri account username and password.
- Create a new project using the **Map** template. Call this project **Module6** and place it in your **C:\GIS\Module6** folder. Ensure that there is not a checkmark in the box next to **Create a new folder for this project**.
- When ArcGIS Pro opens, change the map's name to **Campus**.

  Important Note: Refer to the ParkingMap PDF for the proper locations and orientations of objects. Figure 6.1 shows a visualization of the section of campus used in this module.

- In this module, you'll be using a basemap streamed from ArcGIS Online resources as the source for digitizing (in this case, a high-resolution image from Esri). ArcGIS Online contains imagery for the entire United States (and many other countries as well). To get your bearings, you'll want to look only at the portion around YSU and create data at that scale. To get this process started, add the **Campusbound** shapefile to the Contents pane. It's a polygon showing an estimated boundary of campus. You'll see that the extent of the Campus map view adjusts to the spatial dimensions of the Campusbound shapefile as well as the background topographic map.

- Next, from the **Map** tab, within the **Layer** group, click the **Basemap** button. From the options that appear, choose **Imagery**. Give ArcGIS Pro a minute or two to stream the imagery to you over the Internet. The high-resolution imagery should snap into place, centered around your shapefile of campus. If the imagery does not appear, right-click on **Campusbound** in the Contents pane and select **Zoom to Layer** to refocus the Campus map view on the dimensions of the campus boundaries. If you find that the imagery you're using isn't of the best quality, see **Troublebox 6.1** for some suggestions on obtaining better imagery.

- Turn off the Campusbound shapefile but don't remove it from the Contents pane just yet. (So far you have used it only to get the imagery to center on

TROUBLEBOX 6.1

Is there any better-quality imagery available to use as a basemap for digitizing?

Esri makes changes to its available basemaps, including the Imagery basemap, from time to time. However, a separate imagery basemap, called World Imagery Clarity, is available from the Living Atlas (see Module 5), and you can download it into ArcGIS Pro and use it as you would the other basemaps. World Imagery Clarity imagery may not be exactly current, but in many cases it provides crisper imagery to work with.

You can access World Imagery Clarity as you would other Living Atlas items. On the **Map** tab, within the **Layer** group, click the **Add Data** button. In the Add Data dialog box, under the Portal options, choose **Living Atlas**. Then, in the search box, search for **Clarity**. An option for World Imagery (Clarity) should be available (Figure 6.2). Choose it and click **OK**. The new basemap drops into the map and is shown in its Contents pane.

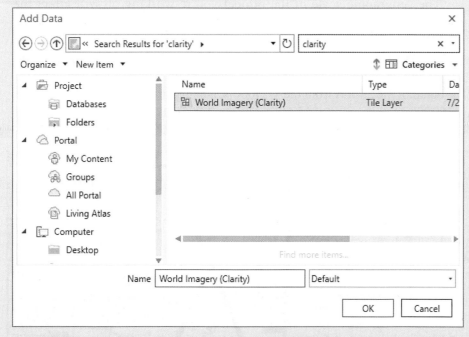

FIGURE 6.2 Obtaining World Imagery (Clarity) from the Living Atlas.

YSU when the basemap was added.) The high-resolution imagery should now be displayed on the screen.

- Right-click on the **Campus** map in the Contents pane and select **Properties** and then select the **General** tab. Select **US Feet** for Display Units and click **OK**. Now you can make measurements, in feet, from the imagery. For more information about measurements and scale, see **Smartbox 6.1**.

SMARTBOX 6.1

Why is scale important when digitizing?

In Module 3, we discussed the importance of scale when making a map. The scale of the map that's used as a digitizing source is critical. For instance, if you're going to be digitizing features of your neighborhood, you need a very large-scale map as a source to capture things such as the dimensions of houses, backyard sheds, or driveways because these kinds of things would not be apparent in smaller-scale maps.

The level of detail available to be digitized changes as the scale of the map source changes. For example, Figure 6.3 shows two topographic maps of McCarran International Airport, with map (a) at a 1:24000 scale and map (b) at a 1:100000 scale. If you were digitizing the airport based on each of these maps, you'd clearly end up with two completely different datasets. If you wanted to capture very detailed features of the airport, the 1:24000-scale map would be the better choice. If you just needed to mark a rough location of the airports and runways, the 1:100000-scale map could be used.

The geographic scale of the project you're working on will affect the choice of the map scale to choose for a digitizing source. For instance, using the 1:24000-scale map of McCarran to digitize the runways would provide you with a lot of detail for analysis. However, you wouldn't be able to digitize individual building or housing footprints at such a scale. Similarly, if you digitized the detail of the airport from the 1:24000-scale map and then the roads from the 1:100000-scale map, you'd have two very different datasets that might cause issues in future analysis.

FIGURE 6.3 The features of McCarran International Airport in Las Vegas, as shown on (a) a 1:24000-scale topographic map and (b) a 1:100000-scale topographic map.

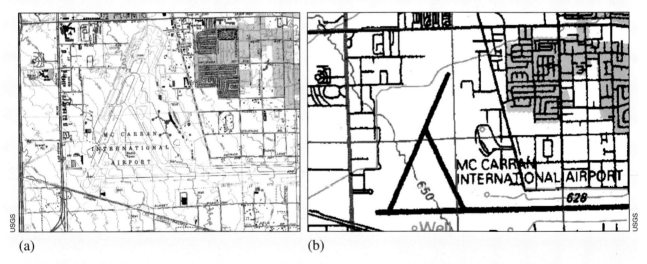

(a)   (b)

Step 6.2 Working with a Geodatabase to Contain GIS Data

- Because you'll be digitizing objects that represent the features of YSU campus—roads, sidewalks, campus and non-campus building footprints, and parking lots—you'll be creating separate feature classes to contain each of these types of data. You therefore need a file geodatabase to contain all these items. When you created the Module6 project, a default file geodatabase, called Module6, was created for you, and you can use that in this module.

- *Note:* To create another file geodatabase to use instead of the default one for a project, you can go to the **Insert** tab, and within the **Project** group, click the **Connections** button and then choose **New File Geodatabase**.
- For more information on geodatabases and how they're used to store GIS data, see **Smartbox 6.2**. Also, you've just used a shapefile to represent the campus boundaries—for more information on shapefiles and other data types available for use in ArcGIS Pro, see **Smartbox 6.3**.

SMARTBOX 6.2

What is a geodatabase, and how does it store geospatial data?

A **geodatabase** is a structure for storing GIS data in ArcGIS Pro. In the Catalog pane, the geodatabase symbol is a gray cylinder, which is a good way of thinking about a geodatabase: It's designed to hold several different types of geospatial (as well as non-spatial) data. In ArcGIS Pro, you use a **file geodatabase**. A file geodatabase has essentially no limit on the number of feature classes or tables it may contain (although each item in the file geodatabase is limited to 1 TB of storage space). It has the file extension .gdb and is stored on your computer as a folder of files. Also, a file geodatabase can be edited and utilized by multiple persons at one time.

A file geodatabase is a single folder that can contain many different types of GIS data. Geodatabases store data as **feature classes**. Each feature class holds a different type of object (such as points, lines, or polygons), so the type of feature class being created in a geodatabase depends on what you're modeling in ArcGIS Pro. For instance, roads or streams would be stored as line feature classes, while building footprints or land parcels would be stored as polygon feature classes.

Beyond the three vector object feature classes, a geodatabase can also store an annotation feature class (for text used for labeling; see Module 3), a dimension feature class (for annotation used for measurements), a multipoint feature class (for storing very large point sets; see Module 17), or a multipatch feature class (used for 3D objects; see Module 18). Geodatabases can also hold tables of non-spatial data (see Module 2) and raster datasets (see Module 12), and other data types such as shapefiles can be imported to become new feature classes. All feature classes are shaded gray in the Catalog pane.

Feature classes can also be grouped together into a **feature dataset**, which can hold several feature classes with the same coordinate system information. For instance, if you have many feature classes related to transportation, you might want

geodatabase An ArcGIS Pro structure for a single item that contains multiple datasets, each as its own feature class.

file geodatabase A geodatabase that can hold an essentially unlimited amount of data and has the file extension .gdb.

feature class A GIS layer stored inside a geodatabase that can contain multiple items of the same data type.

feature dataset A grouping of related feature classes stored within a geodatabase.

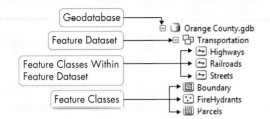

FIGURE 6.4 A file geodatabase containing a feature dataset and several feature classes.

to create a transportation feature dataset within the geodatabase to hold them all for data management or sharing purposes. Feature datasets are also used to collect feature classes together to create things such as network datasets (see Module 11), topologies (see Module 7), terrain datasets (see Module 17), geometric networks (see Module 11), or parcel fabrics. Like feature classes, feature datasets are shaded gray in the Catalog pane. See Figure 6.4 for the organization of feature classes and feature datasets within a file geodatabase.

SMARTBOX 6.3

What other data formats can be used in ArcGIS Pro?

shapefile A series of files (with extensions including .shp, .shx, and .dbf) that make up one vector data layer.

While file geodatabases are very versatile for handling geospatial data in ArcGIS Pro, they're not the only supported Esri data format. Another common format is the **shapefile**, a simple, non-topological file structure for holding GIS data. Shapefiles were introduced in the 1990s with the ArcView software program and continue to be used today. A shapefile can hold objects of a specific type. For example, you could use a polygon shapefile to store property boundaries or a point shapefile for locations of culverts. A shapefile consists of several files with different file extensions, including .shp, .shx, and .dbf. All these many files must be present in the same folder for the shapefile to work properly in ArcGIS Pro. Shapefiles are shaded green in the Catalog pane.

As noted in Module 5, Esri promotes the concept of data interoperability, whereby Esri data formats can be exported for use in other software packages. Likewise, a multitude of software-specific GIS formats can be imported into ArcGIS Pro, including CAD (computer-aided design) datasets and files in the KML (Keyhole Markup Language) format. The data interoperability extension allows for more than 100 different data formats to be used in ArcGIS Pro.

Step 6.3 Converting x/y Coordinates to a Point Feature Class

- One of the simplest ways to create GIS layers is to plot the locations of x and y coordinates as points and then save those points as a feature class. ArcGIS Pro allows you to start with a spreadsheet (such as an Excel file) containing coordinates such as latitude and longitude and convert them to points.

- For this module, you'll be creating a point layer from the coordinates of a group of parking arms on campus. To get started, add the Parkingcoords Excel file to the Contents pane. Choose the **Sheet1$** layer to add it. You may receive an error that ArcGIS Pro cannot read the Excel file; see **Troublebox 6.2** for more information on working with Microsoft Excel in ArcGIS Pro.

TROUBLEBOX 6.2

How can ArcGIS Pro properly work with Microsoft Excel tables?

In order to be able to add Microsoft Excel tables to ArcGIS Pro, you may have to first install a free Microsoft driver on your computer. For full details on what files you need and how to properly install them, see **https://pro.arcgis.com/en/pro-app/help/data/excel/work-with-excel-in-arcgis-pro.htm**. After you install the driver, Excel tables will work like other tables in ArcGIS Pro.

As an alternative, you can use the Excel to Table Geoprocessing tool to convert a Microsoft Excel file into an ArcGIS table format to use so you do not have to directly add the Excel file to the Contents pane. The tool is available in the Conversion Tools toolbox, within the Excel toolset.

- Open the Sheet1$ table. You see the names and locations of two parking lift arms on campus. You can close the table for now.

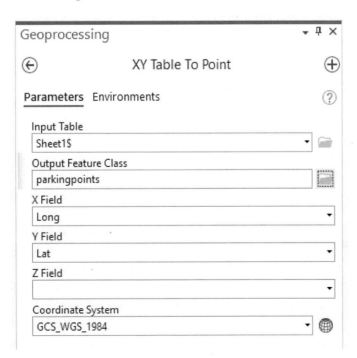

- To convert these coordinates to a layer, from the **Map** tab, within the **Layer** group, click the **Add Data** button and then choose **XY Point Data**. The Geoprocessing pane opens with the XY Table to Point tool:
  - For Input Table, choose **Sheet1$**.
  - For Output Feature Class, click the **browse** button. By default, ArcGIS Pro wants to name the new shapefile Sheet1_XYTableToPoint, but you should give it a new name. Navigate to the C:\GIS:\Module6 folder, and within the Module6 geodatabase, call the new feature class **Parkingpoints**.
  - For X Field choose **Long**.
  - For Y Field choose **Lat**.
- For Coordinate System, use **GCS_WGS_1984**. This should be the default option; if it isn't, click the **Select coordinate system** button (the gray sphere next to the Spatial Reference entry) and use it to choose that coordinate system (it will be WGS 1984 with WKID of 4326).

    Important Note: If you plot your own coordinates that were not measured in WGS84, you can use the Spatial Reference Properties dialog to select the proper coordinate system for the input coordinates in your own Excel file.
- Click **Run** when all settings are correct.
- Parkingpoints is added to the Contents pane. Change the symbology of the Parkingpoints feature class to something more distinctive.
- Look in the Catalog pane at the Module6 geodatabase and expand it until you can see the Parkingpoints feature class successfully added to it.

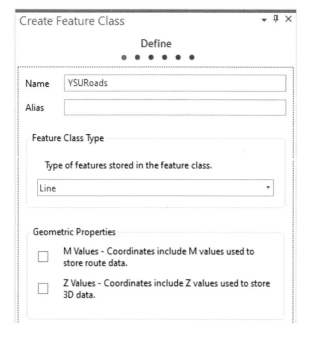

Step 6.4 Creating New Feature Classes to Hold Data

- You'll be digitizing several different types of things (roads, sidewalks, parking lots, and building footprints), and each of these items can be stored in its own feature class inside the Module6 geodatabase. The first thing you'll be digitizing is the roads around campus, so you need to create a new feature class called YSURoads in the geodatabase where you want to store this information. In the Catalog pane, right-click on the **Module6** geodatabase and select **New** and then select **Feature Class**. The Geoprocessing pane reappears, now containing the Create Feature Class dialog. You can see six dots at the top of the dialog, representing the six steps you need to take to build the empty feature class in ArcGIS Pro.
- In the first step, you define the basics of the new feature class you'll be creating. For Name, type **YSURoads**.

- For Feature Class Type, choose **Line** (because you will be digitizing roads as lines).
- Ensure that there is not a checkmark next to the box **M Values**. (M features are linear referencing, which you are not using here.)
- Ensure that there is not a checkmark next to the box **Z Values**. (Z features add a third dimension. You'll learn about this in Module 15.)
- Click **Next**.

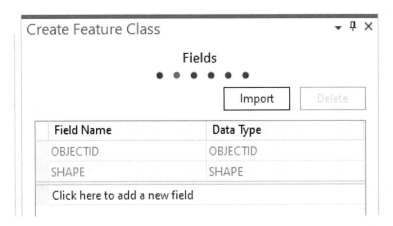

- In the second step, you add any new fields to the feature class as you're creating it. For this YSURoads feature class, you should add a new field that lets you define the name of the line/road as you're digitizing it, so click on the part of the dialog that says **Click here to add new field**.

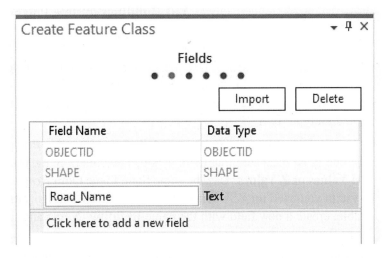

- A new row appears, where you can tell ArcGIS Pro what the field is to be called and what type of field it is supposed to be. Call the new field **Road_Name** and make sure it is of type **Text**.
- Click **Next**.
- In the third step, you select the coordinate system that you want your features to be digitized into. For this module, you should create data in the State Plane Coordinate System for Ohio North. Under the option for Layers, select **NAD 1983 StatePlane Ohio North FIPS 3401 (US Feet)**. This system is an available option here because one of the layers in your Module6 project

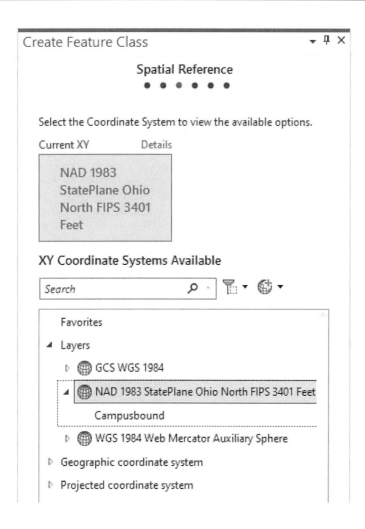

(Campusbound) has this coordinate system. To use a different coordinate system for your new feature class, you would select it here instead.

- Under Current XY, you should see the proper selected coordinate system. Click **Next**.
- In the fourth step, you change tolerances for coordinates. For this part, just click **Next**.
- In the fifth step, you change resolution. For this part, just click **Next**.
- In the sixth step, you change settings related to the database storage configuration. For this part, make sure **Default** is chosen and then click **Finish**.
- Return to the Catalog pane again, and you see that a new feature class called YSURoads has been added, and it now accepts lines.
- Repeat the six steps to create four other feature classes and store them in your Module6 geodatabase as follows:
  - Create a feature class called **YSUBuildings** to store **Polygon** for the geometry type. Ensure that there is not a checkmark in the box next to **M Values** or **Z Values**. Use the same State Plane Ohio North coordinate system you used for YSURoads. In the YSUBuildings attribute table, create a new field with the name **Building_Name** and the data type **Text**.
  - Create a feature class called **NonYSUBuildings** to store **Polygon** for the geometry type. Ensure that there is not a checkmark in the box next to **M Values** or **Z Values**. Use the same State Plane Ohio North coordinate system you used for YSURoads. In the NonYSUBuildings attribute table, create a new field with the name **Building_Name** and the data type **Text**.
  - Create a feature class called **ParkingLots** to store **Polygon** for the geometry type. Ensure that there is not a checkmark in the box next to **M Values** or **Z Values**. Use the same State Plane Ohio North coordinate system you used for YSURoads. In the ParkingLots attribute table, create a new field with the name **Lot_Name** and the data type **Text**.
  - Create a feature class called **Sidewalks** to store **Line** for the geometry type. Ensure that there is not a checkmark in the box next to **M Values** or **Z Values**. Use the same State Plane Ohio North coordinate system you used for YSURoads. Because this class will

hold lines of sidewalks, there is no need to add any attribute fields to this layer.

- Check the Catalog pane. At this point, you should have one geodatabase (Module6) that has one feature class you've already created (Parkingpoints) and five new empty feature classes (YSURoads, YSUBuildings, NonYSUBuildings, ParkingLots, and Sidewalks). Double-check to make sure they are all there and that each is set up to hold objects of the appropriate type (polygons for YSUBuildings, NonYSUBuildings, and ParkingLots and lines for YSURoads and Sidewalks) before proceeding.

- Add all five of these new, empty feature classes to the Contents pane.

Step 6.5 Heads-Up Digitizing of Lines

FIGURE 6.5 The section of YSU campus in which you will be digitizing features.

- It's now time to get some data into the empty feature classes. For this module, you'll be digitizing the features on just a small part of campus. Specifically, you'll be working with the southern portion of campus, the block bounded by Wick Avenue, Walnut Avenue, Spring Street, and Rayen Avenue. (See Figure 6.5 for the area and also refer to the ParkingMap PDF.) Pan and zoom the view so that you're looking at only that one-block area.

- In this section of campus, you'll want to digitize all the campus buildings, non-campus buildings, parking lots, roads, and sidewalks. A good strategy is to set a bookmark in ArcGIS Pro at the full extent of the study area you'll be digitizing. That way, you can easily reset the view to show only the area you need if you zoom in too closely or zoom out too far. Once you have the view set up to show only the block of campus you'll be digitizing, do the following: On the **Map** tab, within the **Navigate** group, click on the **Bookmarks** button and then choose **New Bookmark**. In the Create Bookmark dialog, call the new bookmark **YSU Study Area** and click **OK**.

- Now, when you click the **Bookmarks** button on the ribbon, you have a new option called YSU Study Area. Selecting it resets the view to the boundaries and scale that you currently see on the screen.

- You can now digitize roads that will be stored in the YSURoads feature class in the Contents pane. Note that, even with this feature class turned on, nothing is displayed because it contains no data (yet). Change the color and width of YSURoads to something that will distinguish it from the colors of the features on the basemap.

- On the **Edit** tab, within the **Features** group, click the **Create** button.

- You now see a new Create Features pane open on the right side of the screen, and you can see each of your feature classes stored there (as "templates," in ArcGIS Pro terminology). Whenever you want to digitize a new feature, you can select its template from this Create Features pane. If you find that some or all of your data layers are not available as templates, see **Troublebox 6.3** for a solution.

TROUBLEBOX 6.3

What if some templates are missing from the Create Features pane?

There are a few reasons you may not have a template available in the Create Features pane. First, if your layers are not displayed (either if they have not been added or if they are not shown with a checkmark next to them) in the Contents pane for the map you're digitizing, those layers will not be shown as templates. To ensure that you have a template available for a layer, be sure it is displayed.

Second, if a layer has not been marked as "editable," it will not have a template available in the Create Features pane. To determine whether a feature is editable, in the Contents pane, choose List By Editing (see Module 1). If there is a checkmark next to a layer, you know it is editable; layers without checkmarks do not have available templates.

Finally, a template may not have been created for a layer by ArcGIS Pro. To create a specific template, go to the **Edit** tab, and within the **Features** group, click on the **Manage Templates** dialog launcher (the little arrow in the corner of the Features group on the ribbon). The Manage Templates pane (Figure 6.6) opens; it allows you to specifically add, delete, duplicate, or examine the properties of a template.

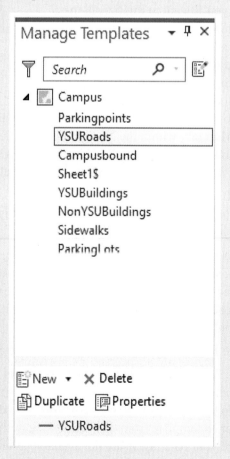

FIGURE 6.6 The Manage Templates pane in ArcGIS Pro.

- See **Smartbox 6.4** for more about the digitizing process in ArcGIS Pro, along with the various digitizing options you have. To begin, in the Create Features pane, click on **YSURoads**. You see it highlighted in blue, and you see a new set of options appear. Click on the **right-pointing arrow** next to YSURoads.

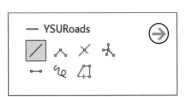

SMARTBOX 6.4

How is digitizing performed in ArcGIS Pro?

When you're digitizing in ArcGIS Pro, you're basically drawing a sketch on the screen, using a source (such as a map or a remotely sensed image) as your backdrop. A sketch consists of one or more **segments** that you draw. Multiple segments are connected by a **vertex** (Figure 6.7). For instance, when you digitize the outline of a parking lot as a polygon, the sketch consists of several line segments connected by vertices, and the segments connect back to the first vertex, thus closing the polygon. **Point digitizing** refers to placing a vertex at each spot where you click the mouse.

segment One part of a line or polygon object created through digitizing.

vertex A defining point placed at the start or end of a segment during digitizing.

point digitizing A digitizing method in which a vertex is placed with each click of a mouse.

When a sketch is complete, it becomes a feature (and, thus, geospatial data). Each feature is then saved as a separate record in the layer's attribute table and thus has a real-world length or area. For instance, if you sketch the outlines of five houses in a subdivision, the polygon layer will have five records, and you will be able to access information about the area and perimeter of each object. If you then digitize five line segments into a "driveways" feature class, it contains five records, each measuring the real-world length of the driveways you just sketched.

When you are digitizing lines, you have several different options available to you (Figure 6.8):

- The Line option allows you to digitize a straight line from vertex to vertex.

- The Right-Angle Line option allows you to digitize a line at a right angle from another line (which would be useful for digitizing features like roads that turn at a right angle).

- The Split Line option allows you to digitize a line across another line and split the line into multiple segments where the lines cross (which would be useful for digitizing intersecting roads).

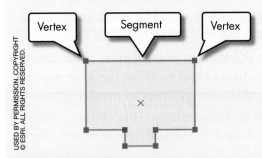

FIGURE 6.7 The segments and vertices that make up a sketch.

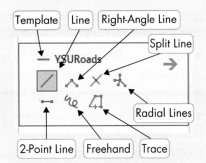

FIGURE 6.8 Options for digitizing line/polyline features in ArcGIS Pro.

- The Radial Lines option allows you to digitize multiple lines from a common vertex (which would be useful for digitizing several pavement paths that start at a single location).

- The 2-Point Line option allows you digitize two points, and then ArcGIS Pro will be connecting those two points with a line (which would be useful for digitizing utilities or sightlines).

- The Freehand option allows you to digitize lines without clicking the mouse at a particular point (which would be useful for digitizing shorelines or water boundaries).

- The Trace option allows you to digitize lines along the path of previously digitized lines (which would be useful if you had multiple layers that all had to follow the same paths).

When you are digitizing polygons, you have several different options available to you (Figure 6.9):

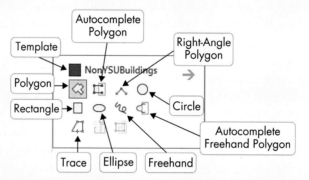

FIGURE 6.9 Options for digitizing line/polyline features in ArcGIS Pro.

- The Polygon option allows you to digitize a polygon shape by inserting a vertex each time you click the mouse.

- The Autocomplete Polygon option allows you to digitize a polygon shape by inserting a vertex each time you click the mouse, and ArcGIS Pro automatically completes a polygon if some of its edges are made up of existing polygons. (This would be useful for digitizing parcel boundaries where the edge of one parcel makes up part of the polygon being digitized.)

- The Right-Angle Polygon option allows you to digitize the edges of a polygon at a right angle from a previous edge (which would be useful for digitizing features like buildings that have corners that turn at a right angle).

- The Circle, Rectangle, and Ellipse options allow you to digitize those shapes without clicking the mouse to insert vertices (which would be useful for digitizing shapes of a certain size, such as a rectangular gravesite or a park fountain).

- The Radial Lines option allows you to digitize multiple lines from a common vertex (which would be useful for digitizing several pavement paths that start at a single location).

- The Freehand option allows you to digitize a polygon without clicking the mouse at a particular point (which would be useful for digitizing water bodies).

- The Autocomplete Freehand polygon option allows you to digitize a polygon without clicking the mouse at a particular point, and ArcGIS Pro automatically completes a polygon if some of its edges are made up of existing polygons. (This would be useful for digitizing soil boundaries where the edge of one patch makes up part of the polygon being digitized.)

- The Trace option allows you to digitize polygons along the path of previously digitized polygons (which would be useful if you had multiple layers that all had to follow the same paths).

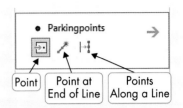

FIGURE 6.10 Options for digitizing point features in ArcGIS Pro.

When you are digitizing points, you have several different options available to you (Figure 6.10):

- The Point option allows you to digitize a point each time you click the mouse.

- The Point at End of Line option sets up a temporary construction line, and you can place a point at the end of that line (which would be useful if you needed to place a point a certain distance away from an object, such as a proposed location for digging a water well a distance away from a site).

- The Points Along a Line option allows you to follow along an already-made line feature (which would be useful for placing points representing light poles in a straight line along a sidewalk).

Before you begin digitizing a feature, ArcGIS Pro prompts you to fill in the data for any attributes about that object. For instance, if you were going to be digitizing lines representing roads, and the feature class the roads would go in had an attribute field called Name, you could type the name of the road first before digitizing the line object. You can also add attribute content after digitizing through the Attribute pane, which can be accessed by going to the **Edit** tab, and within the **Selection** group, clicking the **Attributes** button.

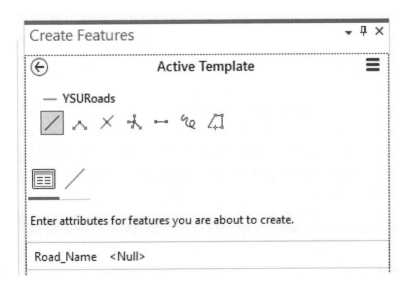

- The Create Features pane changes to give you the Active Template options. (Because YSURoads is the template you chose, that is now the active template, and all the choices and options you have for digitizing will reflect that.) The first road that you will be digitizing is Wick Avenue, so type **Wick Avenue** next to Road_Name (the attribute you created earlier).

- Choose the Line tool to begin digitizing. You should start at the intersection where Wick Avenue meets Rayen, so closely zoom into this area.

- Your cursor turns into a crosshairs. Click once at the Wick and Rayen intersection, and a red dot appears; this is the first vertex in the line. Drag the cursor up to the intersection of Wick and Spring and double-click. The ending node of the line is digitized, and the line segment turns a solid light-blue color. The line has now been digitized, and a line object has been created in the YSURoads feature class.

- You should also see a set of controls at the bottom of the ArcGIS Pro screen to give you some additional options in digitizing.

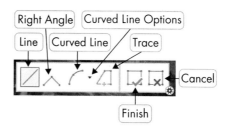

- If you're satisfied with the line you made, click the **Finish** button in the new controls (or press the F2 key on the keyboard). If you're not satisfied, click **Cancel** and digitize the line again.

    Important Note: You should make sure to frequently save your edits so that all your changes are saved to the feature class in the geodatabase. On the **Edit** tab, within the **Manage Edits** group, click the **Save** button. When ArcGIS Pro asks whether you want to save, click **Yes**. If you don't do this, your digitizing won't be saved.

- Use the templates in the Create Features pane and continue to digitize the roads of this section of the YSU campus and update their road names. Refer to the ParkingMap PDF (and the other online maps, if needed) for the locations and names of roads in this section of campus. Consider the following notes to help out:

    - Be sure to add the Road_Name attribute before you begin digitizing. You can add it after you begin if you prefer but make sure that each road segment has the correct name before clicking Finish and moving to the next one.

    - You will likely want to have your roads connect to each other at right angles.

    - Remember that you are only digitizing the roads in this one small section of campus. Four major roads that border this section, and there are also some smaller roads that come into it as well.

    - When digitizing, you may see some unconnected lines. You will learn all about editing vector in Module 7, but for now, you just need to know these simple techniques for editing your work as you digitize:

        - On the **Edit** tab, within the **Features** group, click the **Modify** button. The Modify Features pane opens.

- There are a lot of options, but under the Reshape heading, click on **Vertices** and then click on a line you digitized. The vertices that make up the line appear in green. You can grab and drag these vertices to help unconnected roads better line up with each other.

- When you're done, in the Create Features pane, click the **left-pointing arrow** next to Active Template to save your changes and return to the main Create Features pane options so that you can digitize other layers.

- Each line segment you digitize is stored as a separate record within the attribute table. You can open the attribute table for the YSURoads feature class to see the number of lines you have digitized and what their names are.

- Be sure to save your edits: On the **Edit** tab, within the **Manage Edits** group, click the **Save** button. Answer Question 6.1.

> **Question 6.1** How many road line segments were digitized, and which roads do they represent?

Step 6.6 Heads-Up Digitizing of Polygons

- Now that the roads have been digitized, you can digitize both the campus and non-campus buildings.

- From Create Features, choose the **YSUBuildings** feature class. It becomes the new active template, and you get some new options for drawing tools. (Refer to Smartbox 6.4 on page 157 for descriptions of these options.) You also see the additional tools (right angles, curves, finish, cancel, and so on) at the bottom of the screen. Click on the **Polygon** tool and zoom in closely to the first building you are going to digitize: Meshel Hall. In the Building_Name attribute field, type **Meshel Hall**.

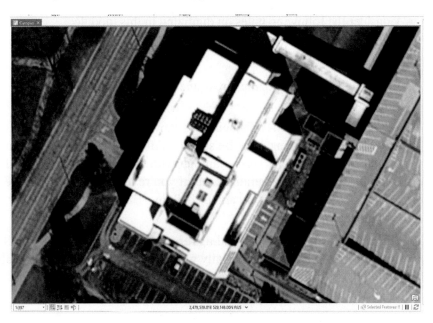

- When digitizing polygons, keep in mind the following:
  - When digitizing a building, you should zoom in very closely on the image so you can see the building dimensions. You want to be close enough to the building that you're not digitizing a crude blob (and doing a poor job in this module). The accompanying graphic shows the boundaries of Meshel Hall (a YSU building) at a good zoom level for digitizing. (You might want to zoom in even more closely for some areas.)
  - When digitizing a polygon shape, click the crosshairs once at the spot where you want to begin creating the starting point. As you click around the outline of the building, ArcGIS Pro begins to show you the full shape of the polygon you're creating as a translucent colored overlay. Be sure to sketch all the dimensions of the building. When you're done, double-click the mouse, and ArcGIS Pro completes the polygon.
  - More options for drawing the lines of polygons—for example, digitizing at right angles, which would be especially important when creating building footprints, or creating curved lines, which might be useful for the semicircular areas on Bliss Hall or the McDonough Museum of Art—are available from the options at the bottom of the screen.
  - Change the colors of your polygon shapes to differentiate them from one another. You can do this, for example, to more easily differentiate a YSU building from a non-YSU building or a parking lot.
  - Just as with the lines, you can supply the name of the polygon before you begin digitizing.
  - Be careful about which layer is the active template when you're digitizing; for example, when digitizing a YSU-owned building, be sure that YSUBuildings is your active template, and when digitizing a non-YSU building, be sure that NonYSUBuildings is your active template.
- Consider the following notes on digitizing in this module:
  - For the purposes of this module, consider the M30 parking deck a YSU building.
  - You will not be digitizing any of the features of the Courtyard Apartments.
  - There are some non-YSU buildings shown in this area that are not listed on the parking map. Be sure to digitize them and add their names to their attributes. (If you're familiar with the area, be sure to use the proper names of the buildings, but if you're unfamiliar with the area, just use names like "non-campus building 1" or something similar to complete your digitizing.)
  - Carefully check out which buildings are on the basemap to digitize. Save your edits as you go and especially when you're done: On the **Edit** tab, within the **Manage Edits** group, click the **Save** button. Answer Questions 6.2 and 6.3 when you've finished digitizing the buildings.

> **Question 6.2**
> How many YSU campus building polygons were digitized, and which buildings are they? (Be careful to check one or more of the maps to be sure you have digitized all campus buildings.)

> **Question 6.3**
> How many non-YSU campus building polygons were digitized, and which buildings do they represent? (Be sure you found and identified all non-campus buildings.)

Step 6.7 Finishing Up Digitizing

- Digitize the polygon shapes representing the parking lots into the Parkinglots feature class. Zoom in as you did while digitizing the buildings to make sure you accurately capture the lot dimensions. Update their attributes during the digitizing process by completing the Lot_Name attribute (for example, M16 or V4). If a lot is not a YSU parking lot, give it a descriptive name, such as "Church Parking."

- Be sure to save your edits: On the **Edit** tab, within the **Manage Edits** group, click the **Save** button. Answer Question 6.4.

> **Question 6.4**
> How many parking lot polygons were digitized, and which ones were they?

- Digitize the lines making up the sidewalks in the area; you'll probably have to zoom in to see the walking paths for the area. Save these lines in the Sidewalks feature class.

  Important Note: With the sidewalks, all you have to do is digitize them; you don't have to update any attributes.

- When you are finished digitizing everything, be sure to save your edits: On the **Edit** tab, within the **Manage Edits** group, click the **Save** button. Also save your Module6 ArcGIS Pro project.

Step 6.8 Measuring the Lengths of Digitized Objects

- As noted in **Smartbox 6.4**, when you're digitizing, you're not simply drawing sketches. Because each object you digitize represents geospatial data, polygons measure real-world areas just as lines measure real-world lengths. ArcGIS Pro can compute the real-world measurements of your sketches through a calculation process. In this module, you'll be computing the total length (in miles) of all the sidewalk segments that you have digitized.

- Open the attribute table of the Sidewalks feature class. You should see a field called Shape_Length, which represents the real-world length of each digitized sidewalk segment in the feature class in the units used (such as feet). You need to compute the total distance, in miles, of all the digitized sidewalk segments. To begin, click on the **Analysis** tab, and within the

Geoprocessing group, click the **Tools** button to open the Geoprocessing pane. From within the Toolboxes section, click on the **Data Management** toolbox and then the **Features** toolset and then click on the **Add Geometry Attributes** script.

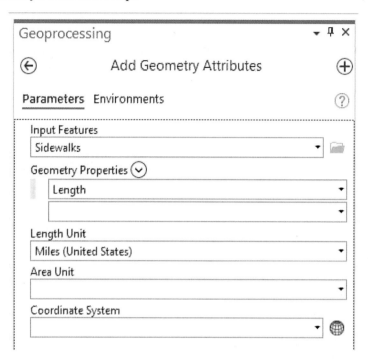

- In the Add Geometry Attributes tool, for Input Features choose **Sidewalks**.
- For Geometry Properties, choose **Length**.
- For Length Unit, choose **Miles (Unites States)**.
- Leave the other entries blank and click **Run**.
- Go to the Sidewalks attribute table to see the results of the tool: A new field called LENGTH has been added to the table that represents the length of each segment, calculated in miles. Answer Question 6.5.

Question 6.5 What is the total length (in miles) of all the digitized sidewalk segments?

Step 6.9 Printing or Sharing Your Results

- Close any open attribute tables.
- Display your YSU buildings, non-YSU buildings, and parking lot polygons, with each category in a different color.
- Also display your roads and sidewalks, using different colors and line thicknesses.
- Turn off the Imagery basemap and change the display so that all of your digitizing can be seen on the screen.
- Turn on the labels for the roads, YSU buildings, non-YSU buildings, and parking lots, using the names you gave them when you assigned their attributes during digitizing.

- Save your final Module6 project.
- Finally, either print a layout (see Module 3) of the final version of your digitized campus map (including all of the usual map elements and the design for a layout) or share your results as a web map through ArcGIS Online (see Module 4).

Closing Time

This module describes various methods of creating your own data to use in ArcGIS Pro. Creating point layers from x/y coordinates and creating features through digitizing are two common and simple ways of generating firsthand GIS data. However, these are not the only ways to create GIS data. Readings obtained using the Global Positioning System (GPS) can measure points and coordinates, and using a GPS receiver or a mobile version of GIS can allow you to directly measure points, lines, and polygons in the field. See *Related Concepts for Module 6* for more about using mobile GIS and GPS to create geospatial data.

Even with careful digitizing or data input, some errors may creep into the process. For example, a road may be digitized at an incorrect length, a parking lot may be digitized with the wrong shape, or a sidewalk may be incorrectly created in the roads layer. Module 7 describes how to assess the quality of datasets you create (or obtain from others) as well as how to edit and fix possible errors in the data.

RELATED CONCEPTS FOR MODULE 6

Using GNSS and Mobile GIS Apps for Data Creation

A common method of data creation is to simply go out into the field and collect the data yourself. For example, a historian working on plotting out troop movements at Civil War battlefields would probably find little in the way of premade GIS data. Rather, she would likely end up on site on battlefield land, where she could try to make measurements. Also, instead of trying to locate spots on a basemap or aerial imagery and digitize them, she would likely be identifying locations while out in the field. Much GIS data is created in this way, and several tools are available to aid this process.

Field data are often collected via the **Global Positioning System (GPS)**, a series of satellites that broadcast signals to Earth. A GPS receiver—whether a handheld unit, a device in your car, or your smartphone or tablet—can pick up these signals (from a minimum of three satellites but commonly four or more) and use the information in them to determine the user's location on Earth's surface. These coordinates are given in GCS using the WGS84 datum, but often the device can translate the coordinates to another datum or projected coordinate system. (See Module 2 for more about GCS and WGS84.)

Data collected from GPS are often used in other programs in the **GPX** data format, a standard that is readable by many software programs, including ArcGIS Pro (which allows you to convert data from GPX format to feature classes). For example, several smartphone apps can track your movements (whether you're hiking, running, or driving) using the GPS capabilities of your phone and produce a

Global Positioning System (GPS) A technology that uses signals broadcast from satellites for position determination on Earth.

GPX The standard format for data collected by a GPS receiver.

GPX file of your path, which can then be added to ArcGIS Pro or ArcGIS Online as a GPX layer.

The name "GPS" refers to the Navstar GPS, which was set up by the U.S. Department of Defense and is operated by the U.S. Air Force, but there are other satellite navigation systems available as well, including GLONASS (the Russian version of GPS), Beidou-2 (the Chinese version, also called Compass), and Galileo (the European Union version, currently in development). These satellite navigation systems are referred to with the blanket name **Global Navigation Satellite Systems (GNSS)**.

Global Navigation Satellite System (GNSS) An overall term for the technologies that use signals from satellites for finding locations on Earth's surface.

ArcPad An Esri software product for mobile devices that allows for GPS integration and field data collection.

Collector for ArcGIS A free Esri app that allows you to collect location-based data in the field with a smartphone or tablet and upload the data to ArcGIS Online.

Survey 123 A free Esri app that allows you to collect form-based data along with location data using a smartphone or tablet and upload the data to ArcGIS Online.

GNSS receiver capability is a common function of mobile devices such as smartphones and tablets, and some GNSS receivers are capable of running GIS software. Esri products are also designed to run on mobile devices; for example, the **ArcPad** program, which combines GIS and GPS capabilities, is designed for field data collection. Aerial imagery can be loaded onto ArcPad, and a GPS receiver can plot your location on the image. As you move, your positions can be saved directly into a shapefile or feature class, allowing you to create point, line, or polygon shapefiles based on your locations in the field. You can then edit these feature classes or add them directly to ArcGIS Pro rather than having to convert or import data.

Esri also distributes several free apps for Android and iOS devices for use in collection of field data. The first of these is **Collector for ArcGIS**, which allows you to use the GPS capabilities of a smartphone or tablet to collect point, line, or polygon data in the field and then upload the data directly to ArcGIS Online. To use Collector for ArcGIS, you first create a new feature layer in ArcGIS Online that can hold the objects for which you want to collect data out in the field (such as street poles). You could also create an editable layer in ArcGIS Pro and publish it to ArcGIS Online as a feature web layer. Then, in ArcGIS Online, you would assemble a web map of these layers. Throughout this process, you would, in essence, create feature templates to populate with data similar to what you did before you began digitizing. Then, when you go out in the field, the Collector for ArcGIS app allows you to use your mobile device's GPS receiver to mark locations of street poles, update the attributes, and publish data to the appropriate template in ArcGIS Online (Figure 6.11). Then a web map of those points can be created, as described in Module 4.

Note that Collector for ArcGIS is the current and updated version of this app. A previous, widely used version of Collector for ArcGIS, referred to as Collector Classic, remains available for download and use as well. Collector for ArcGIS and Collector Classic can be downloaded for free from the Apple App Store or Google Play. For more information about Collector for ArcGIS, see **https://www.esri.com/en-us/arcgis/products/collector-for-arcgis/resources**.

Another free Esri mobile app is **Survey 123**. This app allows you to collect GPS locations in conjunction with form-based information. Then, as with Collector for ArcGIS, you can transfer those data to ArcGIS Online and build web maps. To begin with Survey 123, you first create the form (called a *survey*) of the attributes or information that need to be assessed and entered in the field. For instance, when doing a field assessment of blighted neighborhoods, the survey would have questions to be answered by the person collecting the field data, such as the number of stories of a house, the condition of the home, and other related information. You can also attach photos of the house as part of the survey. Survey 123

FIGURE 6.11 Using Collector for ArcGIS Pro on a smartphone for mobile field data collection.

can be downloaded for free from the Apple App Store or Google Play. For more information about Survey 123, see **https://www.esri.com/en-us/arcgis/products/survey123/overview**.

Key Terms

digitizing (p. 143)
heads-up digitizing (p. 143)
geodatabase (p. 149)
file geodatabase (p. 149)
feature class (p. 149)
feature dataset (p. 149)

shapefile (p. 150)
segment (p. 157)
vertex (p. 157)
point digitizing (p. 157)
Global Positioning System (GPS) (p. 165)

GPX (p. 165)
Global Navigation Satellite System (GNSS) (p. 166)
ArcPad (p. 166)
Collector for ArcGIS (p. 166)
Survey 123 (p. 166)

7

How to Edit Data with ArcGIS Pro

Learning Outcomes

After studying this module, you should be able to:

- Explain the difference between accuracy and precision.
- Describe the usage and importance of five quantitative elements of assessing data quality.
- Explain how the snapping process is used for editing GIS data.
- Describe what undershoots and overshoots are.
- Explain the concept of topology and describe how topology can be used for editing GIS data.

accuracy How closely a measurement matches up with its real-world counterpart.

precision How consistent or exact a measurement is.

ArcGIS Pro Skills

In this module, you will learn how to do the following in ArcGIS Pro:

- Use the Edit tab and Modify Features pane functions to correct multiple types of errors.
- Apply a variety of interactive editing techniques (cutting polygons, splitting lines, reshaping objects, moving objects) to edit errors in feature classes.
- Update and change attribute information for a feature class.
- Use the snapping environment for editing.

 Introduction

When you create GIS data, errors may creep into the data. Digitizing is often a time-consuming process that requires a lot of attention to detail, and it's easy to make mistakes when tracing building footprints or stream patterns from a basemap. Similarly, attribute information (such as a name or a number) can be entered incorrectly. GPS isn't perfectly accurate all the time, and coordinates can sometimes be off by several meters. Incorrect projection or datum information, poor field conditions, and out-of-date basemaps can all lead to flawed final GIS data. All data quality issues must be addressed before GIS data can be useful to you. This module looks at several issues of data quality in GIS; it also teaches you how to edit GIS data using the many editing tools available in ArcGIS Pro.

The terms *accurate* and *precise* are often used interchangeably with regard to the quality of data and measurements. However, in terms of GIS data quality, they have distinct meanings. **Accuracy** describes how closely a measurement reflects the actual value being measured. **Precision** describes how consistent or exact the measurements are, or the level of detail (or significant digits) used in making a measurement. For instance, if you're shooting arrows at the bull's-eye of a target and you strike the bull's-eye dead center, your shot was accurate. If you put three more arrows tightly clustered around the first, your shooting was also precise. However, if all of your shots hit far away from the bull's-eye, but they are still tightly bunched together, your shooting was not accurate, but it was precise: You hit the wrong location but you did it with every shot (Figure 7.1). You can think of GIS data quality in the same way: If your digitizing of a hiking trail followed all of the curves and bends of the path, but you consistently placed the path

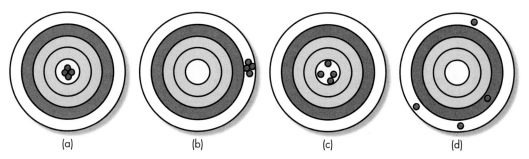

FIGURE 7.1 Four archery results: (a) accurate and precise, (b) not accurate but precise, (c) accurate but not precise, and (d) not accurate and not precise.

10 feet north of where the trail should actually be, your GIS line data would be precise but not accurate.

The metadata that accompanies a layer (see Module 5) should contain several metrics that allow you to assess the quality of the data. One of these metrics is the data's **lineage**, which describes the original source used in creating the data, along with the steps and actions taken during the data creation process. For instance, if streams were digitized from a 1992 USGS 1:100000 topographic map, that information would be contained in the lineage and would aid you in assessing the quality of the data layer. Metadata should also include five other quantitative measures of data quality: temporal accuracy, positional accuracy, attribute accuracy, completeness, and logical consistency. We'll examine each of these measures in this module.

lineage The data sources and processes used for creating a dataset.

 ## Module Scenario and Applications

In this module, you'll continue in the role of an event planner for a university campus (the same role you had in Module 6). The new campus map has been completed (as in Module 6), but now it needs to be checked for accuracy; any errors must be fixed before the map can move to production. It's your job to assess the quality of the new campus map and fix any problems related to the roads, sidewalks, parking lots, and university buildings.

The following are additional examples of other real-world applications of this module's theory and skills:

- Public utilities offices need their datasets to be as up-to-date and error-free as possible. GIS analysts can use editing tools to fix any errors in infrastructure-related datasets.

- A historian receives data related to the locations of generals and troop movements during Civil War battles. She would want to edit the data to be sure that all spatial and non-spatial qualities are as accurate as possible before proceeding with her study.

- A park manager needs the online web map of the park to be updated to reflect the opening of a new set of trails and the seasonal closing of others. He would use the GIS editing tools in ArcGIS Pro to make the appropriate changes to the park data and then share the new map as a map service.

Study Area

- For this module, you will be examining a portion of the Youngstown State University (YSU) campus in Youngstown, Ohio.

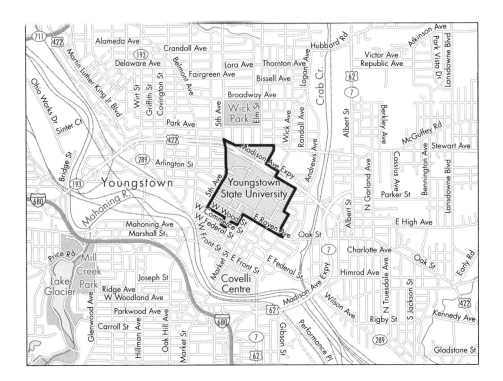

 ## Data Sources and Localizing This Module

The data in this module focus on features and locations within the YSU campus. However, you can easily modify this module to use data from your own campus or local area instead. Because you will be creating the data directly from the Imagery basemap, you can simply find your own campus area that you're familiar with on the basemap and digitize data for feature classes (roads, sidewalks, parking lots, campus buildings, and non-campus buildings) and edit these datasets to contain similar errors that require editing like the ones used in this module.

Step 7.1 Getting Started

- This module's hands-on applications use the data folder called Module7. Your instructor will be able to supply you with this data, or you can download it directly from this book's website at **https://www.macmillanlearning.com/college/us/product/Discovering-GIS-and-ArcGIS-Pro/p/131923075X**. The text in this module assumes that you have this Module7 folder in a computer location referenced as C:\GIS; if you have it somewhere else (for instance, in a flash drive referenced as G:\GISClass), substitute that location and path to the Module7 folder throughout this module.

- The Module7 folder contains the following:
  - ParkingMap: a map of the YSU campus in PDF format. You won't be able to see this in ArcGIS Pro because it's just a PDF file. Open this file in Windows and use it as a reference for this module.
  - Edittests: a file geodatabase that contains four feature classes:
    - Testbuilds: a pre-digitized set of building polygons
    - Testcourts: a pre-digitized set of tennis court polygons
    - Testlots: a pre-digitized set of parking lot polygons
    - Testroads: a pre-digitized set of road lines
  - YSUEdits: a file geodatabase that contains five feature classes:
    - YSURoads: a pre-digitized set of road lines
    - Sidewalks: a pre-digitized set of sidewalk lines
    - ParkingLots: a pre-digitized set of parking lot polygons
    - YSUBuildings: a pre-digitized set of polygons of YSU building footprints
    - NonYSUBuildings: a pre-digitized set of building footprints of buildings not owned by YSU
- Start ArcGIS Pro.
- Sign in with your Esri account username and password.
- Create a new project using the **Map** template. Call this project **Module7** and place it in your **C:\GIS\Module7** folder. Ensure that there is not a checkmark in the box next to **Create a new folder for this project**.
- When ArcGIS Pro opens, change the map's name to **Campus**.
- *Important Note:* Refer to the ParkingMap PDF for the proper locations and orientations of objects. Figure 7.2 shows a visualization of the section of campus used in this part of the module.
- Add all four of the feature classes from the Edittests geodatabase to the Contents pane and display them. (Leave the feature classes from the YSUEdits geodatabase alone for now; you'll get to them in Step 7.7.)
- Zoom in to the area of campus that has the digitized feature classes. You'll be working with a section of the northern portion of campus, roughly bounded by Wick Avenue, Elm Street, and Service Road. (See Figure 7.3 and refer to the ParkingMap PDF.)
- Change colors and symbology for the layers so that you're able to easily see them and distinguish them from one another. (For example, you should be able to easily see the differences among the three polygon layers.) Some or all of these layers contain several editing errors that you will be correcting in this module.

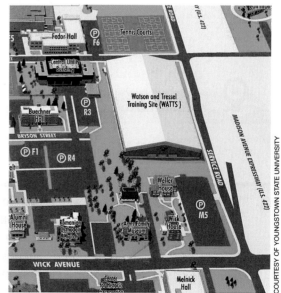

FIGURE 7.2 A section of the YSU campus.

FIGURE 7.3 The portion of YSU campus used in this part of the module.

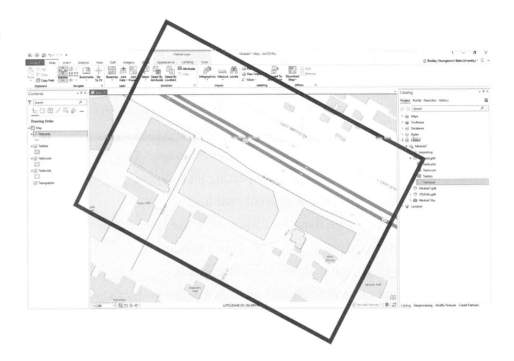

- You need to switch to a different basemap instead of the default topographic one; see either Module 5 or 6 for how to do this. Add the **Imagery** basemap; give ArcGIS Pro a minute or two to stream it in from the Internet. The imagery for the campus area should load, centered around your four feature classes. If the imagery is not of the best quality, see **Troublebox 6.1** in Module 6 for suggestions on obtaining better-quality imagery (that is, World Imagery Clarity) to work with. Due to the nature of the editing, it's highly recommended that you use World Imagery Clarity for this module.

- Note that the imagery represents one snapshot in time. The Esri imagery of the Youngstown area is fairly recent, but it doesn't represent what the campus exactly looks like on today's date. See **Smartbox 7.1** for more information about how such time lags affect data quality assessment.

SMARTBOX 7.1

What is temporal accuracy?

temporal accuracy The time period and currentness of a dataset.

The time period associated with the data you're working with is an indicator of the data's quality. **Temporal accuracy** is a measure of how current the data are, as well as the time period to which the data correspond. For instance, if you're digitizing a map of your campus with the goal of planning campus events, using current (or the most recent) imagery as a source would provide very good temporal accuracy. However, if you were using an image from 1999 to digitize a campus map, it is likely that several changes have occurred since that year (such as new or expanded buildings or different road patterns) that would reduce the temporal accuracy. However, not every dataset requires present-day measurements. For example, land cover data (see Module 12) and digital elevation models (see Module 15) may have been created a few years ago, but because large-scale land cover or terrain changes so slowly, they will likely still be useful (and temporally accurate).

- You may also find it useful to make the polygon feature classes semi-transparent so that you can see the basemap imagery underneath the feature classes for editing purposes. To do so, choose a layer on the Contents pane and then, on the ribbon, go to the **Appearance** tab, and then, within the **Effects** group, you can see the layer transparency slider; you can move this slider back and forth to change the transparency of a layer, or you can just type a new value (such as 50%) in the box next to the slider. Adjust the transparency value as you see fit so that it is semi-transparent, and you can see both the layer and the basemap underneath it. Do this for each polygon layer (Testlots, Testbuilds, and Testcourts).

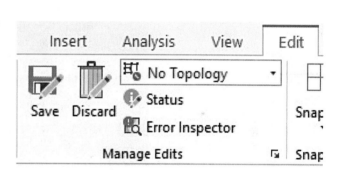

- *Important Note:* In this module, you'll be doing a lot of editing: changing features, moving them, manipulating them, and so on. When you want to save your work, it's not enough to just save your project; you must save your edits to write these changes to the feature classes in the geodatabase. If you don't save your edits, the changes you've made will not be kept. To save your edits at any time, go to the **Edit** tab, and within the **Manage Edits** group, click the **Save** button.

Step 7.2 Reshaping a Feature

- You can easily edit the shapes and lengths of objects. For instance, examine the digitized version of Service Road and note that its eastern edge does not extend to the intersection with Wick Avenue. You need to change the length of it. To do so, in the Contents pane, click on the **Testroads** layer. Then, on the **Edit** tab, within the **Features** group, click the **Modify** button.

- The Modify Features pane opens, and it contains a wide variety of editing options. To change the length of the line, under the Reshape heading, choose **Vertices**.

- The Modify Features pane changes to give you options to edit vertices, and it first prompts you to make a selection. Your cursor changes into a selection cursor that you can use to click on the line you want to reshape. The line segment turns a default cyan color to indicate that it's now selected.

- Click on the line, and you see the vertices that were created to give the line its shape. In addition, the Modify Vertices pane gives you information about the line feature you selected as well as coordinates for each of the vertices.

- Click on the selected line. It turns into a dashed outline, and the vertices are visible. Next, the cursor changes into a four-pointed symbol; by holding down the left mouse button, you can drag that vertex (and, thus, the whole line connected to it) over to the intersection of Wick and Service. When you have the vertex positioned where you want it, release the mouse button. In the new set of controls at the bottom of the screen, click the **Finish** button (the button with the green checkmark). The vertex (and thus the entire line segment) is extended and moved to its new location. If you don't like where you positioned the line, in the new set of controls at the bottom of the screen, click the **Cancel** button (the button with the black X) to cancel the editing and try it again.

- Save your edits (see page 173 for how to do this), and the new length and location of the line are updated in the Testroads feature class. Then, to clear the selection, on the **Edit** tab, within the **Selection** group, click the **Clear** button.

- Now you can use the same types of techniques to reshape a polygon. For instance, the Watson and Tressel Training Sites (WATTS) building has not been digitized properly: One corner has not been digitized, and one side is misshapen. Use the **Explore** tool to reposition the map so that you can see the entire polygon for the building. To begin reshaping, in the Contents pane, click on the **Testbuilds** layer. Then, on the **Edit** tab, within the **Features** group, click the **Modify** button.

- The Modify Features pane opens again. To change the dimensions of a polygon, under the Reshape heading, choose **Vertices**.

- The Modify Features pane again changes to give you options to edit vertices, and it prompts you to make a selection. Your cursor once again changes into a selection cursor. Click on the polygon for the WATTS building you want to reshape, and the polygon outline segment turns a default cyan color to indicate that it's now selected.

- To reshape the polygon, click on the polygon, and the vertices that were created to give the polygon its shape appear. You also see that the Modify Vertices pane gives you information about the polygon feature you selected, as well as coordinates for each of its vertices.

- The cursor changes into a four-pointed symbol. By holding down the left mouse button, you can drag that vertex (and, thus, the whole line connected to it) over to the corner of the WATTS building so that the corner is in the correct position. When you have the vertex positioned where you want it, release the mouse button.

- In the new set of controls at the bottom of the screen, click on the **Finish** button (the button with the green checkmark). The vertex (and thus the entire polygon) is reshaped and moved to its new location. If you don't like how you reshaped the polygon, in the new set of controls at the bottom of the screen, click the **Cancel** button (the button with the black X) to cancel the editing and try it again.

- Save your edits when you've reshaped the polygon (see page 173 for how to do this) and clear the selected polygons.

- Note that you can reshape a feature only by manipulating its vertices. Sometimes you need to add vertices to a feature before you can properly reshape it. For example, to keep reshaping the WATTS building, reopen the Modify Features pane and, under the Reshape heading, choose **Vertices**. Then click on the object you want to add vertices to—in this case, the WATTS building. It is selected in cyan, its outline appears as a dashed line, and the position of the available vertices is shown. At the bottom of the screen is the new set of controls you've been using to finish your reshaping. There are two other buttons here that you can use: the button with the cursor and the plus is the **Add** button, which you click to add a vertex to a line or polygon, and the button with the cursor and the minus is the **Delete** button, which you click to remove a vertex from a line or polygon.

- Click on the **Add** button, then double-click on the polygon outline of where you want to add a vertex. A vertex is placed at that location, and if you continue to reshape the WATTS building, it gives you a new vertex to work with. Click the **Finish** button to cause that vertex to stay there as a new addition to the polygon.

- Remember that whenever you reshape a feature, you're changing its spatial features. For more information about how reshaping affects the measurement of data quality, see **Smartbox 7.2**.

SMARTBOX 7.2

What is positional accuracy?

positional accuracy How closely the spatial features of a dataset match their real-world locations.

Positional accuracy is a measure of how closely the spatial features of a dataset match their real-world equivalents. For instance, if all of the digitized campus buildings are in their proper geospatial locations, the dataset has high positional accuracy. Several factors can affect positional accuracy, including (a) the accuracy of the source map from which the data were derived and (b) the accuracy of the GPS used to obtain data from the field. If the initial source contained errors, then those errors will be carried through to the final product and affect the data's overall positional accuracy.

Step 7.3 Cutting Polygons and Splitting Lines

- During editing, you can split a polygon into two or more separate polygons. For instance, examine the polygon covering the tennis courts adjacent to the WATTS building. Turn the Testcourts layer on and off, and you see that, instead of two polygons (for each separate court complex) being digitized, only one polygon was created. Instead of reshaping the polygon down to cover one court complex and then creating a new second polygon, you could instead cut the single polygon into two smaller polygons.

- Reopen the Modify Features pane and, from the options under Divide, choose **Split**.

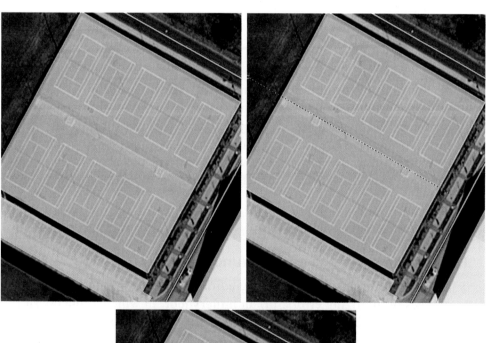

- To divide the single polygon into two, click on the Testcourts polygon to select it and then choose the location off to one side of the polygon that you want to use as a starting point for cutting. Click the mouse at this location, and a red marker appears.

- Use the mouse to draw a line across the length of the polygon, past the other side; this line signifies where the two polygons will be cut.

- Double-click to end the line past the other side of the polygon. The polygon splits into two smaller polygons along the boundary you just created, and the two polygons can then be treated as separate objects.

- In the new set of controls at the bottom of the screen, click on the **Finish** button (the button with the green checkmark). The polygon split finishes, and you have two separate polygons. If you don't like how you split the polygon, in the new set of controls at the bottom of the screen, click the **Cancel** button (the button with the black X) to cancel the editing and try it again.

- Save your edits and open the attribute table for the Testcourts feature class. It now contains two records instead of just one. Close the attribute table when you're done and clear any selected polygons.

Step 7.4 Using Snapping for Editing

snapping Linking two vertices together to become a single vertex.

- **Snapping** (sometimes also referred to as "node snapping" in GIS) is a useful editing technique for joining features together (such as connecting the ends of two lines). With snapping, you can connect lines to one another without having any dangling vertices, or you can link new polygons to existing ones. See **Smartbox 7.3** for more information about the snapping process and the various snapping options available in ArcGIS Pro.

SMARTBOX 7.3

How do the snapping options function in ArcGIS Pro?

Snapping is a very useful tool during editing. Figure 7.4 shows two common types of digitizing errors: (a) an **undershoot**, in which a line doesn't extend far enough to an intersection, and (b) an **overshoot**, in which a line has been extended too far. While you can reshape the line easily enough to lengthen or shorten it, getting the line to the proper length to touch the intersection point would be very difficult. If these lines were not connected, there would be a substantial error in the dataset. For example, if these unconnected road lines were used in putting a network together, ArcGIS Pro wouldn't see that the two roads joined and thus couldn't properly provide a route or directions to allow a turn to occur at the intersection. With snapping, ArcGIS Pro connects the two lines, leaving no gaps between them.

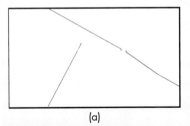
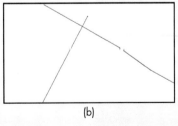

FIGURE 7.4 Two roads, digitized with (a) an undershoot and (b) an overshoot.

undershoot A digitizing error in which a vertex of a line falls short of its target location.

Snapping is done manually. You grab the vertex you want to be connected and drag it within a certain distance of the feature you want to snap it to. ArcGIS Pro then joins the items together. The **snapping tolerance** is the minimum distance around the feature that the vertex must be dragged to in order for snapping to occur. Snapping tolerance can be defined as a real-world distance or a certain number of pixels. Once a vertex is dragged within this distance, ArcGIS Pro tries to automatically "jump" or "snap" to the connection point.

What the vertex connects to is defined by what snapping type you have selected. See Figure 7.5 for an example of this. When a vertex is dragged within the snapping tolerance, it can connect to several different positions on the line. It can link to (a) the endpoint of the line, (b) the edge of the line, or (c) another vertex along the line. Each of these results in a different outcome, and the lines are connected differently, depending on the option you have chosen. In ArcGIS Pro, you should first select the snapping type you need before dragging the vertex within the snapping distance. For instance, if the two roads should intersect perpendicularly, choose the Edge Snapping option.

overshoot A digitizing error in which a vertex of a line goes beyond its target location.

snapping tolerance How closely a vertex must be placed within a feature to activate snapping.

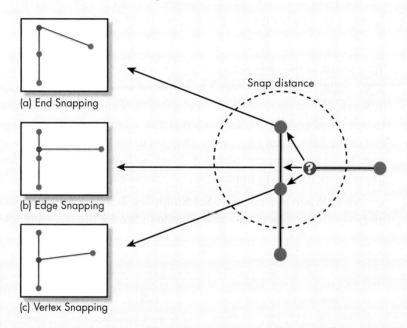

FIGURE 7.5 Three different outcomes for snapping, depending on the snapping type chosen.

In ArcGIS Pro, you can toggle the snapping on and off, so that you can control when you use snapping and when you do not. You have seven different snapping options (Figure 7.6):

- **Point Snapping:** Choosing this option allows you to snap a vertex to a point feature (not a vertex but a separate point object in a point feature class).

- **End Snapping:** Choosing this option allows you to snap a vertex to the vertex at the end of a line of another feature.

- **Vertex Snapping:** Choosing this option allows you to snap a vertex to a vertex in another feature.

- **Edge Snapping:** Choosing this option allows you to snap a vertex to a location you choose along a line or the side of a polygon in another feature.

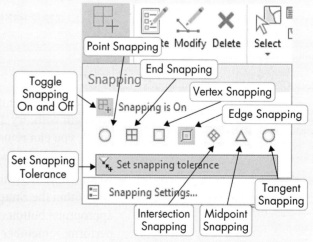

FIGURE 7.6 The various snapping settings in ArcGIS Pro.

- **Intersection Snapping:** Choosing this option allows you to snap a vertex to the intersection to two or more lines.
- **Midpoint Snapping:** Choosing this option allows you to snap a vertex to the middle of a line or the side of a polygon in another feature.
- **Tangent Snapping:** Choosing this option allows you to snap a vertex to a curved line or polygon.

When a snapping option is chosen, it is highlighted in blue; for instance, in Figure 7.6, the Edge Snapping option is shown selected. Snapping can only be done with the options that are chosen, so if you want to snap to an edge, be sure the Edge Snapping option is turned on. It's also a good idea to have activated only the option that you want to use; if you want to snap to the end of a line, for example, be sure the End Snapping option is activated and that no other options are chosen. If you have too many options chosen, ArcGIS Pro may snap the vertex somewhere you don't want it to simply because it has multiple options enabled.

- Take a look at the intersection of Elm Street and Service Road. The line representing Elm Street should be extended to connect with the intersection at Service Road. Instead of just reshaping the Elm Street line, you want it to link together with the Service Road line. In this case, the vertex at the end of the Elm Street line can be snapped to the edge of the Service Road line. To set the snapping tolerance, from the **Edit** tab, within the **Snapping** group, click the **Snapping** button and choose **Snapping Settings…**.

  Note: When you click the **Set Snapping Tolerance** button, ArcGIS Pro allows you to interactively draw a circle to use for the snapping tolerance.

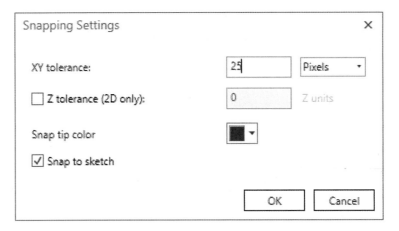

- You can select the number of pixels that make up the snapping distance. To begin with, try 25 pixels, and if this number turns out to be too small, you can return and change it to a larger number. After setting the tolerance, click **OK**.
- Next, to choose which type of snapping you want to use, in the **Edit** tab, within the **Snapping** group, click the **Snapping** button. Click the appropriate button on the toolbar for the type of snapping you want to perform; remember that when the button has a blue box around it, that type of snapping is enabled. In this case, select the **Edge Snapping** button. Also make sure that Edge Snapping is the only option that is enabled.

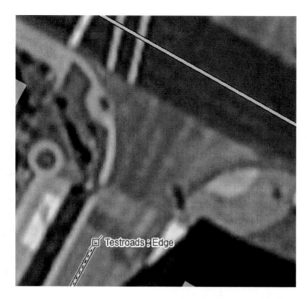
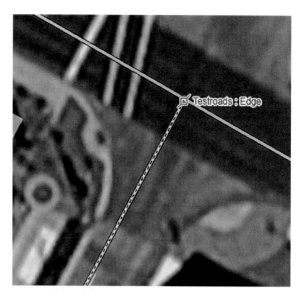
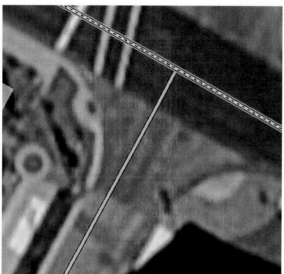

- Bring up the Modify Features pane, and from the Reshape options, choose **Vertices**.
- To reshape the line and snap it, click on the Elm Street line to select it.
- Use the mouse to grab hold of the vertex at the end of the line and drag it within the distance you specified in the snapping tolerance. The vertex "jumps" to try to "snap" to the edge since you selected the Edge Snapping option.
- Double-click the mouse to lock the snapped vertex into place.
- In the new set of controls at the bottom of the screen, click on the **Finish** button (the button with the green checkmark). ArcGIS Pro finishes the reshaping, and you see the Elm Street line snapped to the Service Road line. If you don't like where you joined the lines, in the new set of controls at the bottom of the screen, click the **Cancel** button (the button with the black X) to cancel the editing and to try it again.
- Save your edits and clear any selected lines.

Step 7.5 Updating Attributes During Editing

- You can update or alter the attributes for a feature in a manner similar to how you originally created them in Module 6. For instance, on the **Edit** tab, within the **Selection** group, click the **Select** button and then click on the building polygon that represents Weller House. Then, on the **Edit** tab, within the **Selection** group, click the **Attributes** button.

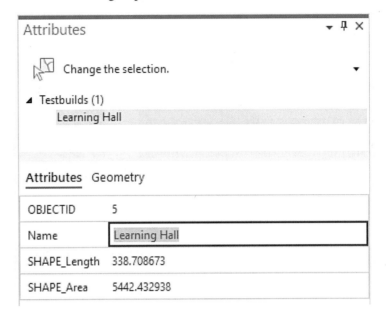

- The Attributes pane opens on the right-hand side of the screen. You can see that the Name attribute of the building has been incorrectly called "Learning Hall" instead of "Weller House." Under the **Attributes** heading in the pane, type the proper building name into the Name field and press the **Enter** key on the keyboard.

- The building name is now correct, so save your edits. Open the Testbuilds attribute table, and you see that the polygon now has the correct name. Having correct attributes for each object is an important part of data quality. See **Smartbox 7.4** for more information about attribute accuracy.

SMARTBOX 7.4

attribute accuracy How closely the non-spatial features of a dataset match their real-world counterparts.

What is attribute accuracy?

Just as the spatial features of a layer can contain errors, so can the non-spatial attributes. **Attribute accuracy** refers to how well the non-spatial parts of a dataset match their real-world equivalents. For instance, one of the attributes in the YSUBuildings layer is Name. If all of the buildings in the layer have the same names as the actual campus buildings, then attribute accuracy (for that field) is high. However, if building names are different or not up-to-date, the attribute accuracy is low. Attribute accuracy doesn't just apply to names but to other non-spatial attributes as well. Land use codes, parcel ID numbers, and tax assessment values are examples of other non-spatial data for which high attribute accuracy is particularly important.

Step 7.6 Deleting, Moving, Rotating, and Creating Features

- During editing, you can easily delete or move features; you can also create new features. To delete a feature, you first have to select it. To do so, on the **Edit** tab, within the **Selection** group, click the **Select** button and then click on the feature you want to delete. You can then remove it in one of two ways:
  - On the **Edit** tab, within the **Features** group, click the **Delete** button.
  - In the View, right-click on the feature you want to delete and then choose **Delete** from the pop-up menu.
- To move a digitized point, line, or polygon, select the feature. Then, on the **Edit** tab, within the **Selection** group, click the **Select** button and then click on the feature you want to move.
- Bring up the Modify Features pane and, under the Alignment options, choose **Move**.
- Using the mouse, click and drag the feature to a new location, release the mouse, and then click the **Finish** button at the bottom of the screen. The object is moved and, when you save your edits, the object's new location is saved in ArcGIS Pro.
- To rotate a feature so it is pointing in a different direction, select the feature, and in the Modify Features pane, under the Alignment options, choose **Rotate**.
- Using the mouse, turn the selected object in a circular path until it gets to the position you want it to be in and then release it.
- Click the **Finish** button at the bottom of the screen. The object is rotated, and when you save your edits, the object's new position is saved in ArcGIS Pro.
- Creating new features (to add missing points, lines, or polygons) is done in the same way as heads-up digitizing. Follow the steps from Module 6 to select the proper template and the correct tool to digitize any new features you might need.
- When editing, you might find that features that belong in one layer have been improperly created or placed in a different layer. For instance, in the feature classes you're working with now, you have two separate feature classes: one for parking lots and another for buildings. Take a closer look at the F82 parking lot, and you can see that it's been incorrectly digitized as part of the Testbuilds feature class when that polygon should be in the Testlots feature class. The brute-force way of fixing this problem would be to delete the F82 parking lot polygon from the Testbuilds feature class and then create a new polygon in the Testlots feature class. However, ArcGIS Pro provides an easier way of making this change: You can simply cut and paste objects from one feature class into another.
- Select the object you want to cut: On the **Edit** tab, within the **Selection** group, click the **Select** button and then click on the object.

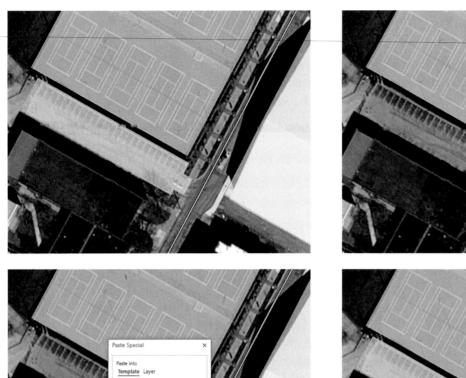

- On the ribbon, in the **Edit** tab, within the **Clipboard** group, click the **Cut** button. The object is cut and vanishes, but it is still stored internally on the clipboard.
- On the ribbon, in the **Edit** tab, within the **Clipboard** group, click the **arrow** under the Paste button and choose **Paste Special**.
- In the Paste Special dialog box that appears, select the feature class you want the object to go into. Choose **Paste Into: Layer**. In this case, select the **Testlots** feature class. Also put a checkmark in the box next to **Keep source attribute values**. Then click **OK**. You now have an F82 polygon of the same spatial dimensions in the same location, but it is in the Testlots feature class.
- Save your edits and clear any selected polygons.
- Ensuring that you haven't inadvertently added objects that shouldn't be there or left out objects that *should* be there is key to evaluating data quality. See **Smartbox 7.5** and **Smartbox 7.6** for more information related to these two types of data quality assessment metrics.

SMARTBOX 7.5

What is completeness?

A dataset should be complete. That is, it should contain all the necessary features but no unnecessary data. In terms of data quality, **completeness** can be judged in two ways:

completeness A measure of the wholeness of a dataset.

error of omission An error in which items that should be part of a dataset are left out of it.

error of commission An error in which extra items that should not be part of a dataset are added to it.

- **Errors of omission**: A dataset is not complete if features have been left out. For instance, if there are 55 buildings on campus, and the "buildings" feature class contains only 53 polygons, then at least two buildings on campus are not properly represented ("at least two" because some of the 55 polygons could possibly be in error), and errors of omission have occurred.

- **Errors of commission**: A dataset is not complete if additional features have been added to it. For instance, if the campus suddenly has an additional building entry for "Stately Shellito Hall" (which doesn't really exist), then an error of commission has occurred. Errors of commission can also occur when some features are incorrectly included in a dataset. For example, while there may be only 55 official named campus buildings, there could be additional storage sheds, maintenance areas, or restroom facilities that could be captured as polygons in a "campus buildings" dataset but not really be considered "official" buildings in a campus directory or map.

SMARTBOX 7.6

What is logical consistency?

logical consistency A measure of whether the same rules were used throughout a dataset.

Another factor of data quality is **logical consistency**, which is a measure of whether the same rules were followed throughout the creation of the dataset. For instance, if more than one person is mapping the trails of a metropark, is everyone following the same rules and methods of data creation and attribute coding? Logical consistency is commonly related to how topological rules are implemented throughout a dataset. (See *Related Concepts for Module 7* at the end of this module for more information about topology.)

Step 7.7 Locating and Editing Errors in a GIS Dataset

- At this point, you can remove the Edittests geodatabase from the Contents pane. In its place, add the five feature classes from the YSUEdits geodatabase. Change their symbology so that each layer is easily distinguished from the others. You may find it useful to make the polygon layers semi-transparent (as you did to the Edittests data in Step 7.1). You now have the same area that you digitized in Module 6, except that intentional errors have been added into the various feature classes.

- Zoom in to the same area you digitized in Module 6—the southern portion of campus, the block bounded by Wick Avenue, Walnut Avenue, E Spring Street, and Rayen Avenue. (See Figure 7.7 for the area and refer to the ParkingMap PDF.)

186 CHAPTER 7 How to Edit Data with ArcGIS Pro

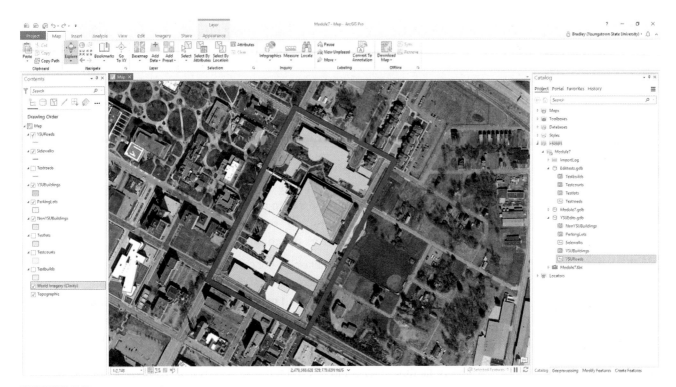

FIGURE 7.7 The section of YSU campus used for editing.

- Within the five feature classes, there are 11 notable errors in the digitizing. Note that these are *significant* errors with the attributes, polygons, and lines—not just minor or insubstantial errors.

- For each of the errors, describe what the problem was and how you fixed it (for example, "Wick and Lincoln lines were overshot; fixed by snapping nodes together," or "YSU building was digitized as a parking lot; switched the building into the parking lots feature class"). Among the significant errors for which you're looking:

  - Six are polygon errors
  - Two involve overshoots or undershoots (and will need to be fixed with snapping)
  - One is some other type of line error
  - Two are attribute errors

- *Important Note.* If everything is correct about a layer—it is digitized properly, it is digitized into the proper feature class, and so on—but the only thing incorrect about it is an attribute, it is considered an attribute error. If there is something wrong with its location—such as its placement or the fact that it is digitized into the wrong feature class—then it is considered a line or polygon error.

- Keep in mind the following hints when looking for errors to fix:

  - The M30 parking deck should be a YSU building, not a parking lot.
  - The bridge going over Wick Avenue and its related features are not YSU buildings, non-YSU buildings, or parking lots.
  - The road that loops around in front of the Courtyard Apartments is not intended to be digitized because the whole road does not fit into the study area.

- The Courtyard Apartments are not part of the study area either.
- For the purposes of this module, the drive-up loop in front of McDonough Museum is not a parking lot.
- Roads will be named or digitized as roads, not as entrances to parking lots.
- Note that some of the features that appear on the paper map do not appear on the Imagery basemap, and vice versa, so stick to what's on the basemap for the locations and dimensions of objects.
- *Important Note:* Refer to the ParkingMap PDF for the proper locations and orientations of objects. Figure 7.8 shows a visualization of the section of campus used in this part of the module.

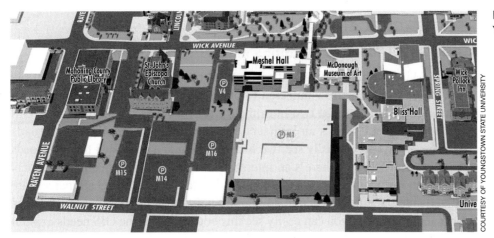

FIGURE 7.8 A section of the YSU campus.

Step 7.8 Printing or Sharing Your Results

- Save your Module7 project.
- Turn off the Imagery basemap. Display your final versions of the YSU buildings, non-YSU buildings, and parking lot polygons all in different colors. Also, display your roads and sidewalks in different colors and line thicknesses.
- Turn on the labels for the roads, YSU buildings, non-YSU buildings, and parking lots, using the name of each for the labeling.
- Finally, either print a layout (see Module 3) of your final version of your edited campus map (including all the usual map elements and design for a layout) or share your results as a web map through ArcGIS Online (see Module 4).

Closing Time

The module examines the basics of editing data in ArcGIS Pro as well as concepts surrounding data quality. Because geospatial technologies and GIS are so widely used, high-quality data is essential to ensuring that the results of analyses or models are reliable. Error-filled or low-quality data will likely produce inaccurate or unreliable products. For example, county auditors use GIS to examine things like property boundaries, land valuations, and property tax assessments. If the geospatial features (such as the dimensions of a land parcel) or the non-spatial features (such as the zoning code) are incorrect, numerous problems can arise.

How objects in a dataset connect and relate to one another is an important aspect of GIS data. When editing, rules for the types of connections that exist between features (independent of their coordinates) can be used in the editing process. See *Related Concepts for Module 7* for more about these topological relationships and how they're used in GIS.

The next module begins to look at how all of these geospatial objects relate to one another. For instance, once you have a well-digitized and edited map of campus, you can begin using that data for analysis, such as to measure the distances between buildings, determine the proximity of parking lots to buildings, and estimate the number of potential students who live within a mile of campus.

RELATED CONCEPTS FOR MODULE 7

Topology and Topological Editing in ArcGIS Pro

When working with vector data, it's important that the GIS understands that the various pieces of data are connected to one another independent of their coordinates. For instance, when two lines that represent roads are digitized, the GIS needs to know that there's an intersection at the place where those two roads cross. Visually, you can easily see two lines crossing at a point, but the GIS needs to be told explicitly that there's an intersection there; this information is critical for applications that provide vehicle routing. Similarly, when two land parcels are digitized side by side, they share a common polygon segment between them as their boundary. Again, you can easily see this boundary, but the GIS needs explicit information telling it that these two polygons are adjacent.

Topology refers to how objects relate or connect to one another independent of their coordinates. Topology establishes three factors for data:

- Adjacency: how one polygon relates to another polygon, such as a common property boundary
- Connectivity: how two lines connect or intersect, as in the intersecting roads example
- Containment: how all locations are situated within a polygon boundary

In ArcGIS Pro, a separate item can be created for a geodatabase feature dataset, simply referred to as a *topology*, sometimes also called a **geodatabase topology**. In this case, a topology is a set of rules that can be applied to the feature classes within the feature dataset. This set of rules can be used to establish connectivity, adjacency, or containment for the objects in those feature classes. For example, a topology rule may state that no polygons may overlap (and an error will be generated whenever two or more polygons overlap) or that lines may not overlap themselves (and errors will be flagged wherever this occurs). Different rules may be applied to different feature classes, so a line feature class representing walking paths may be allowed to have dangling nodes, while a line feature class representing paved driving paths may not.

See Figure 7.9 for examples of some topology rules. For a complete list of the 32 available geodatabase topology rules, see the topology rules poster available in PDF format via the Resource Center, at **https://pro.arcgis.com/en/pro-app/help/editing/pdf/topology_rules_poster.pdf**.

> **topology** How objects relate or connect to one another, independent of their coordinates.
>
> **geodatabase topology** A set of rules applied to the feature classes of a feature dataset in order to remove errors.

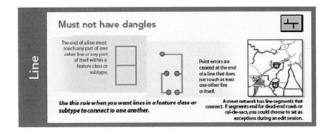
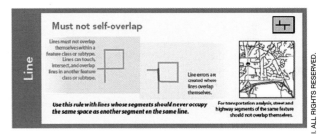

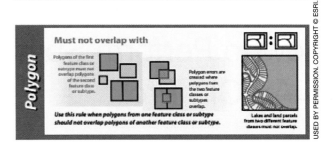

FIGURE 7.9 Examples of four geodatabase topology rules available in ArcGIS Pro.

When one or more rules are chosen for a feature class, those rules are then applied to the objects within that feature class (in what is referred to as the *validation process*). For instance, if the "must not have dangles" topology rule were chosen for a line feature class, all line segments within a certain distance of one another (a value referred to as the *cluster tolerance*) would be automatically edited to remove the dangling vertices. The lines will usually be extended or trimmed accordingly, and the dangling vertex will be snapped to the nearest line or vertex.

Often, however, there will be certain vertices that are not within the cluster tolerance. These will have to be located and edited by hand. The Error Inspector—available on the **Edit** tab, within the **Manage Edits** group—can be used to inspect an entire feature class. It highlights any lines and any vertices that violate the rule. You can then visit each of the problematic vertices in turn, manually selecting how each will be edited and adjusted to comply with the rule. You can also mark a rule violation as an exception to the rule and leave it as is. (***Note:*** These geodatabase topology functions are available to use in ArcGIS Pro only in ArcGIS Pro Standard and ArcGIS Pro Advanced, not in ArcGIS Pro Basic.)

Key Terms

accuracy (p. 168)
precision (p. 168)
lineage (p. 169)
temporal accuracy (p. 172)
positional accuracy (p. 176)
snapping (p. 178)

undershoot (p. 178)
overshoot (p. 179)
snapping tolerance (p. 179)
attribute accuracy (p. 182)
completeness (p. 185)
errors of omission (p. 185)

errors of commission (p. 185)
logical consistency (p. 185)
topology (p. 188)
geodatabase topology (p. 188)

8

How to Perform Spatial Analysis in ArcGIS Pro

Learning Outcomes

After studying this module, you should be able to:

- Define what spatial analysis is.
- List at least five ways that a target layer and a source layer can spatially interact with each other.
- Describe the operation of a Summarize calculation in spatial analysis.
- Explain how a spatial join operates, how it is different from a regular join, and what the output of a spatial join is.
- Explain how a selection by location process operates.
- Describe the usage of pattern analysis in spatial analysis procedures.

ArcGIS Pro Skills

In this module, you will learn how to do the following in ArcGIS Pro:
- Make distance measurements with the Measure tool.
- Use the Summarize Within tool for counting objects within polygons.
- Perform two different types of spatial joins.
- Perform spatial analysis using the Select Layer by Location tool.

Introduction

Using GIS to work with geospatial data means more than simply making a map or performing a query to obtain attribute information about certain features. When you have two or more layers of data, you usually use GIS to answer questions about how they relate to one another spatially. For example, these are some spatial questions that a real estate agent might ask:

- "What school district is this house in?"
- "How many rental properties are within 1 mile of this house?"
- "Which wastewater treatment plant is closest to this house?"

A law-enforcement officer might ask these spatial questions:

- "How many robberies have occurred within a 1-mile radius of this location?"
- "How many registered sex offenders are living within 500 feet of this school?"
- "How many 911 calls have been made from each neighborhood within the city?"

Each professional can answer all these questions by using GIS to examine how features relate to one another with regard to their spatial locations or their proximity. **Spatial analysis** refers to examining the spatial characteristics of features and how they relate to one another across distances. In many ways, spatial analysis lies at the heart of GIS; once you have geospatial representations of the real world's features or phenomena, you can begin to examine how they interact.

spatial analysis The process of examining the characteristics or features of spatial data or how features spatially relate to one another.

In ArcGIS Pro, there are a number of different ways (referred to as *relationship rules*) that two or more GIS layers can relate to each other spatially. We will refer to one of these layers as the **source layer** (which we will be using as the starting point of the relation) and the other as the **target layer** (which we want to examine as the result of the spatial analysis). For instance, the earlier question "How many rental properties are within 1 mile of this house?" would have the layer representing the house as the source layer (what we are starting with) and the layer representing the rental properties as the target layer (what we want to see as the result). At the end of this spatial query, you would have chosen a subset of rental properties that met the criteria of being within 1 mile of the house.

Similarly, the question "How many registered sex offenders are living within 500 feet of this school?" would have the school layer as the source and the locations of registered offenders as the target. At the end of this spatial query, you would have chosen a subset of registered offenders that met the criteria of being within 500 feet of the school. Table 8.1 lists and describes the types of relationship rules that are available in spatial analysis in ArcGIS Pro.

source layer (spatial analysis) The layer that serves as the starting point for a spatial analysis operation.

target layer (spatial analysis) The layer that contains the features that will be examined in the end result of a spatial analysis operation.

Spatial Relationship of the Target Layer with the Source Layer	What ArcGIS Pro Calls This	What Is Chosen as a Result of Spatial Analysis
Intersecting the source layer feature	Intersect	Features from the target layer that overlap or occupy any of the same space with features from the source layer (either a complete overlap or a partial overlap)
Intersecting (3D) the source layer feature	Intersect 3D	Features from the target layer that overlap or occupy any of the same space with features from the source layer (either a complete overlap or a partial overlap)—but only with three-dimensional features (see Module 18)
Within a distance of the source layer feature	Within a distance	Features from the target layer that fall inside a user-defined buffered distance away from the features of the source layer
Within a (geodesic) distance of the source layer	Within a distance Geodesic	Features from the target layer that fall inside a user-defined buffered distance (as measured with geodesic distance) away from the features of the source layer
Within a (3D) distance of the source layer feature	Within a distance 3D	Features from the target layer that fall inside a user-defined buffered distance away from the features of the source layer—but only with three-dimensional features (see Module 18)

Table 8.1 The various relationship rules available for spatial analysis techniques in ArcGIS Pro

Spatial Relationship of the Target Layer with the Source Layer	What ArcGIS Pro Calls This	What Is Chosen as a Result of Spatial Analysis
Containing the source layer feature	Contains	Features from the target layer that hold some or all of the source layer's features within their boundaries
Completely containing the source layer feature	Completely contains	Features from the target layer that hold all the source layer's features within their boundaries—but if the target layer's features touch the boundaries of the source layer's features, they will not be chosen
Containing (Clementini) the source layer feature	Contains Clementini	Features from the target layer that hold all the source layer's features within their boundaries—but if the target layer's features are completely on the boundaries of the source layer's features instead of holding them, they will not be chosen
Within the source layer feature	Within	Features from the target layer that are inside the boundaries of the source layer
Completely within the source layer feature	Completely within	Features from the target layer that are entirely within the boundaries of the source layer—but if the target layer's features touch the boundaries of the source layer's features, they will not be chosen
Within (Clementini) the source layer feature	Within Clementini	Features from the target layer that are inside the boundaries of the source layer—but if the target layer's features are completely on the boundaries of the source layer's features instead of within them, they will not be chosen
Identical to the source layer feature	Are identical to	Features from the target layer that have the exact same spatial features and dimensions as features in the source layer—but if the target layer is of a different vector type (point, line, or polygon) than the source layer, nothing will be chosen
Touching the boundary of the source layer feature	Boundary touches	Features from the target layer that share some or all of the same boundaries with the features of the source layer

Table 8.1 Continued

Spatial Relationship of the Target Layer with the Source Layer	What ArcGIS Pro Calls This	What Is Chosen as a Result of Spatial Analysis
Sharing a line segment with the source layer feature	Share a line segment with	Features from the target layer that have at least one piece of line (from vertex to vertex) in common with the features of the source layer
Crossed by the outline of the source layer feature	Crossed by the outline of	Features from the target layer that have their boundaries passed through by the boundaries of the source layer—but features from the target layer that have a common boundary with the source layer will not be chosen
Have their center in the source layer feature	Have their center in	Features from the target layer that have their weighted center location inside the boundaries of the source layer's features

Table 8.1 Continued

When working with spatial analysis, a good rule of thumb is to first figure out what type of question you want to answer and then to move on to the nuts and bolts of which tool or method and parameters to use. As you'll see in this module and later ones, there are many different types of spatial analysis tools available in ArcGIS Pro, so before picking a tool and using it, first think through the question you want to answer. For instance, if you have a dataset of hundreds of historic bridges and you want to know which township each one is located within, your spatial analysis question becomes "Which county is each bridge within?" Then, you can move to deciding which tool, method, and parameters are best used to properly answer that question. In this module, we'll use ArcGIS Pro to answer these kinds of spatial questions.

Module Scenario and Applications

In this module, you will take the role of a developer working for the state of Ohio. You are looking to find viable places for construction of a new airport. You'll first conduct an assessment of the locations and distributions of regional and international airport complexes in the state. (For purposes of this module, you'll not be working with smaller airstrips or runways.) Next, you'll identify the Ohio counties with populations that are sufficient to justify the expense of a new airport. You'll then look at the places in these counties (a "place" in this sense is an incorporated or non-incorporated location within the state) to find those that are near a major road (such as an interstate or a U.S. highway) but are also sufficiently far enough away from an existing airport complex so the new site won't be in competition with another airport.

194 CHAPTER 8 How to Perform Spatial Analysis in ArcGIS Pro

The following are additional examples of real-world applications of this module's skills:

- A law-enforcement officer is beginning to investigate geographic patterns of crime. As a first step, she wants to know how many houses made 911 calls in proximity to a certain location. She can quickly determine the answer to this question by performing a Select By Location query.

- An urban planner has two different geospatial layers, one of housing locations and another of U.S. Census blocks. He wants to determine the number of abandoned housing units within each census block. He can perform a spatial join of the two layers to determine this information.

- A researcher for the Centers for Disease Control and Prevention is provided with a map of reported dead bird locations within a county. She wants to locate the closest house to each of these places for potential disease monitoring. A spatial join of the two layers will allow her to conduct her analysis.

 ## Study Area

- For this module, you will be working with data for counties, places, and airport complexes for the state of Ohio.

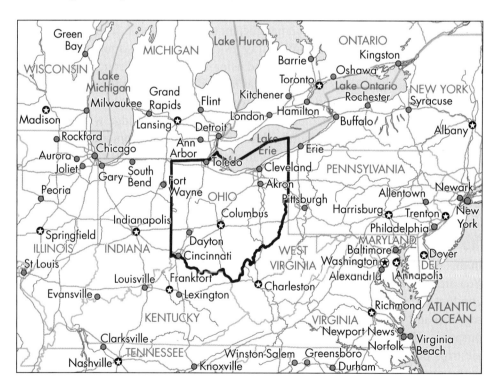

 ## Data Sources and Localizing This Module

The data in this module focus on datasets from within the state of Ohio. However, you can easily modify this module to use data from your own state instead. For instance, if you were going to perform this module's activities in California, the same types of data would be available. The counties dataset was downloaded (for free) and extracted from the Esri USA Counties dataset available on ArcGIS Online (and also available for download from https://www.arcgis.com/home/item.html?id=a00d6b6149b34ed3b833e10fb72ef47b). By querying the USA Counties layer for "STATE_NAME = 'California'," you could select all California counties and export them to a new feature class.

The roads and places (incorporated and unincorporated) datasets were downloaded and extracted from The National Map and are available for all states. Use the Trans_RoadSegment feature class as the source for your state's roads and the GU_IncorporatedPlace and GU_UnincorporatedPlace feature classes for your state's places. The airports data were also downloaded and extracted from The National Map (using the TransAirport_Point feature class). Only airports labeled as complexes (Ftype = 'Complex') that were international or regional airports (Airport_Class = 'Regional Airport' or 'International Airport') were separated out for use in this module.

Step 8.1 Getting Started

- This module's hands-on applications use the data folder called Module8. Your instructor will be able to supply you with this data, or you can download it directly from this book's website at **https://www.macmillanlearning.com/college/us/product/Discovering-GIS-and-ArcGIS-Pro/p/131923075X**. The text in this module assumes that you have this Module8 folder in a computer location referenced as C:\GIS; if you have it somewhere else (for instance, in a flash drive referenced as G:\GISClass), substitute that location and path to the Module8 folder throughout this module.

- The Module8 folder contains the following:
    - Ohioselect: a file geodatabase containing four feature classes:
        - Ohiocountiesproj: a polygon feature class of the 88 counties of Ohio
        - Ohiomajrdsproj: a line feature class of the major roads in Ohio
        - Ohioairportsproj: a point feature class of international and regional airport complexes in Ohio
        - Ohioplacesproj: a polygon feature class of Ohio's incorporated and nonincorporated places
    - Start ArcGIS Pro.
    - Sign in with your Esri account username and password.
    - Create a new project using the **Map** template. Call this project **Module8** and place it in your **C:\GIS\Module8** folder. Ensure that there is not a checkmark in the box next to **Create a new folder for this project**.
    - When ArcGIS Pro opens, change the map's name to **Ohio**.

- Add the Ohiocountiesproj layer to the Contents pane.

- The Ohiocountiesproj layer (as well as the other layers being used) was projected from its initial coordinate system to another one for use in this module. The coordinate system used in this module is **WGS 84 Web Mercator (auxiliary sphere)**, with **meters** for the map units. Check the Ohio map properties under the Coordinate Systems heading to verify that this system is being used.

Step 8.2 Making Basic Measurements in ArcGIS Pro

- Add the Ohioairportsproj and Ohioplacesproj layers to the Contents pane and arrange them so that they lie on top of the counties and you can clearly see all three layers. (Change their symbology as necessary.)
- Zoom in on Fulton and Lucas Counties in northwestern Ohio.
- There is an airport in each county: Fulton County Airport in Fulton County and the Toledo Express Airport in Lucas County. Use the Explore tool to determine which airport is which.

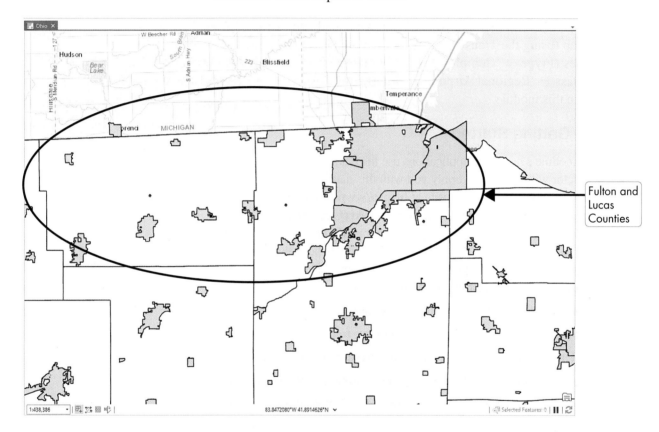

- The Measure tool in ArcGIS Pro allows you to make basic measurements of features or the distances between features. For example, you could use the Measure tool to answer a simple question such as "How far is Fulton County Airport from Toledo Express Airport?" To use the Measure tool like this, on the **Map** tab, in the **Inquiry** group, click on the **Measure** button and choose **Measure Distance**.
- A new box is added to the corner of the Ohio map view; this is the Measure tool. From the **Measure** tool's pull-down menu, choose **Miles** to show the measurements in miles instead of feet.
- Also, click on the **gear icon** on the right-hand side of the Measure tool and from the available options, choose **Geodesic**. This will allow you to measure geodesic distance rather than planar distances.
- Move the cursor into the Ohio map view, and you see it turn into a measuring tool with a crosshairs. You also see a concentric circle appear around the point, in essence "anchoring" the crosshairs to one point to allow you to

Step 8.2 Making Basic Measurements in ArcGIS Pro   197

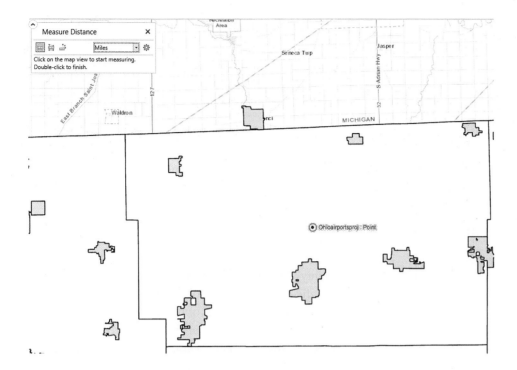

more easily make measurements. If this circle option does not appear when you click with the Measure tool, see **Troublebox 8.1** for how to activate it.

TROUBLEBOX 8.1

Why doesn't the Measure tool snap to a point?

The functions of the Measure tool are influenced by the current settings of the snapping environment. (See Module 7 for more about using snapping in ArcGIS Pro.) To set up the snapping options, on the **Edit** tab, in the **Snapping** group, click the **Snapping** button. When the Snapping tool opens, make sure the button to turn on snapping is selected (so you see "Snapping is On"). Next, make sure that only the Point Snapping option is selected (highlighted in blue), as shown in Figure 8.1. Now, when you use the Measure tool to start measuring from a point, your measurements begin snapped to a point and then also end snapped to a point.

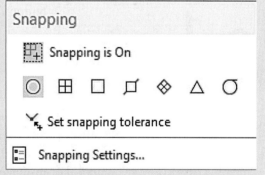

FIGURE 8.1 Choosing the Point Snapping option from the Snapping toolbar.

- Zoom in closely, so you can see both airports, and click the crosshairs on the Fulton County Airport. Drag the mouse toward the Toledo Express Airport, and a line appears. Click the mouse on the Toledo Express Airport, and a distance measurement appears in the Measure dialog. The values for the segment involve the distance between two clicks of the mouse; if you were to measure the distance between Toledo Express and another airport, that new distance would become the value for the segment, while the length would be the sum of all segments combined. Answer Question 8.1.

Question 8.1 What is the geodesic distance between Fulton County Airport and Toledo Express Airport?

- Minimize the Measure tool when you're done.

Step 8.3 Using Spatial Analysis for Summarizing Features

- The Measure tool is useful for making simple measurements of lengths or areas, but you need to use several other ArcGIS Pro tools for answering more in-depth spatial questions. For instance, part of your scenario for this module is to assess the locations and distributions of Ohio's airport complexes and to answer spatial questions like "How many airports are in each county?" To do so, you'll need to begin using more specific geoprocessing tools, such as the Summarize Within tool. For more information about using Summarize tools, see **Smartbox 8.1**. There are two ways to access some commonly used geoprocessing tools:

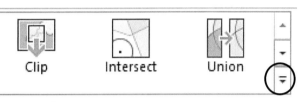

Tools

1. Open the Geoprocessing pane (on the **Analysis** tab, in the **Geoprocessing** group, click the **Tools** button). The tool you will be using in this section is the Summarize Within tool, and you can either search for it or, from the **Analysis** toolbox, within the **Statistics** toolset, choose the **Summarize Within** script.

2. On the **Analysis** ribbon tab, in the **Tools** group, click the **Analysis Gallery** button (the small down arrow button in that group), and then from the many tool options that appear, click the **Summarize Within** button.

SMARTBOX 8.1

How are Summarize tools used in ArcGIS Pro?

When you want to answer spatial analysis questions concerning how many features are within a feature, or when you want to compute some sort of statistic of the attributes of the features within another feature, you can use a **Summarize tool**. For instance, if you had a set of polygons representing county boundaries and a set of points representing the locations of car washes, a Summarize tool would count the number of car wash points within each county polygon. Similarly, if each car wash had an attribute reflecting the amount of money earned by that car wash, the Summarize tool could compute statistics about that attribute—such as calculating the sum of all earnings of the car washes in a county or the average value of the earnings of all car washes in a county.

In ArcGIS Pro, the Summarize Within tool is used to answer these types of questions. When you run Summarize Within, you designate one feature class that has the boundaries (such as the counties layer in the car wash example) and a second feature class that has the features you want to summarize (such as the car washes layer). Summarize Within creates a new feature class of the original

Summarize tool A tool to calculate a statistic (such as the sum) of the objects or their attributes within an area.

features (such as the counties), but it will contain new attribute fields of the summarized feature class (such as a field called Count of Points, whose value indicates the number of points inside each polygon, or a field called Count of Lines that indicates the number of lines inside each polygon). Summarize Within can also calculate statistics about attributes of those summarized features, including sum, mean, minimum, maximum, standard deviation, variance, and range.

Another type of spatial analysis question that can be answered using a Summarize tool concerns the count of features or statistics of attributes of features that are within a certain distance of other features. For instance, if you wanted to know how many car washes are within 1 mile of a freeway off-ramp, or the total or average revenue of those car washes within this distance, you could find out this information by using the Summarize Nearby tool. Summarize Nearby works like Summarize Within, but it creates its own set of polygons showing a distance around a feature and then counts or calculates statistics of features within these newly created polygons. For instance, using Summarize Nearby on a set of points representing the freeway off-ramps would create polygons showing a 1-mile radius around them and then would count the number of car wash points or compute statistics about the car wash attributes within these polygons.

Also, through using credits from ArcGIS Online (see Module 4), Summarize Nearby can not only compute circular polygons at a certain distance but can also create drive-time polygons. These polygons represent how far one could travel by vehicle on a road during a certain amount of time—for instance, a 2-minute driving time or a 5-minute driving time from a point. Thus, you could create a 2-minute drive-time polygon from each freeway off-ramp, and see how many car washes are within this time-based distance. The drive-time polygons are computed using ArcGIS Online resources and then added to ArcGIS Pro for you.

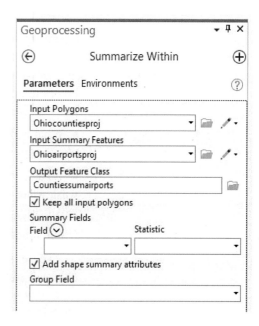

- The Summarize Within tool opens in the Geoprocessing pane.
- Under Input Polygons (the feature that will define the boundaries you will use), choose **Ohiocountiesproj**.
- Under Input Summary Features (what you want to count up), choose **Ohioairportsproj**.
- Under Output Feature Class, click the **browse** button and navigate to your **C:\GIS\Module8** folder, then go within the **Module8** geodatabase, and name the output **Countiessumairports**.
- Leave the rest of the defaults alone and run the tool. A new layer called Countiessumairports is added to the Contents pane. Open this new layer's attribute table and scroll across it to the far right side. You should see that a new field called Count of Points has been added; the values in this field are the sums of the number of airport point features that are within the polygon boundaries of each county. Answer Questions 8.2 and 8.3.

Question 8.2 How many airport complexes are in Cuyahoga County?

Question 8.3 How many airport complexes are in Mahoning County?

- When you're done, close the Countiessumairports attribute table and turn off that layer in the Contents pane.

Step 8.4 Using Spatial Joins for Analysis

- Summarizing is an example of spatial analysis that involves how the points representing airports relate to the polygons representing the county boundaries. However, many more types of analysis can be done. For instance, part of your scenario for this module is to assess the locations and distributions of Ohio's airport complexes and to answer spatial questions like "Which county is each airport in?" or "What is the closest airport to each population center?" In ArcGIS Pro, you could use spatial joins as a way to analyze and answer these questions. See **Smartbox 8.2** for more about spatial joins.

SMARTBOX 8.2

How are spatial joins used in ArcGIS Pro?

In Module 2, we discussed how two attribute tables are joined together: One of the table's fields is appended to the other table on the basis of both tables sharing a common field. This is a good method for linking non-spatial data to geospatial features, but you can perform other kinds of joins as well. In a **spatial join**, the attributes of one layer are joined to another layer not because they have a common field but rather on the basis of their geospatial locations.

spatial join A join in which attributes from one layer are appended to another based on the spatial relationship of the layers.

Spatial joins are versatile joins for spatial analysis. Each of the three vector object types (points, lines, and polygons) can have their attributes spatially joined to any of the other types (that is, points may be joined to polygons, lines may be joined to points, and so on). Spatial joins can be used to relate two GIS layers in any of the ways listed on Table 8.1. Three common types of spatial joins answer these questions:

- What is the closest feature to another feature? This could be used, for example, to find what is the closest culvert to each stream in an area.

- What is within a feature? This could be used, for example, to identify which wastewater treatment plants are within each township.

- What intersects a feature? This could be used, for example, to identify which roads cross a township boundary.

These types of spatial joins can be used with any of the three vector objects, although some types of joins are just not possible; for example, you can't compute how many polygon features are within a point layer or how many lines contain polygons. Also, the attributes of the joined features may be aggregated together in different ways, depending on what settings you use in the spatial join.

For example, as part of a study of Ohio's higher education facilities, say that you want to know which county each college is in. You have a point layer showing the name and location of each university and a separate polygon layer of Ohio counties that contains the name, population, and spatial dimensions of a subset of five counties. These two layers don't have a common field, so you can't use the normal joining of the counties' attribute information to the records of the colleges' layer (as you did in Module 2).

What you *can* do is perform a spatial join that appends the attributes of each county to the college that lies within the county's boundaries. In this case, the counties layer will be spatial joined to the colleges layer. The end result will be a new point layer (of colleges), where each record now has a county name and population matched to it. This matching occurs on the basis of the spatial relationship between college point locations and county polygon boundaries (Figure 8.2). This spatial join acts like a one-to-many join, in which a county's attributes can potentially be matched up with several different universities, if more than one university lies within a county.

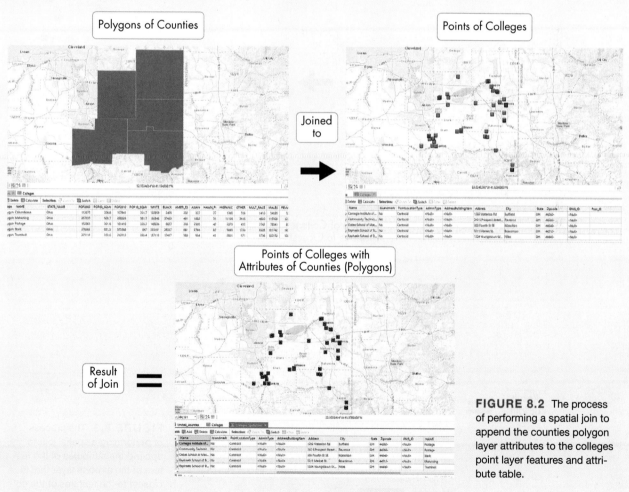

FIGURE 8.2 The process of performing a spatial join to append the counties polygon layer attributes to the colleges point layer features and attribute table.

You can use another type of spatial join to join attributes of one layer to another based on the proximity of the closest feature in the layer to be joined. For example, if you wanted to know which airport is closest to each of the colleges, you could join the airport point layer and its attributes to the colleges points layer. By performing this spatial join, you can determine which airport is the closest to a college point. ArcGIS Pro can then append the record (and its attributes) for that airport to the record for the point to which it is closest. It also adds to the table a new field (in this example, Disttoair) that represents the straight-line distance from each point to its closest airport (using the units of the coordinate system of the features). The end result of this spatial join is a new point layer with the attributes of the closest airport appended to each record (Figure 8.3).

An important thing to remember when performing a spatial join is the direction of the join (that is, which layer gets joined to which). For instance, in the

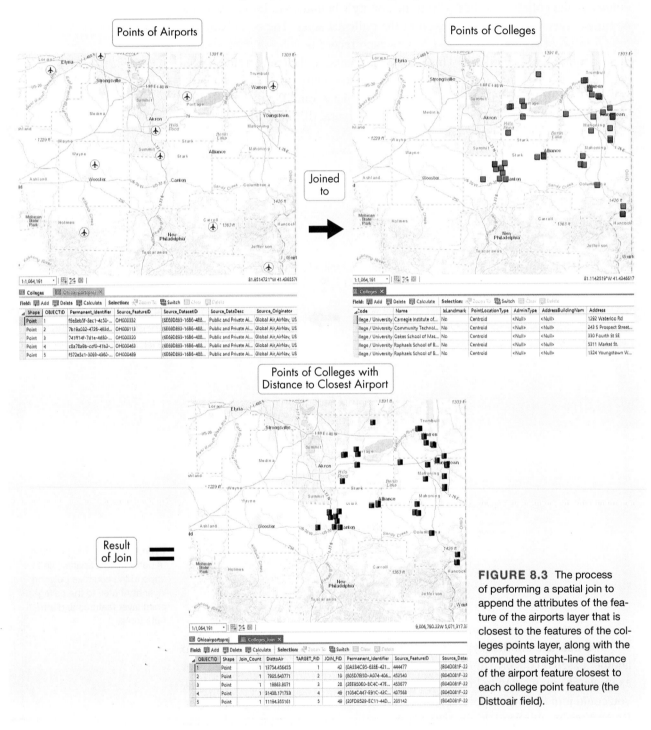

FIGURE 8.3 The process of performing a spatial join to append the attributes of the feature of the airports layer that is closest to the features of the colleges points layer, along with the computed straight-line distance of the airport feature closest to each college point feature (the Disttoair field).

counties and colleges example, you joined counties to colleges, not the other way around. When you opened up the newly joined colleges point layer, each college point had the attribute information of each corresponding county. If you had done the join the other way, each county polygon would have had the information about each college point, which is not what you wanted. It's easy to get the direction of the joins confused. A good rule of thumb to use is that the layer whose attribute table you will be opening to get the information you need is going to be the target layer you join *to*. In this case, you would open the newly joined colleges layer's attribute table to find out county information about each college point.

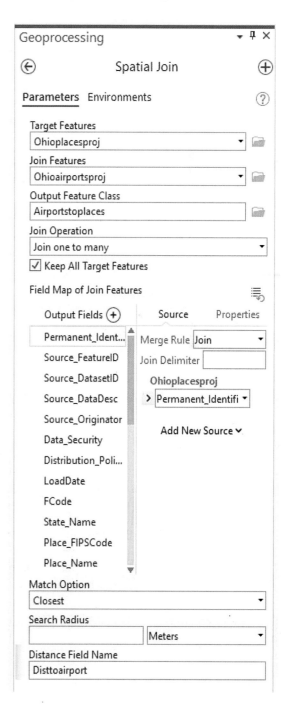

- The first spatial join you will do is to answer the question "For each populated place, what is the closest airport complex (and how far away is that complex)?" Using some common geoprocessing tools, there are two ways of doing this:

  1. Open the Geoprocessing pane, and on the **Analysis** tab, in the **Geoprocessing** group, click the **Tools** button. The tool you will be using in this section is the Spatial Join tool, so either search for it or from the **Analysis** toolbox, in the **Overlay** toolset, choose the **Spatial Join** tool.

  2. On the **Analysis** tab, in the **Tools** group, click the **Spatial Join** button. The Geoprocessing pane opens or switches to the Spatial Join tool.

- Under Target Features (the layer you want to join to), choose **Ohioplacesproj**.

- Under Join Features (the layer you will be joining from), choose **Ohioairportsproj**.

- Under Output Feature Class (the end result of the join), click the **browse** button and navigate to the **C:\GIS\Module8** folder, then go within the **Module8** geodatabase and name the output **Airportstoplaces**.

- Under Join Operation, choose **Join one to many**.

- Put a checkmark in the box next to **Keep All Target Features**.

- For Merge Rule, choose **Join**.

- Under Match Option (which is how the join will be performed), choose **Closest**.

- Under Distance Field Name (what the newly calculated closest distance field will be called), type **Disttoairport**.

- When all settings are correct, run the Spatial Join tool.

- A new layer called AirportstoPlaces is added to the Contents pane. This is the same polygon layer of the places, but it now has a new set of fields joined to it; these fields deal with the distance of each place to the closest point of the Ohioairportsproj layer, based on their spatial locations.

- Open the AirportstoPlaces layer to see these new fields. The fields from the Ohioairportsproj layer that correspond with the spatial locations of the places are joined for each record in the places layer. A new field called Disttoairport is added near the left side of the table; this is the computed distance (in meters) that each populated place is away from the closest airport complex.

- However, this field has the distance value in meters (as that was the units of the Ohioplacesproj field), and you want to measure things in miles. To do so, you can add a new field to the Airportstoplaces attribute table and calculate a value to convert this closest distance measurement from meters to miles. To begin, in the AirportstoPlaces attribute table, click the **Add** button. In the new Fields: AirportstoPlaces tab that appears, give the new field the name **DistMiles** and set its data type to **Float**. On the **Fields** tab, in the **Changes** group, click the **Save** button to save this new field. Close the Fields: Airportstoplaces tab.

- A new field appears in the AirportstoPlaces attribute table called DistMiles, but each record contains the value <null> because you haven't specified what value is to be placed in each record of the new field. What you want to do is take the value in the Distance field (which is in meters) and multiply that value by 0.00062137 to convert that value to miles. To do this, click on the name of the DistMiles field and then, in the attribute table, click the **Calculate Field** button.

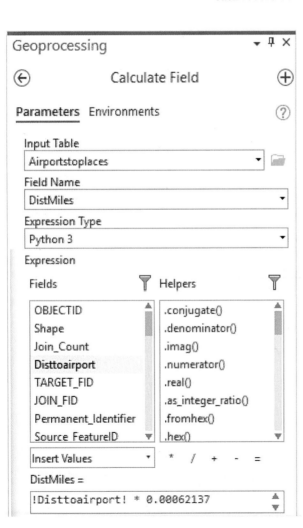

- The Geoprocessing pane switches to the Calculate Field tool.

- Under Input Table, choose **Airportstoplaces**.

- Under Field Name, choose **DistMiles**.

- Under Expression Type, choose **Python 3**.

- You'll be building an expression to compute a new value for the DistMiles field that will be equal to the value in the Disttoairport field multiplied by 0.00062137. To do so, in the Fields box, double-click on **Disttoairport**. The expression changes to DistMiles = !Disttoairport!.

- Next, click the * icon to add a multiplication symbol to the expression.

- Complete the expression box by typing **0.00062137**.

- Your completed expression should read **DistMiles = !Disttoairport! * 0.00062137**. When the expression is complete, run the tool.

- The DistMiles field now shows how far away, in miles, the closest airport is from each place. Answer Questions 8.4, 8.5, and 8.6.

> **Question 8.4** If you lived in Uniopolis, Ohio, what airport complex would be the closest to you, and how far away would it be in miles?

> **Question 8.5** What populated place is the farthest away from an airport complex, and how far away is it in miles?

> **Question 8.6** Why (specifically) was a value of 0 computed for the distance from Richmond Heights to an airport complex? (*Hint:* You might want to locate Richmond Heights on the map and zoom in closely to examine it to help answer this question.)

- Turn off the AirportstoPlaces layer and close its attribute table.

Step 8.5 Using Select By Location for Analysis

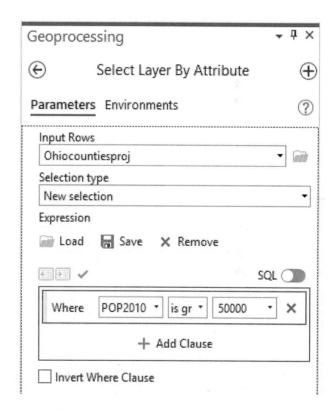

- Now that you have information about the airports and their distribution, you want to determine the location for the new airport complex. Keeping in mind the criteria from the *Module Scenario and Applications* section, you want the new airport to be located in a county with a large population (more than 50,000 persons). You can find this information with a simple query. On the **Map** tab, in the **Selection** group, click the **Select By Attributes** button.

- The Geoprocessing pane switches to the Select Layer By Attribute tool. Use this tool to build an SQL query to select all records in the Ohiocountiesproj layer that have the values of the POP2010 field greater than 50000. (See Module 2 for how to construct a simple SQL query using Select By Attributes.) Run the Select Layer By Attribute tool when you have the query ready.

- The counties meeting this criterion are now selected by ArcGIS Pro (and outlined in the cyan color representing selected features). Answer Question 8.7.

> **Question 8.7** How many counties in Ohio have a population greater than 50,000 persons?

- Next, you want to select all the populated places within the borders of these counties. You can do this by performing a Select Layer By Location operation. (See **Smartbox 8.3** for more about Select Layer By Location in ArcGIS Pro.)

SMARTBOX 8.3

How does Select Layer By Location work in ArcGIS Pro?

There are many different means of examining the spatial relationship of GIS layers. For instance, Module 2 explains how to build a query to select records from an attribute table and how to extract the corresponding features to their own layer. This type of query isn't unique to the analysis of geospatial data; many types of database or spreadsheet programs could perform the same task. What sets GIS apart is its ability to perform a **spatial query** that selects features based not on their attributes but rather on the basis of their location. For instance, Figure 8.4 shows roads and airports of northern Ohio. The records that comprise the length of Interstate 80 have been selected (in cyan). This spatial query was posed to ArcGIS Pro: "Which airports are within 5 miles of Interstate 80?" ArcGIS Pro showed the airports matching this spatial query in the default cyan color used to denote selected features.

spatial query A query that selects records or objects from a layer based on their spatial relationships with other layers rather than their attributes.

FIGURE 8.4 The results of a spatial query asking "Which airports in Ohio are within 5 miles of the selected features of Interstate 80?"

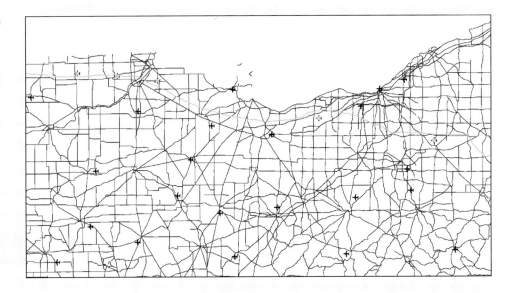

Select Layer By Location An ArcGIS Pro tool used to select features from one or more layers based on their spatial relationship to another layer.

The **Select Layer By Location** tool in ArcGIS Pro allows you to perform spatial queries. With Select Layer By Location, you can pose a spatial query, and ArcGIS Pro will return the records/features that match that query; however, the query conditions are based on spatial features rather than attributes. The selection process is the same as for the Select Layer By Attribute query (see Module 2); selected features are highlighted in the default cyan color, along with the corresponding records in the layer's attribute table. These selected records can be extracted to their own layers or used as the basis for new queries. The same five types of selections can be performed as in Select Layer By Attribute: select features from, add to the currently selected features, remove from the currently selected features, select subset from the current selection, and switch the selection.

With a Select Layer By Location query, two distinct types of layers have to be specified:

- The Input Features layer: This is the target layer and the layer from which the feature will be selected.

- The Selecting Features layer: This is the source layer that specifies the layer to be used as the layer from which to query.

For example, if you want to select all airports within 5 miles of an interstate, the airports layer is the Input Features layer (from which the features/records are selected), and the interstate layer is the Selecting Features layer; you look for features in the Input Features layer that are 5 miles away from the features in the Selecting Features layer. Similarly, if you want to select all cities that are within a certain congressional district, the cities layer is the Input Features layer (and these would be the features that are selected), and the congressional district boundaries layer would be the Selecting Features layer (because you will be selecting features that fall within the spatial dimensions of the district).

Figure 8.5 shows the results of a Select Layer By Location query that searches for all airports (the Input Features layer) that are completely within the boundaries of a populated place polygon (the source). You can see that, of the subset of the airports and places shown in Figure 8.5, only one airport is selected because it is the only airport completely within the populated place boundaries (and thus the only feature that meets the criteria of the query).

FIGURE 8.5 The results of a Select By Location query to find all features of the Input Features layer (the airports) that are completely within the boundaries of a Selecting Features layer (the polygons representing populated places).

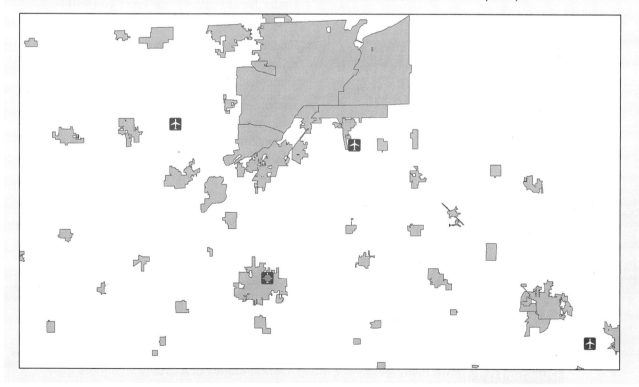

Many different types of spatial queries can be posed using the Select By Location tool, and it's important to be sure you're choosing the right one. (See Table 8.1 on pages 191–193 for all the available options for selecting by location and note that the target here is the Input Features layer and the source is the Selecting Features layer.) For example, say you're involved in a real estate study to determine which houses in a city are prone to flooding. You have two polygons—one representing the residential land parcels and the other representing the floodplain. By treating the parcels layer as the Input Features (or target) layer and the floodplain layer as the Selecting Features (or source) layer, you can use Select Layer By Location to select the parcels on the floodplain.

However, depending on the type of selection you do, you'll likely end up with different sets of selected parcels. If you're trying to select the parcels of land that are at least partially on the floodplain, you should use the "intersect" the source layer feature option. If you're trying to select the parcels of land that are entirely inside the floodplain, you should use the "are completely within" the source layer feature option. If you're trying to select the parcels of land that are at least partly on the floodplain but at its edges, you should use the "crossed by the outline of" the source layer feature option.

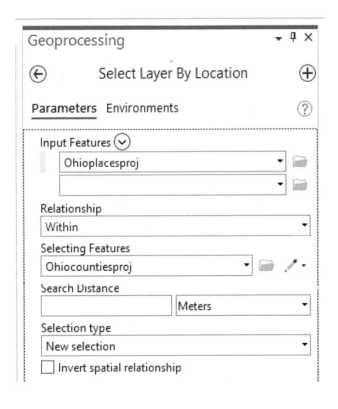

- On the **Map** tab, in the **Selection** group, click the **Select By Location** button. The Geoprocessing pane switches to the Select Layer By Location tool.
- Under Input Features (the layer you will be selecting features from), choose **Ohioplacesproj**.
- Under Relationship, choose **Within** for the relationship rule you want to use (refer to Table 8.1).
- Under Selecting Features (what you will be using as the basis for this selection), choose **Ohiocountiesproj**. Note that as you have already selected some features from this layer with the previous Select By Attributes query, only those selected features will be used rather than all of the features.
- Under Selection type, choose **New Selection**.
- When all the settings are correct, run the Select Layer By Location tool. You see the populated places that were within a previously selected county highlighted in cyan, as they are now selected as well. Answer Question 8.8.

Question 8.8 How many places are within the boundaries of a county with more than 50,000 persons?

- The new airport will have to be built in one of these selected populated places. The second criterion is that the places must be near a major road (within 0.25 miles). To find what places meet this criterion, add the **Ohiomajrdsproj** layer to the Contents pane. You can use Select Layer By Location again to choose these places, so return to this tool in the Geoprocessing pane.
- Under Input Features, choose **Ohioplacesproj**.
- Under Relationship, choose **Within a distance**.
- Under Selecting Features, choose **Ohiomajrdsproj**.
- For Search Distance, type **0.25**.
- From the pull-down box next to this, choose **Miles**.
- For Selection type, choose **Select subset from the current selection**. Note that here you are choosing this option as you already have some features selected in the Ohioplacesproj layer and want to only choose from those previously selected features, not from all the features in that layer.
- When all the settings are correct, run the Select Layer By Location tool. You see the populated places that met your criteria highlighted in cyan (and, thus, were selected). There should be fewer of them this time because your newly selected records were chosen from the previously selected bunch. Answer Question 8.9.

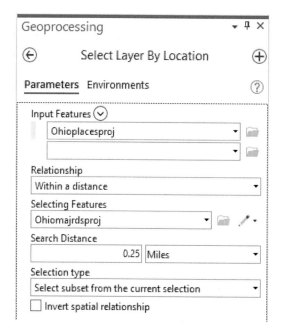

Question 8.9 How many places are within an Ohio county with more than 50,000 persons and also within 0.25 miles of a major road?

- The last criterion is that the chosen places must be far away (35 miles) from an airport complex. You can use Select Layer By Location one more time to determine this, so return to the Geoprocessing pane and the Select Layer By Location tool.
- Under Input Features, choose **Ohioplacesproj**.
- Under Relationship, choose **Within a distance**.
- Under Selecting Features, choose **Ohioairportsproj**.
- For Search Distance, type **35**.
- From the pull-down box next to this, choose **Miles**.
- For Selection type, choose **Remove from the current selection**. Note that here you are choosing this option because the currently selected group of places has met the first two criteria, and you want to start with only these places and take out all the places near an existing airport complex.

- When all the settings are correct, run the Select Layer By Location tool. You see the populated places that met your criteria highlighted in cyan (and, thus, were selected). There should be far fewer of them this time because you just removed many of them from the previously selected bunch. Answer Question 8.10.

Question 8.10 How many places are within an Ohio county with more than 50,000 persons, within 0.25 miles of a major road, and more than 35 miles away from an existing airport complex?

- One of these places (your answer to Question 8.10) will be the site of the new airport complex. You can create a new layer of only these final places for continued analysis (as you did in Module 2). In the Contents pane, choose the **Ohioplacesproj** layer, and then, on the **Data** tab, in the **Export** group, click the **Export Features** button. The Feature Class to Feature Class tool opens in the Geoprocessing pane.

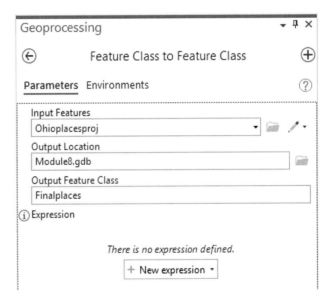

- Under Input Features, choose **Ohioplacesproj**.
- Under Output Location, use the default, **Module8.gdb**.
- Under Output Feature Class, type **Finalplaces**.
- Leave the other options as they are and run the tool. This new Finalplaces layer is added to the Contents pane. Turn off all the layers except Finalplaces, Ohiocountiesproj, and Ohioairportsproj so that you can see your final sites in relation to the airports.

Step 8.6 Using Spatial Joins and Spatial Queries for Further Analysis

- One of the locations in the Finalplaces layer will be the candidate for a new airport complex. Examine the geographic distribution of the places and answer Questions 8.11 and 8.12.

> **Question 8.11** What are the names of the places in the Finalplaces layer? Which of them are classified as cities, and which are classified as villages?

> **Question 8.12** In what county is each of these places located?

- Next, you'll determine which of the existing airports each of these five places is closest to. As in Step 8.4, you can do this with a spatial join. Bring up the Spatial Join tool in the Geoprocessing pane.
- Under Target Features, choose **Finalplaces**.
- Under Join Features, choose **Ohioairportsproj**.
- Under Output Feature Class (the end result of the join), click the **browse** button and navigate to the **C:\GIS\Module8** folder, then go within the **Module8** geodatabase and name the output **Finalclosestairports**.
- Under Join Operation, choose **Join one to many**.
- Put a checkmark in the box next to **Keep All Target Features**.
- For Merge Rule, choose **Join**.
- Under Match Option, choose **Closest**.
- For Distance Field Name, type **Distancetoairports**.
- When all settings are correct, run the Spatial Join tool.
- A new layer called Finalclosestairports is added to the Contents pane. This is the same polygon layer of the final five places, but it now has a new set of fields joined to it, specifying the distance of each place to the closest point of the Ohioairportsproj layer, based on their spatial locations. Answer Question 8.13.

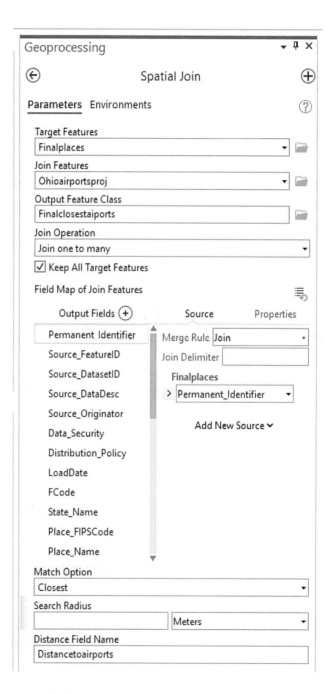

> **Question 8.13** What is the closest airport to each of the potential places, and what is the distance, in miles, that each airport is away from the place?

- Keep the Finalclosestairports layer open. You see a field indicating the population value of each of these final places. The place with the highest population value will serve as the site of the new airport complex. Select only that one record. (See Step 2.4 in Module 2 for one way of selecting only a single record in a table.)

- Export the one selected feature into its own feature class (as you did in Step 8.5). Call this new feature class **chosensite** and save it in your **Module8** geodatabase. The chosensite class should consist of a single polygon of your final site.

- With your site selected, you can do some final analyses of the areas around this site. You need to do two separate Select Layer By Location operations. First, use the Select Layer By Location tool to see how many airports (in the Ohioairportsproj layer) are within 100 miles of the chosen site and answer Question 8.14. Then use Select Layer By Location a second time to find out how many places (from the Ohioplacesproj layer) are within 35 miles of the chosen site and answer Question 8.15. (*Hint:* Be especially careful with the selection methods you're using in each of these two spatial queries; ensure that you choose the correct layers for Input Features and Selecting Features, and make sure you use the correct Select Layer By Location action. You may want to clear all selected features before performing each of these selections.)

Question 8.14 How many Ohio airport complexes are within 100 miles of the chosen site?

Question 8.15 How many populated places are within 35 miles of the chosen site?

Step 8.7 Printing or Sharing Your Results

- Save your Module8 project.
- Zoom in on your chosen site and show it, the surrounding counties, the nearby places, and any surrounding airports. Make sure your chosen site is labeled, along with the counties and airports.
- Either print a layout (see Module 3) of the final version of your airport site (including all the usual map elements and design for a layout) or share your results as a web map through ArcGIS Online (see Module 4).

Closing Time

This module introduces different methods of spatial analysis included with ArcGIS Pro. Joining layers and their attributes based on their spatial location and performing spatial queries are powerful tools for answering many location-based questions. Determining the closest feature or how many objects are within a certain distance from other objects allows you to delve more deeply into geospatial data. Some additional types of analysis methods related to patterns and clusters are discussed in *Related Concepts for Module 8*.

You'll use many kinds of different spatial analysis methods in later modules. For instance, two expansions of the concepts you learned in this module are creating a buffer zone polygon around a feature and combining two or more layers into a single layer. In the next module, you'll begin examining the geoprocessing techniques for doing some of these kinds of tasks.

RELATED CONCEPTS FOR MODULE 8

Analysis of Patterns and Clusters in ArcGIS Pro

Both spatial queries and spatial joins allow you to determine information related to distance, whether you are seeking the closest feature or selecting the objects within a certain distance of other objects. However, these types of distance measures don't tell you anything about how clustered various items might be, based on their distance apart or based on the values associated with the features. For instance, a police analyst could compute a summation of the number of car break-ins occurring in each neighborhood in a city, but this summation will not tell her if these break-ins are happening near one another, if high numbers are grouped together in certain locations, or if these break-ins are happening in random locations. In ArcGIS Pro, a set of complex techniques called *spatial statistics* is used to measure and assess patterns and clusters of different phenomena. Many types of spatial statistical tools are available in ArcGIS Pro; this section presents a brief overview of a few of them.

One useful tool is the Spatial Autocorrelation (Global Moran's I) tool in the Analyzing Patterns toolset of the Spatial Statistics toolbox; it examines the distribution of feature locations and their values (for instance, the distance between city parking lot locations and the number of car break-ins at each one) to determine the extent to which items are clustered together, evenly dispersed, or simply distributed randomly. The degree of clustering is referred to as the **spatial autocorrelation** of the features. The results of Moran's I allow you to examine the spatial autocorrelation of the locations and their values. Figure 8.6 shows an example of different patterns of polygons and their values, which range from dispersed to clustered.

spatial autocorrelation A measure of the degree of clustering of objects and their data values.

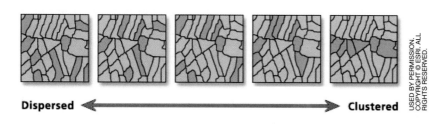

FIGURE 8.6 A set of polygons with associated values (denoted by color) that range from dispersed to clustered.

Another tool for cluster analysis is the High/Low Clustering (Getis Ord General G) tool in the Analyzing Patterns toolset of the Spatial Statistics toolbox. This tool allows you to determine the degree of clustering (for high values or low values within a dataset). ArcGIS Pro uses related types of techniques to examine clusters of values (for example, with the Hot Spot Analysis Getis Ord Gi* tool).

Hot spots in the data are locations where clusters of high values are found (while cold spots are locations where clusters of low values are found).

This kind of analysis can also be performed in ArcGIS Pro using the Optimized Hot Spot Analysis tool. For instance, police analysts would use **hot spot analysis** to determine where spikes in certain types of crimes are occurring. Figure 8.7 shows the mapped result of hot spot analysis to determine hot spots or high levels (in shades of red and orange) and cold spots or low levels (in shades of blue) of parking enforcement in Vancouver, British Columbia.

hot spot analysis A technique used to determine spatial clusters of high values and low values.

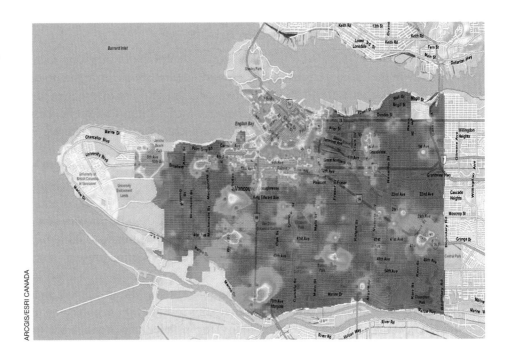

FIGURE 8.7 A map of hot spots and cold spots of parking enforcement in Vancouver, British Columbia.

Key Terms

spatial analysis (p. 190)
source layer (p. 191)
target layer (p. 191)

Summarize tool (p. 198)
spatial join (p. 200)
spatial query (p. 206)

Select Layer By Location (p. 206)
spatial autocorrelation (p. 213)
hot spot analysis (p. 214)

How to Perform Geoprocessing in ArcGIS Pro

ArcGIS Pro Skills

In this module, you will learn how to do the following in ArcGIS Pro:
- Establish geoprocessing environment settings for a project.
- Merge multiple feature classes together into a single feature class.
- Interactively select and extract objects to their own layer.
- Indicate which layers can have objects selected from them.
- Dissolve the boundaries of a set of polygons and combine them.
- Extract a subset of the features in a layer by using the shape of another layer.
- Build a buffer around features.
- Overlay two polygons to create a new third polygon feature.
- Perform point-in-polygon overlay.

Learning Outcomes

After studying this module, you should be able to:
- Describe how two or more layers are merged together in GIS.
- Explain how a dissolve operation is performed.
- Explain what a buffer is in GIS.
- Describe the concept of polygon overlay.
- List four types of polygon overlay operations and explain how they are performed.
- Describe the output of point-in-polygon and line-in-polygon operations.

Introduction

Many types of GIS analysis involve performing an action on a layer and then getting a new output layer in return. **Geoprocessing** is the general term used for this type of task. For example, you could start with a point layer representing culvert locations and create a new polygon layer that represents a 1-mile area around each point, which you could then use for examining the land cover characteristics in each of these 1-mile areas. In ArcGIS Pro, geoprocessing covers a wide range of tasks that are used for multiple applications. You've already used numerous geoprocessing tools in the previous modules; in this one you will use a lot more.

Geoprocessing is often used to create zones of spatial proximity around features or to combine similar polygons. Another common form of geoprocessing is an **overlay**, which involves combining two (or more) layers that have some of the same spatial boundaries but different properties or attributes to create a new layer. Think of creating an overlay as the process of placing one layer over top of another layer, determining where these layers match up and combine, and then producing a new layer as a result (Figure 9.1).

For instance, to determine what sites are available for developing new coastal properties, a polygon representing a buffer around a lake could be overlaid with a

geoprocessing The process of taking an action on a dataset that results in a new dataset being created.

overlay The combination of two or more layers in GIS.

FIGURE 9.1 An example of overlaying two layers and the resultant third layer that combines the spatial and attribute characteristics of the two inputs.

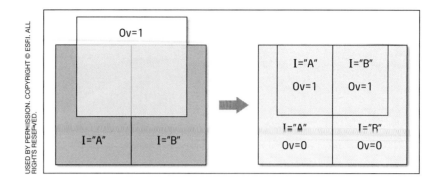

second polygon layer representing the parcel boundaries that are zoned for commercial use. The resultant overlay would show only those polygons within the buffer zone that are also available for commercial usage.

Module Scenario and Applications

This module puts you in the role of an Ohio teacher examining the distribution of schools in one of Ohio's regions as well as access to those schools. As part of your study, you will be trying to determine which schools are located within place boundaries (whether an incorporated place or an unincorporated place), as well as which schools are near major roads. You want to figure out which schools are in which places, as well as which schools are not in a place and not near a major road.

The following are additional examples of real-world applications of this module's skills:

- An urban planner wants to determine the distribution of residential land parcels on a floodplain. He can use polygon overlay to find out which homes are in the floodplain.

- A county engineer needs to examine which areas of the county cannot hear tornado sirens. He can create a buffer around each siren's location to denote the extent of the siren and then determine which populated areas are not reached by the sirens.

- A developer wants to see the potential effect of the placement of a new shopping center and its parking area. She can overlay the digitized footprint of the proposed development onto the existing land use to gauge the amount of wetlands and forested areas that would be removed by such a project.

Study Area

- For this module, you will be starting with data for the entire state of Ohio and then working with only a subset of four counties representing Ohio education State Support Team Region #5: Ashtabula, Trumbull, Mahoning, and Columbiana counties.

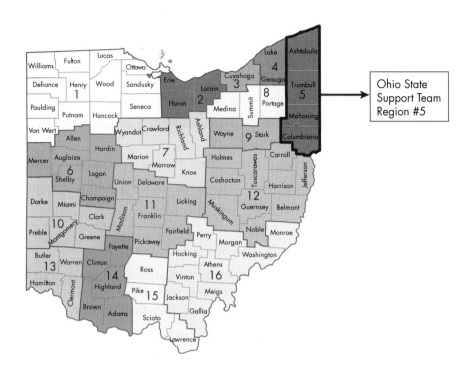

Data Sources and Localizing This Module

The data in this module focus on datasets from within the state of Ohio. However, you can easily modify this module to use data from your own state instead. For instance, if you were going to perform this lab's activities in Missouri, the same sets of data would be available. The Ohio counties dataset was downloaded (free) and extracted from the USA Counties dataset available on ArcGIS Online (and also available for download from **https://www.arcgis.com/home/item.html?id=a 00d6b6149b34ed3b833e10fb72ef47b**). By querying the USA Counties layer for "STATE_NAME = 'Missouri'," you could select all Missouri counties and export them to a new feature class. From there, you could choose the counties within Missouri that correspond to the educational region you wish to examine.

The roads and places (incorporated and unincorporated) datasets were downloaded and extracted from The National Map and are available for all states. Use the Trans_RoadSegment feature class as the source for your state's roads and the GU_IncorporatedPlace and GU_UnincorporatedPlace feature classes for your state's places. The schools data were also downloaded and extracted from The National Map (using the Struct_Point feature class and using FType = '730 - Education').

Step 9.1 Getting Started

- This module's hands-on applications use the data folder called Module9. Your instructor will be able to supply you with this data, or you can download it directly from this book's website at **https://www.macmillanlearning.com/college/us/product/Discovering-GIS-and-ArcGIS-Pro/p/131923075X**. The text in this module assumes that you have this Module9 folder in a computer location referenced as C:\GIS; if you have it somewhere else (for instance, in a flash drive referenced as G:\GISClass), substitute that location and path to the Module9 folder throughout this module.

- The Module9 folder contains the following:
  - Ohiogeo: a file geodatabase containing five feature classes:
    - Ohiocountiesproj: a polygon feature class of the 88 counties of Ohio
    - Ohiomajrds: a line feature class of the major roads in Ohio
    - Ohioschools: a point feature class of elementary and secondary schools in Ohio
    - Ohioincplaces: a polygon feature class of Ohio's incorporated places
    - Ohiononincplaces: a polygon feature class of Ohio's unincorporated places
- Start ArcGIS Pro.
- Sign in with your Esri account username and password.
- Create a new project using the **Map** template. Call this project **Module9** and place it in your **C:\GIS\Module9** folder. Ensure that there is not a checkmark in the box next to **Create a new folder for this project**.
- When ArcGIS Pro opens, change the map's name to **Ohio**.
- Add the Ohiocountiesproj layer to the Contents pane.
- The Ohiocountiesproj layer (as well as the other layers being used) was projected from its initial coordinate system to another one for use in this module. The coordinate system being used in this module is **WGS 84 Web Mercator (auxiliary sphere)**, with **meters** for the map units. Check the Ohio map properties under the Coordinate Systems heading to verify that this system is being used.
- Before proceeding, you'll want to make sure your geoprocessing **environment settings** are correct for the work you'll be doing in this module. When performing geoprocessing in ArcGIS Pro, several settings affect all the actions that you take during a project. For example, you can specify the extent or coordinate system that all results should have. Several upcoming modules will instruct you to fix these settings in place before starting the module. To set up the environment settings for this module, on the **Analysis** tab, in the **Geoprocessing** group, click the **Environments** button to open the Environments dialog box.
- In the Workspace section, for Current Workspace, use **Module9.gdb**. This should be the default, but if it is not, click the **browse** button and navigate to **C:\GIS\Module9** and choose the **Module9** file geodatabase.
- In the Workspace section, for Scratch Workspace, use **Module9.gdb**. This should be the default, but if it is not, click the **browse** button and navigate to **C:\GIS\Module9** and choose the **Module9** file geodatabase.
- In the Output Coordinates section, for Output Coordinate System, choose **Ohiocountiesproj**. This choice tells ArcGIS Pro to use the same coordinate system as the Ohiocountiesproj layer for all of its output; ArcGIS Pro then updates this choice to the WGS_84_Web_Mercator projection.
- In the Processing Extent section, for Extent, choose **Ohiocountiesproj**. This choice tells ArcGIS Pro to use same extent as the Ohiocountiesproj

environment settings
Geoprocessing options that control activities in a map document, such as the extent of outputs, the coordinate system of outputs, or the cell size of outputs.

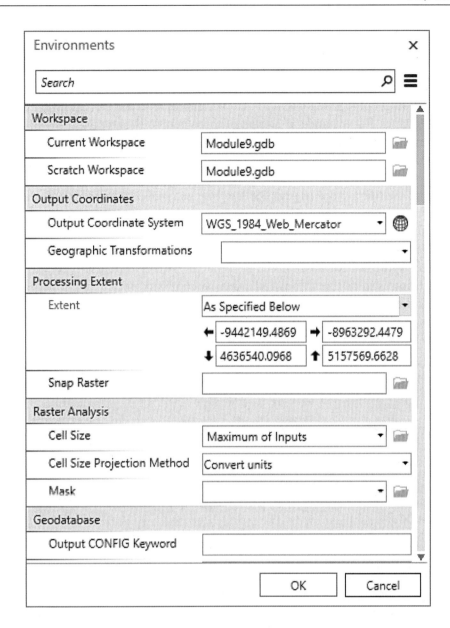

layer for all output; ArcGIS Pro then updates this choice to the actual coordinates for the extent.

- Click **OK** when all the environment settings are in place.

Step 9.2 Merging Multiple Layers

- Add the Ohioincplaces and Ohiononincplaces layers to the Contents pane and position them so that you can see both of them on top of the Ohiocountiesproj layer.

- The National Map separates areas into two categories: incorporated places and unincorporated places. This module does not make a distinction between these two. Thus, you want to combine these two layers into a single layer rather than work with two different layers. In ArcGIS Pro, you can do this by using the merge process. (See **Smartbox 9.1** for more information.)

SMARTBOX 9.1

How can multiple layers be merged in ArcGIS Pro ?

A common geoprocessing operation is to combine several different layers into a single layer. For instance, if a team of three people is mapping the trails of a large park, each person will be producing feature classes of one-third of the park. Ultimately, you do not want three separate layers of trails but a single layer of trails that combines the geography of all three. You can accomplish this with the Merge tool in ArcGIS Pro. **Merge** combines the geospatial features and attributes of two or more vector layers into a single new vector layer. A related geoprocessing tool in ArcGIS Pro is **Append**, which combines one or more layers with an already existing feature class.

See Figure 9.2 for an example of a merge. Two different polygon layers (with features in different locations) can be combined into a single polygon layer that contains all features (and attributes) from both layers. If attributes are present in one layer and not another, the value "null" is assigned to the missing values. Keep in mind that only features of the same type can be merged; for instance, two line feature classes can be merged into a single line feature class, but a point feature class and a line feature class cannot be merged into a single feature class.

Merge A geoprocessing tool that combines two or more layers of data together into a single layer.

Append A geoprocessing tool that combines one or more layers of data together with an already existing layer.

FIGURE 9.2 Using the Merge tool to combine two layers into one.

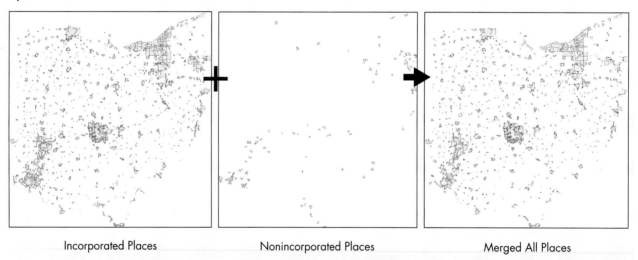

- The tools that you will be working with in this module (including Merge) are available on the **Analysis** tab, in the **Tools** group. Clicking the **Tools** down-arrow button opens the Analysis Gallery, from which you can choose commonly used geoprocessing tools.

- When the Analysis Gallery opens and expands, click the **Merge** button inside it. The Merge tool opens in the Geoprocessing pane.

- Under Input Datasets, first choose **Ohionincplaces**. When another pull-down menu appears, allowing you to choose the additional layers to merge, choose **Ohioincplaces**.

- Under Output Dataset, click the **browse** button and navigate to your **C:\GIS\Module9** folder and go within the **Ohiogeo** geodatabase. Call this output dataset **Ohioplaces**.

- Leave the other options at their defaults and **run** the Merge tool. A new layer called Ohioplaces is added to the Contents pane; it has both of the layers combined in it. At this point, you can remove both the Ohioincplaces and Ohionincplaces layers from the Contents pane since you now have a new merged layer to work with.

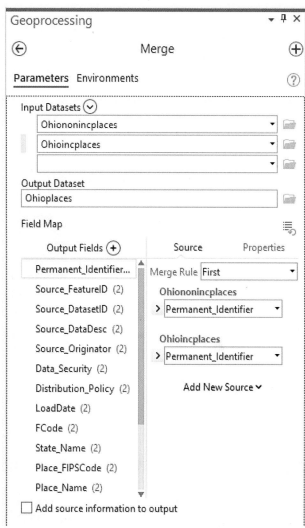

Step 9.3 Interactively Selecting and Extracting Data Layers

- The focus of this module is on the four counties in northeastern Ohio that make up State Support Region #5: Mahoning, Trumbull, Columbiana, and Ashtabula. Instead of working with all 88 Ohio counties, you can select and extract only the 4 counties of interest. For this module, you can select those counties and extract them to their own layer. (See Module 2 for more about selection and selected features in ArcGIS Pro.) You want to select features only from the Ohiocountiesproj layer (and not from the Ohioplaces layer), and you need to communicate this to ArcGIS Pro. In the Contents pane, choose the List By Selection option to display the layers this way. (See Module 1 for more about the Contents pane and different ways to list layers in it.) ArcGIS Pro now shows you the layers from which you can and cannot select.

- By default, each layer has a checkmark next to it to indicate that each layer can have features selected from it. Remove the checkmark from the Ohioplaces layer so that only the Ohiocountiesproj layer has a checkmark. Now, Ohiocountiesproj is the only layer in the Contents pane from which you can select things.

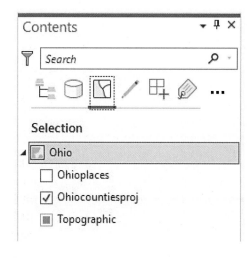

- On the **Map** tab, in the **Selection** group, click the **Select** drop-down arrow and choose **Lasso**. The other options allow you to draw particular shapes (such as a rectangle or circle) and select features that touch those shapes, but the lasso option gives you much more flexibility in terms of the shapes that you draw.

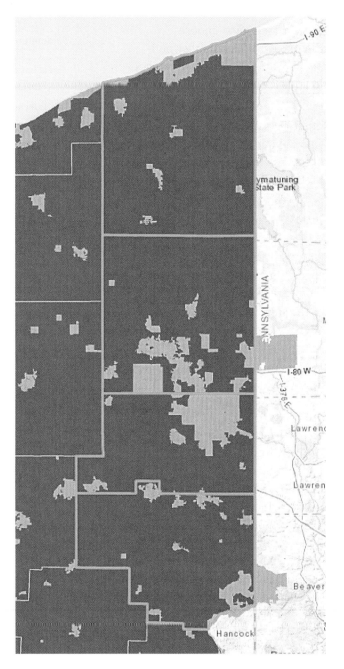

- The cursor changes to the selection cursor. Click once within the boundary of the northernmost county that makes up the four counties you're using in this module (see the graphic in the *Study Area* section on page 217 for their locations). Because you indicated to ArcGIS Pro that only features within Ohiocountiesproj would be selected, none of the corresponding features in Ohioplaces that fell within the rectangle were selected. Continue drawing using the lasso tool until you have a completed polygon that intersects or touches a boundary of each of the four counties. When you release the mouse button, you see the four counties outlined in cyan, indicating that they have been selected.

Important Note: Select only the four counties. If you find that you've selected other counties, you can deselect them by holding down the **Shift** key on the keyboard and clicking on the counties you want to deselect to remove them from the selection. If you want to clear all selected counties and start over, in the Map tab of the ribbon, in the Selection group, click the **Clear** button.

- Next, change how you're viewing items in the Contents pane to **List By Drawing Order**.
- The next step is to create a new feature class containing just the four selected counties. (See Modules 2 and 8 for how to export data into a new feature class in a geodatabase.) Export the selected features of the Ohiocountiesproj layer to a new layer called **Fourcounties** and save it as a feature class in the **Module9** geodatabase. After you export this layer, the Fourcounties layer is added to the Contents pane.
- At this point, turn off the Ohiocountiesproj layer and zoom in on the four counties that you are using in this lab.
- Switch the Fourcounties polygons symbology to **Black Outline 2 pts** so that you can see through the polygons and so they also have a thick black outline. (See Module 1 for how to change the color or symbology of a layer.)

Step 9.4 Dissolving Polygon Boundaries

- The next step in defining the four-county region is to dissolve the borders between the counties so that you can use the entire region—rather than four separate counties—as a boundary. (See **Smartbox 9.2** for more about using Dissolve for geoprocessing.) You will use this new "region" in further analysis.

SMARTBOX 9.2

How does Dissolve operate in ArcGIS Pro?

The **Dissolve** function in ArcGIS Pro combines multiple objects in a feature class into a single object when these objects have the same attributes. For example, a polygon feature class of U.S. counties might have an attribute indicating the state to which the counties belong (such as a FIPS code or state name). When Dissolve is used, the boundaries of the counties' polygons are removed, and they are combined into a single polygon, as long as those polygons have the same attribute (for example, all counties have the same state FIPS code or state name). Figure 9.3 shows how Dissolve can be used to remove county boundaries and create a larger object of the geography of the combined area. By dissolving polygon boundaries, you can create regions of data that have the same attributes. For instance, a polygon feature of parcels might contain an attribute indicating how that parcel is zoned. You can dissolve the polygon boundaries according to their zoning type so that you can create larger polygons of areas zoned residential, commercial, or industrial.

Dissolve A geoprocessing tool that combines polygons with similar attributes.

FIGURE 9.3 Using the Dissolve tool to combine several features in a single layer into one feature.

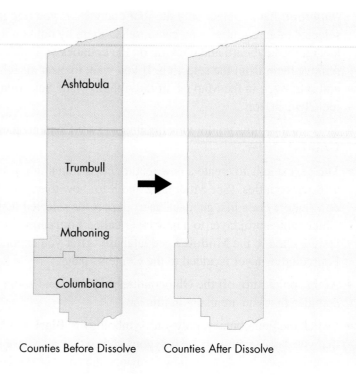

Counties Before Dissolve   Counties After Dissolve

- On the **Analysis** tab, in the **Tools** group, open the **Analysis Gallery** again and then click the **Dissolve** button. The Dissolve tool opens in the Geoprocessing pane.

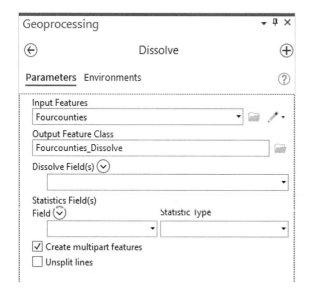

- Under Input Features, use **Fourcounties** as the feature class you wish to dissolve.
- Under Output Feature Class, click the **browse** button and navigate to your **C:\GIS\Module9** folder, go to the **Module9** geodatabase, and name it **Fourcounties_Dissolve**.
- Don't choose anything under Dissolve Field or Statistics Field. When all options are set, run the Dissolve tool.
- You see a "borderless" four-county region layer called Fourcounties_Dissolve added to the Contents pane. Change its symbology to **Black Outline 2 pts** and remove the original Fourcounties layer from the Contents pane.

Step 9.5 Clipping Feature Classes

- Add the Ohioschools and Ohiomajrds layers to the Contents pane. Right now, these two layers and the Ohioplaces feature class contain information about all of Ohio. You want to focus only on schools and major roads that are within the boundaries of the four-county region. To select only those areas, you can use the Clip operation in ArcGIS Pro. (For more information about Clip, see **Smartbox 9.3**.)

SMARTBOX 9.3

How does Clip operate in ArcGIS Pro?

The **Clip** tool in ArcGIS Pro uses the shape and geometry of one feature class to extract objects out of another feature class. In Figure 9.4, the roads layer covers the entire state, but you want to use only the roads that fall within the multi-county polygon boundaries. The roads can be clipped so that a new dataset consisting of only the roads within the geographic area of the polygon can be extracted for use. Using Clip is like using a feature class (such as a polygon) as a "cookie cutter"; all features from another layer that fit within this cookie cutter are removed to a new layer for use. Polygons, lines, and points can all be clipped using this operation. For instance, if you have a land-cover map of your entire state but will be focusing your environmental analysis on a single county, you can use the polygon boundaries of the county to clip out the land-cover polygons that are within that single county.

Clip A geoprocessing tool that extracts objects from one layer based on the geometry of a second layer.

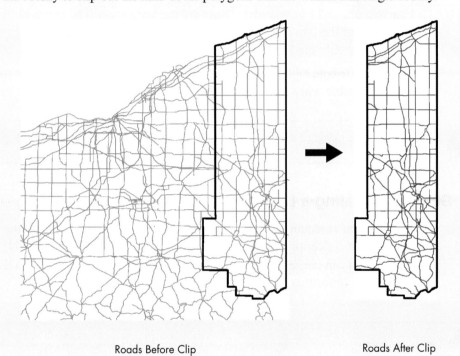

FIGURE 9.4 Using the Clip tool to create a new feature class.

Roads Before Clip   Roads After Clip

- On the **Analysis** tab, in the **Tools** group, open the **Analysis Gallery** again and then click the **Clip** button. The Clip tool opens in the Geoprocessing pane.

- Under Input Features, select **Ohioplaces** from the Ohiogeo geodatabase. This is the feature class that you want to clip.

- Under Clip Features, select **Fourcounties_Dissolve**. This is the feature class that will define the cookie cutter shape used for clipping.

- Under Output Feature Class, click the **browse** button and navigate to your **C:\GIS\Module9** folder, go to the **Module9** geodatabase, and name it **Fourplaces**.

- Run the Clip tool. You now have a new feature class called Fourplaces added to the Contents pane that shows only the places inside the four-county region.

- Use the Clip tool again to clip the Ohioschools layer to the boundaries of the Fourcounties_Dissolve layer to create a new output feature class called **Fourschools** and place this output feature class in your **Module9** geodatabase.

- Use the Clip tool again to clip the Ohiomajrds layer to the boundaries of the Fourcounties_Dissolve layer to create a new output feature class called **Fourmajrds** and place this output feature class in your **Module9** geodatabase.

- You now have a lot of layers in the Contents pane. Remove all layers from the Contents pane except for the dissolved county boundaries (the Fourcounties_Dissolve layer) and the three clipped layers (Fourplaces, Fourschools, and Fourmajrds). These are the layers you'll be using for the remainder of the lab. **Zoom to layer** with the Fourcounties_Dissolve layer to center on only the four-county region.

- Open the attribute table for the Fourschools layer. Answer Question 9.1. (Close the table when you're done.)

Question 9.1 How many schools are within the four-county region?

Step 9.6 Creating a Buffer Around Features

- Because you're examining access to schools, the next step is to examine information concerning areas that are near the region's major roads. To do this, you can create a half-mile buffer around all the major roads in the region. See **Smartbox 9.4** for more information on buffers.

SMARTBOX 9.4

How do buffers operate in ArcGIS Pro?

buffer A zone of spatial proximity around a feature or set of features.

A **buffer** is a region of spatial proximity created around a set of objects. In ArcGIS Pro, a buffer is a new feature class created by calculating a distance around a set of objects in another layer (whether points, lines, or polygons). See Figure 9.5 for an example. For this example, a feature class contains a set of lines representing roads, and a half-mile buffer is created around these roads and saved as a polygon feature class. Distances and polygons are calculated around each road, and the borders between those polygons have been dissolved in order to create a single new object that is the buffer.

Creating buffers is a common geoprocessing operation. For instance, engineers may create a buffer around a road that needs to be widened to see which sections of neighboring properties will be affected, or emergency planners might create buffers around tornado sirens that represent the distance that the sirens'

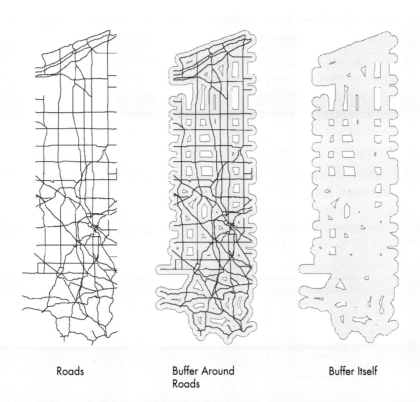

FIGURE 9.5 Creating a half-mile buffer around the roads of a feature class.

Roads   Buffer Around Roads   Buffer Itself

sound reaches. Overlapping buffers may also have the boundaries between them dissolved to create larger buffer regions for future analysis rather than having multiple individual overlapping buffers available. Dissolving the area between the overlap of the buffers of each road segment is also shown in Figure 9.5.

Note that creating a buffer is different from using a Select Layer By Location query (see Module 8). In Select Layer By Location, one of the options allows you to select how many objects are within a certain distance of other objects. The Buffer tool creates a new polygon object that represents this distance, allowing you to work with the actual dimensions of the buffer zone as a separate layer.

- On the **Analysis** tab, in the **Tools** group, open the **Analysis Gallery** again and then click the **Buffer** button. The Buffer tool opens in the Geoprocessing pane.
- Under Input Features, select **Fourmajrds**. This is the layer whose features you want to create a buffer around.
- Under Output Feature Class, click the **browse** button and navigate to your **C:\GIS\Module9** folder, go to the **Module9** geodatabase, and name it **Fourmajrds_halfmilebuffer**.
- For Distance [value or field], select **Linear unit** from the drop-down. Type **0.5** and select **Miles** from the pull-down menu to create a half-mile buffer around the road features.
- The next four options help define the shape of the buffer. Under Side Type, select **Full**. Under End Type, select **Round**. Under Method, select **Planar**. Under Dissolve

228 CHAPTER 9 How to Perform Geoprocessing in ArcGIS Pro

Type, select **Dissolve all output features into a single feature** (to dissolve any boundaries between overlapping buffers).

- Run the Buffer tool. A new layer called Fourmajrds_halfmilebuffer is added to the Contents pane. Place the Fourschools layer on top of the Fourmajrds_halfmilebuffer buffer layer in the Contents pane to eyeball just how many schools are (and are not) within a half mile of a major road. Zoom in and pan around the map to see what the dimensions of the buffer are and how the schools fit in the buffer.

- Counting points in the buffer by hand would be inefficient and likely inaccurate. Instead, you can do a spatial query operation (as in Module 8) to select only those schools within the buffer (and will thus not select schools not inside the buffer). From the **Map** tab, in the **Selection** group, click the **Select By Location** button.

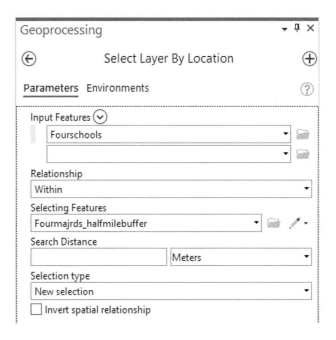

- Use the Select Layer By Location tool (see page 206 in Module 8) to select features from the Fourschools layer that are within the half-mile buffer (the Fourmajrds_halfmilebuffer layer). Answer Questions 9.2 and 9.3.

Question 9.2 How many schools in the four-county region are within half a mile of a major road?

Question 9.3 How many schools in the four-county region are not within half a mile of a major road?

- Clear your selected features.

Step 9.7 Performing Polygon Overlay

- Knowing about the proximity of schools to roads is only part of the analysis you need to do. You also want to know which schools are within the boundaries of an incorporated or unincorporated place and also near a major road. You want to identify areas on the ground that are in the place boundaries and also near a road, and a spatial query cannot do this for you. What you can do is create a new feature class that shows the boundaries of the buffer and the boundaries of the places together—but there are several different ways you can overlay these two types of polygons. See **Smartbox 9.5** for more information about overlaying polygons.

SMARTBOX 9.5

What are the different types of polygon overlays in ArcGIS Pro?

When two polygons are overlaid, the two layers are combined to create a new, third, layer that has qualities retained from the two polygons used to create it. This new layer combines not only the geospatial qualities and geometry of the two polygons but also their attributes. For an example, see Figure 9.6, where two polygon layers (a set of land parcels and the floodplain) are overlaid to create a new, third layer that has the geographic characteristics of both inputs. However, where a land parcel polygon is split into two polygons (with one on the floodplain and one not on the floodplain), the attribute table is updated to include two separate records: One polygon has the attributes of being on the floodplain (FID_flood = 3), and the other polygon has the attributes of not being on the floodplain (FID_flood = −1).

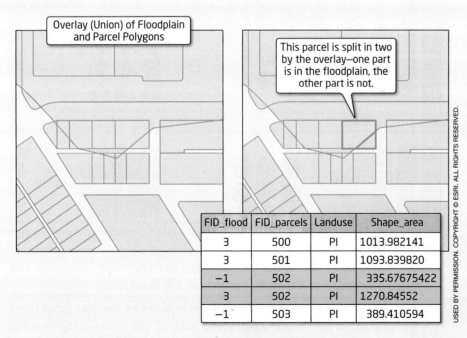

FIGURE 9.6 An overlay of two polygon layers (land parcels and the floodplain) to create a new, third, layer. Note how the attribute table of the new layer combines the two layers.

Overlay operations are used for multiple types of applications. Figure 9.7 shows some of the overlay methods available in ArcGIS Pro:

- **Intersect**: In this type of overlay, only the features that the two input layers have in common are retained in the output layer. If the land parcels and the floodplain are overlaid using an Intersect operation, the

Intersect A type of GIS overlay that retains the features that are common to both layers.

FIGURE 9.7 Six different types of polygon overlay operations—Intersect, Identity, Symmetrical Difference, Union, Erase, and Update.

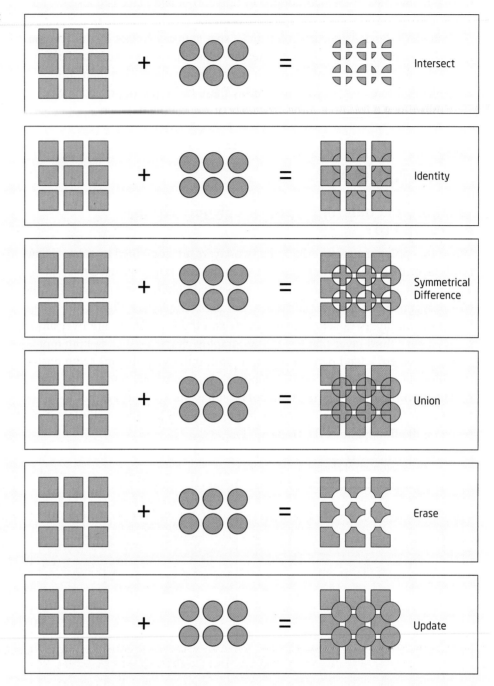

Identity A type of GIS overlay that retains all the features from the first layer along with the features it has in common with the second layer.

new layer shows only areas that have land parcels on the floodplain, not any other parcels or any other sections of the floodplain.

- **Identity**: In this type of overlay, all the features of the first input layer are retained in the output layer, along with all the intersecting features of the second layer. If the land parcels and floodplain are overlaid using an Identity operation, the new layer shows all available land parcels plus the area of the floodplain covering these parcels (but no other parts of the floodplain).

- **Symmetrical Difference**: In this type of overlay, the output layer retains all features of the two input layers except for those areas they have in common. If the land parcels and floodplain are overlaid using a Symmetrical Difference operation, the new layer shows all land parcels and all areas of the floodplain except for those parcels that were on the floodplain.
- **Union**: In this type of overlay, the output layer retains all features of the two layers. If the land parcels and floodplain are overlaid using a Union operation, the new layer shows all land parcels as well as all areas of the floodplain combined together.
- **Erase**: In this type of overlay, the output layer retains all features of the first input layer except for those areas that the first layer has in common with the second input layer. If the land parcels and the floodplain are overlaid using an Erase operation, the new layer shows all land parcels except those on the floodplain.
- **Update**: In this type of overlay, the output layer retains all features of the first input layer with all features of the second input layer placed on top of them, sort of like a "cut and paste" operation. All the features of the first layer that are in common with the second layer are covered over by the second layer. If the land parcels and the floodplain are overlaid using an Update operation, the new layer shows all of the floodplain and any parts of the land parcels not covered by the floodplain.

Symmetrical Difference A type of GIS overlay that retains all the features from both layers except for the features that they have in common.

Union A type of GIS overlay that retains all the features from both layers.

Erase A type of GIS overlay that retains all the features from the first layer except for what they have in common with the second layer.

Update A type of GIS overlay that retains all features from the second layer as well as those features from the first layer that are not in common with the second.

- To show the areas that are both near a road and also within a place boundary, use the Intersect option. Answer Questions 9.4 and 9.5.

> **Question 9.4** Why did you use the Intersect operation and not the Union operation? What would the resulting overlay be if Union were used?

> **Question 9.5** Why did you use the Intersect operation and not the Symmetrical Difference operation? What would the resulting overlay be if Symmetrical Difference were used?

- On the **Analysis** tab, in the **Tools** group, open the **Analysis Gallery** again and then click the **Intersect** button. The Intersect tool opens in the Geoprocessing pane.
- Under Input Features, choose **Fourmajrds_halfmilebuffer**. Another pull-down menu appears, so you can choose the next layer to intersect this with; in this case, choose **Fourplaces**. These are the features that you want to overlay using an Intersect operation.
- Under Output Feature Class, click the **browse** button and navigate to your **C:\GIS\Module9** folder, go to the Module9 geodatabase, and name it **Placesandbuffer**. This is the name of the new layer created from the overlay.

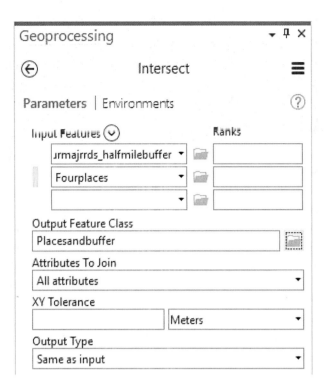

- Under Attributes To Join, choose **All attributes**.
- Under Output Type, choose **Same as input**.
- Leave the other options alone and run the Intersect tool. The Placesandbuffer layer is added to the Contents pane. Turn off the Fourplaces and Fourmajrds_halfmilebuffer layers to see only the new overlay.
- Use the Select Layer By Location tool to determine which of the schools in the Fourschools layer are within the new Placesandbuffer layer. Answer Question 9.6.

Question 9.6 How many schools are within both half a mile of a major road and also within the boundaries of a place?

- In the Fourschools attribute table, you can switch the selection to select the records (see Module 2) that did not meet the selection criteria. Do so and answer Question 9.7.

Question 9.7 How many schools do not meet the selection criteria of Question 9.6?

- Export these points (for the schools that do not fall within the Placesandbuffer layer) to a new feature class in your Module9 geodatabase called **Fourschoolsoutside**.
- Examine the attribute table of the Fourschoolsoutside layer. Answer Question 9.8.

> **Question 9.8**
>
> Of the schools not near a road and not within a place, how many of them are considered to be affiliated with the city of Youngstown? How many of them are considered to be affiliated with the city of Warren? (*Hint:* You might want to use Select Layer By Attribute queries on the Fourschools-outside layer to help determine the answers to these questions.)

- Clear your selected features.

Step 9.8 Performing Point-in-Polygon Overlay

- For the last step of analysis, for those schools that are near a major road and within the boundary of a place, you want to identify which place they are in. You can do this by overlaying the points of the Fourschools layer with the polygon boundaries of the Placesandbuffer layer as a point-in-polygon overlay. (See **Smartbox 9.6** for more information.)

SMARTBOX 9.6

How is a point-in-polygon overlay performed?

Overlay operations can be performed with more than just polygons. As its name implies, a **point-in-polygon overlay** operation overlays points onto polygons. The end result is a new point layer, in which each point contains the attributes of the polygon that it was overlaid in (a process similar to a spatial join; see Module 8). Figure 9.8 shows an example of a point-in-polygon overlay. The figure shows a set of polygons representing land parcels in one layer and a set

point-in-polygon overlay A type of GIS overlay that results in a new point layer where the points have the attributes of the polygons within which they are.

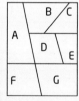 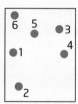 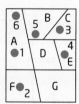

FIGURE 9.8 The layers and output of a point-in-polygon overlay operation.

Land Parcel	Owner
A	Smith
B	Jones
C	Roberts
D	Hill
E	James
F	Young
G	Cross

Parcel polygons layer

Wells
1
2
3
4
5
6

Wells points layer

Wells	Owner
1	Smith
2	Young
3	Roberts
4	James
5	Jones
6	Smith

Points with polygon attributes layer

of points representing water wells in a second layer. You want to determine which wells are on which parcels of land. Because you want to see where these two layers intersect, you can perform the point-in-polygon overlay by using the Intersect tool. Like the Intersect overlay operation, it determines which points and which polygons are in common with one another.

The output layer is a new point layer in which each point now also has the attributes of its corresponding polygon parcel. For instance, in Figure 9.8, point 2 is in polygon F. In the resultant new point layer, point 2 will have the attributes of polygon F (that is, the record for point 2 will now contain information saying that it is in a parcel owned by Young).

A similar operation is a **line-on-polygon overlay**, which overlays lines onto polygons. The result of this operation is a new line layer in which each line is assigned the attributes of the polygon through which it passes. If a single line were to pass through three polygons, that line would be split into three separate lines (and, thus, the attribute table would have three records instead of one), and each of these segmented lines would have the attributes of the corresponding polygon through which it passed.

line-on-polygon overlay
A type of GIS overlay that results in a new line layer where the lines have the attributes of the polygons within which they are.

- Out of your available geoprocessing operations, an Intersect operation will do the job, so on the **Analysis** tab, in the **Tools** group, open the **Analysis Gallery** again and then click the **Intersect** button. The Intersect tool opens in the Geoprocessing pane.

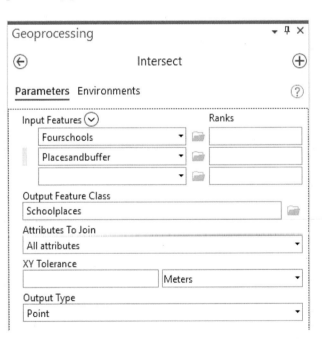

- Under Input Features, choose the **Fourschools** points layer. Another pull-down menu appears, allowing you to choose the next layer to intersect this with; in this case, choose the **Placesandbuffer** polygon layer.

- Under Output Feature Class, click the **browse** button and navigate to your **C:\GIS\Module9** folder, go to the **Module9** geodatabase, and name it **Schoolplaces**.

- Under Attributes To Join, choose **All attributes**.
- Under Output Type, choose **Point**. This indicates to ArcGIS Pro that because you are intersecting points and polygons, your output will be a new point layer (and thus you are performing a point-in-polygon overlay).
- Leave the other options alone and run the Intersect tool. The Schoolplaces layer is added to the Contents pane. Turn off the Fourschools and Fourschoolsoutside layers to see only the new overlay. Using the information from the Schoolplaces layer, answer Questions 9.9, 9.10, 9.11, and 9.12. (*Hint:* You may want to use Select Layer By Attribute queries to help find the answers.)

Question 9.9 In which incorporated or unincorporated place is Cardinal Mooney High School located?

Question 9.10 In which incorporated or unincorporated place is John F. Kennedy High School located?

Question 9.11 In which incorporated or unincorporated place is Saints John and Paul Elementary located?

Question 9.12 Which three schools are considered to be in Edgewood?

Step 9.9 Printing or Sharing Your Results

- Save your Module9 project.
- Turn off all layers except the Placesandbuffer and the Schoolplaces layers and change the symbology so that the overlaid areas and the points can be clearly seen.
- Finally, either print a layout (see Module 3) of your final version of your layers (including all the usual map elements and design for a layout) or share your results as a web map through ArcGIS Online (see Module 4).

Closing Time

Geoprocessing operations are common and versatile methods for performing spatial analysis in GIS. Buffers, overlay operations, merge, and dissolve operations are all highly useful. In ArcGIS Pro, geoprocessing tools can be incorporated into more complex projects. You can, for example, chain together several commands and have the various options and settings for the tools already established. See *Related Concepts for Module 9* for information on how tasks can be used to work with these actions in ArcGIS Pro.

You'll continue using geoprocessing tools throughout future modules. In the next module, you'll keep working with other GIS applications, such as turning a table of addresses into a layer of geospatial data. Once you've created GIS data through this geocoding process, you can use spatial analysis techniques to examine the data.

RELATED CONCEPTS FOR MODULE 9

Workflows and Tasks in ArcGIS Pro

The various steps that you went through in this module—from applying the environment settings, to merging the two places layers, to interactively selecting the various counties, all the way through the buffers and Select Layer By Location analysis—were a specific set of actions done in a specific order. This series of steps to complete an overall operation is referred to as a **workflow** in ArcGIS Pro. In order to complete the workflow for this module, you had to do a number of different things to finish the analysis for the four-county area of Region #5. However, from the map shown in the *Study Area* section on page 217, you can see that there are 16 different education regions, and if you were to repeat all of this module's actions as a workflow for the other 15 regions, it would be a very repetitive process with plenty of room for error; accidentally using the wrong selection option, typing the wrong number for the buffer size, or setting a tool option incorrectly could result in errors or the wrong analysis of the data.

workflow A sequence of actions performed using GIS layers and tools.

task A preset interactive series of actions used to establish the steps, settings, and tools in a workflow.

task item An overall container that can hold many tasks and can also be exported to a task file and imported to another project.

task group A collection of tasks inside a task item.

Fortunately, ArcGIS Pro gives you a set of tools called **tasks** that allow you to set up a workflow once, and then you can execute all the steps in that workflow in the same way each time you work with it. When you convert all the workflow steps into a task, you can be confident that when you do that same analysis on each of the other 15 regions, you're not missing a step or a setting, and you know you're carrying out the same actions in the same way each time. Tasks allow for uniformity when doing the same workflow over and over. They also allow you to carry out these workflows more quickly, as ArcGIS Pro knows what needs to be done each time. By creating a task, you can turn the execution of the workflow over to someone else, and that person will be able to carry out the workflow in the same way that you would. Also, a task is more than an automated process; unlike a model (see Module 20), a task can be designed to be interactive at each step.

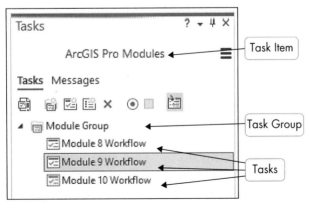

FIGURE 9.9 The hierarchy of the task item, task group, and tasks.

In ArcGIS Pro, the first thing that is created is a **task item**, which you can think of as an overall container for holding all the tasks that you want to work with in a project. A task item can hold several different tasks within it. Inside the task item, similar tasks can be organized together within a **task group**. Figure 9.9 shows this setup. The heading ArcGIS Pro Modules is the task item, and it contains one task group called Module Group, which in turn contains multiple tasks (Module8 Workflow, Module9 Workflow, and Module10 Workflow). A task item and everything in it is saved as part of a project in ArcGIS Pro. A task item may be exported to become a task file, and then the task file may be opened in another project, so that you can easily move tasks from project to project.

A task consists of a series of **steps**, which are the actions that ArcGIS Pro takes when the task is run. Each step consists of running a single ArcGIS Pro command or geoprocessing tool that is part of the workflow. Figure 9.10 shows an example of a task comprised of a set of steps; each of these is a separate step you took in this module's workflow.

When a task is created, at each step you can select what tool or command ArcGIS will open for you (such as the Environments dialog box or the Merge tool in the Geoprocessing pane), and you can also set up any specific parameters. For instance, when setting up step 9 (Buffer Roads) in Figure 9.10, you would be able to preset the half-mile radius of the buffer, as well as the other settings of the tool. Thus, when someone runs the task and reaches that step, all of those settings and options will already be in place for the user.

step A single action (either a command or geoprocessing tool) that is executed when running a task.

Task Designer The tools used to edit a task item, task group, task, or step.

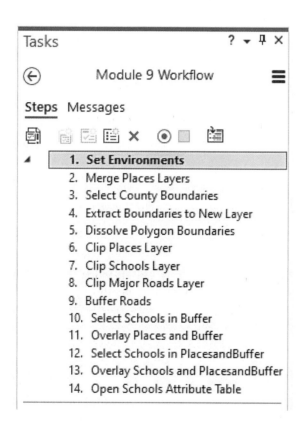

FIGURE 9.10 A task containing a set of 14 steps. When the task is run, these steps are all activated, in order.

Each step of the task is set up individually, using the **Task Designer**, where you can make adjustments to each step. (Note that the Task Designer is also used to make edits to task items, task groups, and tasks themselves.) In the Task Designer pane, you can add any additional information for the user and instructions about what will happen in this step, and you can choose what happens when this step is reached (see Figure 9.11). You can choose whether the user needs to manually advance to the next step, whether the step will run and then proceed onward, or whether the step runs and advances without input from the user. When designing steps, you can input the parameters manually, but ArcGIS Pro also gives you the option to "record" the process as you design a

FIGURE 9.11 An example of the settings used in the Task Designer when editing a step in a task.

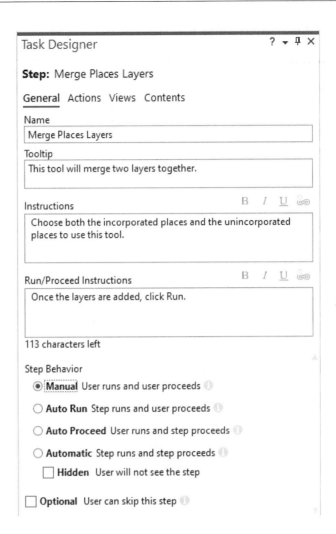

step, so that you are free to go through the process of the workflow, and ArcGIS Pro automatically translates your actions in the workflow into the various steps of the task.

Key Terms

geoprocessing (p. 215)
overlay (p. 215)
environment settings (p. 218)
Merge (p. 220)
Append (p. 220)
Dissolve (p. 223)
Clip (p. 225)
buffer (p. 226)

Intersect (p. 229)
Identity (p. 230)
Symmetrical Difference (p. 231)
Union (p. 231)
Erase (p. 231)
Update (p. 231)
point-in-polygon overlay (p. 233)
line-on-polygon overlay (p. 234)

workflow (p. 236)
task (p. 236)
task item (p. 236)
task group (p. 236)
step (p. 237)
Task Designer (p. 237)

10

How to Perform Geocoding in ArcGIS Pro

ArcGIS Pro Skills

In this module, you will learn how to do the following in ArcGIS Pro:
- Construct and use an address locator.
- Perform geocoding by matching street addresses from a table to their locations on a road layer.
- Perform interactive rematching.
- Fix address information in the geocoding process.
- Use geocoded points in basic spatial analysis.

Learning Outcomes

After studying this module, you should be able to:
- Define the term *geocoding*.
- Describe the necessary components in a reference database to perform geocoding.
- Explain what an address locator is and why it is important for geocoding.
- Explain how address parsing and grammar are used in geocoding.
- Describe the steps involved in address matching in GIS.
- Explain at least two reasons geocoded results may not be correct.

Introduction

A common application of geospatial technologies is locating an address on a map. When you're using a service like MapQuest or Google Maps, you type in an address. The service converts that string of letters and numbers into a location plotted on a map. This matching of an address with its corresponding geospatial location is called **geocoding**. In ArcGIS Pro, you can easily plot a single address on a map, much as you would in Google Maps. ArcGIS Pro also allows you to perform **batch geocoding**, in which multiple addresses are plotted at the same time and then mapped as a point feature class (which can then be used for further analysis).

The key to geocoding is matching addresses to good reference data. With street addresses, the **reference data** take the form of a geospatial data layer representing roads or streets. In GIS, a road is not represented as one long line; for example, a road like Fifth Avenue in New York City does not consist of one line/object and a single record in an attribute table. Instead, it is split into multiple line **segments** (likely at each intersection with another street or road), and each segment has different attributes. Thus, a road like Fifth Avenue might be represented with hundreds of different records in the line layer. See Figure 10.1 for another example: In Mahoning County, Ohio, Mahoning Avenue is a major road that runs across the county. In the roads layer, however, Mahoning Avenue is represented not as 1 line but as 173 segments.

When a road is represented as a series of segments, each segment can have different attributes assigned to it, such as the name of the road, the range of addresses on the left and right sides of the road, the zip codes on the left and right sides of

geocoding The process of using the numbers and letters of an address to plot a point at the appropriate location.

batch geocoding The process of matching a group of addresses together at once.

reference data The base street layer used as a source for geocoding.

segment A single part of a line that corresponds to one portion of a street.

the road, the type of road, and the speed limit for each stretch or the measured length of the road. Separating sections of a road into smaller bits, each with a narrow range of addresses assigned to it, enables the GIS to use more information when attempting to match an address.

The result of a geocoding process is a new set of points representing the matched locations of each address found using this kind of reference data. Because geocoding has converted a set of addresses in a table into geospatial data, each point can also be assigned coordinates (such as the latitude and longitude) that represent the location, and the point layer can be used alongside your other GIS layers. Several steps occur in the geocoding process, and in this module, you'll see how they work in ArcGIS Pro.

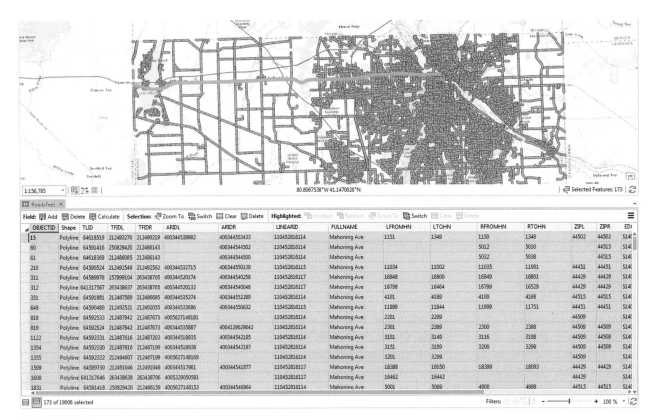

FIGURE 10.1 Mahoning Avenue shown selected in ArcGIS Pro—one road, but 173 segments.

Module Scenario and Applications

This module puts you in the role of a county librarian, working to coordinate several library services, such as deliveries and community outreach. Before you can begin coordinating these services, you need to have a GIS layer of the libraries' locations. Starting with an Excel file of library names and addresses, you'll create a new GIS layer of the libraries' locations and perform some initial spatial analysis of the libraries in relation to the county's main roads.

The following are additional examples of real-world applications of this module's skills:

- An operator at a 911 call center needs to quickly determine a caller's address so that she can convey this information to emergency services. She would use geocoding to determine the location from which the call was placed.

- A marketing department for a retail store is putting together a targeted advertising campaign for certain customers. The marketing department can use geocoding to create a map of where these customers live and can then examine the demographics of these neighborhoods.
- A police detective is investigating a series of break-ins at houses. He can use geocoding to plot the locations of the break-ins as a starting point for analyzing their locations and determining possible patterns.

Study Area

- For this module, you will be working with data from Mahoning County, Ohio.

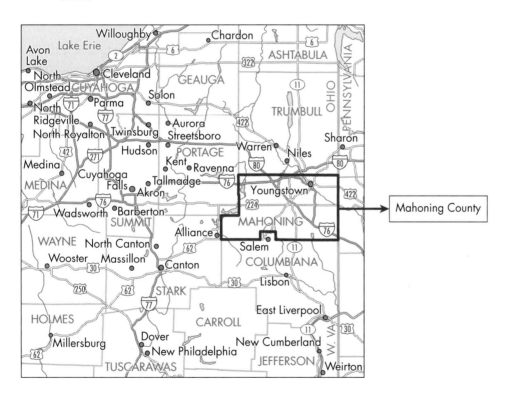

Data Sources and Localizing This Module

The data in this module are related to library addresses and roads in Mahoning County, Ohio. However, you can easily modify this module to use the streets and libraries of your own county. The address information for the county libraries comes from the Mahoning County Public Library's website, at **https://www.libraryvisit.org**. Using your own county's library website (or the phone book), you can put together a Microsoft Excel file of library names and addresses (for instance, the addresses of libraries in Johnson County, Iowa, are online at **https://www.icpl.org/public-libraries-johnson-county**). If you do not have a cluster of libraries in your area, use the addresses of something else like high schools, gas stations, coffee shops, or drugstores.

The roads layer used in this module is a TIGER file downloaded from the U.S. Census Bureau, and the same kind of data are available for all U.S. counties. For example, if you were examining roads in Johnson County, Iowa, you would download the TIGER file for that particular county. TIGER files are available for each county for free download from **https://www2.census.gov/geo/tiger/TIGER2018/ADDRFEAT/**. From this website, you can download the shapefile that corresponds with your county. The TIGER files are organized by FIPS code, and for Mahoning County, Ohio, the downloaded shapefile was the one with the filename ending in 39099. Data for each county in the United States is available on the website by FIPS code. For instance, Johnson County, Iowa, would be the file whose name ends in 19193. A full list of FIPS codes for each state and county is available from the National Weather Service, at **https://www.weather.gov/pimar/FIPSCodes**. This TIGER file was also projected into the State Plane Coordinate System using US Feet (for Mahoning County, this was in Ohio North).

Step 10.1 Getting Started

- This module's hands-on applications use the data folder called Module10. Your instructor will be able to supply you with this data, or you can download it directly from this book's website at **https://www.macmillanlearning.com/college/us/product/Discovering-GIS-and-ArcGIS-Pro/p/131923075X**. The text in this module assumes that you have this Module10 folder in a computer location referenced as C:\GIS; if you have it somewhere else (for instance, in a flash drive referenced as G:\GISClass), substitute that location and path to the Module10 folder throughout this module.

- The Module10 folder contains the following:
  - libraries.xlsx: an Excel file containing address information for Mahoning County libraries
  - Geocode: a file geodatabase that contains a line feature class called RoadsFeet, with distances measured in feet

- Start ArcGIS Pro.

- Sign in with your Esri account username and password.

- Create a new project using the **Map** template. Call this project **Module10** and place it in your **C:\GIS\Module10** folder. Ensure that there is not a checkmark in the box next to **Create a new folder for this project**.

- When ArcGIS Pro opens, change the map's name to **Mahoning**.

- Add the RoadsFeet feature class to the map.

- The RoadsFeet layer was projected from its initial coordinate system to another one for use in this module. The coordinate system used in this module is **NAD 83 Ohio North State Plane**, with **feet** for the map units.

Check the Mahoning map's properties under the Coordinate Systems heading to verify that this coordinate system is being used. Also, in the Mahoning map's properties, under the **General** tab, change Display Units to **Feet**. By making these changes, you ensure that any measurements you make will be done in feet.

- Add the libraries.xlsx table to the Contents pane. (See Module 2 for more about adding a table to the Contents pane.) To do so, you need to first expand the table in the Catalog pane and then add the file called **Sheet1$**.

Step 10.2 Using a Streets or Roads Layer for Geocoding

- Open the attribute table for RoadsFeet and answer Question 10.1. Each of the records represents a segment of a road, not a complete road. Scroll across the attribute table, and you see several attributes assigned to each of the segments that will be used in the geocoding process. This roads layer will serve as your reference data for the geocoding process in this module. See **Smartbox 10.1** for more about using road layers in the geocoding process.

> How many records (segments) are in the RoadsFeet layer? **Question 10.1**

SMARTBOX 10.1

How is a streets layer used as a reference database for geocoding in ArcGIS Pro?

A geospatial layer that represents streets or roads divides a street into a series of segments, each containing attributes that help in identifying the address-related information for a particular segment. The exact attributes and their names vary depending on the source of the streets layer, but the following are some common attributes:

- The name of the street
- The range of address numbers on each side of the street (usually separated into four categories—the starting and ending addresses on the left side and the starting and ending addresses on the right side)
- The state and city of the segment
- The zip code of the segment (and sometimes the zip code on the left side of the street and the zip code on the right side of the street)

Additional attributes that may sometimes be found, including the following, add further detail for each segment:

- The prefix of the street (such as North or West)
- The suffix of the street (such as NW or East)
- The type of street (such as Avenue, Blvd, or Road)

street centerline A file containing line segments representing roads.

With this level of detail assigned to each segment, the streets layer can serve as reference data for the geocoding process, which will search for a particular segment to which to match an address.

These types of street layers are often referred to as **street centerline** files. They are line layers used in GIS, created to model each street as a series of line segments. Street centerline files are common forms of GIS data and are available from a variety of sources. Because they are used for geocoding, companies may have their own proprietary street centerline files commercially available, but there are several free alternatives as well. For instance, the roads file you're using in this module is from the U.S. Census **TIGER/Line** files. (See Module 2 for more about U.S. Census GIS data.) TIGER (Topologically Integrated Geographic Encoding and Referencing) files for streets are line layers that model streets as segments with attributes similar to those described above; they are sources of reference data for geocoding. TIGER/Line files are created by the U.S. Census Bureau and are freely available online at **https://www.census.gov/geographies/mapping-files/time-series/geo/tiger-line-file.html**. Also, your local county or state GIS division may have street centerline files available for you to download or use.

TIGER/Line A file produced by the U.S. Census Bureau that contains (among other things) the line segments that correspond with roads in the United States.

- In ArcGIS Pro, open the libraries.xlsx table (use **$Sheet1**). You should see several fields containing information about the address of each library. In the next step, you'll create an address locator to help match up the data in this table with the proper locations, using the reference data.

Step 10.3 Using an Address Locator

- In the libraries table, you'll see 15 addresses, with their information broken into separate fields called Name, Address, State, City, and Zip. (You may have to expand the far-right side of the table to see all attributes and the full five-digit zip code.) Close the table when you're done examining it.

- The first step in geocoding is to set up an address locator to use for matching. See **Smartbox 10.2** for more about what an address locator is and how it operates.

SMARTBOX 10.2

address locator ArcGIS Pro functionality that specifies the reference data, geocoding style, and necessary attributes for geocoding.

What is an address locator and how does it work in ArcGIS Pro?

An **address locator** is the key component of geocoding in ArcGIS Pro; it establishes the parameters for how geocoding will be performed. An address locator includes information on what layer is used as the reference data, what address locator style is used, and which specific attributes from the reference data layer are used (and how they will be used) during the geocoding process. In other words, creating an address locator is a "preprocessing" step you must undertake before you can perform the geocoding. Once you have created a single address locator, it will be established, and you will not need to re-create it if you perform further address matching operations.

When creating an address locator, you need to select which address locator style to use for matching addresses. An **address locator style** determines the criteria that are used for matching an address to a segment in the reference data (and thus also determines what attributes need to be present in the reference data for use in geocoding). An example of an address locator style is the *Dual Ranges* style (which you will be using in this module). Dual Ranges is used to find the location of a house on a specific side of a street. To function properly, it needs the following attributes to be present for the segments in the reference data: the name of the street, the low value for the addresses on the left side of the street, the high value for the addresses on the left side of the street, the low value for the addresses on the right side of the street, the high value for the addresses on the right side of the street, the zip code on the left side of the street, and the zip code on the right side of the street. If your reference data don't have these attributes, then Dual Ranges cannot be used.

Several other address locator styles are available, as there may be times when you don't need to match an address to a particular house on a street. Retail sales analysts, for example, may target marketing or mailings to customers in a particular zip code; thus, they may need to determine the location of customers only by zip code instead of house address. The *US Address - Street Name* style requires only the name of the street, the city, and the state—but the geocoding matches only to a particular street, not to a particular address on that street. Similarly, the *US Addresses - Zip 5-Digit* style requires only the zip code of an address—but it geocodes points only to a zip code, not to a street address. In short, your choice of address locator style determines not only the kind of output data that will be generated by the geocoding process but also the kind of attributes needed in the reference data.

In addition, sometimes a street is known by several different names; for instance, in a town, a road called Main Street may also be known as US Route 224 and also as Boardman-Canfield Road. You can create an address locator that uses an alternate street name table, which will allow you to find an address based on either the primary reference data or the alternative table that contains different names.

address locator style The format that will be used for geocoding addresses.

- To create your own address locator to use for geocoding, open the Geoprocessing pane (by selecting the **Analysis** tab and then the **Geoprocessing** group and then clicking the **Tools** button). The tool you want to use is Create Address Locator, and you can either search for it or find it in the **Geocoding** toolbox. The tool opens in the Geoprocessing pane.

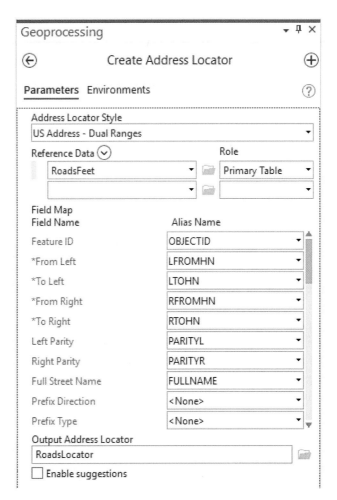

- Under Address Locator Style, select **US Address - Dual Ranges**.
- Under Reference Data, choose **RoadsFeet**.
- Under Role, choose **Primary Table**.
- In the Field Map pull-down menus, specify the specific attributes in the RoadsFeet layer that you will use for performing the geocoding. ArcGIS Pro fills in some default options, and you need to manually fill in the following:
  - For From Left, choose **LFROMHN** as the starting address range on the left side of the street for each segment.
  - For To Left, choose **LTOHN** as the ending address range on the left side of the street for each segment.
  - For From Right, choose **RFROMHN** as the starting address range on the right side of the street for each segment.
  - For To Left, choose **RTOHN** as the ending address range on the right side of the street for each segment.
  - For Full Street Name, choose **FULLNAME** as the name assigned to each street segment.
  - For Street Name, also choose **FULLNAME** as the required street name for each segment.
  - For Left Zip Code, choose **ZIPL** as the zip code the left side of the street for each segment.

- For Right Zip Code, choose **ZIPR** as the zip code the right side of the street for each segment.
- Under Output Address Locator, click the **browse** button and navigate to your **C:\GIS\Module10** folder and name the locator **RoadsLocator**.
- Run the Create Address Locator tool. Return to the Catalog pane and expand the option for **Locators**. In addition to some default Esri locators, you see your RoadsLocator now created and available in this project.
- You can also make some adjustments to the settings for the locator at this point. In the Catalog pane, expand the options under **Locators**. Then right-click on **RoadsLocator** and choose **Locator Properties**.

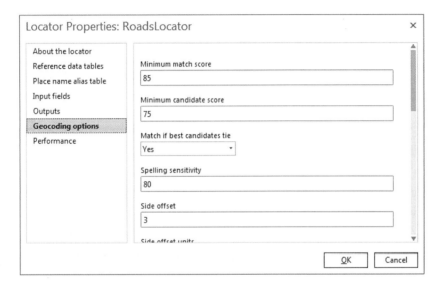

- In the Locator Properties dialog box, click on the **Geocoding options** tab. If the following options are different from what you see on your screen, change your dialog to match these values:
  - Minimum match score is **85**.
  - Minimum candidate score is **75**.
  - Spelling sensitivity is **80**.

  These three settings are used to adjust the minimum ranking for an address to be matched, chosen as a possible matching candidate, and also how ArcGIS Pro accounts for possible spelling errors.
- Click on the **Outputs** tab. For both the Write X and Y coordinates and Write DisplayX and DisplayY coordinates, choose **Yes**. With these options chosen, ArcGIS Pro creates new fields with the coordinates of the geocoded points.
- Leave the other options set at their defaults and click **OK**.

Step 10.4 The Geocoding Process

- Now you can begin geocoding the addresses in your libraries.xlsx file. There are two ways to initiate the process:
  - In the Contents pane, right-click on the table you want to geocode (in this case, **Sheet1$**) and choose **Geocode Table**.

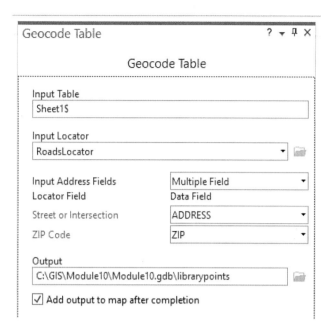

- In the Contents pane, click once on the table you want to geocode (in this case, **Sheet1$**) and on the **Data** tab, in the **Geocoding** group, click the **Geocode Table** button.
- The Geocode Table pane opens. At the bottom of the pane, click on **Go to Tool** to get started.
- For Input Table, enter **Sheet1$**.
- Under Input Locator, choose **RoadsLocator**.
- For Input Address Fields, choose **Multiple Field**.
- For Street or Intersection, choose **ADDRESS**.
- For Zip Code, choose **ZIP**. The Street or Intersection and Zip Code options reflect the names of the fields in the Excel table to match the items used by the address locator. (If they were called something different in the Excel table, you'd have to choose different field names for Street or Intersection and also Zip Code at this point.)
- Under Output, click the **browse** button and navigate to your **C:\GIS\Module10**, go to the **Module10** geodatabase, and name the geocoded output **librarypoints**.
- When all the settings are correct, run the Geocode Table tool.
- You receive a report indicating how many of the 15 library addresses were matched, how many were tied, and how many were unmatched. You should see that 13 of the libraries were matched but 2 were not. A lot happened behind the scenes to geocode these addresses according to the settings you gave ArcGIS Pro. For more information on what happens during the geocoding process, see **Smartbox 10.3**.

SMARTBOX 10.3

How does the geocoding process work in ArcGIS Pro?

The geocoding process involves converting the string of letters and numbers in an address into a point matching a location referenced by a line segment. How? First, the addresses are parsed into the various components that make up the address. **Parsing** means breaking a string of numbers and characters (such as an address) into smaller components, separating the street number, the street prefix, and the street name into individual items. Once an address is parsed, its pieces are matched with the segments in the reference data, as indicated by the address locator style. See Table 10.1 for an example of parsing. The second column of the table shows the street address, and the columns to its right show how pieces of the address are parsed into separate headings.

Once an address is parsed, multiple representations of the components of the address are assigned. This is necessary because there are often several

parsing Breaking an address into its component parts.

Location	Address	Prefix Direction	Number	Street Name	Street Type	Suffix Direction
White House	1600 Pennsylvania Avenue NW		1600	Pennsylvania	Avenue	NW
Smithsonian Air and Space Museum	600 Independence Ave SW		600	Independence	Ave	SW
U.S. Government Printing Office	732 North Capitol St NW	N	732	Capitol	St	NW

Table 10.1 Three Washington, D.C., addresses parsed into their component pieces for geocoding

ways of representing some of the pieces of an address. For instance, in a street address, "avenue" could be written out as "avenue" or shortened to "ave" or "av," but the geocoding should be able to recognize all three of these representations. As Table 10.1 shows, the White House is on Pennsylvania Avenue, and the Smithsonian Air and Space Museum is on Independence Ave. Both "Avenue" and "Ave" are used in writing these addresses, and the geocoding should recognize that both of these strings of characters mean the same thing. Similarly, the address of the U.S. Government Printing Office could be written as "North Capitol" street or "N Capitol" street, indicating the north section of Capitol Street, or the street itself could simply be called "North Capitol" and not indicate a prefix direction. The geocoding process will thus set up multiple representations of this address to cover these conditions.

Next, the segments are searched according to the address locator style, and the various attributes are used to find the best options for segments to match the address. For instance (using the Dual Ranges address locator style), if the reference data used to match the address for the White House contain a segment with its name equal to "Pennsylvania Ave," its address range on one side of the street from 1522 to 1608, and its address range on the other side of the street from 1523 to 1609, that segment would likely be one of the best matches, but there will likely be some other potential candidates as well.

ArcGIS Pro assigns a ranking of 0 to 100 to each potential candidate segment; higher numbers indicate a greater likelihood that the particular segment is the correct match. ArcGIS Pro then selects the best candidate and chooses that segment for the matching. Before beginning geocoding in the Locator Properties dialog box, you can adjust the settings for the minimum score that a candidate should have to be considered a match, as well as the minimum acceptable score that a candidate should have for an address to be matched to it.

Finally, ArcGIS Pro plots a point corresponding to the address somewhere along the selected segment. If using address ranges (as with the Dual Ranges style), ArcGIS Pro uses the ranges of the addresses to determine the side of the street (right or left) on which the geocoded point ends up, as well as how far along the segment the point should be placed. For instance, if an address is at 150 Shady Street, and the range of addresses for a Shady Street segment

linear interpolation A method used in geocoding to place an address location among a range of addresses.

extends from 100 to 198, the point for 150 will be placed at about the halfway point of the segment. If the address is for 120 Shady Street, it will be placed about one-fifth of the way from the low end of the segment. This method for placing a point at an approximate distance along a segment is referred to as **linear interpolation**. See Figure 10.2 for an example of plotting a geocoded point in ArcGIS Pro using linear interpolation.

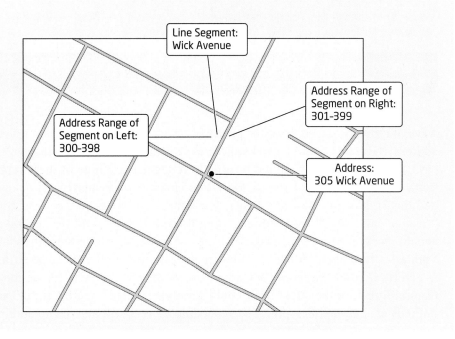

FIGURE 10.2 Plotting a geocoded point for an address in ArcGIS Pro.

Step 10.5 Rematching Addresses by Editing

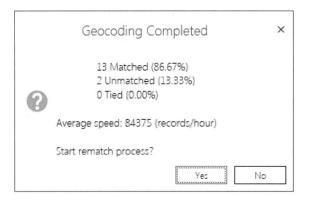

- Two of your library addresses did not match. To attempt to rematch them (and to see what was wrong and figure out why they did not match the first time around), click on **Yes** in the Geocoding Completed dialog box. This brings up the Rematch Addresses pane. If you closed the box and need to get back to Rematch Addresses, see **Troublebox 10.1** for assistance.

TROUBLEBOX 10.1

How can I return to the Rematch Addresses pane in ArcGIS Pro?

Regardless of whether you answer ArcGIS Pro's "Start rematch process?" question by clicking Yes or No, you can always get back to rematching addresses. For instance, if you later want to return and rematch an address, you can easily access the Rematch Addresses pane for a layer. In the Contents pane, right-click on the layer for which you want to rematch addresses (for instance, the librarypoints layer), choose **Data**, and then choose **Rematch Addresses**.

- In the Rematch Addresses pane, click on **Unmatched**. You see that you have two records that were not matched in the process.
- The first library (listed as 1 of 2) that was unmatched (Greenford Library) is located at address 7441 W South Range Road. This address just has a minor problem that can be fixed with some simple editing. In the Zip Code box, delete the value 44422 and instead type in **44460** and press the **Enter** key on the keyboard.

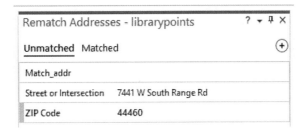

- Click **Apply**. Several new options appear as potential matching candidates for this address. Answer Question 10.2.

> **Question 10.2** How many candidates are listed as potential matches for the Greenford address?

- At the bottom of the Rematch Addresses pane you have some potential candidates to choose from. You also see the locations of the candidates shown on the map (with Candidate A shown on the map at a point labeled A and so on). In the Rematch Addresses pane, select the candidate with the highest score (which should be the candidate at the top of the list), and it is highlighted in cyan both in the pane and on the map.

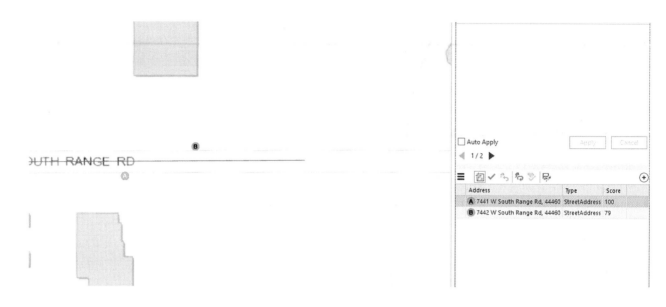

- After selecting the candidate you want to match, in the Rematch Addresses pane, click the **Match** button (the green checkmark). You'll see that address is now matched to that point's location. Answer Question 10.3.

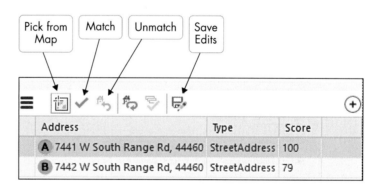

Question 10.3 Why did changing the zip code from 44422 to 44460 allow this address to be rematched, when 44422 is the correct, listed, address for the library? (*Hint:* The problem is not with the original address or the geocoding process itself. Carefully check the attribute table of the reference data layer to determine the answer to this question.)

Step 10.6 Rematching Addresses by Picking Addresses from the Map

- In the Rematch Addresses pane, you see that you have one more point to geocode. ArcGIS Pro shows you on the map the point to which it wants to match that address and also shows that its best choice for matching is the address 306 Wick Ave. However, the Main Library (at 305 Wick Ave) is actually on the other side of the street, and just making edits to the address won't change anything about how ArcGIS Pro is trying to geocode this point. What you're going to have to do is select the proper location interactively from the map and, in essence, tell ArcGIS Pro "put the point at this location." ArcGIS Pro refers to this process as *Pick address from map*, and this is how you can geocode the second unmatched library. When picking an address from a map, you need to find its specific location. The spot that you click on becomes the matched location. Thus, you need to make sure you know where Main Library is located. To get some context for choosing the location, select the **Map** tab and then the **Layer** group and click the **Basemap** button and add the **Imagery** basemap.

- Pan the map around to see the unmatched point and the area across the street from it. The building on the right-hand side of the Wick Avenue street is the Main Library, so you'll want to move the point to match up with it. Before proceeding, though, answer Question 10.4.

> **Question 10.4** Where will you place the geocoded point for the Main library? Why?

- Return to the Rematch Addresses pane and click the **Pick from Map** button.
- Click on the Mahoning map view where you want to place the geocoded point. A point appears, shown selected in cyan, and on the Rematch Addresses pane, a second candidate with x/y coordinates is available for geocoding. ArcGIS Pro is now giving you two options: either geocode the address to the point you just clicked or geocode it to the x/y coordinates.

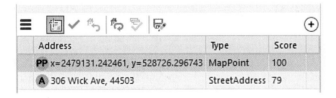

- In the Rematch Addresses pane, make sure the new point you picked is selected and click the **Match** button. The point is matched, and you see that you now have 15 matched addresses.
- At this point, click the **Save Edits** button in the Rematch Addresses pane. If ArcGIS Pro asks whether you want to save edits, click **Yes**.

Step 10.7 Examining Geocoded Results

- The fact that a result is returned to you as "Matched" doesn't necessarily mean that its point has been placed in absolutely the correct geocoded location. Points are sometimes placed in a location that's not correct yet may be flagged as "matched" by a geocoding process (and not just in ArcGIS Pro). For more information about why geocoding results may contain inaccuracies, see **Smartbox 10.4**.

SMARTBOX 10.4

Why are geocoded results sometimes inaccurate in ArcGIS Pro?

Geocoded results are only as good as the reference data used for performing the geocoding. If the reference data are out of date (for instance, if they don't contain streets that correspond to the new subdivision to which you're trying to match an address), you may end up with strange results as a point is placed on a different street with a similar name. Similarly, if the reference data contain errors or missing data in address ranges, zip codes, or other essential attributes, the accuracy of the geocoded results may be compromised. Sometimes, when a street address cannot be located, the center of its zip code is used as the next best source for matching; in such a case, the point is "matched" not to a house address but to the zip code.

Another source of inaccuracy in geocoded results is linear interpolation, which uses the range of addresses as a guide for where to place the geocoded point along the segment. Linear interpolation results in an approximation of where the point should go, and it adheres to the values in the reference data's address ranges, not to the distribution of houses in real life. For instance, in the reference data, the segment for Shady Street may have values between 100 and 148, and a house at 126 Shady Street would have its geocoded point placed at about the middle of the segment. However, if the actual position of 126 Shady Street in the real world were closer to the end of the street, the position of the geocoded point would be significantly off. If the address range for the same segment stretched from 100 to 198, the geocoded position of 126 Shady Street would be placed about one-quarter of the way down the segment, with the geocoded address shown even farther off from its real-world position (Figure 10.3).

FIGURE 10.3 The location of 126 Shady Street in the real world and its geocoded position (via linear interpolation) using two different sets of reference data.

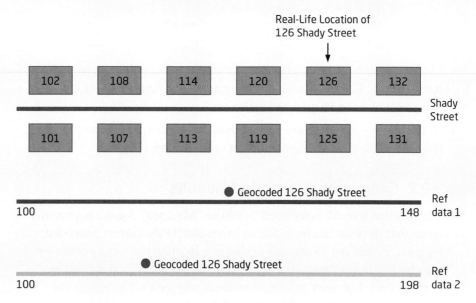

- In the Rematch Addresses pane, you see the first matched address with its information, and you also see its corresponding matched point shown on the Mahoning map view. As a way of checking your geocoded results, you can compare the point that you geocoded (using the Dual Ranges address locator style and the free TIGER/Line file) against the results generated by a geocoding service. (See *Related Concepts for Module 10* for more about using geocoding services in ArcGIS Pro.) To check your results, select the **Map** tab and then the **Inquiry** group and click the **Locate** button.

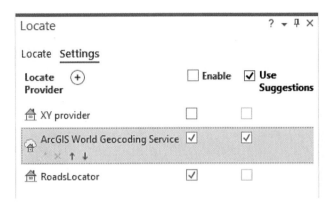

- In the Locate pane, select the **Settings** tab, and under the Enable heading, put checkmarks in the options for **ArcGIS World Geocoding Service** and **RoadsLocator**. Remove the checkmark for **XY provider**.

ArcGIS World Geocoding Service An ArcGIS Online service that allows you to geocode addresses using Esri resources in the cloud.

- In the Locate pane, select the **Locate** tab and enter the address for the first library to examine, Austintown Library: **600 S Raccoon Rd, Youngstown, OH 44515**. Then click the **Enter** key on the keyboard.

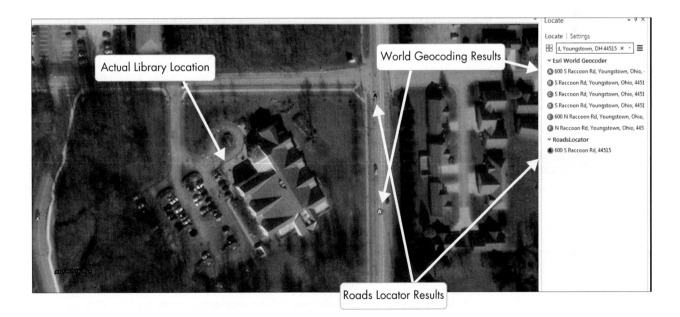

- The results you get back in the Locate pane are the geocoding results from both ArcGIS World Geocoding Service (Esri's online geocoding tools) in green and RoadsLocator (the address locator you created from the TIGER file) in purple. You also see the two geocoded points on the map. With the Imagery basemap turned on, you can see the real-world location of the Austintown library as well. Answer Question 10.5.

Question 10.5 Where are your geocoded results of the Austintown Library located in relation to the location provided by the ArcGIS World Geocoding Service? You should also be able to see the library on the Imagery basemap. Where are the two locations positioned in relation to the actual library?

- Continue through the other 14 library addresses in Table 10.2, comparing the results of your geocoding through the RoadsLocator address locator with that of the ArcGIS World Geocoding Service locations and then in relation to the actual real-world location of the library as you can view it on the basemap. As you complete each one, answer Question 10.6.

Library Name	Address
Boardman Library	7680 Glenwood Ave, Youngstown OH 44512
Brownlee Woods	4010 Sheridan Rd, Youngstown OH 44514
Campbell	374 Sanderson Ave, Campbell OH 44405
Canfield	43 W Main St, Canfield OH 44406
East	430 Early Rd, Youngstown OH 44505
Greenford	7441 W South Range Rd, Greenford OH 44460
Main Library	305 Wick Ave, Youngstown OH 44503
Newport	3730 Market St, Youngstown OH 44507
Poland	311 S Main St, Poland OH 44514
Sebring	195 W Ohio Ave, Sebring OH 44672
Springfield	10418 Main St, New Middletown OH 44442
Struthers	95 Poland Ave, Struthers OH 44471
Tri Lakes	13820 Mahoning Ave, North Jackson OH 44451
West	2815 Mahoning Ave. Youngstown OH 44509

Table 10.2 The 14 remaining libraries and their addresses

Question 10.6 For each of the libraries, describe the position of both results (your geocoded points using RoadsLocator and the results provided by the ArcGIS World Geocoding Service) in relation to each other and the library building as you can observe it in the Imagery basemap. Be very specific in your answers.

Step 10.8 Basic Spatial Analysis of Geocoded Points

- In this part of the module, you will be conducting some analysis of your geocoded points in relation to the roads layer to answer some spatial questions.

    Important Note: Before proceeding, clear any selected features (such as libraries selected from the geocoding procedure) by selecting the **Map** tab and then the **Selection** group and clicking the **Clear** button.

- First, you need to examine how many libraries are near a major road (Market St) in Mahoning County. To begin, use the Select Layer By Attribute tool (see Module 2) and build an SQL query to find all the road segments that comprise Market Street. Build your query to find all the road segments (records) in the **RoadsFeet** layer with the attribute **FULLNAME is equal to Market St**.

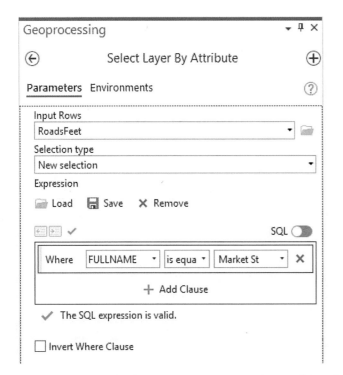
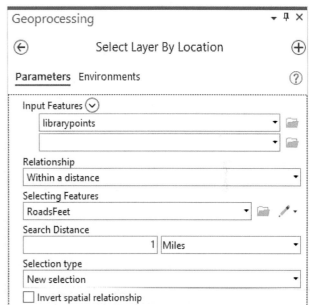

- Next, use the Select Layer By Location tool (see Module 8) to select all libraries within 1 mile of these selected features of the RoadsFeet layer that comprise Market St.
- Answer Question 10.7.

Question 10.7
What are the names of the libraries that are within 1 mile of Market Street?

- Clear all of your selected features. Next, use Select Layer By Attribute to find all the road segments for the street named Mahoning Ave and then use Select Layer By Location to determine which libraries are within 1 mile of Mahoning Avenue. Answer Question 10.8.

Question 10.8
What are the names of the libraries that are within 1 mile of Mahoning Avenue?

- Clear all of your selected features.
- You're now ready to do one last set of queries. One of the main roads that runs the east–west length of the county is State Route 224. However, it changes names several times as it passes through many different townships

and populated places. You need to select all the road segments of State Route 224, but this means finding several differently named segments (as opposed to all segments of the same road having the same name, such as Market St or Mahoning Ave). Use Select Layer By Attribute to find all segments with the following names:

1. US Hwy 224
2. Akron Canfield Rd
3. W Main St
4. E Main St
5. Boardman Canfield Rd
6. Boardman Poland Rd
7. W McKinley Way
8. E McKinley Way
9. Main St (but note that there are two Main Sts in the county, and the ones that correspond with State Route 224 have the zip code 44406, so be sure to select only these segments and not any other segments with the name Main St)

- When these are all selected, you see the whole of State Route 224 stretched across the county with all road segments selected. Next, use Select Layer By Location to find all libraries within 1 mile of State Route 224. Answer Question 10.9.

Question 10.9 What are the names of the libraries that are within 1 mile of State Route 224?

Step 10.9 Sharing or Printing Your Results

- Save your work as a map document and include the usual information in the Map Document Properties pane.
- Next, display only the RoadsFeet layer and the geocoded librarypoints layer. Turn on the labels for librarypoints by using the USER_NAME field. You may have to adjust the font or symbol to make the names stand out in relation to the lines of the roads layer.
- Finally, either print a layout (see Module 3) of the final version of your geocoded libraries (including all the usual map elements and design for a layout) or share your results as a web map through ArcGIS Online (see Module 4).
- If you are sharing your data, you might also want to share your address locator via ArcGIS Online so that others can use the locator you created. You can share your locator by creating a *locator package* (.gcpk file) that contains the address locator. This is done through the Package Locator tool, available in the Geoprocessing pane (in the Data Management toolbox, within the Packaging toolset). By using this tool, you can create a

single-file package containing the components of the locator. To later unpack the locator package, you need to use the Extract Package tool, which is also located in the Data Management toolbox, within the Packaging toolset.

Closing Time

This module demonstrates how to match a set of addresses to their geographic locations by using a street layer and geocoding techniques. Geocoding is a powerful tool that you likely will use often, whether you're finding a destination with an online mapping website, a mapping app on your phone or tablet, or a vehicle navigation system in your car. ArcGIS Pro moves beyond simple address location, providing the tools needed to geocode multiple addresses and then use the plotted data for spatial analysis. Through ArcGIS Pro resources, Esri provides other types of geocoding options. See *Related Concepts for Module 10* for more information about cloud-based geocoding services.

geocoding service An online utility that allows for geocoding one or more addresses.

A lot more can be done with street data and geocoded results. For instance, a service like Google Maps allows you to geocode an origin and a destination and then shows you what it computes as the shortest driving route between those two points. ArcGIS Pro allows you to compute shortest paths between multiple points on a street network, and you'll use the same data from this module to do just that in Module 11.

RELATED CONCEPTS FOR MODULE 10

Geocoding Services and ArcGIS Pro

One thing you've probably noticed is that, when you use an online system for geocoding a single address (such as Google Maps—**https://maps.google.com**) or batch geocoding multiple addresses (such as BatchGeo—**https://batchgeo.com**), you don't have to go through many of the steps outlined in this module, such as setting up an address locator or examining ranking results. These activities happen behind the scenes on the website. The advantage to using one of these systems (referred to as **geocoding services**) is that the steps in the geocoding process (from setting up the reference database to selecting the type of address locators to the matching itself) have already been implemented, leaving you to focus on the results of the process.

In ArcGIS Pro, a geocoding service is available through the ArcGIS World Geocoding Service via ArcGIS Online. In this module, you used the ArcGIS World Geocoding Service in the Locate pane; when you typed an address, ArcGIS Pro used the cloud to find a matching geocoded location for that point. All of the reference data (that is, the roads layer) used for this geocoding is also located in the cloud.

You can also use the ArcGIS World Geocoding Service as a source for batch geocoding. When you use the tools in the Geocode Table pane, you can use the ArcGIS World Geocoding Service as a locator instead of using your own homemade address locator and downloaded street centerline file (see Figure 10.4). However, use of this service requires an ArcGIS Online organization connection

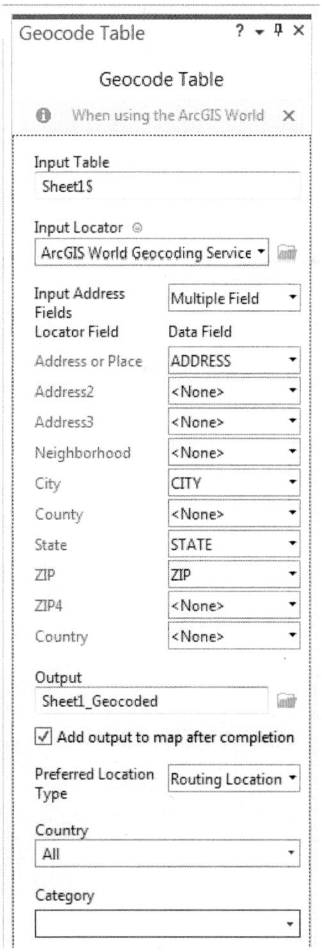

FIGURE 10.4 Using the ArcGIS World Geocoding Service as a locator when geocoding addresses in ArcGIS Pro.

and consumes several credits from your organization's ArcGIS Online account, depending on how many addresses you are batch geocoding. (See Module 4 for more about using credits and the ArcGIS Online organization.) For instance, at the time of writing, doing batch geocoding of a group of 1000 addresses would consume about 40 credits.

Key Terms

geocoding (p. 239)
batch geocoding (p. 239)
reference data (p. 239)
segment (p. 239)
street centerline (p. 244)

TIGER/Line (p. 244)
address locator (p. 244)
address locator style (p. 245)
parsing (p. 248)
linear interpolation (p. 250)

ArcGIS World Geocoding Service (p. 255)
geocoding services (p. 259)

How to Perform Network Analysis in ArcGIS Pro

ArcGIS Pro Skills

In this module, you will learn how to do the following in ArcGIS Pro:
- Activate ArcGIS Pro extensions.
- Access the features of the Network Analyst extension.
- Create routes with different parameters between a set of stops.
- Incorporate barriers into route analysis.
- Create service areas using Network Analyst.
- Use the Closest Facility tools of Network Analyst.

Learning Outcomes

After studying this module, you should be able to:
- Describe what a network consists of in GIS.
- Explain the importance of connectivity in a network.
- Define what transit cost is in a network.
- Explain how Dijkstra's algorithm can find the shortest path between stops on a network.
- Explain how the shortest path between multiple stops on a network can be found.
- Explain what a service area is.
- Describe how a closest facility operation is performed.

Introduction

Once you have a starting point and a destination geocoded, you may want to find the shortest driving distance between these points. For instance, if you're going out to dinner at a new restaurant, a mapping app on your phone or a website like Google Maps can determine the shortest route from your house to the restaurant. However, network analysis in GIS allows for more than just finding the shortest driving path between two points. ArcGIS Pro tools allow you to determine the shortest driving distance between two or more locations, find the closest facility (via a transportation system) to a location, and create buffers that are constrained by driving distance instead of a circular radius. Emergency management workers, delivery companies, and real estate agents are among the many professionals who use these types of network analysis.

In Module 10, we discussed how GIS usually represents a street: as a line layer (such as a street centerline file) with roads split into segments, with each segment having various attributes for that small piece of the street. In this module, you'll begin with a streets layer and use ArcGIS Pro to set it up as a **network**, a connected set of segments that can be used for analyzing routes. In its most basic form, a network consists of two items: a set of **edges** (the line segments that make up the streets and have attributes assigned to them) and a set of **junctions** (the nodes that connect the edges). In a road network, edges would be the lengths of the roads, and junctions would be the intersections of two roads or places where you could turn from one road onto another (Figure 11.1).

network A series of junctions and edges connected together for modeling concepts such as streets and routes.

edges The links of a network.

junctions The nodes of a network (or the places where edges come together).

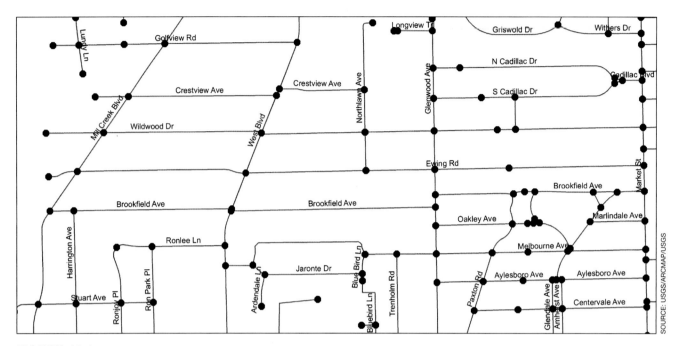

FIGURE 11.1 A road network of Mahoning County, Ohio, consisting of junctions and edges.

Networks aren't limited to roads. Other types of transportation networks can also be modeled using ArcGIS Pro, such as railroads or walkways. The tools for creating and working with transportation networks are in a separate extension for ArcGIS Pro, called Network Analyst, which you'll be using in this module. While you'll be working only with a road network in this module, other types of networks, including non-transportation networks, can be modeled in ArcGIS Pro, such as utilities or power lines, but such networks are better handled with a different set of network tools; see *Related Concepts for Module 11* for more information.

Module Scenario and Applications

This module puts you in the role of a local librarian who is coordinating several tasks for a county library system. Your jobs include overseeing book deliveries to libraries as well as identifying the high schools most likely to make use of a library's resources. You'll be creating a network in ArcGIS Pro, determining the shortest driving distance between libraries to cut down on delivery times, and using the network applications to connect schools and libraries.

The following are additional examples of real-world applications of this module's skills:

- A dispatcher for a messenger service needs to route a courier on the quickest path to a destination through an area where several roads are heavily congested. He can use the routing tools explained in this module to determine the shortest route to take.

- A 911 operator needs to determine the closest hospital to an accident location so that EMTs can reach the hospital as quickly as possible. She can use the closest facility tools in ArcGIS Pro to do so.

- A real estate agent showing potential homes to clients wants to find the optimal driving route between houses so that she can show as many homes as possible within a certain amount of time. The route tools of Network Analyst can set up her best driving routes.

Study Area

- For this module, you will be working with data from Mahoning County, Ohio.

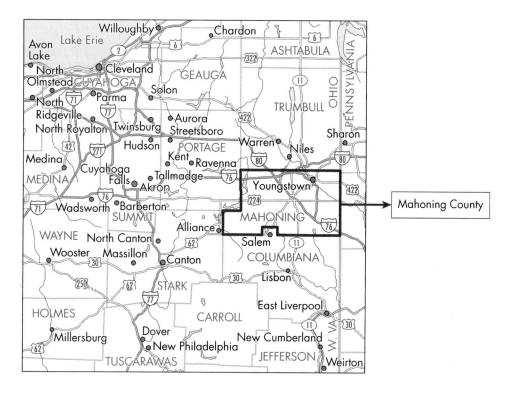

Data Sources and Localizing This Module

The data in this module are related to library addresses, school addresses, and streets in Mahoning County, Ohio. However, you can easily modify this module to use the libraries, schools, and streets of your own county. The address information for the county libraries comes from the Mahoning County Public Library's website, at **https://www.libraryvisit.org**. Using your own county's library website (or the phone book), you can put together a Microsoft Excel file of library names and addresses (for instance, Johnson County, Iowa's library addresses are online at **https://www.icpl.org/public-libraries-johnson-county**). This module also focuses on the locations of high schools in the county. The address information for the Mahoning County high schools comes from **https://www.publicschoolreview.com/ohio/mahoning-county**. Similar information for other U.S. counties is also available from this website; for instance, Johnson County, Iowa, school information is available at **https://www.publicschoolreview.com/iowa/johnson-county**. Using your own county's library and high school resources, create a Microsoft Excel file of library names and addresses and a second Excel file of school names and addresses and geocode them into their own point layers. (See Module 10 for more about geocoding.)

The roads layer used in this module was downloaded from The National Map, and the same kind of data are available for all U.S. states. For example, if you were examining Johnson County, Iowa, you would download the Transportation data for Iowa from The National Map, use the Trans_RoadSegment feature class, and clip out the features of Johnson County. For purposes of this module, the roads file was converted from NAD 83 GCS to NAD 83 Ohio North State Plane (to keep network measurements in a projected coordinate system). The Add Geometry Attributes tool was used in ArcGIS Pro to calculate a new field in Trans_RoadSegment called LENGTH, which computed the length of each road segment, in miles.

Step 11.1 Getting Started

- This module's hands-on applications use the data folder called Module11. Your instructor will be able to supply you with this data, or you can download it directly from this book's website at **https://www.macmillanlearning.com/college/us/product/Discovering-GIS-and-ArcGIS-Pro/p/131923075X**. The text in this module assumes that you have this Module11 folder in a computer location referenced as C:\GIS; if you have it somewhere else (for instance, in a flash drive referenced as G:\GISClass), substitute that location and path to the Module11 folder throughout this module.

- Module11 contains a file geodatabase called CountyNet that contains the following items:

  - Roads: A feature dataset that contains a network dataset (called Roads_ND) and five feature classes. A feature dataset is necessary here to contain the network dataset used in this module. The five other feature classes in this feature dataset are:

    - Roads_ND_Junctions: a point feature class of junctions in the road network, which is part of the Roads_ND network dataset
    - RoadsMiles: a line feature class of Mahoning County roads that was used to construct the Roads_ND Network Dataset
    - librarypoints: a point feature class containing the locations of 15 libraries in Mahoning County. This represents geocoded and corrected library addresses from the Mahoning County Public Library website.
    - construction: a point feature class containing the locations of simulated construction blocking a road
    - schoolpoints: a point feature class containing the locations of 30 high schools in Mahoning County

- Start ArcGIS Pro.
- Sign in with your Esri account username and password.
- Create a new project using the **Map** template. Call this project **Module11** and place it in your **C:\GIS\Module11** folder. Ensure that there is not a checkmark in the box next to **Create a new folder for this project**.
- When ArcGIS Pro opens, change the map's name to **Mahoning**.
- You will be using the ArcGIS Pro Network Analyst extension in this module. For more information about extensions in ArcGIS Pro, see

Smartbox 11.1. If you don't have access to the Network Analyst extension, see Troublebox 11.1 for more information on how to get it set up in ArcGIS Pro.

SMARTBOX 11.1

What is an extension in ArcGIS Pro and how is it used?

Several ArcGIS Pro network functions and tools are part of **extensions** that must first be activated before they can be used. Extensions are separately purchased ArcGIS Pro components that enable a wide range of features; they do not automatically come included with the core system. There are numerous extensions available for specialized functions in ArcGIS Pro, including the Business Analyst and Defense Mapping extensions. In this book, you'll make use of Spatial Analyst, 3D Analyst, and, in this module, **Network Analyst**, which enables you to work with different network functions that are not available in ArcGIS Pro.

When an extension is licensed and activated, you have several new tools and options available to use. For instance, there are several Network Analyst tools in ArcGIS Pro that require an active Network Analyst extension to run. If the licensed extension is not active, those tools appear with a lock symbol next to them, and they are unavailable until the extension is activated.

extension An add-on set of functions for ArcGIS Pro.

Network Analyst An ArcGIS Pro extension used with transportation networks.

TROUBLEBOX 11.1

How can I activate an extension in ArcGIS Pro?

Extensions are individually licensed to users via their ArcGIS Online accounts, much the same way a license to run ArcGIS Pro is tied to an account. (See Smartbox 1.1 in Module 1.) When the administrator for your organization assigns your ArcGIS Online account a license to run ArcGIS Pro, he or she can also assign you licenses for available extensions. For instance, this module requires a license for the Network Analyst extension to be assigned to your ArcGIS Online account, and for future modules you will need licenses for the Spatial Analyst and 3D Analyst extensions. Contact your ArcGIS Online organization administrator to assign these extension licenses to your account so that when you sign in to ArcGIS Pro with your account, all those extension functions will be active and working properly.

To determine what extensions you have licensed and assigned to the Esri account that you use to sign in to ArcGIS Pro, go to the **Project** tab of the ribbon and click **Licensing**. You can then see which extensions are licensed and assigned to your account, along with when those licenses expire.

Step 11.2 Creating a Network Dataset

- In the Catalog pane, navigate to your **C:\GIS\Module11** folder and expand it so you can view its contents. Expand the **CountyNet** file geodatabase and also the **Roads** feature dataset it contains. You should see the network dataset called **Roads_ND** that has already been created for you. At the time of writing, the ability to create network datasets will be added to ArcGIS Pro 2.5, so Roads_ND was already created in ArcMap. Instructions for creating network datasets in ArcMap and ArcGIS Pro are available on this book's website, at **https://www.macmillanlearning.com/college/us/product/Discovering-GIS-and-ArcGIS-Pro/p/131923075X**.

- Add the **Roads_ND** network dataset to the map. For more information about the elements of a network dataset in ArcGIS Pro, see **Smartbox 11.2**.

SMARTBOX 11.2

What are the elements of a network dataset in ArcGIS Pro?

network dataset An ArcGIS Pro structure that is a series of connected junctions and edges, commonly used for transportation-related problems.

turns Information used by ArcGIS Pro to determine information about valid flows along a network.

connectivity The linkages between edges and junctions of a network.

In ArcGIS Pro, a network is stored in a **network dataset**, which uses the files stored as a feature dataset in a geodatabase. The initial streets layer that contains the segments (which will be used as edges in the network) must be one of the files in the feature dataset. If other files are used, they are stored in the feature dataset as well.

A network dataset is not simply a layer of lines that represent road segments; a network allows you to design routes and do analysis with them. For this reason, a network has to contain much more information than just a layer of line segments. First, a network can contain information about **turns**. That is, parts of the network can have information about items such as where left turns cannot be made, where U-turns can be made, and what streets are one way. This kind of information is important because you don't want to create routes that allow vehicles to travel the wrong way on a one-way street or allow them to make turns where (in the real world) the turns would be illegal. Turn information can be set up at a junction and is stored as a separate turn feature class.

Another key component of a network is its **connectivity**, or how all the edges link up with one another. Proper connectivity in a network ensures that items that should link together do, while others that shouldn't link together don't. For instance, ArcGIS Pro needs to understand that the junction where four line road segments come together allows travel to continue from one segment to another. However, the segments of a road and a freeway overpass should not connect in a way that would allow a vehicle to turn from the road onto the freeway at that point.

Similarly, if a roads layer has lines that represent both roads and railroad tracks, the network connectivity should not be set up in a way that allows drivers to continue from a road onto a railroad track. It might sound silly to think that a GIS would instruct a driver to drive on a railroad track, but keep in mind that a roads layer consists only of line segments. Without something to explicitly tell the GIS that the segments of roads and the segments of railroads should not connect, the GIS simply sees edges and junctions that fit together and does not automatically distinguish a road from a railroad. If different parts of a city transportation network (such as roads, railroads, sidewalks, and subways) are modeled separately, ArcGIS Pro uses special connectivity groups that allow the user to specify which of these groups connect. (For example, sidewalks and metro stations connect to each other, and metro stations and subways connect to each other, but sidewalks and subways do not connect.)

A network also needs a means of measuring the travel distance between destinations on a route. A route is made up of a series of edges, and there is a cost associated with traveling across an edge. This travel cost (referred to as **impedance** in ArcGIS Pro, and sometimes also referred to as *transit cost*) is used to determine the overall shortest route. For example, the cost of traveling the length of an edge could be the physical distance of that edge (for instance, a quarter of a mile). This cost could also be expressed as the time it takes to drive that distance (for instance, about 20 seconds, if you were driving at 40 mph). Depending on the type of transportation network being used, this cost could reflect the length of the segment and its assigned speed limit; it might also have additional impedance due to heavy traffic conditions, or it could reflect the amount of walking time it takes to travel the segment.

If a route consists of 10 segments, the total cost of that route will be the sum total of the travel costs of all the segments. If the impedance value being used is the actual distance you had to travel for each segment, the overall cost of the route would be the 10 distance values added together. If the impedance value is the time it takes to drive each segment, the overall cost of the route would be the 10 drive time values added together. Thus, the overall cost of taking a route reflects whatever impedance values are used in the network.

> **impedance** A value that represents how many units (of time or distance) are used in moving across a network edge. Also referred to as *transit cost*.

Step 11.3 Creating a Route by Visiting Stops in Order

- You can use the Roads_ND network dataset for calculating routes between points. You can start by finding shortest driving distance route among all the libraries. To begin, add the **librarypoints** feature class to the Contents pane.

- In the Mahoning map tab's Contents pane, right-click on the name of the map itself (**Mahoning**) and choose **Properties**. In the Map Properties dialog box, choose the **General** tab and set Display Units to **Feet**. (Leave Map Units set to **Feet**.)

- On the **Analysis** tab, in the **Tools** group, click the **Network Analysis** button and choose **Route**. You see a new item called Route added to the Contents pane, featuring symbols for a number of different route elements (stops, point barriers, routes, line barriers, and polygon barriers). With this new Route item chosen in the Contents pane, a new tab called Route is available on the ribbon.

- The Route item can be treated like any other layer. In the Contents pane, right-click on it and choose **Properties**. In the Layer Properties: Route dialog box that opens, click on the **General** tab and rename the Route layer **Libroute1**. Then click **OK** to close the dialog. In the Contents pane, you now see the Route item called Libroute1.

- In your role in this module as a local librarian, you are helping a delivery person who has to make a set of deliveries to the libraries. To help the delivery person, you have to locate the shortest driving

stops Destinations to visit on a network.

distance route between all libraries. In this case, the library locations will be considered **stops** along the network. Your next step will be to add the points representing the libraries as stops to be visited. In the Contents pane, choose the **Libroute1** layer and then, on the ribbon, select the **Route** tab and then the **Input Data** group and click the **Import Stops** button. The Add Locations tool opens in the Geoprocessing pane.

- Under Input Network Analysis Layer, choose **Libroute1**.
- Under Sub Layer, choose **Stops** to tell ArcGIS Pro you want to add the library points as stops (and not any other option) along the network.
- Under Input Locations, choose **librarypoints** to tell ArcGIS Pro to use the librarypoints layer for the stops.
- For Field Mappings, choose **Use Geometry**.
- Leave the other settings alone and run the Add Locations tool.
- You see that Libroute1 now has 15 stops added to it; these stops will be displayed in the map view. Answer Question 11.1. Keep in mind that the order in which the libraries are listed is their assigned order shown in the librarypoints attribute table, and thus that will also be the numerical order ArcGIS Pro assigns to them. When you're done examining Libroute1, you can turn off the librarypoints layer, but be sure to keep the Stops layer turned on.

Question 11.1 Use the Explore tool to familiarize yourself with each of the 15 stops/libraries. Briefly sketch out the locations of the 15 points, as well as their names (Austintown, Boardman, Brownlee Woods, and so on) and the number initially assigned to each one. (You are doing this for your own reference for future questions.)

- You can now create the best (shortest) route among these libraries. In the Contents pane, choose **Libroute1**, and then, on the **Route** tab, in the **Travel Settings** group, under the **Sequence** pull-down menu, choose **Use Current**. This tells ArcGIS Pro to use the current arrangement of stops when computing its route; for instance, it will start at Stop 1, compute the shortest route to Stop 2, compute the shortest route from Stop 2 to Stop 3, and so on.
- On the **Route** tab, in the **Analysis** group, click the **Run** button. Your "best route" between these 15 stops appears. For further information on how these shortest routes are calculated, see **Smartbox 11.3**.

SMARTBOX 11.3

How is a shortest route calculated in ArcGIS Pro?

As described in Smartbox 11.2, the cost of traveling a route is determined by summing up the impedance costs for all the segments of the route. However, when you're traveling between an origin and a destination on a road network, there are multiple different routes you can take. For instance, if you're driving from your house to a friend's house on the other side of town, there are likely numerous routes you could take. Some of them may be very direct, and others would take you well out of your way but would eventually get you there. A system (such as an online website or a vehicle navigation device) that's finding your shortest route needs to have a way of determining which of the routes is the shortest (based on the impedance value being used, such as the shortest overall driving time or the shortest overall driving distance). ArcGIS Pro uses an **algorithm**, or a set of steps used in a process, to figure out which route, of all the possible ways of traveling between two points, is the shortest one.

algorithm A set of steps used in a process.

Dijkstra's algorithm A mathematical process that determines the shortest path along a network from an origin node to the other nodes on the network.

ArcGIS Pro uses a shortest path algorithm called **Dijkstra's algorithm**, with some modifications depending on the setup of the network and destinations. Dijkstra's algorithm is designed to determine the shortest path between an origin point and other destinations on a network. Figures 11.2 through 11.9 provide simple examples of how Dijkstra's algorithm finds the shortest path from an origin point to locations on a network.

Figure 11.2 shows an initial network of four locations (nodes or junctions) that are connected by edges. Node a is the origin node; you want to find the shortest path from node a to node d. Each of the edges has a cost assigned to it for traveling across that edge (the units of the impedance value), and you see that some of the edges have different costs, depending on the direction being traveled. Note that only the origin node (node a) has a value assigned to it (a value of 0 because it's the starting point); right now, you have no idea how much the cost will be to travel to any of the other nodes.

In the first step of the algorithm (Figure 11.3), you will travel each of the edges from the starting node a to the nodes to which it directly connects (node b and node c), and you will assign a value to each of those two nodes for the total cost of travel to each of the nodes. For instance, the travel cost of moving from node a to node b is 5 units, so node b is assigned a value of 5, while the cost of moving from node a to node c is 9 units, so node c is assigned a value of 9. Node a is now colored red to signify that you've already evaluated the costs for traveling away from that node (and you don't have to do that again).

Since you've already searched from node a to all directly connected nodes, you need to select another node from which to search.

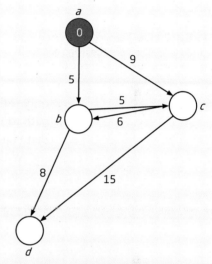

FIGURE 11.2 A sample network of four junctions and six edges.

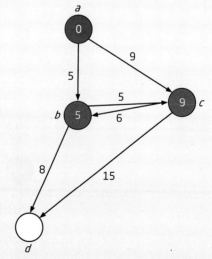

FIGURE 11.3 Searching from the origin node (node a) to the two nodes directly connected to it (node b and node c).

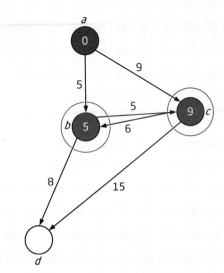

FIGURE 11.4 Selecting the new node to begin searching from again (the node with the lowest value—either node *b* or node *c*).

Figure 11.4 shows this step. The algorithm looks at the values of all nodes (except for the ones from which you've already searched) and tries to decide which node you should use to begin searching from again. The algorithm chooses the node with the lowest value. In this case, the value of 5 in node *b* is lower than the value of 9 in node *c*. Thus, node *b* becomes the new node from which to begin searching.

You will now travel from node *b* to all directly connecting nodes and update the values of those nodes (Figure 11.5). Node *b* connects directly to node *c* and node *d*. It costs 8 units to travel from node *b* to node *d*, so node *d*'s value is updated to 13 (you take node *b*'s value of 5 and add the additional 8 units it would take to travel from node *b* to node *d*). The cost of traveling from node *b* to node *c* is 5 units, so under this logic, node *c* would be updated to a value of 10 (the value of 5 in node *b* is added to the other 5 units it would take to travel from node *b* to node *c*). However, node *c* has already been assigned a cost to travel to it (a value of 9). Because node *c*'s value of 9 is lower than the travel cost from node *b* to node *c* (a value of 10), it is not updated to the higher value. A node's value is updated only if the new value is lower than the existing value (indicating that a shorter path to that node has been found). Node *b* is also now colored red, indicating that you can no longer search for paths from it.

With node *b* now exhausted, you need to find a new node from which to search. As before, the algorithm looks at the values of the nodes that haven't been searched from yet and selects the lowest one (Figure 11.6). Node *c* has a value of 9, and node *d* has a value of 13, so node *c* becomes the new node from which to search.

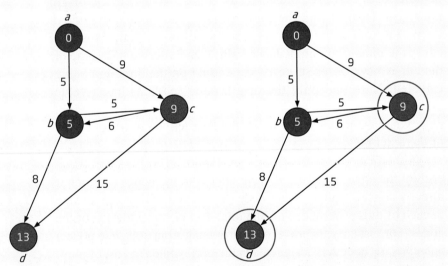

FIGURE 11.5 Searching from node *b* and updating the values for the nodes it directly connects to (node *c* and node *d*).

FIGURE 11.6 Determining the new node to begin searching from (the node with the lowest value, either node *c* or node *d*).

Node *c* connects directly to node *b* and node *d*. The total cost of traveling from node *c* to node *b* would be 15 (the value of 9 for node *c* added to the cost of 6 for traveling from node *c* to node *b*). The value of 15 is higher than node *b*'s value of 5, so node *b* is not updated. Similarly, the cost of traveling from node *c* to node *d* would be 24 (the value of 9 for node *c* added to the cost of 15 for traveling from node *c* to node *d*). Because this value is higher than the current value of node *d* (13), node *d* is not updated (Figure 11.7). Node *c* is now exhausted, so it is colored red (and you can no longer search from it).

The algorithm now needs to select a new node from which to begin searching. Only one node has not yet been exhausted (node *d*), so it is selected as the new search node (Figure 11.8).

However, there are no nodes for it to directly connect to, so no comparisons of values are made, and no updating is done. Node *d* is colored red because its searching is exhausted (Figure 11.9). Because there are no more nodes from which to search, the algorithm is completed, and all shortest paths have now been found. The shortest path to travel from node *a* to node *d* costs a total of 13 units, and the shortest route is to travel from node *a* to node *b* and then to node *d*. Any other possible routes are longer than this.

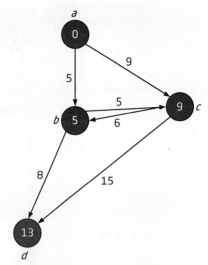

FIGURE 11.7 Updating the values for the nodes directly connected to node *c* (node *b* and node *d*). Both nodes already have lower values, so they are not updated.

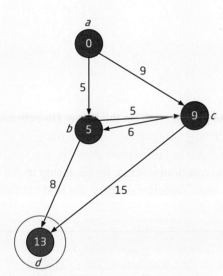

FIGURE 11.8 Determining the new node to begin searching from; only node *d* remains as a choice.

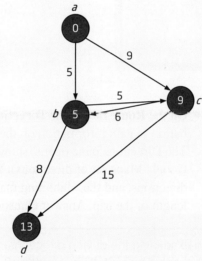

FIGURE 11.9 Node *d* does not directly connect to any other nodes, so searching is complete, and the algorithm has determined all shortest paths.

ArcGIS Pro uses the same type of algorithm to find the shortest distance between an origin and a destination, with some adjustments to allow locations to be found at locations other than junctions (nodes) and also to allow for settings in the network, such as turns.

- Use the Explore tool to click on the route associated with Libroute1. In the pop-up box that appears, the route is called Greenford-West because the route starts at Greenford Library and ends at West Library. You see general information about the route, including the length of the route. Answer Question 11.2.

Question 11.2 How long is the best route between the stops, as measured in miles? (*Important Note:* The length reported back from ArcGIS Pro is in the units of the network, which in this case is feet.)

- On the **Route** tab, in the **Directions** group, click on the **Show Directions** button to gain information on the driving directions along the route. The Directions pane opens, showing driving directions (similar to a Google Maps set of directions). You can double-click on each step in the directions, and the Mahoning map view switches to show the route at that length of the trip. Answer Question 11.3.

Question 11.3 In examining the driving directions, do you see any particular directions that seem non-intuitive to a driver? (*Hint:* Chances are you're unfamiliar with driving throughout Mahoning County, Ohio. However, look for any directions that drivers may be unlikely to take, such as switching to an Interstate for very short distances, or similar kinds of activities.) Why would ArcGIS Pro provide these kinds of directions?

Step 11.4 Creating a Route by Rearranging Stops

- Libroute1 shows the shortest path on the network among the 15 stops, but if you look at it, you see that the route does a lot of crisscrossing and backtracking in order to visit the 15 stops in the numerical order specified. Some of the libraries (such as Tri Lakes and Sebring) are in more remote areas of the county, while others (such as Brownlee Woods, Struthers, and Poland) are relatively close together. To save time, it would make more sense for a driver to visit the closer libraries one after the other instead of having to drive far away to a remote location and then back again. What you will do next is compute a new delivery route that rearranges the order of the stops. For more information about reordering stops to compute best routes, see **Smartbox 11.4**.

SMARTBOX 11.4

How does reordering stops affect the shortest routes in ArcGIS Pro?

By changing the order of the stops you visit, you can adjust the overall cost of traversing a route. For instance, Figure 11.10 shows two different routes used to visit five different stops (public libraries in Mahoning County, Ohio). Figure 11.10a shows the route generated by visiting the stops in a predetermined order; Dijkstra's algorithm is used to find the shortest route between library number 1 and library number 2, then the shortest route between library number 2 and library number 3, and so on. You can see that an awful lot of backtracking is being done to find the shortest path between stops in this particular order. In contrast, Figure 11.10b shows another shortest route between the five stops, but this time the order of the stops is more optimal. The route begins at one end, visits the stops nearest one another, and ends at the furthest stop.

FIGURE 11.10 Two different shortest routes between stops (public libraries in Mahoning County, Ohio)—visiting stops in a predetermined order (11.10a) and visiting stops in a rearranged, more optimal order (11.10b).

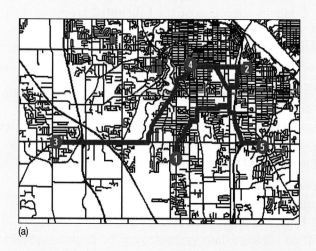
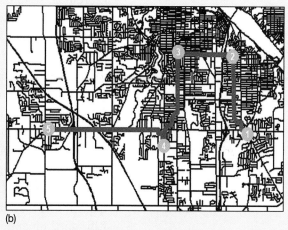

(a) (b)

Reordering stops gives you more flexibility when trying to determine the best route for visiting all destinations. In this case, the Traveling Salesman Problem (**TSP**) can help. The TSP is a mathematical process for determining the optimal configuration of stops on a network. Think of it this way: A traveling salesman must visit several destinations to make sales and needs to find the shortest (or least-cost) route among all these destinations. The TSP examines the different combinations of how to reorder all the destinations to find the best route. (For instance, is the lowest overall cost incurred by visiting the five stops as 1–2–4–5–3, by visiting them as 2–3–4–5–1, and so on?) While algorithms can provide decent solutions to the problem (and ArcGIS Pro uses heuristics to come up with good solutions), the only way to really know which of the different combinations of ordering the stops is the best is to work through all the combinations, which would be impossible with a large number of stops.

TSP Traveling Salesman Problem, a mathematical process that involves determining the optimal configuration for rearranging a series of stops on a network.

You can place additional constraints on the reordering of stops. You might want to keep the first stop and the last stop fixed and shuffle only the order of stops in between; for instance, if you have errands to run, you will leave work—a fixed stop—and then find the best order to run your errands, then arrive home at the end—a second fixed stop. You may also want to keep the first stop the same as the last stop; for example, if you were planning a new bus route, the initial stop and the final stop would both be the location of the bus depot in order to simulate the bus leaving and returning to its point of origin each day. There might be times when you are not able to shuffle the order of the stops and will have to visit locations in a predetermined order; for example, these constraints would apply to a shipping company vehicle that has to make time-sensitive deliveries.

- On the **Analysis tab**, in the **Tools** group, click the **Network Analysis** button and choose **Route**. You see that a new item called Route is added to the Contents pane. Right-click on **Route** and choose **Properties**. In the Layer Properties: Route dialog box that opens, click on the **General** tab and rename the Route layer **Libroute2**. Then click **OK** to close the dialog. Next, you can load the library points as stops for this new Libroute2, so in the Contents pane, choose the **Libroute2** layer and then select the **Route** tab and the **Input Data** group and click the **Import Stops** button. The Add Locations tool opens in the Geoprocessing pane. Load the locations of the stops as you did with Libroute1 so that you have the 15 libraries loaded as stops.

- To make changes to how the route will be computed for Libroute2, on the **Route** tab, in the **Travel Settings** group, for Sequence, choose **Preserve First and Last Stop**. This means that the first and the last stop are both preserved, or fixed (that is, you must start at stop #1 and must end at stop #15).

- Then on the **Route** tab, in the **Analysis** group, click the **Run** button. Your "best route" between these 15 stops appears. Answer Questions 11.4 and 11.5. You might find it helpful to turn Libroute1 and Libroute2 on and off and change their symbology to be distinctive from one another when examining them.

Question 11.4 How long is this new best route (Libroute2) between the libraries (in miles)?

Question 11.5 How does Libroute2 differ from Libroute1? Rather than answering "it's shorter" or "it's longer," how specifically has the route itself changed in relationship to the stops and the computed path?

- Make a third new route and name it **Libroute3**.
- Load the **librarypoints** layer as stops for Libroute3.
- To make changes to how the route will be computed for Libroute3, on the **Route** tab, in the **Travel Settings** group, for Sequence, choose **Find Best**.

This means that the first and the last stop are no longer fixed (that is, you no longer need to start at stop #1 and end at stop #15, and ArcGIS Pro will figure out the best order in which to visit these stops).

- On the **Route** tab, in the **Analysis** group, click the **Run** button to see your "best route" between these 15 stops. Answer Questions 11.6, 11.7, and 11.8.

> **Question 11.6** How long is this new best route (Libroute3) between the stops (in miles)?

> **Question 11.7** How is the arrangement of stops different among the three routes? (That is, which stop is visited first, which is visited second, and so forth in each of the three routes? List them as #1 through #15 for each route.)

> **Question 11.8** You have three shortest paths among the libraries. What is so special about the specific route taken by Libroute3 to make it the shortest of the three?

Step 11.5 Creating Routes with Barriers on the Network

- Turn off Libroute1 and Libroute2 but leave Libroute3 on. In this step, you focus only on Libroute3 (the shortest or least-cost of the three paths).
- Add the **construction** feature class to the map and change the symbology so that you can clearly see the points. The construction class consists of seven points representing places where the roads have been blocked due to construction. ArcGIS Pro treats these seven points as absolute barriers, not letting you cross the road at these spots. You're now going to reexamine Libroute3 with this new constraint added.
- Make a fourth new route and name it **Libroute4**.
- Load the **librarypoints** layer as stops for Libroute4.
- Next on the **Route** tab, in the **Input Data** group, click on the **pull-down arrow** next to the small red and white button and choose **Input Point Barriers**. The Add Locations tool opens again; this time you will use it to add point barriers rather than stops.
- Under Input Network Analysis Layer, choose **Libroute4**.
- Under Sub Layer, choose **Point Barriers** to tell ArcGIS Pro you want to add the construction points as barriers (and not any other option, such as stops like the libraries) along the network.
- Under Input Locations, choose **construction** to tell ArcGIS Pro to use the construction layer for the source of the point barriers.
- For Field Mappings, choose **Use Geometry**.
- Leave the other settings alone and run the Add Locations tool. The seven barrier locations are shown in the view as large red dot symbols. You see that with the barriers, four of the library locations are between Main and West, and three of them are near Springfield (on different nearby roads).

276 CHAPTER 11 How to Perform Network Analysis in ArcGIS Pro

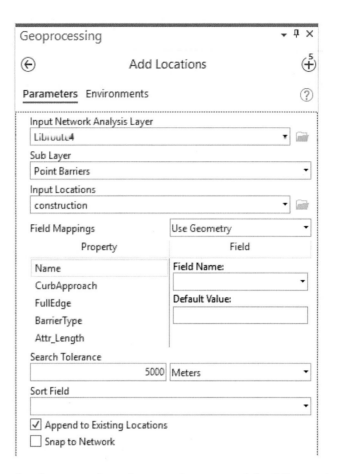

- To make changes to how the route is computed for Libroute4, on the **Route** tab, in the **Travel Settings** group, for Sequence, choose **Find Best** (just as you did with Libroute3). Then on the **Route** tab, in the **Analysis** group, click the **Run** button. You now see your "best route" for Libroute4 between the 15 stops while accounting for the 7 barriers. Keep in mind that Libroute4 has the same problem constraints as Libroute3, except it now contains these construction barriers.

- Answer Questions 11.9 through 11.12. Turn off all four of the routes and the construction points when you're done.

Question 11.9 What is the length (in miles) of Libroute4 with these barriers?

Question 11.10 What changed between Libroute3 and Libroute4 with the addition of the barriers?

Question 11.11 How (specifically) did the path between West and Main change?

Question 11.12 How (specifically) did the path between Springfield and Poland change?

Step 11.6 Working with Service Areas

- Routes are not the only tools that can be created with networks. Another network function is a service area, which shows the parts of the network near (or "served by") a particular point. See **Smartbox 11.5** for more information about service areas.

SMARTBOX 11.5

What are service areas and how are they used in ArcGIS Pro?

A **service area** is used in network analysis to determine which parts of a network are near a particular location (referred to as a **facility**) and are thus "being served" by that location. The service area is a polygon (or series of polygons) defined by the impedance factor being used (such as the actual driving distance or the driving time) for the network. These are sometimes referred to as **drive-time polygons**. Figure 11.11 shows two service areas for fire stations, based on driving time. The service areas in red show which parts of the network can be reached from the fire station within a certain amount of time. (Note that this value can also include the *turnout*, or amount of time needed to scramble and deploy vehicles from a facility, not simply the time or distance of the network.)

Service areas help determine the accessibility of a facility and can be thought of as buffers around the facility point (see Module 9), but the buffer is constrained by the properties of the network rather than just a large circle. For instance, a circular 1-mile buffer around a fire station is likely significantly different from the places on a network that can be reached within 1 mile of driving along city streets. Service areas are used for finding which parts of a network can be reached by emergency vehicles within a certain amount of time, which parts of a network are within a certain distance of a store, or which sections of a network are within a certain walking time of a school (among many other examples).

service area A polygon boundary created around sections of a network to determine which areas of the network are within a certain distance of a location.

facility A location from which service areas are created.

drive-time polygon Another term used for a service area.

FIGURE 11.11 Service areas for fire stations, based on their driving response times.

- On the **Analysis tab**, in the **Tools** group, click the **Network Analysis** button and choose **Service Area**. You see a new item called Service Area added to the Contents pane.

- Right-click on **Service Area** and choose **Properties**. In the Layer Properties: Service Area dialog box that opens, click on the **General** tab and rename the Service Area layer **Libservice**. Then click **OK** to close the dialog.

- Next, you can load the library points as stops for the new Libservice service area. In the Contents pane, choose the **Libservice** layer and then, on **Service Area** tab, in the **Input Data** group, click the **Import Facilities** button. The Add Locations tool opens in the Geoprocessing pane.

- Under Input Network Analysis Layer, choose **Libservice**.

- Under Sub Layer, choose **Facilities** to tell ArcGIS Pro you want to add the library points as the facilities for which to create service areas (and not any other option) along the network.

278 CHAPTER 11 How to Perform Network Analysis in ArcGIS Pro

- Under Input Locations, choose **librarypoints** to tell ArcGIS Pro to use the librarypoints layer for the source of the facilities.
- For Field Mappings, choose **Use Geometry**.
- Leave the other settings alone and run the Add Locations tool.

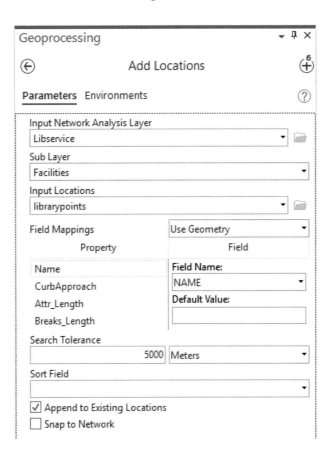

- Next, you need to establish the settings for the Libservice service area. On the **Service Area** tab, in the **Travel Settings** group, for Direction choose **Away from Facilities** to tell ArcGIS Pro you are creating service areas for driving distance away from each facility (as opposed to driving toward a facility).
- Also in the **Travel Settings** group, for **Cutoff**, type **5280** to tell ArcGIS you want to make the breaks for each service area at 5280 feet. (Because there are 5280 feet in a mile, you will be looking at 1-mile service areas.)
- On the **Service Area** tab, in the **Analysis** group, click the **Run** button. A new set of polygon areas (the service areas) are created 1 mile around each library.
- Answer Question 11.13. Then turn off the service area polygons.

Question 11.13 Are there any overlapping service areas? If so, which libraries have overlapping areas within the 1-mile regions? (Your answer may be something like "Austintown Library and Main Library overlap service areas within 1 mile.")

Step 11.7 Working with Closest Facilities

- The closest facility function can be used to determine which facilities are closest (in terms of network distance) to a set of incidents. (See **Smartbox 11.6** for more information about the closest facility function.) In this step, you will be determining which libraries are closest to which high schools in terms of their driving distance on the network (rather than creating a buffer or using a search radius from Select Layer By Location). This step will help determine which libraries a high school student is most likely to use.

SMARTBOX 11.6

What is the closest facility function, and how does it work in ArcGIS Pro?

The **closest facility** function of ArcGIS Pro determines (using a network and impedance values) which one of several locations on a network (referred to as *facilities*) is the nearest one via the network to a separate location (referred to as an **incident**). For instance, the closest facility function can find which hospitals (facilities) are closest in terms of network distance to an accident location (an incident).

Figure 11.12 provides an example of the closest facility function. The goal is to determine which public libraries are nearest to each high school, via travel along a network, and the impedance value is the actual physical driving distance along the network. For each high school, ArcGIS Pro will identify the library that is closest to it via the driving distance (using Dijkstra's algorithm) and plot out that route. Note that each high school can be closest to only one library, but a single library may be the closest facility for multiple high schools.

closest facility A tool used to determine which locations on a network are nearest (via network distances) to a different set of locations.

incident The locations for which the closest facility function tries to determine the nearest facility.

FIGURE 11.12 The output of the closest facility function, showing the shortest network distance between high schools (the incidents, shown with flags) and the library nearest to the school (the facilities, shown with blue dots).

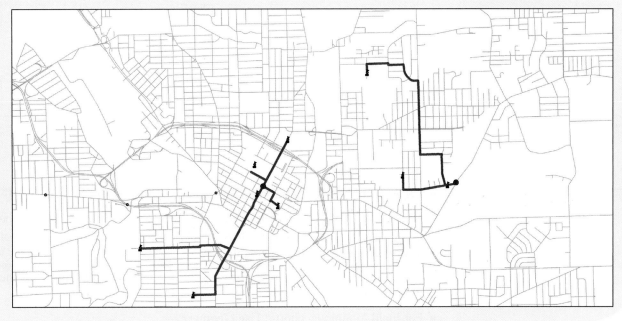

- Add the schoolpoints layer to the Contents pane and change its symbology to something distinctive that you can visualize differently from the library points.

- On the **Analysis** tab, in the **Tools** group, click the **Network Analysis** button and choose **Closest Facility**. You see a new item called Closest Facility added to the Contents pane.

- Right-click on **Closest Facility** and choose **Properties**. In the Layer Properties: Closest Facility dialog box that opens, click on the **General** tab and rename the Closest Facility layer **ClosestLib**. Then click **OK** to close the dialog.

- Next, you can load the library points as stops for the new ClosestLib service area. In the Contents pane, choose the **ClosestLib** layer and then, on the new **Closest Facility** tab, in the **Input Data** group, click the **Import Facilities** button. The Add Locations tool opens in the Geoprocessing pane.

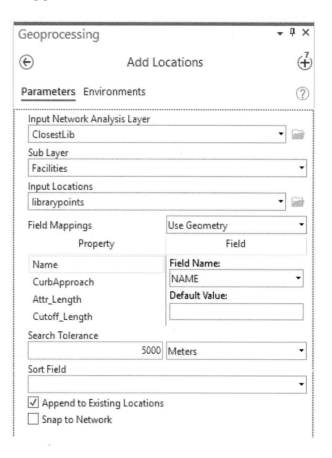

- Under Input Network Analysis Layer, choose **ClosestLib**.

- Under Sub Layer, choose **Facilities** to tell ArcGIS Pro you want to add the library points as the facilities when calculating the closest facility (and not any other option, such as incidents).

- Under Input Locations, choose **librarypoints** to tell ArcGIS Pro to use the librarypoints layer for the source of the facilities.

- For Field Mappings, choose **Use Geometry**.

- Leave the other settings alone and run the Add Locations tool. You see the points representing the libraries appear with the symbology of the facilities in the Contents pane.
- Next, you can add the other half of the closest facility problem: the incidents. On the **Closest Facility** tab, in the **Input Data** group, click the **Import Incidents** button to return to the Add Locations tool.
- Under Input Network Analysis Layer, choose **ClosestLib**.
- Under Sub Layer, choose **Incidents** to tell ArcGIS Pro you want to add the school points as the incidents when calculating the closest facility (as opposed to the facilities).
- Under Input Locations, choose **schoolpoints** to tell ArcGIS Pro to use the schoolpoints layer for the source of the incidents.
- For Field Mappings, choose **Use Geometry**.
- Leave the other settings alone and run the Add Locations tool. You see the points representing the schools.

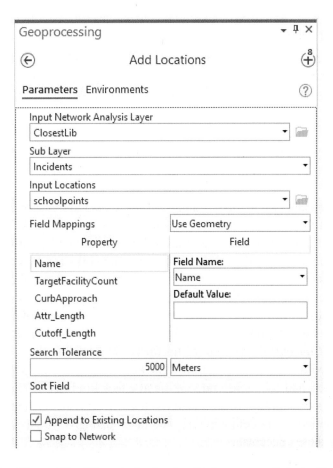

- On the **Closest Facility** tab, in the **Analysis** group, click the **Run** button. A new set of lines is created to establish the closest facility (that is, which schools are assigned to each library).
- On the **Closest Facility** tab, in the **Directions** group, click the **Directions** button. The Directions pane opens again, with information about the distance and driving directions between each school (incident) and its

closest library (facility). Look at which schools connect to which libraries and answer Questions 11.14 and 11.15. *Hint:* You can use the **Explore** tool to click on a closest facility line, a facility, or an incident to get more information about it.

> **Question 11.14** Which library is considered the closest facility for Western Reserve, and how long is the route to that library (in miles)?

> **Question 11.15** Which library is considered the closest facility for Ursuline High School, and how long is that route (in miles)?

Step 11.8 Saving, Printing, or Sharing Your Results

- Save your Module11 project.
- Display only the lines of the network, Libroute3, and the libraries. Turn on the labels for the librarypoints layer, using the Name field. You may have to adjust the font or symbol to make the names stand out in relation to the lines of the roads layer.
- Finally, either print a layout (see Module 3) of the final version of your best route to the libraries (including all the usual map elements and design for a layout) or share your results as a web map through ArcGIS Online (see Module 4).

Closing Time

This module examines a number of applications related to transportation networks and the use of the Network Analyst tools in ArcGIS Pro. Networks are more than just line files representing roads, and several important factors must be taken into account when creating and working with them. Networks can be used for numerous types of spatial analysis, including routing, service areas, and closest facility studies. However, as noted in the introduction to this module, the Network Analyst functions are used with transportation networks. For other types of networks (such as power lines or telephone lines), it is more common to use a utility network. (See *Related Concepts for Module 11* for further information.)

Starting with the next module, we'll be switching gears from the vector data of points, lines, and polygons and looking at a new kind of GIS data representation called *raster data*. There are many features in the real world that vector data may not be suited for modeling in GIS, and raster data give you a different way of representing these concepts.

RELATED CONCEPTS FOR MODULE 11

Utility Networks in ArcGIS Pro

Utility networks are used in ArcGIS Pro to model and analyze non-transportation-related features. There are many things that move along networks that are not vehicles—for example, gas pipelines, sewer lines, and electrical mains. All of

utility network An ArcGIS Pro structure of a series of connected junctions and edges, commonly used for utility-related problems.

these can be modeled with GIS. In these cases, functions such as shortest routes are less applicable than techniques such as tracing the flow of water upstream or determining which electrical transformer is faulty and causing a power outage in a subdivision. Figure 11.13 shows an example of a utility network established to map the components of a gas line (such as compressor stations and regulator stations) and the flow of the gas pipelines.

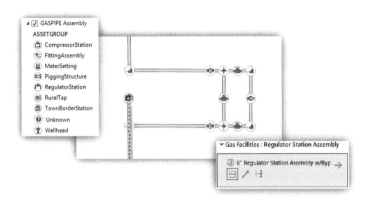

FIGURE 11.13 Using a utility network to examine utility gas lines.

Like a transportation network, a utility network consists of a series of edges and junctions with proper connectivity. Also like a transportation network, a utility network is set up with the proper feature classes present in a feature dataset, but instead of creating a new network dataset, you create a new utility network by using the Create Utility Network tool. Other tools allow you to set up barriers on a utility network as well as place flags on junctions or edges that represent key points in trying to solve problems (for example, placing a flag at the location of a broken transformer or water main).

Traces are used to solve problems in a utility network. ArcGIS Pro comes with a variety of traces. Analysts can trace how the flow along a utility network upstream or downstream will be affected when the power is shut off, or they can find which parts of a network are connected or not connected to a particular point. For instance, a trace might be used to determine the areas on a network that will be affected by a power failure (and this information can be combined with other data to figure out which homes will not have power).

trace A problem-solving technique used with a utility network.

Key Terms

network (p. 261)	connectivity (p. 266)	facility (p. 277)
edges (p. 261)	impedance (p. 267)	drive-time polygon (p. 277)
junctions (p. 261)	stops (p. 268)	closest facility (p. 279)
extension (p. 265)	algorithm (p. 269)	incident (p. 279)
Network Analyst (p. 265)	Dijkstra's algorithm (p. 269)	utility network (p. 282)
network dataset (p. 266)	TSP (p. 273)	trace (p. 283)
turns (p. 266)	service area (p. 277)	

12

How to Use Raster Data in ArcGIS Pro

Learning Outcomes

After studying this module, you should be able to:

- Describe the components of a raster data model.
- Define what grid cell resolution is.
- Explain what a raster zone is.
- Demonstrate the region grouping procedure with raster data.
- List at least two advantages and two disadvantages of the raster data model compared to the vector data model.
- Explain the effect of grid cell resolution on a raster dataset.
- Explain how mosaics of raster datasets are created.
- Describe how a subset of a raster data set is created.

continuous field view A conceptualization of the world in which all items vary across Earth's surface as constant fields, and values are available at all locations along the field.

raster data model A model that represents geospatial data using a series of equally spaced and sized grid cells.

grid cell A single square unit of a raster.

ArcGIS Pro Skills

In this module, you will learn how to do the following in ArcGIS Pro:

- Use geoprocessing settings for raster data.
- Access raster data layer information in the Catalog pane.
- Add raster data to the Contents pane.
- Work with raster data and raster attribute tables.
- Create and analyze regions from raster data.
- Convert vector data to raster format at a variety of raster grid cell resolutions.

 Introduction

The first 11 modules in this book use the points, lines, and polygons of the vector data model to represent real-world items. This system is intuitive; after all, building footprints have definite boundaries that are modeled with polygons, and sidewalks have definite starting and stopping points that are modeled with lines. However, many types of phenomena aren't easily modeled with a set of objects. For instance, every location on Earth has a value for the measured temperature or elevation at that spot. If you were to represent these measurements in GIS using the vector data model, you would need a nearly infinite set of points, each with an assigned value.

A different way of viewing these types of continuously varying phenomena is to think of them as a surface stretching across Earth, with every location on the surface having a value for temperature or elevation at that spot. This way of looking at the world is called the **continuous field view**. In this model, phenomena are modeled as a surface (or field), and every location on the surface can hold a value for each phenomenon, with no gaps in the measurements.

GIS usually uses the **raster data model** to represent these continuous surfaces. This model doesn't use points, lines, or polygons. Rather, it uses a set of evenly distributed square **grid cells**. While the cells themselves are square, an entire raster can be either square or rectangular, and it contains a certain number of cells. For example, a raster with seven rows and five columns would contain 35 grid cells (Figure 12.1). Each cell represents the same sized area on Earth's surface. For instance, each grid cell could represent 9 square meters, or

900 square meters, or 9 square kilometers, depending on the grid cell size being used in the model. In our 7 × 5 raster grid example, if each cell were 900 square meters in size (that is, measuring 30 meters on one side and 30 meters on the other), the entire raster surface would cover a geographic area of 31,500 square meters (35 cells × 900 square meters). All grid cells represent the same size; a single raster cannot have mixed sized cells within it.

Each grid cell contains a single value representing the phenomenon being measured in that area. For example, if our sample raster represented elevation, each cell would have one value representing the height above sea level for the area of the ground covered by the cell. With raster data, it's assumed that the value for the cell is applied to the entire area of the cell. Thus, for our sample grid, all places within the 30 × 30 meter area of a grid cell would be assumed to have the same elevation value. Thus, every location within the raster surface contains an elevation value. In addition to representing elevation, raster data are often used to represent phenomena such as soil type or land cover and are frequently used in delineating watersheds, modeling groundwater, and measuring solar radiation levels on Earth's surface.

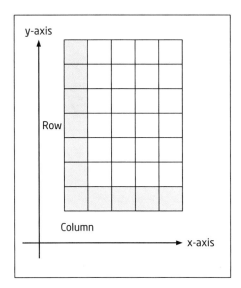

FIGURE 12.1 A sample raster and its grid cells.

 ## Module Scenario and Applications

This module puts you in the role of a county environmental planner, tasked with putting together a map of the county's land cover resources (forests, water bodies, and agricultural areas). Before you can begin the mapping process, you need to acquire and examine land-cover data in ArcGIS Pro.

The following are additional examples of real-world applications of this module's skills:

- As part of a research project about land management policies, a forest manager needs to compile information on the number of deciduous and coniferous forests in an area. A surface of raster land cover provides the basis for him to do so.

- An urban planner needs to assess the amount of impervious surface within a suburban area as part of an analysis of water runoff. Raster data provide an excellent basis for modeling and measuring this area.

- A geologist is undertaking a large-scale study of environmental conditions across a state as these conditions relate to hydraulic fracturing. Utilizing data about elevation, soil conditions, and land cover in a raster format allows her to develop a large-scale dataset within GIS.

 ## Study Area

For this module, you will be working with data from Mahoning County, Ohio.

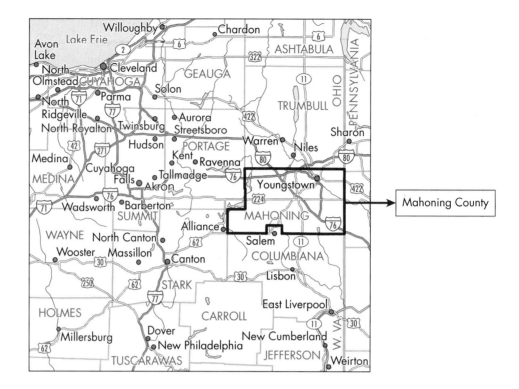

Data Sources and Localizing This Module

The data in this module focus on features and locations within Mahoning County, Ohio. However, you can easily modify this module to use data from your own county or local area instead. The lakes data were extracted from the county-level hydrography dataset available for download from The National Map (see Module 5). Use the NHDWaterbody feature class for the county as the lakes layer.

The NLCD 2011 dataset was downloaded (for free) from the Multi-Resolution Land Characteristics Consortium (MRLC), which is online at **https://www.mrlc.gov/data**. NLCD 2011 data for the entire United States can be downloaded from that site, and the data for your individual county (for example, Delaware County, Indiana) can be extracted using the Extract By Mask tool in ArcGIS Pro. (See page 301 for more about using Extract By Mask.) Alternatively, you can download NLCD 2011 data for a region directly via The National Map and extract your own county's land cover. Note that when using The National Map for downloading NLCD 2011 data, the files are laid out in large sections, and it may take more than one section to cover your entire county.

Step 12.1 Getting Started

- This module's hands-on applications use the data folder called Module12. Your instructor will be able to supply you with this data, or you can download it directly from this book's website at **https://www.macmillanlearning.com/college/us/product/Discovering-GIS-and-ArcGIS-Pro/p/131923075X**. The text in this module assumes that you have this Module12 folder in a computer location referenced as C:\GIS; if you have it somewhere else (for instance, in a flash drive referenced as G:\GISClass), substitute that location and path to the Module12 folder throughout this module.

- The Module12 folder contains the following:
  - Mahonraster: a file geodatabase containing two items:
    - NLCDmahon: a raster dataset of National Land Cover Database (NLCD) with land-use/land-cover classifications for each pixel. It has been clipped for you to the boundaries of Mahoning County.
    - Lakes: a polygon feature class of lakes in Mahoning County
- Start ArcGIS Pro.
- Sign in with your Esri account username and password.
- Create a new project using the **Map** template. Call this project **Module12** and place it in your **C:\GIS\Module12** folder. Ensure that there is not a checkmark in the box next to **Create a new folder for this project**.
- When ArcGIS Pro opens, change the map's name to **Mahoning Land Cover**.
- Several of the raster functions and tools in ArcGIS Pro are part of an extension called **Spatial Analyst** that you need to have licensed to your Esri account and active in order to work with it in this module. If you don't have Spatial Analyst licensed and active, some tools will not work; they will have a lock symbol next to them and will be unavailable. (See Smartbox 11.1 and Troublebox 11.1 in Module 11 for more about extensions and ensuring that you have the Spatial Analyst extension available for use.)
- Add the **NLCDmahon** layer to the Contents pane.
- Before proceeding, you need to make sure your geoprocessing environment settings are correct for the work you'll be doing with raster data in this module. (See **Smartbox 12.1** for more about environment settings and raster data.)

Spatial Analyst An extension that enables several raster functions and tools in ArcGIS Pro.

SMARTBOX 12.1

How do environment settings affect analysis and usage of raster data in ArcGIS Pro?

A raster dataset in ArcGIS Pro is a rectangular grid consisting of values assigned to each cell. When doing conversions or analysis in ArcGIS Pro, several adjustable settings affect all actions that you take in the project. For instance, you can specify the folder to which all raster results should be written or the raster cell size that all analysis results should have. A few environment settings affect only raster data, and several forthcoming modules instruct you on how to put these settings in place, as the settings are crucial when working with raster data. The following settings are particularly important:

- Workspace: This is the location to which raster output will be written. By default this is the file geodatabase that is created along with the project you're working on.
- Scratch Workspace: This is the location for temporary files that are created but not kept (for instance, during the use of ModelBuilder, covered in

Module 20). By default this is the file geodatabase that is created along with the project you're working on.

- Output Coordinate System: This defines what coordinate system is used for any new raster outputs. This is important if you are working with several layers with different coordinate systems: You can specify which coordinate system any new rasters created will match up with.

- Extent: This defines the rectangular boundary that tools and raster functions will operate on. For instance, if you have two different raster grids of land use placed side by side in ArcGIS Pro but Extent is set only to the boundary of one of those two grids, raster tools will only operate on the cells within that boundary and not the other layer.

- Snap Raster: This is used to align raster grids to the extent of another grid. This is typically used when you have grids of different cell sizes and want their output to cover the same area.

- Cell Size: This is used to tell ArcGIS Pro what cell size to use for raster analysis or any new rasters that are created. Setting this saves you time as you will not have to define it each time you use a new raster tool; setting Cell Size also keeps all new rasters you create at a consistent cell size.

- Mask: This defines which cells a tool will operate on. Mask is a different concept from Extent; whereas Extent defines the rectangular boundary of the entire grid, Mask allows you to specify a shape within that boundary, and only the cells in that shape are used. For instance, if the grid represents land cover over an entire region, and you only want to perform an action on the boundaries of a watershed within that region, Extent would set the entire region, and Mask could conform to the shape of the watershed within that region.

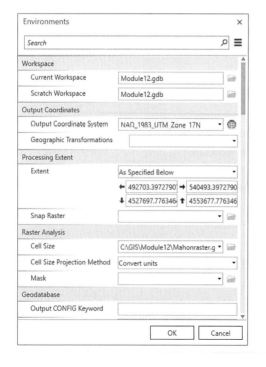

- On the **Analysis** tab, in the **Geoprocessing** group, click the **Environments** button. The Environments pane opens.

- For Current Workspace, use the **Module12.gdb** geodatabase from your C:\GIS\Module12 folder.

- For Scratch Workspace, use the **Module12.gdb** geodatabase from your C:\GIS\Module12 folder.

- For Output Coordinate System, choose **NAD_1983_UTM_Zone_17N**. ArcGIS Pro then updates the coordinate system to that of the NLCDmahon layer.

- For Extent, choose **NLCDmahon**. ArcGIS Pro then updates the x/y coordinates of the lower-left corner and upper-right corner of the grid extent to that of the NLCDmahon layer.

- For Cell Size, choose **NLCDmahon**. ArcGIS Pro then updates the cell size for raster output in this project to that of the NLCDmahon layer (that is, 30 meters).

- Leave the other settings as is and click **OK** to put these geoprocessing environment settings into place.

Step 12.2 Raster Data Basics

- The raster data of the Mahoning County land use should be in the view (because you added the layer in the previous step). Note that ArcGIS Pro assigns a random color to each different cell value when the layer is added. Use the zoom tools to zoom in and out of the preview image to see how some features are represented through the use of grid cells. Answer Question 12.1. For more information on what a raster dataset consists of, see **Smartbox 12.2**.

SMARTBOX 12.2

What does a raster dataset consist of?

A raster dataset in ArcGIS Pro is a rectangular grid consisting of values assigned to each cell. ArcGIS Pro needs to know the geographic coordinates of the entire grid (usually the upper-left or lower-left corner of the grid) in order to properly align it with other datasets. The grid consists of rows and columns, and all the grid cells are the same size. Thus, in its most basic form, a raster consists of a set of coordinates for the grid, the dimensions (rows and columns) of the grid, the size of each cell, and a list of values to populate the grid with (Figure 12.2).

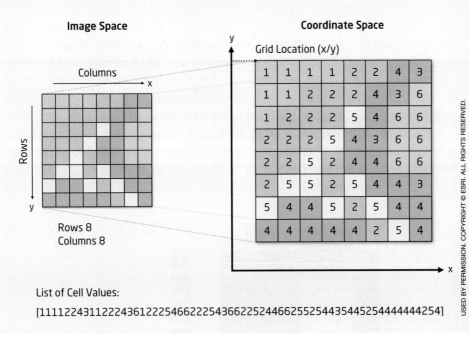

FIGURE 12.2 The basic setup of a raster grid, in which each cell receives a single value, and the coordinates of the grid are known.

List of Cell Values:
[11112243112224361222546622254366225244662552544354452544444444254]

Question 12.1 How are the major roads in the county being modeled with raster data?

- Back in Catalog pane, expand the **Module12** folder, expand the **Mahonraster** geodatabase, right-click on **NLCDmahon**, and select **Properties**. Answer Questions 12.2 and 12.3.

Question 12.2 How many grid cells does this dataset contain? (*Hint:* The rows and columns are available to you in the Properties dialog box.)

Question 12.3 How many square meters does the entire grid cover? (*Hint:* The grid cell size, in meters, is available to you in the Properties dialog box.)

Step 12.3 Working with Zones of Raster Data

- In the Mahoning County land-use raster, you can see several different values assigned to this land-cover raster. Each of these different categories is a zone of raster data. (See **Smartbox 12.3** for more information about zones.) Answer Question 12.4.

SMARTBOX 12.3

What are zones of raster data in ArcGIS Pro?

zone One of several different values assigned to a raster grid cell.

NoData A raster data cell that contains a null value.

In the raster data model, a **zone** refers to each different value assigned to a grid cell. For instance, if a grid cell could contain a value of 1, 2, or 3, then that raster would have three different zones. There can be (and probably will be) multiple grid cells in each zone. A special value that is separate from the zones (but not considered its own zone) is the **NoData** value. As its name implies, NoData contains no value at all (not a value of zero, or a value of infinity; it is simply a null value). NoData values are treated as "holes" in the dataset; sometimes they are errors, but other times they are intentionally placed in the dataset to indicate grid cells that should not be included in the analysis of a raster.

The raster in Figure 12.3 contains four different zones, each assigned a value of 1, 2, 3, or 4. Note that multiple grid cells are assigned to each zone. Two grid cells are assigned a value of NoData; neither of these cells is included in any of the four zones.

FIGURE 12.3 An example of a raster containing four zones and two NoData grid cells.

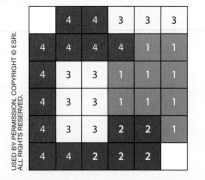

Question 12.4 How many zones are in the NLCDmahon raster grid?

- The values assigned to each grid cell are specific values used as land-cover codes in the National Land Cover Dataset. See **Smartbox 12.4** for more information about the NLCD and what each grid cell value represents.

SMARTBOX 12.4

What is the National Land Cover Dataset?

The **National Land Cover Dataset (NLCD)** is a raster data layer of the various types of land cover of the entire United States (Figure 12.4). Each cell of the raster covers 30 meters by 30 meters, and the value of a cell represents a specific land-cover code. Each code stands for a specific land-cover type, such as open water, deciduous forest, or pasture. See Table 12.1 for a list of the land-cover types represented by each grid cell value.

National Land Cover Dataset (NLCD) A 30-meter raster dataset of land cover for the entire United States.

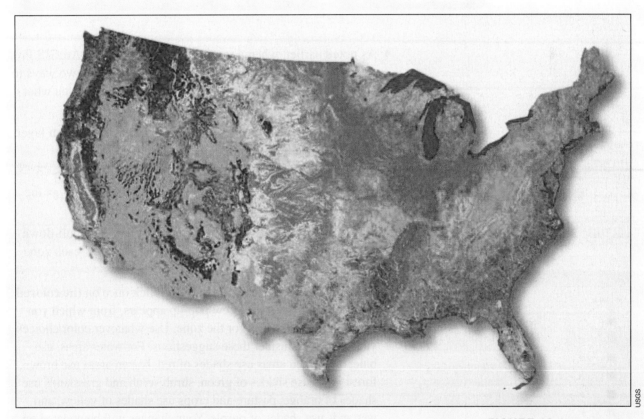

NLCD data are compiled by the Multi-Resolution Land Characteristics (MRLC) Consortium, a group of several U.S. agencies led by the USGS. NLCD data are available in raster layers showing the land cover of the United States circa 2016, 2011, 2006, 2001, and 1992. NLCD data are distributed free, either directly through the MRLC's website (**https://www.mrlc.gov**) or The National Map. In addition, products related to various NLCD data are available for free download, including raster layers of impervious surface, tree canopies, and changes in land cover between time periods.

FIGURE 12.4 The National Land Cover Dataset (NLCD) representing land cover for the United States circa 2011.

Value	NLCD 2011 Classification	Value	NLCD 2011 Classification
11	Open Water	51	Dwarf Scrub
12	Perennial Ice and Snow	52	Shrub Scrub
21	Developed, Open Space	71	Grassland/Herbaceous
22	Developed, Low Intensity	72	Sedge/Herbaceous
23	Developed, Medium Intensity	73	Lichens
24	Developed, High Intensity	74	Moss
31	Barren Land	81	Pasture/Hay
41	Deciduous Forest	82	Cultivated Crops
42	Evergreen Forest	90	Woody Wetlands
43	Mixed Forest	95	Emergent Herbaceous Wetlands

Table 12.1 The assigned values and corresponding classifications for NLCD 2011

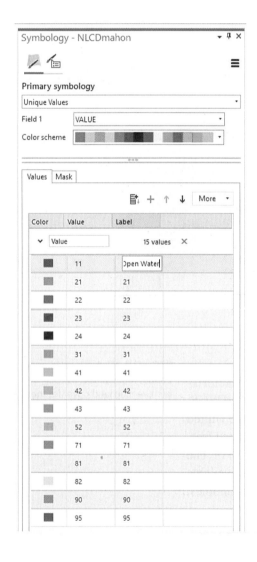

- As noted earlier, when a raster is added to a map in ArcGIS Pro, random colors are assigned to each zone. There are two ways to change the symbology and color choices to better match what those zones represent (such as forests or water):

  1. In the Contents pane, right-click on the **NLCDmahon** layer and choose **Symbology**.

  2. In the Contents pane, click once on the NLCDmahon layer to choose it, and then, on the **Appearance** tab, in the **Rendering** group, click the **Symbology** button.

- The Symbology pane opens. Under the Symbology pull-down menu, choose **Unique Values** so that you can show each zone separately in the raster.

- To change the color scheme of a zone, click once on **the colored box for that zone**, and a new pop-up appears, from which you can choose a new color for the zone. Use whatever color choices you want but consider these suggestions: For water areas, use blue, developed areas use shades of red, barren areas use brown, forest areas use shades of green, shrub scrub and grasslands use shades of orange, pasture and crops use shades of yellow, and wetlands use shades of purple. Your changes will be updated on the map and in the Contents pane.

- Also click in the headings under the Label column to change the displayed text from numbers to text that describes what each zone is really displaying (Open Water, Grasslands, and so on). (See **Smartbox 12.4** on page 291 for the meaning of each land-cover code value.) For instance, you can change the Label field for zone 11 from 11 to Open Water. Change the label for each zone to text of what that land cover type really is. Your changes will be updated on the map and in the Contents pane.

- The next thing you need to do is to open the attribute table of the raster layer. In the Contents pane, right-click on the **NLCDmahon** layer and choose **Attribute Table**. The value of each zone and the count of cells in each zone are displayed. (See **Smartbox 12.5** for further information about raster attribute tables.) Answer Questions 12.5 and 12.6.

SMARTBOX 12.5

How is a raster's attribute table set up in ArcGIS Pro?

In a vector layer's attribute table, each object is stored as its own record, with attributes assigned to that record. Thus, for a polygon layer that represents 200 different lakes in a county, there would be 200 records. However, raster data often use a huge number of grid cells to represent similar data types. (For example, the NLCD layer you're using in this module for land cover of a single Ohio county contains well over a million grid cells.) It would be impossible to navigate an attribute table in which each grid cell was an individual record and the attribute tables for raster layers were set up differently from vector layers.

A raster attribute table consists of two primary columns: a field called Value and another field called Count. The Value field lists each of the raster zones by its designation. For instance, in Figure 12.5, the raster grid cells have numerical values of 1, 2, 3, and 4. Thus, the Value field of the raster's attribute table will contain four entries, one for each number. The Count field is a summation of how many cells comprise each zone. For instance, there are 9 cells in zone 1, so the count for zone 1 will be 9. Additional attributes can then be created in GIS and appended to the raster attribute table for each of the zones (such as the attributes Type, Area, and Code shown in Figure 12.5).

FIGURE 12.5 A raster layer and its corresponding attribute table.

Value	Count	Type	Area	Code
1	9	Forest land	8100	FL010
2	5	Wetland	4500	WL001
3	9	Crop land	8100	CL301
4	11	Urban	9900	VL040

No Data

Question 12.5 How many cells are in each zone? (That is, how many pixels are in the first zone (11), how many are in the second zone (21), and so on?)

Question 12.6 Why is your answer from Question 12.5 so much lower than the sum total of your answers to Question 12.2? What accounts for this discrepancy?

Step 12.4 Regions of Raster Data

- You'll now start using some functions available via the Spatial Analyst extension. In this step, you'll be grouping your cells together to form regions. (See **Smartbox 12.6** for more information.)

SMARTBOX 12.6

What is a region in raster data?

region A set of contiguous grid cells that have the same value.

In ArcGIS Pro, a raster **region** is created by contiguous (touching) cells that have the same value. Each of these groups forms its own region, and a raster may consist of multiple regions. In ArcGIS Pro, the regiongroup process is used to form regions from cells. See Figure 12.6 for an example of regiongrouping. The grid to be regiongrouped is labeled Inputraster. The three cells in the upper-left corner are contiguous and all have the same value. Thus, they form region 1 in the regiongroup output (the grid labeled Grouped). In the Grouped grid, all cells in a region retain the same value; because these cells represent region 1, in the Grouped grid they all have a value of 1. Back in Inputraster, the two cells in the upper-right corner are contiguous and have the same value; thus, they form region 2 in the Grouped grid. Each group of contiguous cells in Inputraster is assigned to a different region in the Grouped grid.

The Grouped raster's attribute table consists of the usual fields Value and Count, where the Value field represents the number assigned to each region—in Figure 12.6, the first region created (value of 1), the second region created (value of 2), and so forth. Count represents how many cells are in each region.

Regiongrouping creates a special field in the output grid called Link, which represents the original grid cell value that each region had in Inputraster. For instance, region 2 originally consisted of grid cells that had a value of 0, so region 2's Link attribute is 0. Similarly, region 6 originally consisted of grid cells that had a value of 100, so region 6's Link attribute is 100.

Regiongrouping is useful in analyzing the size or extent of large contiguous blocks of grid cells that have the same properties. For example, all of the contiguous grid cells in an area that have a land-use code representing a specific type of forested land can be grouped together to create a region representing the mapped boundaries of that type of tree stand. Similarly, several contiguous cells that all have a value representing open water can be grouped together to form a region that represents the extent of a lake.

FIGURE 12.6 An example of the output raster of a regiongroup procedure and its attribute table.

Inputraster:

100	100	0	0
	100	200	200
400	0	0	200
400	0	100	100

Grouped:

1	1	2	2
	1	3	3
4	5	5	3
4	5	6	6

Value = No Data

Grouped Attribute Table

Value	Count	Link
1	3	100
2	2	0
3	3	200
4	2	400
5	3	0
6	2	100

- Open the Geoprocessing pane (on the **Analysis** tab, in the **Geoprocessing** group, click the **Tools** button). The tool you will be using is called Region Group, and you can either search for it or find it within the Spatial Analyst Tools toolbox, in the Generalization toolset. Open the Region Group tool using one of these methods.
- Under Input raster, choose **NLCDmahon**.
- Under Output raster, click the **browse** button and navigate to your **C:\GIS\Module12** folder, go to the **Module12.gdb** geodatabase, and call the output raster **RegionG1**.
- Under Number of neighbors to use, choose **Eight**.
- Under Zone grouping method, choose **Within**.
- Make sure the **Add link field to output** box is checked.
- Run the Region Group tool. A new grid (RegionG1) is added to the Contents pane. Open this new RegionG1 grid's attribute table. Answer Question 12.7.

Question 12.7
How many regions have been created?

- By default ArcGIS Pro assigns the RegionG1 group a graduated color scheme going from black to white; that is, low-numbered regions are shaded black and high-numbered regions are shaded white. This makes it very difficult to discern individual regions. To change this, in the Contents pane, choose the **RegionG1** layer and then, on the **Appearance** tab, in the **Rendering** group, click the arrow under the **Symbology** button and choose **Unique Values**. ArcGIS Pro assigns a different color to each region, making the regions easier to see. (Note that this may take a couple of minutes to fully render because there are several thousand unique values.)
- In RegionG1's attribute table, right-click on the **Count** field (click on the word itself) and choose **Sort Descending** to sort the Count column in order from highest to lowest count.
- Click on the row tab at the far left side of the attribute table on the row that corresponds to the largest region. The entire row is then highlighted in cyan. (In other words, the row and all the grid cells that correspond to it on the map are now selected, just as in Module 2.)
- Examine the map to see the new selected region. Using the information on the map and in the attribute table, answer Questions 12.8, 12.9, and 12.10. Close the RegionG1 attribute table when you're done.

Question 12.8
What land use does the region with the greatest number of grid cells represent?

Question 12.9 How much real-world area does this region take up (in square meters)?

Question 12.10 Redo the regiongroup procedure to create a raster called RegionG2 but instead use the Four option for number of neighbors. (This means that only grid cells that are contiguous in the four cardinal directions will form a group, instead of in all eight directions.) Find the largest region again. How much real-world area does this new region (computed using the Four option instead of Eight) take up (in square meters)?

Step 12.5 Converting Vector Data to Raster Data

- In this portion of the module, you'll be comparing vector data to the raster data. (See **Smartbox 12.7** for more information about the vector data model versus the raster data model.) You can turn off the Region Group raster(s) at this point and leave the NLCDmahon raster turned on.

SMARTBOX 12.7

How do raster data compare with vector data?

The choice between using vector data or raster data to represent geospatial data often depends on what you want to represent in ArcGIS Pro, and there are several advantages and disadvantages to each. For instance, land cover could be mapped with grid cells (as in the NLCD) or with a series of polygons. Figure 12.7 shows a hypothetical landscape composed of four different land-cover types (1 = urban/built, 2 = agricultural, 4 = forested, 5 = water) modeled with polygons (Figure 12.7a) and grid cells (Figure 12.7b).

The raster data model is simpler than the vector data model; a grid of numbers is a simpler structure than several polygons of various shapes and dimensions. However, the vector data model is more compact than the raster data model; the landscape is modeled with only 6 objects in the vector model but with 64 grid cells in the raster model. The vector model can provide more

FIGURE 12.7 A hypothetical landscape modeled using (a) vector data and (b) raster data.

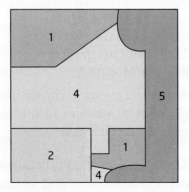

aesthetically appealing output than can the raster model. For example, the border of the "body of water" on the right side of the landscape in Figure 12.7 has naturally curved edges that are more easily represented by the polygons of the vector model than by the square blocks of the raster model.

Vector data are also better for representing topological relationships (see Module 7). It's difficult to represent things like the connectivity of road intersections or the adjacency of shared parcel boundaries with raster data. However, when overlaying two or more data layers, the process is much simpler with raster data (see Module 20) and more difficult with vector data (see Module 9). Finally, raster data are better at representing surfaces or datasets with a high degree of spatial variability. For instance, areas with changes in elevation or temperature are more efficiently represented with raster data than with vector data.

- Add the lakes feature class from your Mahonraster geodatabase folder.

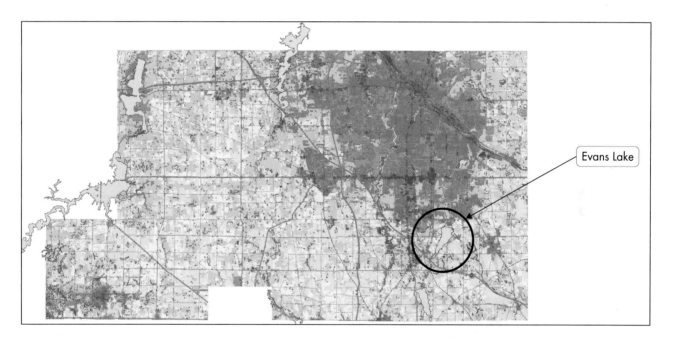

- Zoom in on Evans Lake (in southern Mahoning County). Answer Question 12.11 and then turn off the NLCDmahon raster.

> **Question 12.11** How closely (in terms of the spatial arrangement) does the raster version of Evans Lake match the vector feature class version?

- You'll now convert the vector polygon shapefile into a raster layer. Open the Geoprocessing pane. The tool you will use is Polygon to Raster, so either search for it or find it within the **Conversion** toolbox in the **To Raster** toolset. Open the Polygon to Raster tool using either method.

298 CHAPTER 12 How to Use Raster Data in ArcGIS Pro

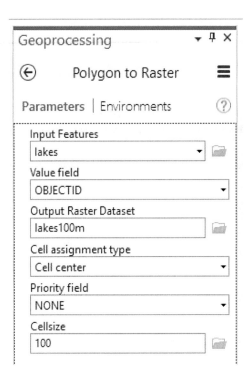

- Under Input Features, choose **lakes**.
- Under Value field, choose **OBJECTID**.
- For Output Raster Dataset, click the **browse** button and navigate to your **C:\GIS\Module12** folder, go to the **Module12.gdb** geodatabase, and call the output raster **lakes100m**.
- For Cellsize, type **100** (for cells of 100-meter resolution).
- Keep the other settings at their defaults and run the Polygon to Raster tool. A new layer called lakes100m is added to the Contents pane; it is a raster version of the lakes layer using grid cells of 100-meter resolution.
- Open the lakes100m attribute table. Answer Question 12.12.

Question 12.12 How many grid cells does Evans Lake consist of when modeled with 100-meter grid cells?

- Examine the vector shapefile of the lakes and your new lakes100m grid; turn each one on and off and compare the two. Answer Question 12.13.

Question 12.13 How does the spatial arrangement of the lakes100m grid compare to the vector feature class at Evans Lake?

- Run the Polygon to Raster tool again but this time create a new output raster dataset called **lakes10m** and for Cellsize enter **10** (for 10 meters). For more about the effect of raster resolution on data representation, see **Smartbox 12.8**. Then Answer Questions 12.14, 12.15, and 12.16.

SMARTBOX 12.8

How does resolution affect the raster dataset?

resolution The ground area represented by a single grid cell in a raster.

The **resolution** of a raster dataset affects not only the size of the raster but also the amount of detail that can be captured in the raster. Remember that each grid cell represents a fixed amount of area on the ground. The larger (or coarser) the raster resolution, the fewer grid cells it takes to cover the surface. However, due to the larger size of the cells, features appear "blockier" and are less detailed. At smaller (or finer) raster resolutions, items appear more detailed, but it takes more grid cells to represent items on the surface.

Figure 12.8 shows three different representations of a lake using raster grid cells. You can see the polygon boundaries of the lake, along with how each of the three raster resolutions matches up with them. The finer-resolution cells (1 meter) can capture more detail of the lake, but the grid requires 256 cells to do so. As resolution becomes coarser, fewer cells are used in the datasets, but the amount of detail declines as well. With 4-meter resolution, only 16 cells are used to cover the same area, but the lake is modeled in a very general way. This trade-off in the size of the raster (and the resulting number of cells being used) and the amount of detail that can be captured in the raster is inherent in the resolution used by the raster.

FIGURE 12.8 The effect of three different raster resolutions on the representation of data.

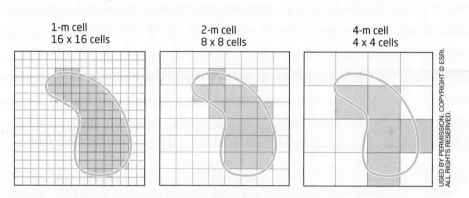

Question 12.14 How many grid cells does Evans Lake consist of when modeled with 10-meter grid cells?

Question 12.15 How does the spatial arrangement of the lakes10m grid compare to the vector feature class?

Question 12.16 What is the advantage of using the lakes100m raster to represent Evans Lake, and what advantage is there to using the lakes10m raster to represent Evans Lake?

Step 12.6 Printing or Sharing Your Results

- Save your Module 12 project. Finally, print a layout (see Module 3) showing the NLCD data for the county (along with the 10-meter lakes grid).

Closing Time

This module examines the basics of the raster data model and compares raster data with vector data. Neither format is inherently superior, but certain problem-solving tasks or models lend themselves more to one format or the other. Raster data are often more associated with representing phenomena that vary continuously, such as precipitation patterns, land-cover types, or elevation, while vector data are used to represent items as series of objects. The raster data model is simple to use and very versatile; see *Related Concepts for Module 12* for more ways to use raster datasets in GIS. You'll work with raster data several more times throughout the book.

One specific use of rasters that this module hasn't discussed is remotely sensed imagery taken from a satellite or aircraft. Remote sensing imagery is a common data source used with GIS, and ArcGIS Pro treats it in much the same way as the raster data you worked with in this module. For instance, Landsat satellite imagery (discussed in Module 13) was used as the initial basis for creating NLCD. In Module 13, you'll do much more with satellite imagery in raster format.

RELATED CONCEPTS FOR MODULE 12

Using Subsets and Mosaics of Raster Data

When you're using raster data, it is always in the form of a rectangular (or square) grid, but this shape may not represent the size or dimensions of the area you're studying. For example, the raster you used in this module of Mahoning County, Ohio, was a full rectangle, even though you were working only with the grid cells within the oddly shaped county boundary. A dataset such as NLCD covers the entire country, but you were working with only a small portion, or **subset** of the data. The rectangular area of the raster didn't include cells with land-cover values from surrounding counties within the raster's area. Instead, ArcGIS Pro was able to extract only the grid cells that fit within the county boundary.

There are a variety of ways of creating subsets of raster datasets in ArcGIS Pro. If you have a raster and want to work with only part of it, you can do this by using the **Clip function** available as a raster function in ArcGIS Pro. Raster functions can be accessed on the ribbon, on the Analysis tab, within the Raster group, by clicking the Raster Functions button. The Raster Functions pane that opens provides a variety of different raster-specific function. The Clip raster function (located in the Data Management options) allows you to either remove the raster grid cells inside the boundary of the input or remove all the grid cells that fall outside the boundary of the input (Figure 12.9). The shape used for the clipping can either be another raster or a polygon feature class.

Important Note: Despite the name, the Clip raster function is not the same as the Clip geoprocessing tool used in Module 9.

subset A raster created by removing a set of grid cells from a larger raster.

Clip function An ArcGIS Pro raster function used for creating subsets of raster layers.

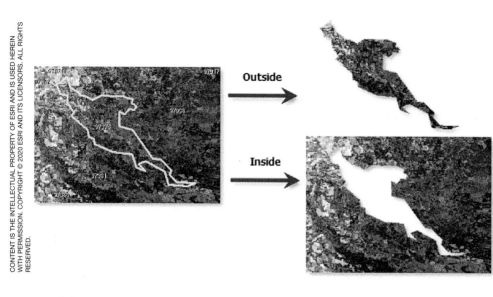

FIGURE 12.9 Using the Clip raster function as both an outside clip and an inside clip.

Subsets can also be created in ArcGIS Pro using a variety of **extraction** geoprocessing tools that allow you to specify the subset of the data you want to work with. For example, you can specify the dimensions of a rectangle or polygon and extract all grid cells within those boundaries to a new raster dataset to create a subset. (This is done by the Extract By Rectangle tool or the Extract By Polygon tool, respectively.) You could also use a polygon boundary as a "cookie cutter" shape to extract all the cells under the digitized shape. (You will do this in Modules 13 and 15.) In essence, you can create a "mask" and extract only the cells under that "masked" region to work with. (You do this with the **Extract By Mask** geoprocessing tool; see Figure 12.10.)

extraction The process of creating a subset raster from a larger raster.

Extract By Mask An ArcGIS Pro tool that performs extraction using a polygon boundary.

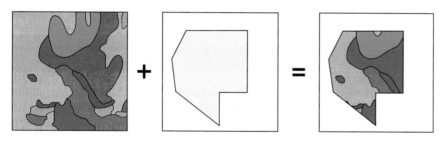

FIGURE 12.10 Extracting a subset of grid cells using the boundaries of a predefined "mask" polygon.

Conversely, you may have multiple rasters that need to be fitted together so that you can manipulate them as a single layer. These rasters may be adjacent to one another (like tiles) or may overlap in certain ways. You need to "knit" the various rasters together to form a single raster **mosaic**. ArcGIS Pro provides several options for doing so through the use of the Mosaic geoprocessing tool (Figure 12.11).

mosaic A raster created by joining several smaller rasters.

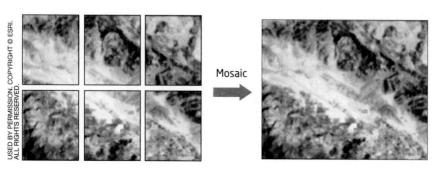

FIGURE 12.11 Six different raster datasets combined into a single mosaicked raster.

mosaic dataset An ArcGIS Pro file structure designed for managing one or more rasters or collections of raster data.

Raster data are also commonly shared by creating a **mosaic dataset** and adding the rasters to it. A mosaic dataset is a special file structure that can be created in the Catalog pane (or through the Create Mosaic Dataset geoprocessing tool); it can be used to work with one or more rasters by pointing to their locations and managing the rules for visualizing and mosaicking the rasters together. (*Important Note:* Despite its name, a mosaic dataset does not actually contain the raster layers themselves; a mosaic dataset is neither a mosaic nor a dataset.) Mosaic datasets are frequently used with very large raster datasets or large numbers of rasters or with several imagery rasters (see Module 13).

Key Terms

continuous field view (p. 284)
raster data model (p. 284)
grid cell (p. 284)
Spatial Analyst (p. 287)
zone (p. 290)
NoData (p. 290)

National Land Cover Dataset (NLCD) (p. 291)
region (p. 294)
resolution (p. 299)
subset (p. 300)
Clip function (p. 300)

extraction (p. 301)
Extract By Mask (p. 301)
mosaic (p. 301)
mosaic dataset (p. 302)

# How to Use Remotely Sensed Imagery in ArcGIS Pro

ArcGIS Pro Skills

In this module, you will learn how to do the following in ArcGIS Pro:
- Add multiband imagery in color to the Contents pane.
- Visually interpret remotely sensed imagery.
- Add NAIP imagery from the Living Atlas.
- Add, visually enhance, and examine Landsat OLI imagery.
- Change band combinations in a multispectral Landsat OLI image.

Learning Outcomes

After studying this module, you should be able to:
- Define the term *remote sensing*.
- Explain how orthoimagery is different from regular aerial photography.
- List seven elements of visual image interpretation.
- Describe the steps in the remote sensing process.
- Define what a band of energy is in relation to remote sensing.
- Explain what a digital number is in relation to remote sensing.
- Describe the characteristics of the Landsat 8 satellite in relation to spatial, spectral, temporal, and radiometric resolution.
- List at least three high-resolution satellite sensors and the spatial resolution associated with each.
- Explain the purpose of georeferencing in relation to imagery in GIS.

Introduction

Thanks to geospatial technology applications such as Google Earth and Google Maps, overhead images of the ground have become commonplace. By examining imagery of a city from different years, it is possible to see how the area has grown and where the city is expanding in terms of new residential, commercial, or industrial developments. Overhead imagery is acquired through **remote sensing** technologies, in which a camera or sensor records information about reflected or emitted energy from targets on the ground. This information is then converted into an image that can be used in GIS.

Remotely sensed imagery comes in two varieties: **satellite imagery** (in which the sensors on a spacecraft capture the image) and **aerial photography** (in which cameras or sensors on an aircraft take the images). Both of these are used extensively with GIS. For instance, the imagery basemap you used in Modules 6 and 7 consisted of remotely sensed information that was processed into images for you. Taking images of the ground from the air has been going on for a long time. The first aerial photograph, of a French landscape, was taken in 1858 by a man in a balloon. Today, equipment on satellites, modern aircraft, and unmanned drones (also referred to as **UAS**, for unmanned aircraft systems) captures digital images of the ground and records energy reflections unseen by the human eye (such as infrared or thermal sources).

The ability of sensors to record energy outside the visible light that we can view opens up many new avenues of analysis. For instance, healthy vegetation (leaves, grass, crops, and so on) reflects a lot of infrared light. If we could see infrared light and looked at a well-manicured, watered, and healthy yard, we would be nearly blinded by its reflection of this type of light. Although infrared light is invisible to us, remote sensing devices have the capability to record it so that we can view this kind of energy in an image. For instance, Figure 13.1 shows the online Esri Landsat Explorer tool, which uses the color red to highlight the areas heavily reflecting infrared light (such as healthy fields, forests, grass). This information can be used for further analysis in GIS.

FIGURE 13.1 A satellite image of Redlands, California, shown in color infrared in the Esri Landsat Explorer app.

A satellite or aerial image represents a "snapshot" in time; it reflects the current conditions at the time the image was acquired. How often you can acquire imagery is dependent on the source. For example, aerial photographs from the National Agriculture Imagery Program (see Smartbox 13.3 on page 311) are acquired every 3 years, while Landsat satellite imagery (see Smartbox 13.6 on page 319) is acquired every 16 days.

Many remote sensing platforms capture the ground below at **nadir**, or the area directly underneath the camera or sensor. When the sensor or camera is looking straight down at the ground below, the result is a **vertical image**. Some satellite platforms have **off-nadir viewing** capability, which allows them to capture images from areas not directly underneath them, thus allowing acquisition of imagery on a more frequent basis (such as every 1 to 3 days, as compared to the 16-day orbital cycle of the Landsat satellite, which captures the ground at nadir). Looking off-nadir often results in capturing **oblique imagery**, with the sensor or camera tilted and viewing the ground at an angle; you often see things like some of the sides of buildings in oblique imagery.

There are many different remote sensing platforms available today, including everything from equipment onboard UAS flown from an app on a tablet to numerous satellites in low Earth orbit. This module provides an overview of how remote sensing works as well as how to utilize remotely sensed data alongside other layers in ArcGIS Pro, but it can't hope to cover every remote sensing platform available.

remote sensing The process of collecting information related to the reflected or emitted electromagnetic energy from a target by a device a considerable distance away from the target onboard an aircraft or spacecraft platform.

satellite imagery Digital images of Earth acquired by sensors onboard orbiting spaceborne platforms.

aerial photography The acquisition of imagery of the ground taken from an airborne platform.

UAS Unmanned aircraft system; drone aircraft flown by pilot on the ground using remote control.

For a more detailed overview of many different types of satellite platforms and how their particular imagery is handled in ArcGIS Pro, see **https://pro.arcgis.com/en/pro-app/help/data/imagery/satellite-sensor-raster-types.htm**.

Module Scenario and Applications

This module puts you in the role of an urban planner for the Cleveland, Ohio, area. You're collecting information regarding the city's waterfront area for potential redevelopment or expansion. Before you begin doing a study in GIS, you'll need current information about many of the properties and their usages (such as sports stadiums, a nature preserve, and a local airport). You'll examine these through different types of satellite imagery and aerial photos.

The following are additional examples of real-world applications of this module's skills:

- A forest ranger is measuring the extent of the burn scar of a wildfire at a national forest. Recent satellite imagery can provide a basis for GIS analysis to reveal not only the spatial dimensions of the burn but also information about the health of the trees and vegetation near the burn area.

- A public utilities worker is in the process of updating a GIS database with information about the placement of new traffic and safety infrastructure items. Recent aerial photographs of the city will aid her in plotting their locations.

- A county auditor is preparing to compute tax assessments on the privately owned properties in the county. As part of this GIS analysis, he needs current high-resolution aerial photographs to help assess whether new additions (such as swimming pools or decks) are now part of the properties.

nadir The location under the sensor or camera in remote sensing.

vertical image A remotely sensed image in which the camera is looking down at the landscape.

off-nadir viewing The capability of a sensor to observe areas other than the ground directly underneath it.

oblique image A remotely sensed image acquired at an angle.

Data Sources and Localizing This Module

The data in this module focus on features and locations within Cleveland, Ohio. However, you can easily modify this module to use data from your own city or local area instead. For example, if you were performing this module's activities in Norfolk, Virginia, you could easily obtain the same kinds of imagery. The Landsat data were downloaded for free from GloVis (**https://glovis.usgs.gov**), processed into an image, and subsetted to show only a section of Cleveland. Similar data are available for Landsat scenes containing Norfolk from GloVis and also Earth-Explorer (**https://earthexplorer.usgs.gov**). In addition, the Landsat imagery from ArcGIS Online used in Step 13.5 is also available for Norfolk (and other areas in the United States and elsewhere around the globe).

The orthoimages used in Step 13.2 were downloaded (for free) from Earth-Explorer, and multiple options for orthoimagery for your own area (such as Norfolk) within the United States are also available. Finally, the NAIP imagery used in Step 13.3 is available free for all states in the United States. If you were examining Norfolk, you would access the Virginia NAIP imagery.

Step 13.1 Getting Started

- This module's hands-on applications use the data folder called Module13. Your instructor will be able to supply you with this data, or you can download it directly from this book's website at **https://www.macmillanlearning.com/college/us/product/Discovering-GIS-and-ArcGIS-Pro/p/131923075X**. The text in this module assumes that you have this Module13 folder in a computer location referenced as C:\GIS; if you have it somewhere else (for instance, in a flash drive referenced as G:\GISClass), substitute that location and path to the Module13 folder throughout this module.
- The Module13 folder contains the following:
  - cle.img: a subset of a Landsat 8 OLI / TIRS satellite image from 9/27/14
  - Rockhall.tif: a high-resolution orthophoto of the Rock & Roll Hall of Fame in Cleveland, Ohio
  - HumphreyPark.tif: a high-resolution orthophoto of Humphrey Park in Cleveland, Ohio
  - Location.tif: a high-resolution orthophoto of an area in Cleveland, Ohio
  - Harbor.tif: a high-resolution orthophoto of a harbor area on the Cleveland waterfront
- Start ArcGIS Pro.
- Sign in with your Esri account username and password.
- Create a new project using the **Map** template. Call this project **Module13** and place it in your **C:\GIS\Module13** folder. Ensure that there is not a checkmark in the box next to **Create a new folder for this project**.
- When ArcGIS Pro opens, change the map's name to **Cleveland Imagery**.
- Ensure that you have the Spatial Analyst extension licensed and activated for your use. (See Module 12 for how to do this.)

Step 13.2 Working with Orthoimagery in ArcGIS Pro

- Aerial photography is an excellent source of remotely sensed imagery for GIS, especially for discerning finer details in an image. Aerial imagery is used in ArcGIS Pro as a raster. In GIS, it's common to use aerial photography as orthoimagery (see **Smartbox 13.1**).

SMARTBOX 13.1

What is orthoimagery?

An aerial photo is, unfortunately, subject to a number of distortions. For one, the photo doesn't have the same scale at all locations covered in the photo. Photo scale is a function of the distance to the camera, and as the plane flies over the landscape below, some items are closer to or farther from the camera, and thus the photo doesn't have the same scale everywhere. In addition, the farther away tall objects are from the center of the photo (called the *principal point*), the more they tend to "lean" away from the center, an effect referred to as *relief displacement*.

A regular aerial photo is affected by these issues, but it can be put through a process called *orthorectification*, in which the effect of relief displacement is removed, the photo is given uniform scale (that is, the same scale at all locations in the photo), and the photo is given real-world coordinates and georeferenced. (See *Related Concepts for Module 13* for further information.) An aerial image that has gone through orthorectification is referred to as an **orthoimage** (or, often, an **orthophoto**). Orthoimagery is often used in GIS because the images align with other geospatial data and have a coordinate system assigned, and thus they can be used for measurements or as a digitizing source because they have uniform scale.

orthoimage (orthophoto) A spatially referenced aerial photo with uniform scale.

- Add the **Rockhall.tif** orthoimage to the map. If prompted by ArcGIS Pro about Pyramids, click Yes. When adding the Rockhall.tif, make sure to add just that one file and not any of the files inside it, such as Band_1, Band_2, or Band_3. You should add the entire image rather than separate bands. After the Rockhall.tif layer is added, you see it listed in the Contents pane as Red: Band_1, Green: Band_2, and Blue: Band_3, each with a different color (red, green, and blue) next to it. You may see, instead, an entry like Rockhall.tif_Band_1 that is in grayscale; see **Troublebox 13.1** for why this may be so and how to fix it.

TROUBLEBOX 13.1

Why is the imagery in grayscale (black and white) instead of color?

Unless an image is intended to be in black and white, a single remotely sensed image is often made up of several layers (or bands). When adding an image in ArcGIS Pro using the Add Data button on the ribbon, if you double-click the filename as usual, you end up in an Add Data dialog box that asks you to select one of the image's bands (Figure 13.2). If you select one band, it is shown in shades of gray from black to white instead of as a color composite image. (See Smartbox 13.5 on page 313 for more information about what these various bands represent.)

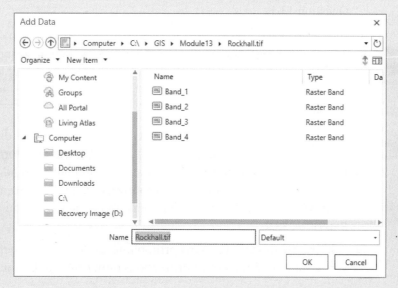

FIGURE 13.2 The multiple bands that make up a remotely sensed image in ArcGIS Pro.

- This orthoimage covers only a small section of the Cleveland waterfront, but it shows that section at a very high level of detail. Zoom in to the Rock & Roll Hall of Fame until you can make out the extent of the building and its nearby surroundings.

- Examine the orthoimage of the Rock & Roll Hall of Fame and answer Question 13.1.

Question 13.1 What features of this building, as seen from the sky in this orthophoto, would help you identify it as the Rock & Roll Hall of Fame (or at least a unique building dedicated to music)?

- Working with remotely sensed imagery in GIS gives you a different perspective when examining objects, structures, and land formations as it means looking at them from above rather than on the ground. It sometimes takes a different way of thinking about what you're looking at to be able to identify items in an image. For more about interpreting images, see **Smartbox 13.2**.

- Add the **HumphreyPark.tif** orthophoto to the map. As before, you want to add the entire orthophoto rather than individual bands.

SMARTBOX 13.2

How is visual image interpretation performed using remotely sensed imagery?

When you try to interpret features in an aerial photo or a satellite image (for instance, objects in developed areas or physical features in a natural landscape), you are like a detective, searching for clues in the image to figure out what you're really looking at. Clues like the size and shape of objects or the texture of glass versus steel can help to determine what you're looking at. **Visual image interpretation** is the process of identifying objects and features in a remotely sensed image based on a number of distinct elements:

- **Pattern:** This is the physical arrangement of objects in an image. How objects are ordered (or disordered) often helps in interpreting an image. A large array of cars set up as if for inspection in a parking lot around a relatively small building will probably be a car dealership rather than some other type of shopping area. Evenly spaced rows of jet airplanes suggest a military base, while a haphazard arrangement of aircraft could be a group of statics on display at an aircraft or military museum.

- **Site and association:** Site deals with the location characteristics of an item, and association relates an object in an image to other nearby features in the image. For example, a football field itself has enough distinctive features to identify it, but the related phenomena you might see in an image (the number of bleachers, the number and distribution of the seats, and perhaps the amount of nearby parking) would help in determining whether you're looking at a field used by a high school team, a 1-AA college team, or a professional NFL team.

- **Size:** This describes the length, width, and area on the ground of objects in an image. The relative sizes of objects in an image can offer good clues in visual image interpretation. For instance, the average length of a car is about 15 feet. If a car is present in an image, you can gain information about other objects by comparing their lengths with the length of the car; you can tell, for instance, if a structure near the car is the size of a house or the size of a shopping center. The sizes of some features remain the same, whatever the image. If baseball diamonds or football fields are present in an image, elements in them (the 90 feet between bases or the 100 yards between goal lines) can be compared with other objects in the image to calculate their relative sizes.

- **Shadow:** This is the dark shape cast by an object with a source of light shining on it. Shadows help provide information about the height or depth of the objects that are casting them. For instance, when looking down at a tall structure like the Space Needle in Seattle, the height of the tower itself would be partially hidden due to the nature of being photographed from directly above, but we could figure out that it must be very tall if we could see the shadow it casts. Shadows can help us identify objects that are virtually unidentifiable from a viewpoint directly above them. From directly above, a set of railings looks like a thin, almost invisible thread around a park or in front of a building, and telephone poles look like dotted lines. However, if you look down at them

visual image interpretation The process of examining information to identify objects in an aerial (or other remotely sensed) image.

pattern The arrangement of objects in an image, used as an element of image interpretation.

site and association Information related to the locations of objects and their related attributes in an image, used as elements of image interpretation.

size The physical dimensions (length, width, and area on the ground) of objects, used as an element of image interpretation.

shadow The dark shapes in an image caused by a light source shining on an object, used as an element of image interpretation.

from directly above late on a sunny afternoon, railings and telephone poles can be identified immediately by their shadows.

- **Shape:** This is the particular form of an object in an image. The distinctive shapes of objects in an aerial photo can help us identify them. A baseball field has a traditional diamond shape, a race track has a distinctive oval shape (and even an abandoned horse race track may still show evidence of the oval shape on the landscape), and the circular shape of crops may indicate the presence of center-pivot irrigation in fields.

- **Texture:** This refers to the differences of a certain shading or color throughout parts of an image. The texture of an object can be identified as coarse or smooth. For instance, different types of greenery can be quickly distinguished by their texture; a forest of trees and a field of grass may have the same tone, but the trees appear very rough in an image, while grass looks very smooth. The texture of a calm lake looks very different from the rocky beach surrounding it.

- **Tone:** This is the particular grayscale (black to white) or intensity of color in an image. The tone of an object can convey important information about its identity. For example, the light-blue color of a swimming pool would make it easy to identify when set off against the pale color of the surrounding concrete deck. Similarly, a wooden walkway and a sandy beach have different tones that help in distinguishing them from one another in a photo.

shape The distinctive form of an object, used as an element of image interpretation.

texture Repeated shadings or colors in an image, used as an element of image interpretation.

tone The grayscale levels (from black to white) or range of intensity of a particular color discerned as a characteristic of particular features present in an image, used as an element of image interpretation.

Although the image covers more than the park itself, you should be able to identify the park. Zoom in to see its entire extent. Answer Questions 13.2 and 13.3.

Question 13.2 What four different outdoor sports does Humphrey Park accommodate? Which elements of image interpretation can you use to determine this, and how do you use them?

Question 13.3 What type of development (residential, industrial, commercial, and so on) is to the immediate north and northeast of Humphrey Park? Which elements of image interpretation can you use to determine this, and how do you use them?

- Add the **Location.tif** orthophoto to the map. As before, you want to add the entire orthophoto rather than individual bands. There is one large prominent complex shown in this orthophoto. Zoom in, closely examine the whole facility, and answer Question 13.4.

Question 13.4 The facility shown in the orthophoto comprises many different things, but they all contribute to one overall place. What kind of facility is shown in this image? Which elements of image interpretation can you use to determine this, and how do you use them?

Step 13.3 Working with NAIP Imagery in ArcGIS Pro

- A good source of remotely sensed imagery is the 1-meter aerial orthoimages from the National Agriculture Imagery Program (NAIP), administered by the U.S. Department of Agriculture's Farm Service Agency. For more information about NAIP, see **Smartbox 13.3**.

SMARTBOX 13.3

What is NAIP imagery?

NAIP imagery, collected since 2003, is provided by the National Agriculture Imagery Program, which is sponsored by the Farm Service Agency, a branch of the USDA (United States Department of Agriculture). Originally on a 5-year cycle of image collection, in 2009 NAIP began a 3-year cycle. NAIP orthoimagery is 1-meter spatial resolution and is distributed free by the USDA. Most NAIP imagery is the usual three-band (red, green, and blue) color imagery, but some imagery is available as color infrared (which uses the NIR, red, and green bands instead).

Instead of obtaining an individual orthoimage, you can obtain NAIP imagery in a special format called a **DOQQ** (Digital Orthophoto Quarter Quad, also often just referred to as a DOQ, or Digital Orthophoto Quad). A DOQQ is a special product in which the orthoimage is compiled into a single image that covers one-quarter of a 1:24000 USGS topographic map quadrangle. (See Module 16 for more information about topographic maps.) The orthoimagery in a DOQQ covers 3.75 degrees of latitude by 3.75 degrees of longitude in area. NAIP imagery is also available as a mosaicked image of an entire county.

NAIP National Agriculture Imagery Program; a program maintained by the USDA.

DOQQ Digital Orthophoto Quarter Quad; a product showing orthoimagery covering 3.75 degrees of latitude by 3.75 degrees of longitude.

- The USDA makes NAIP imagery freely available from a variety of sources, including through Esri's Living Atlas (see Module 5). To access it, on the **Map** tab, in the **Layer** group, click the **Add Data** button.

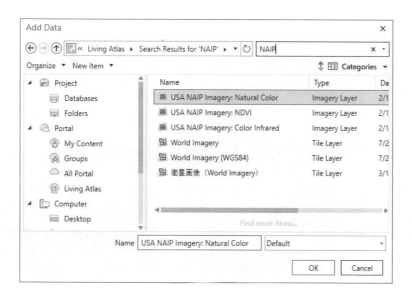

- In the Add Data dialog box, first choose **Living Atlas** from the options on the left. Then, in the search box, type **NAIP** and press **Enter**. Finally, from the choices that appear, choose **USA NAIP Imagery: Natural Color** and click **OK**. A new item called USA NAIP Imagery: Natural Color is added to the Contents pane; this is NAIP imagery for the entire United States.

- Notice that the NAIP imagery doesn't seem as sharp or defined as the individual orthophotos you've been using. This is because the orthophotos are either 1-foot or 0.3-meter spatial resolution, while the NAIP imagery is 1-meter spatial resolution. For more about spatial resolution, see **Smartbox 13.4**.

SMARTBOX 13.4

What is spatial resolution?

spatial resolution The size of the area on the ground represented by 1 pixel's worth of energy measurement.

Each pixel in an image represents a certain amount of size on the ground. The area of a single pixel is the **spatial resolution** of the image. When remote sensing data are collected, the spatial resolution of the sensor is fixed. That is, images with 30-meter resolution can't be "fine-tuned" to obtain 3-meter-resolution images instead.

The spatial resolution helps in determining what kinds of detail can be observed in an image. For instance, in a 30-meter-resolution image, each pixel represents a 30 meter × 30 meter area (900 square meters) on the ground; individual trees and cars are much smaller than that and would not be able to be resolved (seen). So, if you were digitizing tree stands, you would need a much finer spatial resolution in order to see the necessary details for your project. The NAIP imagery, for example, is 1-meter resolution, so each pixel on the screen covers a 1 meter × 1 meter area and, thus, contains much greater detail.

Remotely sensed imagery is acquired at a variety of different spatial resolutions; see Figure 13.3 for examples of the same area on the ground sensed at differing spatial resolutions. For example, most imagery from the Landsat series of satellites is 30-meter spatial resolution. This imagery is often used for larger-scale environmental applications, in which being able to examine fine detail is unnecessary. In many cases, commercial satellite sensors and aerial photography can obtain imagery at sub-meter spatial resolutions that provide crisp detail.

FIGURE 13.3 The same area on the ground sensed at three different resolutions.

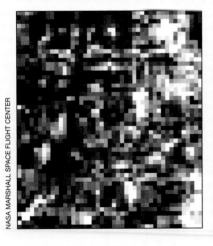

NASA MARSHALL SPACE FLIGHT CENTER

- For a comparison of the difference between the 1-meter NAIP imagery and 0.3-meter individual orthophotos, add the **Harbor.tif** orthophoto to the map. As before, add the image itself rather than the individual bands. In the Contents pane, right-click on **Harbor.tif** and choose **Zoom to Layer** to center the map view on this orthophoto.

- Zoom in on the harbor area until you can see the docked boats. Next, on the **Appearance** tab, in the **Effects** group, click the **Swipe** button. The cursor turns into a solid black triangle. Click on the map and hold down the left mouse button and drag the triangle down past the harbor. You see the NAIP image underneath revealed, and you can drag the orthoimage up and down over it, as if you were opening and closing window blinds. Look carefully at the harbor and the boats in both images and answer Question 13.5.

> **Question 13.5**
> What kinds of details of the harbor and the boats can you see in the 0.3-meter orthophoto that you cannot in the 1-meter NAIP image? Why is this?

- Turn off all four of your individual orthophotos but leave the NAIP imagery on.

Step 13.4 Working with Color Infrared and Multispectral Imagery in ArcGIS Pro

- The NAIP imagery you've already seen in this module is not the only type available. Go through the same process you did before to add NAIP imagery from the Living Atlas, but this time choose the option **USA NAIP Imagery: Color Infrared**. A new type of NAIP imagery is added to the Contents pane. This imagery is similar to the existing NAIP image, but the colors are very different. During the remote sensing process, multiple kinds of data are collected of the same geographic area and are placed together into a single image. For more information about this Color Infrared imagery and what a remotely sensed image is really showing, see **Smartbox 13.5**.

SMARTBOX 13.5

What is a remotely sensed image actually showing?

In the remote sensing process, a sensor on a satellite or an aircraft measures the reflection of the Sun's electromagnetic radiation from objects on Earth. However, there are two exceptions to this. First, *thermal sensors* measure the amount of radiated heat from objects on Earth; they absorb energy from the Sun and re-emit it as heat. Second, *active sensors* generate their own energy, throw it at a target, and record the reflection or backscatter of that energy pulse. (Radar waves are an example of this, as is lidar; see Module 17.) However, sensors typically measure the reflection of energy.

Electromagnetic energy radiates from the Sun through space as waves, and this energy has different properties, depending on the energy's wavelength. For instance, energy with very short wavelengths has the properties associated with x-rays or gamma rays, while slightly longer wavelengths have the properties of ultraviolet or infrared light. Much of this energy gets trapped in Earth's atmosphere by greenhouse gases such as ozone or carbon dioxide, but some of it reaches and penetrates objects on Earth's surface and is reflected. The reflection of energy makes its way upward to the sensor on an aircraft or a satellite, and the sensor records the radiance (Figure 13.4).

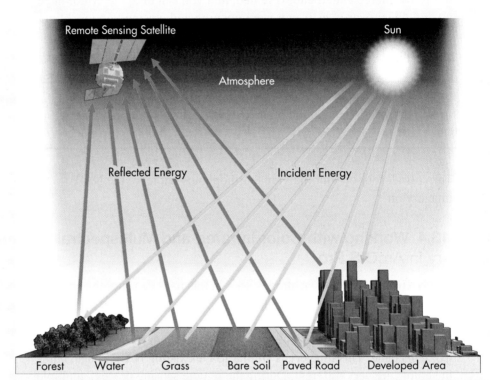

FIGURE 13.4 The remote sensing process.

band A narrow range of wavelengths being measured by a remote sensing device.

multispectral Sensing several bands of energy at once.

hyperspectral Sensing hundreds of bands of energy at once.

panchromatic Black-and-white imagery acquired by sensing the entire visible portion of the spectrum at once.

visible light Electromagnetic energy with wavelengths between 0.4 and 0.7 micrometers.

The sensor often records the radiance of multiple types of energy over the same area at once. For instance, an aircraft sensor could record the three main types of visible light (blue, green, and red) and also a small portion of the wavelengths in the infrared part of the spectrum. A satellite sensor might be tuned to record these four wavelengths and several others, including different, longer wavelengths of infrared energy. These narrow wavelengths of energy are called **bands** of energy. A sensor that records several bands at the same time is called a **multispectral** sensor, and a sensor that can record hundreds of bands at once is called a **hyperspectral** sensor. If a sensor is recording only the range of the entire visible light spectrum at once, it's referred to as a **panchromatic** sensor. Thus, the kind of imagery (multispectral, hyperspectral, or panchromatic) being produced refers to the kind of sensor used to acquire it.

Bands of electromagnetic energy are commonly recorded by sensors from the following wavelength ranges:

• **Visible light** (0.4 to 0.7 micrometers): This is the only range of reflected energy that the human eye can see.

- **Blue** (0.4 to 0.5 micrometers): This is the portion of reflected energy of visible light that the human eye perceives as shades of blue. In remote sensing, the measurement of blue energy is often used for increased penetration into water bodies. Water strongly absorbs energy instead of reflecting it, except for blue light (and wavelengths slightly shorter and longer than blue wavelengths).

- **Green** (0.5 to 0.6 micrometers): This is the portion of reflected energy of visible light that the human eye perceives as shades of green. In remote sensing, green light is used in conjunction with studying the health of vegetation. The more chlorophyll that a plant or leaf has, the more green light it reflects, and thus the healthier that plant or leaf is.

- **Red** (0.6 to 0.7 micrometers): This is the portion of reflected energy of visible light that the human eye perceives as shades of red. In remote sensing, red light is also used in conjunction with studying the health of vegetation—but in a different way. As plants or leaves go through their senescence process, they lose their chlorophyll and begin to grow more stressed and unhealthy. With a lack of chlorophyll, the leaves absorb more green light and in turn reflect more red light.

- **Near infrared (NIR)** (0.7 to 1.3 micrometers): This is reflected energy in the shorter range of infrared wavelengths and is commonly used by both satellite imagery and aerial photography. Healthy greenery, such as trees, leaves, plants, grass, and agriculture, tends to strongly reflect near-infrared energy, and thus it is commonly used in remote sensing for studying the health of vegetation or the amount of biomass in an area.

- **Shortwave infrared (SWIR)** or **middle infrared (MIR)** (1.3 to 3.0 micrometers): This is reflected energy in the medium range of infrared wavelengths. These bands are often used in geologic studies or for assessing water content.

- **Thermal infrared (TIR)** (3.0 to 14.0 micrometers): This is the emitted radiant heat energy from longer portions of the infrared wavelengths.

A sensor records the radiance of several bands and scales them according to a specific range as a record of the amount of radiance measured at that location, with higher numbers indicating a greater amount of radiance and lower numbers indicating a lesser amount. This range is determined by the sensor's **radiometric resolution**, and common scales are 8-bit (in which numbers go from 0 to 255) and 11-bit (in which numbers go from 0 to 2047). Each number is referred to as a **digital number**, or a **brightness value**. The digital number is then assigned to the grid cell, and the size of that cell is based on the sensor's spatial resolution. For instance, the Landsat 8 image you're using of Cleveland has 30-meter spatial resolution, meaning that each cell in the raster is 900 meters square (that is, 30 meters × 30 meters), while the radiometric resolution the image is presented with is 16-bit, indicating that the digital number values in each cell range from 0 through 65,535.

A single band is displayed in shades of gray, where the lowest digital numbers (such as 0 in an 8-bit image) are shown in black, the highest digital numbers (such as 255 in an 8-bit image) are shown in white, and all values in between are dark or light gray (Figure 13.5). In order to show the bands of imagery in color, ArcGIS Pro has three **channels** into which bands can be placed for

---

blue Electromagnetic energy with wavelengths between 0.4 and 0.5 micrometers.

green Electromagnetic energy with wavelengths between 0.5 and 0.6 micrometers.

red Electromagnetic energy with wavelengths between 0.6 and 0.7 micrometers.

near infrared (NIR) Electromagnetic energy with wavelengths between 0.7 and 1.3 micrometers.

shortwave infrared (SWIR) Electromagnetic energy with wavelengths between 1.3 and 3.0 micrometers.

middle infrared (MIR) A term used synonymously with shortwave infrared.

thermal infrared (TIR) Electromagnetic energy with wavelengths between 3.0 and 14.0 micrometers.

radiometric resolution A sensor's ability to determine fine differences in a band of energy measurements.

digital number The energy measured at a single pixel according to a predetermined scale.

brightness value A term used synonymously with digital number.

channel The display of bands of imagery in shades of red, green, or blue.

display: a red channel, a green channel, and a blue channel. When displaying these bands of energy on the screen in color, you're able to display only three of them simultaneously. For instance, you could show the digital numbers from the green band in the blue channel, the near-infrared band in the green channel, and the middle-infrared band in the red channel, and the resulting composite would show the three bands combined, in color, on the screen. The green digital numbers would be shown in shades of blue, the near-infrared digital numbers would be shown in shades of green, and the middle-infrared digital numbers would be shown in shades of red (and all objects would then be displayed in a color composed of whatever shades of blue, green, and red the various types of digital numbers combined to be).

FIGURE 13.5 A single 8-bit raster band with its digital numbers and corresponding grayscale.

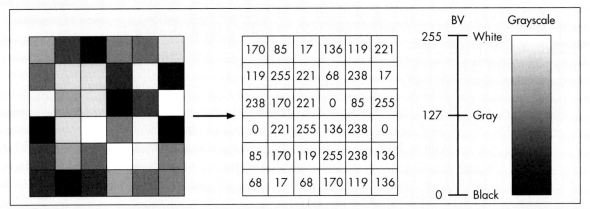

You can show any combination of bands through the three channels at any time and generate a different **color composite** (Figure 13.6). When the display shows a blue band in the blue channel, a green band in the green channel, and a red band in the red channel, the composite is considered a **true color composite** because it represents the image in the way that the human eye sees things (that is, we see reflection of blue light in shades of blue, green light in shades of green, and red light in shades of red). A **false color composite** is generated when the pattern of channels and bands deviates from a true color composite.

color composite An image formed by placing a band of imagery into each of the three channels (red, green, and blue) to view a color image instead of a grayscale one.

true color composite An image arranged by placing the red band in the red channel, the green band in the green channel, and the blue band in the blue channel.

false color composite An image arranged by not placing the red band in the red channel, the green band in the green channel, and the blue band in the blue channel.

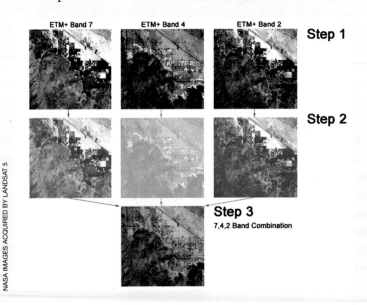

FIGURE 13.6 A sample three-band composite.

False color composites are frequently used for analysis because they enable analysts to view types of energy that are typically invisible to the human eye (such as near infrared or middle infrared) displayed on the screen in shades of blue, green, or red.

A **color-infrared** image is a specific type of false color composite. It was originally achieved with film that was sensitive to the near-infrared portion of the electromagnetic spectrum and blocked the blue portion. As a result, a color-infrared image would show near-infrared reflection in the red channel, the red reflection in the green channel, and the green reflection in the blue channel. Today, color-infrared imagery is commonly collected with aircraft or satellite sensors that can measure at least these three different bands, and it is viewed by placing these three bands (near-infrared, red, and green) into the three channels (red, green, and blue) in that specific arrangement.

Often a remotely sensed image needs to have a **histogram stretch** applied to it to artificially enhance the brightness of the colors that appear on the screen. A histogram stretch expands the ranges of digital numbers that appear on the screen. For instance, in an 8-bit image with a range of digital numbers going from 0 to 255, if the actual range of digital numbers extends from 40 to 150, those correspond to relatively dark colors, and so the image will appear darker on the screen. A histogram stretch artificially adjusts the range of those values so that some or many of them end up at the higher end of the 0–255 range, allowing those pixels to be shown in brighter colors. Note that the imagery still retains the original digital numbers when used for analysis, and this stretching only affects how the imagery is visualized on the screen. There are many different ways of applying such a stretch in ArcGIS Pro to visually enhance an image in different ways.

color-infrared An image arranged by placing the near-infrared band in the red channel, the red band in the green channel, and the green band in the blue channel.

histogram stretch A technique used to artificially enhance the visual brightness and colors of pixels from a remotely sensed image.

- In the Contents pane, place the USA NAIP Imagery: Color Infrared layer so it is on top of the USA NAIP Imagery: Natural Color layer. Also in the Contents pane, you should see the two NAIP images listed with a different band placed into a different channel. When assessing NAIP imagery, the bands and their assigned wavelengths are:
  - Band 1: red wavelengths
  - Band 2: green wavelengths
  - Band 3: blue wavelengths
  - Band 4: NIR wavelengths

- Use the Swipe tool to move back and forth between viewing the two images. Keep in mind that they are the same image, and you're just viewing different band combinations in each one. Zoom out as necessary to see the area surrounding the harbor and how it looks in each image. Then answer Question 13.6.

> **Question 13.6** Why do the trees look green in the NAIP Natural Color image but look red in the NAIP Color Infrared image? *Hint:* Keep in mind which bands are being shown in which of the three channels and the properties of those wavelengths.

- Move back to the waterfront and zoom in on the Cleveland Browns football stadium. Use the swipe tool to closely examine the stadium and the field and then answer Question 13.7.

Question 13.7 Is the field of the Cleveland Browns Stadium real grass or artificial turf? How can you tell just by viewing the two NAIP images?

- Turn off the two NAIP images when you're done.

Step 13.5 Working with Landsat Imagery in ArcGIS Pro

- Add the **cle.img** satellite image to the map. As with the orthophotos, when adding cle.img, be sure to add just that one tile and not any of the files inside of it, such as Band_1, Band_2, or Band_3. You should add the entire image rather than separate bands. After the cle.img layer is added, you see it listed in the Contents pane as Red: Band_1, Green: Band_2, and Blue: Band_3, with a different color (red, green, and blue) next to each of these entries. Instead, the entry in the Contents pane might be something like cle.img_Band_1, and it might be in grayscale; if this is the case, see Troublebox 13.1 on page 307 for why this may be so and how to fix it. Note that cle.img is a Landsat OLI multispectral satellite image of the waterfront area of Cleveland, Ohio. For more information about the Landsat program and the imagery produced from its satellites (including what these various bands represent), see **Smartbox 13.6**.

SMARTBOX 13.6

What is the Landsat program, and what are its capabilities?

The **Landsat** series of satellites is the longest continually running satellite remote sensing program in existence. The first Landsat satellite (Landsat 1) was launched by the U.S. government in 1972, and the most recent satellite (Landsat 8) was launched in 2013. The future of the Landsat program is being designed now, with Landsat 9 tentatively planned for launch at the end of 2020 and Landsat 10 a few years after that. Through the seven satellites in the program, Landsat has provided global imagery coverage for over 40 years. Even better, the U.S. government has recently opened this archive of imagery and made it available to the public for free.

The first five Landsat satellites contained a sensor called the **MSS** (the Multispectral Scanner), which had a spatial resolution of 79 meters and recorded four bands. MSS data over an area was collected every 18 days. Landsat 4, launched in 1982, contained a new instrument called the **TM** (Thematic Mapper). The TM sensor was also onboard Landsat 5 when it launched in 1984. TM imagery had a spatial resolution of 30 meters (the TIR band had a resolution of 120 meters) and sensed in seven multispectral bands (blue, green, red, NIR, two different MIR bands, and a TIR band—see Smartbox 13.5 on page 313 for a review of these bands). Landsat 5's mission continued until 2012, so the satellite provided nearly 28 years of TM data, collected for an area every 16 days.

Landsat 6 did not achieve orbit and was lost in 1993. Landsat 7, launched in 1999, carried a new sensor, **ETM+** (the Enhanced Thematic Mapper). The ETM+ sensor has the same capabilities of TM (all multispectral bands had 30-meter spatial resolution, except for the TIR band, which had 60-meter spatial resolution), with the addition of an eighth panchromatic band (which sensed the entire visible and part of the NIR region as a single band) at 15-meter spatial resolution. Landsat 7 collects data of an area every 16 days. Unfortunately, the Scan Line Corrector aboard Landsat 7 developed an uncorrectable error in 2003, which meant that imagery acquired by ETM+ contained only about 75 percent of the total pixels that should be there. Thus, if you're using Landsat 7 imagery from 2003 onward, you may see diagonal sections of pixels with no data recorded in them.

Landsat 8, the latest mission, contains two different sensors. The first is **OLI** (Operational Land Imager). The OLI sensor monitors bands similar to the six multispectral and the single panchromatic on ETM+, with the addition of two other bands: an "ultra-blue" band for examining coastal areas and another IR band for use with cirrus clouds. The OLI bands are sensed at 30 meters, with the panchromatic at 15 meters. The second instrument onboard Landsat 8, **TIRS** (Thermal Infrared Sensor), senses two different thermal bands at 100 meters (but are resampled to a 30-meter resolution for delivery to the end user). The image of Cleveland you're using in this module is a subset of a Landsat 8 image (with the OLI and TIRS instruments) obtained on September 27, 2014. Table 13.1 summarizes all the Landsat sensor capabilities along with the specific band numbers assigned to each kind of energy being sensed.

Landsat A long-running U.S. remote sensing program that had its first satellite launched in 1972 and continues today.

MSS The Multispectral Scanner onboard Landsats 1 through 5.

TM The Thematic Mapper instrument onboard Landsats 4 and 5.

ETM+ The Enhanced Thematic Mapper instrument onboard Landsat 7.

OLI The Operational Land Imager instrument onboard Landsat 8.

TIRS The Thermal Infrared Sensor onboard Landsat 8.

Satellite and Sensor(s)	Bands Sensed (by band number)	Spatial Resolution
Landsat 1–5 MSS	1: R, 2: G, 3: NIR, 4: NIR	79 m
Landsat 5 TM	1: B, 2: G, 3: R, 4: NIR, 5: SWIR, 6: TIR, 7: SWIR	Multi: 30 m, TIR: 120 m
Landsat 7 ETM+	1: B, 2: G, 3: R, 4: NIR, 5: SWIR, 6: TIR, 7: SWIR, 8: Pan	Multi: 30 m, TIR: 60 m, Pan: 15 m
Landsat 8 OLI/TIRS	1: Ultra-Blue (Coastal), 2: B, 3: G, 4: R, 5: NIR, 6: SWIR, 7: SWIR, 8: Pan, 9: IR (Cirrus), 10: TIR, 11: TIR	Multi: 30 m, Pan: 15 m, TIR: 100 m

Table 13.1 The sensors used by Landsat missions, their specific bands and band numbers, and their spatial resolutions

Landsat scene A single image obtained by a Landsat satellite sensor.

The various Landsat sensors have a swath width of 185 kilometers, which is how much of the ground a sensor can "see" during one pass. An entire **Landsat scene** measures a ground area about 180 kilometers long by 185 kilometers wide. Thus, when you're obtaining Landsat imagery to use in ArcGIS Pro, you'll frequently be downloading files for each of the bands that cover the area defined by a single Landsat scene. If the area of analysis isn't contained in a single scene, you often have to mosaic the images together into a larger image raster (see Module 12).

Imagery from the Landsat archives can be downloaded for free. The USGS has two different utilities online to help you obtain Landsat imagery. The first of these is EarthExplorer (**https://earthexplorer.usgs.gov**), which allows you to access Landsat imagery along with many other types of satellite imagery, aerial photography, and GIS data. (See page 138 in Module 5 for more about EarthExplorer.) The second is GloVis (**https://glovis.usgs.gov**), the Global Visualization Viewer, which allows you to search for Landsat scenes (and other satellite imagery) and download the bands from a scene.

- You need to make some visual adjustments to the image so that it's clearer and brighter. Click once on the cle.img layer in the Contents pane and then, on the **Appearance** tab, in the **Rendering** group, click the **Symbology** button.
- For Primary symbology, choose **RGB**.
- For Red, choose **Band_4**.
- For Green, choose **Band_3**.
- For Blue, choose **Band_2**.
- For Stretch type, choose **Esri**.
- For Statistics, choose **DRA**.
- You should see the cle.img image brighten up considerably so that you can better see the Cleveland waterfront area. Zoom in closely to the Burke Lakefront Airport area on

the shores of Lake Erie. When you zoom in, the image becomes very pixelated, and it's difficult to make out fine details in the image. This is because the spatial resolution is 30 meters; that is, each pixel (or cell) of the raster (see Module 12) represents an area on the ground 30 meters × 30 meters square.

- Examine the areas around the airfield and answer Questions 13.8 and 13.9.

Question 13.8 Why can you not see individual planes at the airport in the cle.img Landsat imagery?

Question 13.9 At the 30-meter resolution of the Landsat imagery, what features of Burke Lakefront Airport can you still identify?

- Right-click in the cle.img layer in the Contents pane and select **Zoom to Layer** to return to the full view of the Landsat image. You can see in the Contents pane that the single cle.img layer is composed of three bands, each placed into a red, green, or blue channel. For instance, this Landsat image actually consists of eleven different bands, each representing a different type of remotely sensed information of the same section of Cleveland.

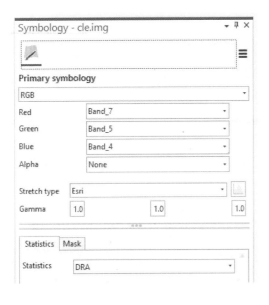

- To change the bands shown in the Landsat imagery, you use the Symbology pane. It can be opened for a remote sensing image in two different ways:

    1. In the Contents pane, right-click on the cle.img layer and then choose **Symbology**.

    2. In the Contents pane, click once on the cle.img layer and then, on the **Appearance** tab, in the **Rendering** group, click the **Symbology** button to open it.

- For Primary symbology, choose **RGB**. When displaying imagery using the RGB (red, green, blue) composite option, you can choose which of the multispectral bands to display in each of the three channels. This is a Landsat 8 image using the OLI and TIRS instruments, so 1 of the 11 bands can be displayed in the red channel, a second band can be displayed in the green channel, and a third band can be displayed in the blue channel. The rest of the bands are not used when drawing the composite on the screen.

- For Red, choose **Band_7**. This is the longer of the two SWIR bands from OLI. The more shortwave infrared reflectance in a pixel, the more red contributes to the color of that pixel.

- For Green, choose **Band_5**. Healthy, natural features (such as healthy trees, grassy fields, or vegetation) reflect a lot of near-infrared light. By choosing to display this band in the green channel, the more near-infrared reflectance in a pixel, the more green contributes to the color of that pixel.

- For Blue, choose **Band_4**. This is the Red band from OLI. Features that would reflect a lot of red light (or visible light in general) have a brighter reflectance in this band. By choosing to display this band in the blue channel, the more red reflectance in a pixel, the more blue contributes to the color of that pixel.

- For Stretch type, choose **Esri**. This is one of several different types of histogram stretches used to enhance the brightness and appearance of the imagery.

- For Statistics, choose **DRA**.

- The image's appearance changes to reflect these three bands in this particular combination in the red, green, and blue channels and stretched and displayed in a certain way to help enhance the brightness of the image. You should see that developed areas (such as cities) are shown in shades of pink and purple, areas with greenery are shown in shades of green, and water bodies are shown in darker blue and black. In the Contents pane, be sure that the cle.img Landsat layer is at the top of the contents and that the USA NAIP Imagery: Natural Color layer is underneath it; then turn on the USA NAIP Imagery: Natural Color layer. You can then use the swipe tool to move back and forth between the two layers.

- In the cle.img Landsat layer, you should be able to see several small bright red areas. Examine them closely in the Landsat layer and also using the swipe tool in the NAIP layer and then answer Question 13.10.

> **Question 13.10**
> What kind of land cover corresponds with these bright red areas in the Landsat image (that is, what do these bright red areas represent)? With this band combination and stretch type, what does this say about the interaction of the three types of energy being shown in the Landsat image's bands and this type of land cover?

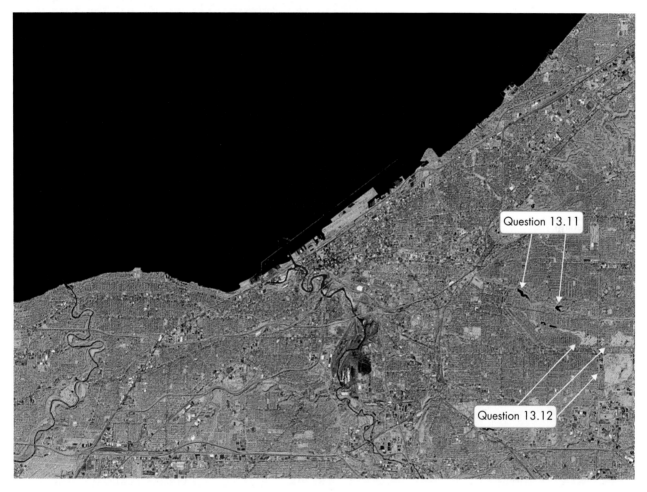

- Next, turn off the NAIP layer and return to just the Landsat imagery. Examine the areas highlighted in the graphic above and answer Questions 13.11 and 13.12.

> **Question 13.11**
> What are these areas on the landscape, and how can you tell? *Hint:* Look at the size and pattern of the areas and think about what kinds of energy reflectance you're examining with this Landsat image.

> **Question 13.12**
> What are these areas on the landscape, and how can you tell? *Hint:* Look at the size and pattern of the areas and think about what kinds of energy reflectance you're examining with this Landsat image.

- Right-click on the cle.img layer in the Contents pane and choose **Zoom to Layer** to return to looking at the whole of the Cleveland region. You can gain a lot of information through satellite imagery, although the spatial resolution of a system like Landsat isn't conducive to examining fine details. Many satellite systems have high-resolution (and multispectral) capabilities, though, and produce very detailed imagery to use in GIS (see **Smartbox 13.7**).

SMARTBOX 13.7

What satellites have high-resolution capabilities?

Many of the available high-spatial-resolution satellite data come from private companies rather than government sources. DigitalGlobe, Airbus Defence and Space, and Planet are among the companies that operate commercial satellites and make the imagery available for purchase. Many commercial satellites have off-nadir capabilities, so their sensors can be tasked with imaging a certain area on the ground during an orbital pass and can then re-image that area within a short time frame, usually one to three days.

If a satellite has a multispectral sensor along with a panchromatic sensor of finer spatial resolution, the panchromatic band is usually used to pan-sharpen the multispectral bands. When **pan-sharpening** occurs, the panchromatic and multispectral bands are fused together in order to refine the image (so that the multispectral bands can be viewed at the resolution of the panchromatic).

Note, however, that prior to 2014, U.S. government regulations indicated that non-government purchasers of commercial imagery would have the imagery resampled to a spatial resolution of 0.5 meters if the sensor resolution was finer than that. Since those restrictions were lifted in 2014, imagery purchased from a commercial provider can be had at its original resolution. Table 13.2 lists a selection of currently operating high-spatial-resolution satellites and their capabilities, and Figure 13.7 provides an example of high-spatial-resolution satellites imagery.

pan-sharpening The technique of fusing a higher-resolution panchromatic band with lower-resolution multispectral bands to improve the clarity and detail seen in an image.

Satellite	Source	Launch Date	Bands Sensed	Spatial Resolution (at Nadir)
GeoEye-1	DigitalGlobe	2008	Pan, B, G, R, NIR	Pan = 0.41 m, Multi = 1.65 m
Pleiades-1A	Airbus	2011	Pan, B, G, R, NIR	Pan = 0.5 m, Multi = 2.0 m
Pleiades-1B	Airbus	2012	Pan, B, G, R, NIR	Pan = 0.5 m, Multi = 2.0 m
SkySat-1	Planet	2013	Pan, B, G, R, NIR	Pan = 0.9 m, Multi = 1.1 m
SkySat-2	Planet	2014	Pan, B, G, R, NIR	Pan = 0.9 m, Multi = 1.1 m
SPOT 6	Airbus	2012	Pan, B, G, R, NIR	Pan = 1.5 m, Multi = 6.0 m
SPOT 7 (Azersky)	Airbus	2014	Pan, B, G, R, NIR	Pan = 1.5 m, Multi = 6.0 m
WorldView-1	DigitalGlobe	2007	Pan	Pan = 0.5 m
WorldView-2	DigitalGlobe	2009	Pan, B, G, R, NIR, Coastal, Yellow, Red Edge, NIR-2	Pan = 0.46 m, Multi = 1.85 m
WorldView-3	DigitalGlobe	2014	Pan, B, G, R, NIR, Coastal, Yellow, Red Edge, NIR-2, 8 SWIR bands, 12 CAVIS bands	Pan = 0.31 m, Multi = 1.24 m, SWIR = 3.7 m, CAVIS = 30.0 m

Table 13.2 A selection of high-spatial-resolution satellites and their capabilities

FIGURE 13.7 A satellite image of the Magic Kingdom in Walt Disney World in Orlando, Florida.

Step 13.6 Printing or Sharing Your Results

- Save your Module13 project. There is no need to produce a layout of the final results.

 ## Closing Time

This module examines the basics of working with satellite and aerial imagery as layers within ArcGIS Pro as well as some basic imagery analysis. When working with imagery like this, always keep in mind that there's a lot of preprocessing done behind the scenes before you get imagery to work with. For instance, in this module, you didn't have to adjust the imagery to make it match up with the basemap; the imagery already contained that information. However, there may be times when you'll need to alter or fit an image (or another raster) to match up with the rest of the data. See *Related Concepts for Module 13* for more about this georeferencing procedure in ArcGIS Pro.

Creating large-scale raster datasets like imagery is easy enough to do with a plane, drone, or satellite sensor, but there may be times when you'll need to work with a raster but will not have that raster data available or will have it only in discrete form. For instance, if you were going to examine patterns of precipitation, you might only have point data available from a set of weather stations, but what you would really need is a raster surface of precipitation for all locations in your study area. The next module examines how to create these types of large-scale rasters through a process called *interpolation*.

RELATED CONCEPTS FOR MODULE 13

Georeferencing an Image

In ArcGIS Pro, **georeferencing** is the process of taking a layer without spatial reference and aligning it with other layers that already have a coordinate system (projected or geographic) assigned to them. Georeferencing is commonly performed on raster images, such as scanned historical maps or scanned aerial photos, so that they can be used as geospatial data along with other layers in GIS. For instance, if you're digitizing old home locations from a 50-year-old aerial photograph, you will want to be sure that the photo has a proper coordinate system assigned and matched to your other layers so that the digitizing results and measurements will be correct. To georeference a raster image, you also need an already spatially referenced source with which to match it, such as an orthoimage or a satellite image. (Vector data layers, such as transportation layers, can also be used as a source.)

Georeferencing is performed by selecting locations that can be clearly identified in both images and placing a **control point** at those spots. At least three control points are necessary to tie the two images together, and often several more are used to get a better fit of the two images. Control points are normally selected at well-delineated locations in both images, such as the intersections of roads or the corners of buildings. When the two images are well aligned, the unreferenced image can undergo a **transformation** to be converted to the new coordinate system. For instance, if the referenced image is in the UTM Zone 17 projected coordinate system, the unreferenced image can be transformed so that its coordinates are assigned the same coordinate system and spatial reference.

The root mean square error (**RMSE**) is used to compare the locations on the newly transformed image with where they match up to the image used as the reference. The lower the RMSE value, the better the alignment between the images. Because the transformation is based on the control point locations, control points are often deleted or moved, or new points are added, to lower the RMSE and achieve better alignment. When this process is complete, the image can be permanently transformed to a new georeferenced dataset in ArcGIS Pro for use in future projects.

> **georeferencing** A process whereby spatial referencing is given to data.
>
> **control points** Point locations where the coordinates are known that are used in aligning an unreferenced image to a source.
>
> **transformation** A process in which data are altered from unreferenced to having spatial reference.
>
> **RMSE** The root mean square error, an error measure used in determining the accuracy of the overall transformation of unreferenced data.

Key Terms

remote sensing (p. 303)
satellite imagery (p. 303)
aerial photography (p. 303)
UAS (p. 303)
nadir (p. 304)
vertical image (p. 304)
off-nadir viewing (p. 304)

oblique imagery (p. 304)
orthoimage (orthophoto) (p. 307)
visual image interpretation (p. 309)
pattern (p. 309)
site and association (p. 309)
size (p. 309)

shadow (p. 309)
shape (p. 310)
texture (p. 310)
tone (p. 310)
NAIP (p. 311)
DOQQ (p. 311)
spatial resolution (p. 312)

band (p. 314)
multispectral (p. 314)
hyperspectral (p. 314)
panchromatic (p. 314)
visible light (p. 314)
blue (p. 315)
green (p. 315)
red (p. 315)
near infrared (NIR) (p. 315)
shortwave infrared (SWIR) (p. 315)
middle infrared (MIR) (p. 315)
thermal infrared (TIR) (p. 315)
radiometric resolution (p. 315)
digital number (p. 315)
brightness value (p. 315)
channel (p. 315)
color composite (p. 316)
true color composite (p. 316)
false color composite (p. 316)
color-infrared (p. 317)
histogram stretch (p. 317)
Landsat (p. 319)
MSS (p. 319)
TM (p. 319)
ETM+ (p. 319)
OLI (p. 319)
TIRS (p. 319)
Landsat scene (p. 320)
pan-sharpening (p. 324)
georeferencing (p. 326)
control points (p. 326)
transformation (p. 326)
RMSE (p. 326)

14

How to Perform Spatial Interpolation with ArcGIS Pro

Learning Outcomes

After studying this module, you should be able to:
- Define what spatial interpolation means.
- Explain the difference between global and local methods of interpolation.
- Explain how the IDW method calculates a value through interpolation.
- Describe the concept of the Spline method of interpolation.
- Explain how zonal statistics are calculated.

spatial interpolation The process of determining an unknown value at a location based on known values at other locations.

surface A raster that contains a value for some phenomena at all locations.

ArcGIS Pro Skills

In this module, you will learn how to do the following in ArcGIS Pro:
- Interpolate a surface from a set of points using the IDW method.
- Interpolate a surface from a set of points using the Spline method.
- Use the Zonal Statistics tool to examine an overall average of cell values within a polygon boundary.
- Use Extract By Mask to clip a raster layer to conform to the boundaries of a polygon layer.
- Use the Swipe tool to visually compare two datasets.

Introduction

When you're planning outdoor activities for the weekend and you look at a weather map, chances are you'll see your location on a color-coded map carrying a value for the high temperature or chance of precipitation. However, weather forecasters don't have access to information about the levels of humidity or amount of rainfall for every location on the map. Rather, weather stations distributed around the region measure this type of data. If you used ArcGIS Pro to plot the locations of these stations, you'd have a series of points around your area but not coverage of every place in your region. To determine the rainfall values that are associated with your location, you'd use the rainfall measurements at nearby weather stations to estimate how much rainfall you should expect at your location. **Spatial interpolation** is the process of figuring out an unknown value at a location based on known values at other locations.

Many phenomena vary continuously across Earth, such as elevation, temperature, barometric pressure, and precipitation. All these are measured using a set of points at selected locations. By using these initial scattered point values, you can use spatial interpolation to create a raster that shows a value for the phenomenon at all locations. In ArcGIS Pro, this raster is referred to as a **surface** because it contains values for each location as they vary continuously. For instance, precipitation, elevation, and distance can all be measured at a nearly infinite number of points, but they are represented as a raster surface that attempts to model their continuous variation. Figure 14.1 shows an example of an initial set of point measurements (representing annual precipitation values) and the corresponding interpolated surface (of precipitation).

FIGURE 14.1 Comparing a set of statewide point measurements of annual precipitation to an interpolated surface of annual precipitation.

There are several different approaches to performing spatial interpolation using GIS. The first distinction is whether a global method or local method is used for interpolation. A **global interpolation** method attempts to use every known value in the dataset when approximating an unknown value. For instance, if you live in northeastern Ohio, and you are trying to interpolate a value for precipitation at your location, a global interpolation method would use the values from all available weather stations in the state (including points in the far south and west) to estimate the precipitation level at your position. As such, global interpolation methods are inexact and produce extremely general results that are often used for determining **trends** (very general patterns or overriding processes that affect the measurements) in a dataset. For instance, a trend may show a greater amount of precipitation as one moves south to north.

For more exact results, a **local interpolation** method can be used. In local interpolation, only a small subset of point measurements near the location are used to approximate the unknown value at that location. See Figure 14.2 for an example: When trying to interpolate the value at the unknown location (shown in yellow), only the values of the closest five points (those shown in red within the radius around the unknown value point) are used. The values at all the other known points (shown in blue) are not used because they are too distant to have a realistic impact on the value at the chosen location. This module introduces you to two different local interpolation methods in ArcGIS Pro for creating raster surfaces from an initial set of points.

global interpolation The process of using all known values in a dataset when approximating an unknown value.

trend A very general pattern or overriding process that affects measurements.

local interpolation The process of using a subset of known values when approximating an unknown value.

FIGURE 14.2 An example of using a subset of points for a local interpolation method.

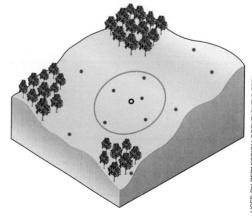

 ## Module Scenario and Applications

This module places you in the role of a climatologist trying to create a statewide map of annual precipitation for Ohio. You also want to be able to determine the average annual precipitation for entire counties as part of a larger climate-related study. You have a set

of points representing precipitation values taken at multiple places across the state, but your goal is to create a continuous surface representation that covers all locations.

The following are additional examples of real-world applications of this module's skills:

- An environmental scientist is attempting to create a map of pollution levels for her state. She has samples of CO_2 (carbon dioxide) levels from a number of different measurement areas around the state, and she needs to use these point samples to create a statewide pollution index using spatial interpolation techniques.

- An engineer is attempting to create a model of the bottom of a section of Chesapeake Bay to help identify the nesting sites of endangered species of sea turtles. He has a set of sample bathymetric points obtained by dredges as they move through the bay. The engineer can use spatial interpolation methods to create a continuous bathymetric surface from these individual points.

- A geologist has measured a sampling of groundwater points based on drilling wells. She wants to create a map showing the depth to groundwater for the entire area and can use spatial interpolation methods to create a surface showing this information based on the samples from the wells.

Study Area

For this module, you will be working with data from the entire state of Ohio, along with areas near the Ohio borders.

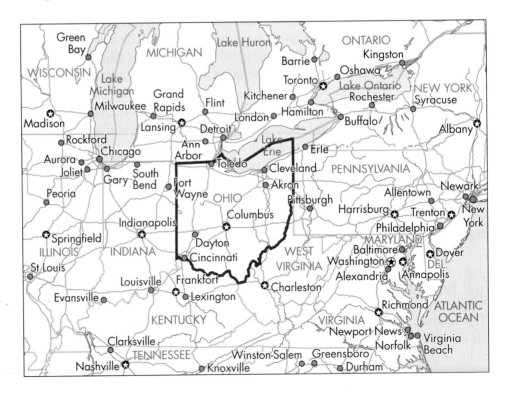

Data Sources and Localizing This Module

The data in this module focus on datasets from within the state of Ohio. However, you can easily modify this module to use data from your own state instead. For instance, if you were doing this analysis in Wisconsin, you could easily obtain the same kind of data. The Ohio county dataset was downloaded (for free) and extracted from the USA Counties dataset available on ArcGIS Online (and also available at **https://www.arcgis.com/home/item.html?id=a00d6b6149b34ed3b833e10fb72ef47b**); you can download and extract data the same way for Wisconsin counties.

The airports data were downloaded and extracted from The National Map (the trans_airportpoint layer from the Structures geodatabase), and airports labeled as complexes that are international or regional airports were separated out for use in this module.

The Ohioprecippoints layer was downloaded from the National Weather Service Advanced Hydrologic Prediction Service (located online at **http://water.weather.gov/precip/download.php**), which provides precipitation data that can be downloaded (for free) for the entire United States. The points were originally derived from gridded datasets, and each point has an attribute that represents the total observed precipitation amount for the year 2011. Keep in mind that, although the dataset used for this module is a gridded set, the interpolation techniques used here can be used (and commonly are used) with non-gridded data.

The values used for the airport precipitation readings in Step 14.6 were obtained from information culled from values on the Wunderground (Weather Underground) website, at **http://www.wunderground.com**. Observed precipitation amounts for airport weather stations for 2011 (for various airports throughout the United States) are available from the "History and Almanac" or "Historical Weather" information for individual stations at airports. A caveat: Some observed values being used to verify the interpolated results may have been used in creating the original Ohioprecippoints data.

Step 14.1 Getting Started

- This module's hands-on applications use the data folder called Module14. Your instructor will be able to supply you with this data, or you can download it directly from this book's website, at **https://www.macmillanlearning.com/college/us/product/Discovering-GIS-and-ArcGIS-Pro/p/131923075X**. The text in this module assumes that you have this Module14 folder in a computer location referenced as C:\GIS; if you have it somewhere else (for instance, in a flash drive referenced as G:\GISClass), substitute that location and path to the Module14 folder throughout this module.
- The Module14 folder contains the following:
  - Ohioprecipdata: a file geodatabase containing three items:
    - Ohioairports: a point feature class of the locations of Ohio airport complexes
    - Ohiocountiesalbers: a polygon feature class representing the Ohio county borders (projected to the Albers Equal Area Conic projection)

- Ohioprecippoints: a point feature class representing total observed precipitation amounts for Ohio and nearby surrounding areas in 2011

- Start ArcGIS Pro.
- Sign in with your Esri account username and password.
- Create a new project using the **Map** template. Call this project **Module14** and place it in your **C:\GIS\Module14** folder. Ensure that there is not a checkmark in the box next to **Create a new folder for this project**.
- When ArcGIS Pro opens, change the map's name to **Ohio Precipitation**.
- Several of the interpolation tools in ArcGIS Pro are part of the Spatial Analyst extension, which you must have licensed to your Esri account and active in order to work with it in this module. If you don't have Spatial Analyst licensed and active, some tools will not work; they will have a lock symbol next to them and will be unavailable. See Smartbox 11.1 and Troublebox 11.1 in Module 11 for more about extensions and ensuring that you have the Spatial Analyst extension available for use.
- Add the **Ohiocountiesalbers** feature class to the Contents pane.
- Add the **Ohioprecippoints** feature class to the Contents pane.
- On the **Analysis** tab, in the **Geoprocessing** group, click the **Environments** button and then set the following geoprocessing environment settings for this module:
  - For Current Workspace, use the **Module14.gdb** geodatabase within your C:\GIS\Module14 folder.
  - For Scratch Workspace, use the **Module14.gdb** geodatabase within your C:\GIS\Module14 folder.
  - For Output Coordinate System, choose **Ohiocountiesalbers**. ArcGIS Pro then updates the coordinate system to that of the Ohiocountiesalbers layer.
  - For Extent, choose **Ohioprecippoints**. ArcGIS Pro then updates the x/y coordinates of the lower-left corner and upper-right corner of the extent to those of the Ohioprecippoints layer.
  - For Cell Size, type **100**. ArcGIS Pro then updates the cell size for raster output in this project to 100 meters.
- Leave the other settings alone and click **OK** to put these geoprocessing environment settings into place.
- The Ohioprecippoints layer was projected from its initial coordinate system to another one for use in this module. The coordinate system being used in this module is **North America Albers Equal Area Conic (NAD 83)**, with meters as the map units. Check the Ohio precipitation map properties (on the Coordinate System tab) to verify that this projection system is being used.

Step 14.2 Examining the Data Points

- Rearrange the layers so that you can see all the available precipitation points placed on top of the Ohio counties. (You might want to make the counties fill color Black Outline so that you can see only the outlines of the counties and leave the counties themselves transparent.) This will give you an indication of the distribution of points relative to the borders of Ohio.

- Answer Question 14.1.

> **Question 14.1**
> How many points are in the Ohioprecippoints layer? (That is, how many precipitation sample points are you dealing with in this exercise?)

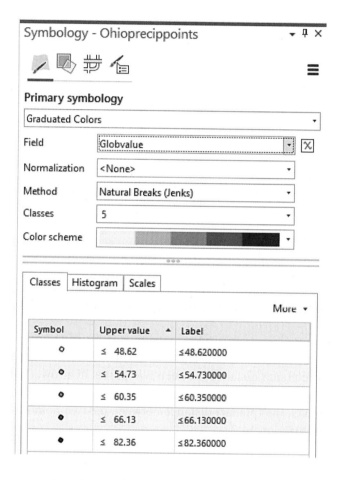

- Next, you need to change the symbology of the Ohioprecippoints layer to visually examine the distribution of precipitation levels in (and around) Ohio in 2011. In the Symbology pane, use the following settings:
  - For Primary symbology, choose **Graduated Colors**.
  - For Field, choose **Globvalue**.
  - For Normalization, choose **<None>**.
  - For Method, choose **Natural Breaks (Jenks)**.
  - For Classes, choose **5**.
  - Select a color scheme that you feel makes the variations in the precipitation levels stand out.
- You should now see the points classified by the amount of precipitation for 2011 (the Globvalue attribute). Answer Question 14.2.

Question 14.2 Based on visual examination, where were the greatest concentrations of rainfall amounts in Ohio in 2011?

- When the National Weather Service puts these data together, it does so in "gridded" format—meaning that the sampling points for precipitation are evenly spaced out from one another. You'll now look at how far apart these precipitation point measurements are. Zoom in closely on a set of points in northern Ohio. On the **Map** tab, in the **Inquiry** group, click the **Measure** button and then choose **Measure Distance**.

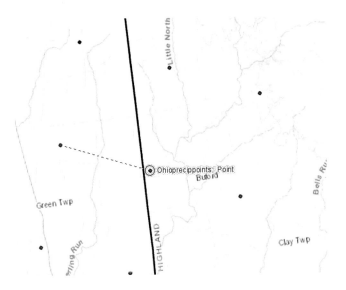

- Your cursor changes to a ruler shape; this is the Measure tool. (See Module 8 for more about using the Measure tool.) Use the Measure tool to find the planar distance between two points in a row. (You will be measuring slightly diagonally to follow a row.) In the Measure dialog that appears, the (planar) distance between the two points is calculated. Answer Question 14.3.

Question 14.3 How far apart are the precipitation points (in feet)?

Step 14.3 Spatial Interpolation with IDW

- As you determined for Question 14.3, the sample points are far apart. Say that you want to know how much precipitation occurred at the places between them, not just at these points. To generate this information, you have to interpolate a surface of total rainfall. The first method you'll use is inverse distance weighted (IDW). See **Smartbox 14.1** for more information about IDW and how it operates. To open the Geoprocessing pane, on the **Analysis** tab, in the **Geoprocessing** group, click the **Tools** button. The tool you will be using is IDW, and you can either search for it or find it directly in the **Spatial Analyst Tools** toolbox, in the **Interpolation** toolset. Using either method, open the IDW Spatial Analyst tool.

SMARTBOX 14.1

How does inverse distance weighted (IDW) interpolation function?

Inverse distance weighted (IDW) is a local interpolation method based on the one premise that the closer a known value point is to an unknown value location, the more important that known point is to determining the unknown value. Conversely, the farther away a known value point is from an unknown value location, the less that known point contributes to determining the unknown value. In IDW, the values of the points within the local search radius are averaged to determine the value at the unknown location. However, the value of each known point is weighted according to how far away that point is (and closer points carry a higher weight than points that are farther away). For instance, the amount of snowfall in areas closer to you is going to have a stronger influence on the amount of snow you get at your location than the amount of snowfall occurring at areas farther away. While IDW is a good and fast interpolation method, it has a tendency to create bull's-eye effects around known value points, especially if there are no other points nearby.

Mathematically, the formula for determining the interpolated value at the unknown point ($F(x,y)$) is the sum of all the values of the known points (F_i) multiplied by a weight (W_i) for each known point value:

$$F(x,y) = \sum_{i=1}^{N} W_i F_i$$

The weight of each point is calculated by taking each known point and measuring the distance (d) from the known point to the unknown point and dividing by the sum total of all distance values. A separate value for power (p) is used in the weighting formula as well:

$$W_i = \frac{d_{i0}^{-p}}{\sum_{i=1}^{N} d_{i0}^{-p}}$$

The value used for power directly affects the equation. If a high value is used for power, the points at closer distances will be weighted even more heavily. The default value for power used in ArcGIS Pro is 2.

inverse distance weighted (IDW) An interpolation process that involves assigning a higher weight to the values of known points closer to the location being interpolated and lower weights to the values of known points that are farther away.

- In the IDW tool, for Input point features, choose **Ohioprecippoints**.
- For Z value field, select **Globvalue** (the attribute field that will be the source of the values you will be interpolating).
- For the Output raster, click the **browse** button and navigate to your **C\GIS:\Module14** folder, go to the **Module14.gdb** file geodatabase, and name the output raster **IDWsurface**.
- For Output cell size, enter **100** (for 100 meters, which you previously set in the Geoprocessing pane).
- Leave the other settings at their defaults and run the IDW tool. In the lower-right corner of the ArcGIS Pro window, you see the progress of the IDW calculation (which may take a couple minutes to run).

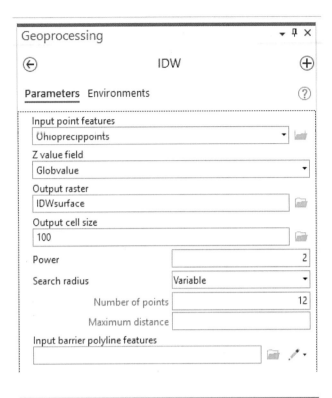

- At this point, the IDWsurface layer shows an interpolated surface for all of the data points, but you're only interested in looking at those within Ohio. To do so, you can create a subset of the raster data—that is, only those cells of the IDWsurface layer that are within the boundary of Ohio. (See page 300 in Module 12 for more about subsets.)

- In the Geoprocessing pane, the tool you want to use is **Extract by Mask**. Either search for it or find it directly in the **Spatial Analyst** toolbox, within the **Extraction** toolset. The Extract by Mask tool opens in the Geoprocessing pane.

- For Input raster choose **IDWsurface**.

- For Input raster or feature mask data, choose **Ohiocountiesalbers**.

- For the Output raster, click the **browse** button and navigate to your **C:\GIS\Module14** folder, go to the **Module14.gdb** file geodatabase, and then name the output raster **IDWOhio**.

- Run the Extract by Mask tool. IDWOhio is added to the Contents pane. It is a new grid with 100-meter resolution, with cell values reflecting the interpolated precipitation at that location.

- In the Symbology pane, change the symbology of the IDWOhio grid as follows:
  - For Primary symbology, choose **Classify**.
  - For Method, use **Geometric Interval**.
  - For Classes, use **10**.
  - For Color Scheme, choose a suitable and intuitive color ramp.

- Turn off the IDWsurface layer at this point as you will be working with only the IDWOhio layer.

- Zoom in to Cuyahoga County in northern Ohio and adjust the view so that you can see the whole county. Arrange the layers in the Contents pane so that you can see the Ohioprecippoints layer on top of the interpolated IDWsurface grid, all framed by the county borders. Examine the precipitation points and their rainfall values in relation to the interpolated surface. Answer Question 14.4.

> **Question 14.4** Why do some points have a circular set of grid cells centered around that point, while others do not?

- Turn off the Ohioprecippoints layer.

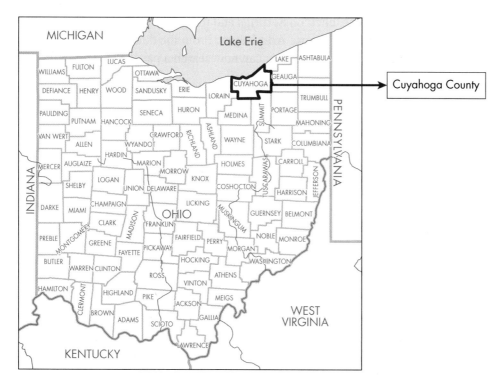

- Add the Ohioairports feature class to the Contents pane so that it sits on top of the other layers. Zoom out so that you can see the entire state again.
- At this point, you need to locate six airports across the state to determine the interpolated precipitation value at each location. These are the airports you'll be examining:

Airport Name	Text in Attribute: Name
Akron Canton KCAK	Akron-Canton Regional Airport
Cleveland Hopkins KCLE	Cleveland-Hopkins International Airport
James M. Cox Dayton KDAY	James M. Cox Dayton International Airport
Mansfield-Lahm Regional KMFD	Mansfield Lahm Regional Airport
Port Columbus International KCMH	Port Columbus International Airport
Youngstown-Warren Regional KYNG	Youngstown-Warren Regional Airport

- To find the correct airport, go to the **Map** tab, and in the **Inquiry** group, click the **Locate** button. (See Module 5 for more about the Locate tool.) The Locate pane opens.

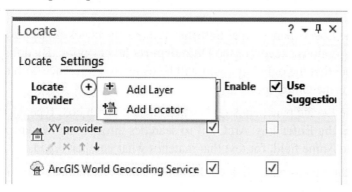

- In the Locate pane, click on **Settings** and then click on the **+** button next to Locate Provider and choose **Add Layer**. From the dialog box that opens next, choose **Ohioairports**. You can now search a particular field in the Ohioairports layer.

- Under Field Options, in the pull-down menu next to Name, choose **Contains**.

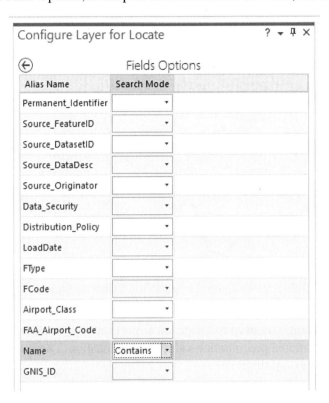

- Click the **left arrow button** at the top of the pane to return to the Locate pane.

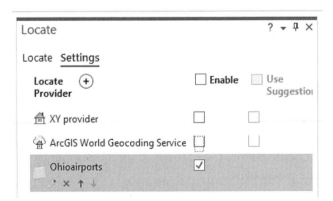

- In the Locate pane, under **Settings**, remove the checkmarks next to all other options except for the **Ohioairports layer** option. By doing so, you ensure that the only thing you will be using as a reference with the Locate tool is the Ohioairports layer.

- In the Locate pane, click on **Locate**. In the Search box, type **Akron and press the Enter key**. ArcGIS Pro searches through the Ohioairports layer, in the Name field, for text that matches what you just typed.

- For the first airport, click on the option that corresponds with Akron Canton Regional Airport. (See the table on p. 337 for the names of the airports and the corresponding value of each one in the Name field.) ArcGIS Pro zooms in to this point.

- Go to the **Map** tab, and in the **Navigate** group, click on the arrow under the **Explore** tool. From the available options, choose **Visible layers**. By doing so, you ensure that when you click on a point on the screen, ArcGIS Pro will return to information about all visible layers at that point. Click on the point representing Akron-Canton Regional Airport.

- Zoom in very closely to the first airport and use the **Explore** tool on the Map tab to click on the Airport point and get information about the name of the airport. (See Module 5 for more about the Explore tool.)

- In the pop-up that opens, you should have the information from three visible layers: Ohioairports, Ohiocountiesalbers, and IDWOhio. You can click the left and right arrows at the bottom of the pop-up to move through these results one by one.

  - Use the Locate and Explore tools for all six airports, carefully selecting the point for each airport and examining the amount of precipitation calculated for the 100-meter grid cell that point falls on.

  Important Note: You need to closely zoom in to the airport points to be certain you are selecting the correct locations. Answer Question 14.5.

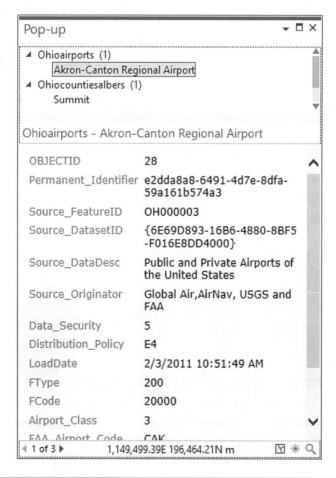

Question 14.5 What is the estimated total precipitation (in inches) at each of the six airport sites (according to the IDW interpolation)?

Step 14.4 Working with the Zonal Tools

- Rather than examine the interpolated precipitation value at a single point, you will now look at an average of many precipitation values for an entire area. The zonal tools in ArcGIS Pro allow you to do this. For more information about zonal tools, see **Smartbox 14.2**.

SMARTBOX 14.2

How do the zonal tools operate in ArcGIS Pro?

The zonal tools in ArcGIS Pro allow you to obtain information about a set of raster grid cells that are spatially within a boundary (which can be defined by a polygon feature class or another raster layer). In a zonal function, the raster for which you want to calculate information serves as the input, and the boundary serves as the "zones." If a polygon feature class is used as the source of the zones, ArcGIS Pro internally converts it to a raster in performing a zonal calculation. An example of a zonal function is the **Zonal Statistics** tool, which allows you to calculate a particular statistic (for example, the mean, the sum, the standard deviation, the median, the value that occurs most often, the value that occurs least often, the number of unique values, the highest value, the lowest value, the range between the highest and lowest values) for the grid cells of the input raster that fall within the boundaries of the zones.

Figure 14.3 shows how the Zonal Statistics function can be used to find the average precipitation for each Ohio county. A raster surface contains interpolated precipitation values for each grid cell. The polygon feature class (Ohiocountiesalbers) designates the boundaries of each zone, with each of the 88 Ohio counties as a separate zone. The Zonal Statistics tool computes the mean of all grid cell values that fall within the boundaries of each of the counties. A new output raster is created, and all grid cells that were within each zone are assigned the same value (specifically, the mean of all interpolated values in that zone).

Zonal Statistics A tool for calculating statistical values based on the raster grid-cell values that fall within a boundary.

FIGURE 14.3 An interpolated raster surface of precipitation values, a polygon feature class defining zones as county boundaries, and a Zonal Statistics output layer showing the average precipitation value calculated for each zone.

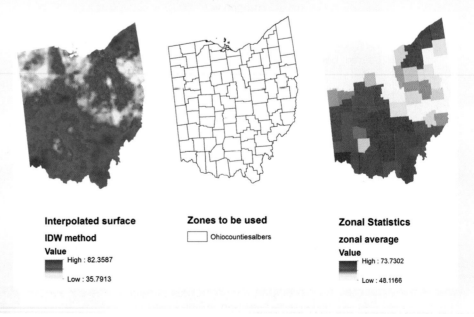

- For this module, you'll be calculating the average amount of precipitation for each county in Ohio. ArcGIS Pro will compute the average of all the interpolated grid cell values that are within the boundaries of each county. Return to the Geoprocessing pane. The tool you will use is Zonal Statistics, and you can either search for it or find it directly in the Spatial Analyst **Tools** toolbox, within the **Zonal** toolset. Open the **Zonal Statistics** Spatial Analyst tool. The Zonal Statistics tool opens in the Geoprocessing pane.

- For Input raster or feature zone data, select the **Ohiocountiesalbers** feature class from the Ohioprecipdata geodatabase. This is the layer that will define the zones.

- For Zone field, use the **NAME** attribute.

- For Input value raster, use **IDWOhio**. This is the raster with the values you want to examine in relation to the zones.

- For Output raster, click the **browse** button and navigate to your **C:\GIS\Module14** folder, go to the **Module14.gdb** file geodatabase, and then name the output raster **CountyIDWave**.

- For Statistics type, choose **Mean** (because you want the average interpolated value for each zone).

- Place a checkmark in the **Ignore NoData in calculations** box.

- Run the Zonal Statistics tool. ArcGIS Pro calculates the mean value of all the grid cells in the IDWOhio raster that fall within the boundaries of the Ohio counties, as modeled in the Ohiocountiesalbers feature class. A new raster grid called CountyIDWave is created, with each cell showing the average amount of rainfall per county in Ohio. Examine this new grid and answer Questions 14.6 and 14.7.

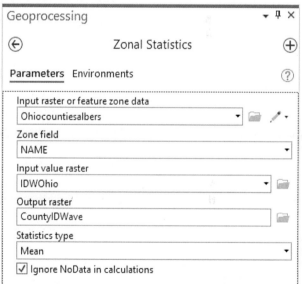

Question 14.6 What is the average amount of precipitation for the entirety of Cuyahoga County (according to the IDW interpolation)?

Question 14.7 What is the average amount of precipitation for the entirety of both Geauga and Lorain Counties (according to the IDW interpolation)? Give a value for each county.

Step 14.5 Spatial Interpolation with Splines

- IDW isn't the only deterministic method of interpolating a surface from a set of points. In this portion of the module, you'll use a method called Spline to create a new interpolated surface. See **Smartbox 14.3** for more information about using the Spline method for spatial interpolation.

SMARTBOX 14.3

How do Spline functions operate in spatial interpolation?

Spline An interpolation method that involves using a mathematical process to fit a surface exactly through known points.

regularized A Spline method that results in an overall smoother surface but that may also result in values that are beyond the range of the known point values.

tension A Spline method that results in a less smooth surface than a regularized Spline but that contains values closer to the range of the known point values.

The **Spline** method of interpolation involves using a mathematical process to fit the surface exactly through the known points and also fit a set of input points in the local search area. Here is one way to conceptualize splines: You have a set of points with known values for some phenomenon (such as precipitation). Instead of thinking of them as dots on a map, think of them as poles that are being raised above a flat surface to a height represented by their value. For instance, if one point has a value of 50 and another has a value of 100, think of these as two poles—one that is 50 units high and another that is 100 units high. When you have all your points set up as poles, you have to adjust a rubber sheet (that can be stretched in various directions) so that it touches the top of each pole. In the end, you'll have a smooth surface that's been stretched this way and that.

The Spline process is a mathematical one that tries to fit the surface to the known points as well as minimize the overall curvature. Two types of Spline methods are available in ArcGIS Pro. The first one is **regularized**, and it results in an overall smoother surface but may also result in values that are beyond the range of the known point values. The second method, called **tension**, generates a less smooth surface than the regularized method, but it contains values closer to the range of the known point values.

The end result of Spline interpolation is a smooth surface. The Spline method is best used for interpolating surfaces that do not have abrupt variations in their values (such as elevation measurements or temperature readings). See Figure 14.4 for an example of a regularized Spline interpolated surface compared to an IDW-generated surface. The two surfaces appear visually similar, but the Spline method produces a smoother overall surface. However, the range of interpolated values is differs greatly between the two methods.

FIGURE 14.4 A comparison of a Spline-generated surface and an IDW-generated surface.

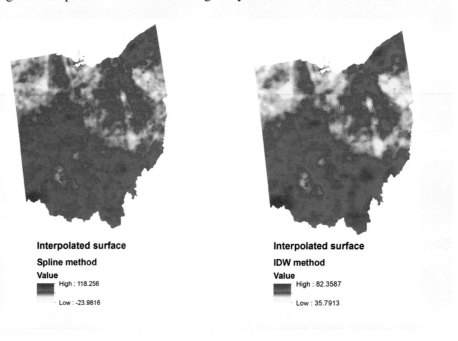

- Return to the Geoprocessing pane. The tool you will use is Spline, and you can either search for it or find it directly in the **Spatial Analyst Tools** toolbox, within the **Interpolation** toolset. Using either method, open the **Spline** Spatial Analyst tool. The Spline tool opens in the Geoprocessing pane.
- For Input point features, select **Ohioprecippoints**.
- For Z value field, select **Globvalue**.
- For Output raster, click the **browse** button and navigate to your **C:\GIS\Module14** folder, go to the **Module14.gdb** file geodatabase, and then name the output raster **splinesurface**.
- For Output cell size, enter **100** (for 100 meters, which you previously set in the Geoprocessing pane).
- For Spline type, choose **Regularized**.
- Leave the other options at their defaults and run the Spline tool. In the lower-right corner of the ArcGIS Pro window, you see the progress of the Spline calculation (which may take a couple of minutes to run).
- Splinesurface is added to the Contents pane. It is a new grid with 100-meter resolution, with the cell values reflecting the new interpolated precipitation at that location.

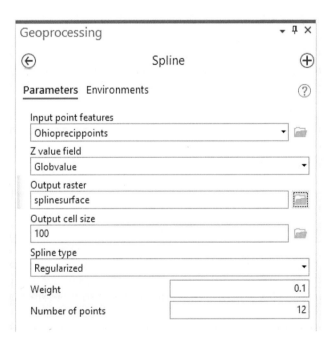

- As you did earlier, use the Extract by Mask tool to create a subset of the Splinesurface raster called **SplineOhio**.
- The new SplineOhio layer is added to the Contents pane. It is a new grid with 100-meter resolution, with the cell values reflecting the interpolated precipitation at that location. In the Symbology pane, change the symbology of the SplineOhio grid as follows:
  - For Primary symbology, choose **Classify**.
  - For Method, choose **Geometric Interval**.

- For Classes, use **10**.
- For Color Scheme, choose a suitable and intuitive color ramp.
• Turn off the Splinesurface raster as you will be using only the SplineOhio subset layer.
• Examine the SplineOhio layer in and around Cuyahoga County, comparing it to the surface generated for the IDWOhio layer. Answer Questions 14.8 and 14.9.

Question 14.8 Why does the spline surface appear "smoother" (that is, with less of the circular effect) than the IDW surface?

Question 14.9 Use the Spline tool again but this time create a tension spline surface. How do the numerical values of the regularized spline compare to the numerical values of the tension spline? Why is this? (You can delete this tension spline after you complete this question.)

• As you did before, use the Locate and Explore tools to determine the amount of precipitation interpolated for each of the six airports used in Step 14.3 but this time examine the SplineOhio layer. Answer Question 14.10.

Question 14.10 What is the estimated total precipitation (in inches) at each of the six airport sites (according to the Spline interpolation)?

• Again use the Zonal Statistics tool but this time calculate the mean value for all of the SplineOhio grid cells that fall within an Ohio county boundary. Answer Questions 14.11 and 14.12.

Question 14.11 What is the average amount of precipitation for the entirety of Cuyahoga County (according to the Spline interpolation)?

Question 14.12 What is the average amount of precipitation for the entirety of both Geauga and Lorain Counties (according to the Spline interpolation)? Give a value for each county.

Step 14.6 Evaluating Interpolated Surfaces

• At this point, you have two interpolated surfaces of total precipitation for the state of Ohio. You need to examine both surfaces to see which of them comes closer to predicting the actual amounts of precipitation in the state of Ohio. First, you can visually compare the two surfaces. Turn off all of your layers except for SplineOhio and IDWOhio and place the SplineOhio layer on top of the IDWOhio layer in the Contents pane. Next, in the Contents pane, click once on **SplineOhio** so that it is highlighted. Then, on the **Appearance** tab, within the **Effects** group, click the **Swipe** button. (See page 313 in Module 13 for more about using Swipe.)

- When you choose the Swipe tool, your cursor becomes a black triangle. Click the mouse on the image and hold and drag the triangle from left to right or up and down; the Splinesurface layer is replaced with the IDWsurface layer underneath, as if you were opening and closing window blinds. Use Swipe to compare the two layers and zoom in on areas where the layers appear similar and areas where the layers appear different. Also pay close attention to the values of the classifications, as seen for both layers in the Contents pane (because they will likely have different values displayed for each surface). Answer Question 14.13.

> **Question 14.13** How do the two surfaces visually compare (in terms of patterns and smoothness of the surfaces)?

- By examining both surfaces (as well as your answers to Questions 14.5 and 14.10), you can see that, while the two surfaces have some similarities, they also contain different interpolated values for the same locations. By adjusting the parameters of both the IDW and the Spline processes, you could generate other similar-but-different interpolated surfaces of precipitation. With different surfaces, you need a way to determine which method provides the most accurate interpolation result.

- The table below lists the observed precipitation values for the six Ohio airports that you've looked at in this module. This information was culled from values on the Wunderground (Weather Underground) website, at **http://www.wunderground.com**.

Airport Name	Name in Feature Class	Observed Precipitation
Akron Canton KCAK	Akron-Canton Regional Airport	58.38 inches
Cleveland Hopkins KCLE	Cleveland-Hopkins International Airport	65.32 inches
James M. Cox Dayton KDAY	James M. Cox Dayton International Airport	56.72 inches
Mansfield-Lahm Regional KMFD	Mansfield Lahm Regional Airport	56.68 inches
Port Columbus International KCMH	Port Columbus International Airport	54.96 inches
Youngstown-Warren Regional KYNG	Youngstown-Warren Regional Airport	54.01 inches

- Compare the Wunderground observed precipitation values to those that you interpolated from each surface and use the Swipe tool to examine both surfaces around these six locations. Answer Question 14.14.

> **Question 14.14** How do the interpolated values for the IDW surface and the Spline surface compare to the six actual values? Which values are too high, which are too low, and which nearly match in each dataset? Based solely on these six points, which surface is more representative of the actual observed values, and why?

Step 14.7 Printing or Sharing Your Results

- Save your Module14 project. You can then present your results in one of two ways:
  - Print a layout (see Module 3) showing the IDWOhio and SplineOhio layers side by side. In order to do this, you need to display the IDWOhio layer in the Ohio precipitation map you've been working with and then create a second new map and place the SplineOhio layer into it. When printing a layout, you can insert two maps into it and position them in the layout so that they are shown side by side, with a separate legend and scale bar for each map. (See Module 3 for more on how to use more than one map frame in a layout.)
  - Share your results as a tiled map service through ArcGIS Pro Online (see Module 4). If you're sharing your data, you might want to create a web application using a template that has the Swipe tool available to allow others to interactively examine both surfaces (as you've been doing in this module) or one of the Compare templates.

Closing Time

This module looks at how to start with an initial set of point measurements and create a raster surface from them to "fill in the gaps" where point measurements were unavailable. Spatial interpolation methods are used to create surfaces to model continuous phenomena, including precipitation, groundwater, and pollution levels. While commonly used, IDW and Spline are not the only spatial interpolation methods available in ArcGIS Pro. *Related Concepts for Module 14* examines another powerful set of interpolation tools.

Elevation is a phenomenon that varies continuously across Earth and is commonly modeled as a raster surface. An elevation surface acts as the interpolated surfaces you've examined in this module (that is, at each x and y location, a z-measurement for elevation can be determined). You will work with these kinds of "functional" surfaces in the next module.

RELATED CONCEPTS FOR MODULE 14

Geostatistical Methods for Spatial Interpolation

deterministic An interpolation method that does not take spatial variation into account.

Both the IDW and Spline techniques are **deterministic** interpolation methods, in that they rely solely on the location of the known points and mathematically fit the surface through these points. As such, they do not rely on information regarding how the values of the known points relate to one another. For instance, if you have several known points with similar values located close together, how will this similarity of values influence how the values for nearby unknown points are calculated? Likewise, if the values for the points are very different from one another but very far apart, how will these conditions affect the new values to be interpolated for other locations? These kinds of spatial variations in the values of the points or the similarities of values of nearby points are not considered in deterministic methods like IDW (which deals only with the distances between points, not their values) or Spline (which seeks to fit the surface through points and minimize the curvatures of the surface). However, a separate set of **geostatistical** interpolation methods take into account spatial autocorrelation (see Module 8) and similarities over space between points when calculating interpolated values for a surface.

Kriging is a commonly used geostatistical interpolation method that takes into account the variation or similarity of the known values as well as their spatial arrangement when predicting what the new interpolated values should be. To model these factors, you first need to know something about the variance between the values of the known points. In theory, points with similar values (that is, a low variance) should be found closer together, while points with dissimilar values (that is, a high variance) should be found farther away from one another. In reality, though, these conditions are not always met, so a model of the variance of the points' values as it relates to their distance apart can be generated. This model can then be used to determine the variance associated with points to be interpolated that are a certain distance apart. This model is called a **semivariogram**, and properly setting it up in relation to the values of the known points is key to its use in the prediction process (Figure 14.5).

geostatistical An interpolation method that accounts for spatial variation.

Kriging A local geostatistical interpolation method.

semivariogram A model of the variance between points and their distance apart.

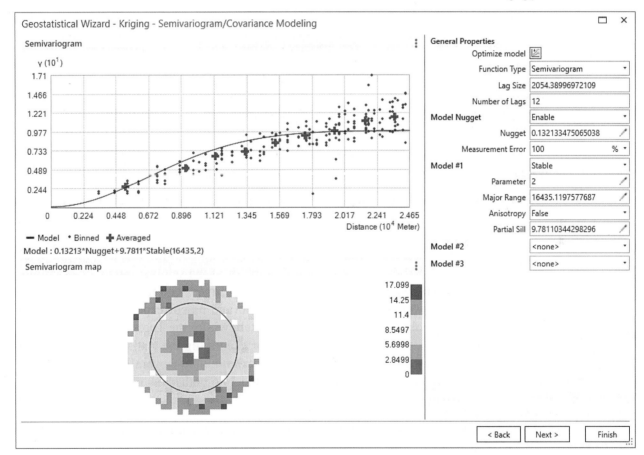

FIGURE 14.5 Fitting a semivariogram for use in the Kriging process.

Kriging is a local interpolation method that uses (a) the variation between the subset of known points being used to determine the unknown value and (b) the variation between the known points and the unknown location. Each point is weighted through a mathematical process that takes into account the variance as well as a computed margin of error. Weights are assigned to each of the points being used in the calculation of the location's unknown value (Figure 14.6).

Several types of Kriging are used in interpolation. A common method is **Ordinary Kriging**, which assumes an unknown mean for the known point

Ordinary Kriging A Kriging method that assumes an unknown mean and a lack of trend in the data.

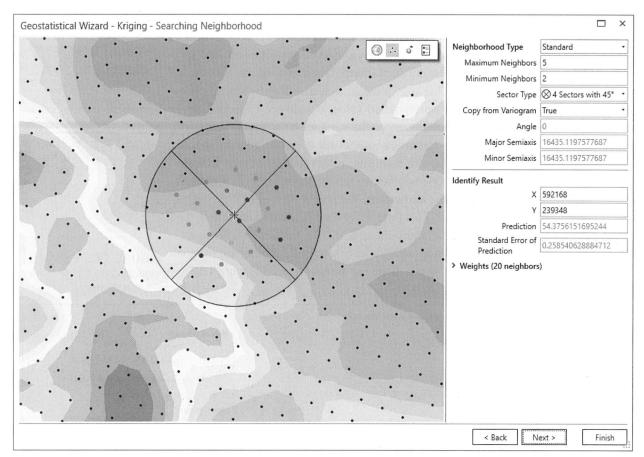

FIGURE 14.6 The assigned weights for the 20 points being used to determine the value for a location using Kriging.

Universal Kriging A Kriging method that assumes a trend within the data (and that removes the trend in order to perform interpolation).

Geostatistical Analyst An ArcGIS Pro extension that contains several functions related to examining and working with spatial statistics.

values as well as the absence of a trend in the data. Another is **Universal Kriging**, which assumes the presence of a trend and removes that trend from the data prior to performing interpolation. A Kriging tool is available in the Interpolation toolset of the Spatial Analyst toolbox, but further in-depth analysis with Kriging (such as adjusting the fit of the semivariogram or examining the weights assigned to points) can be performed with the functions of ArcGIS Pro's **Geostatistical Analyst** extension (and its corresponding Geostatistical Wizard tools, located on the Analysis tab, within the Tools group). This extension provides modeling techniques for examining the statistical nature of spatial data, as well as statistical tools for examining the overall accuracy and fit of an interpolated surface. Figures 14.5 and 14.6 both show the use of Geostatistical Analyst functions.

Key Terms

spatial interpolation (p. 328)
surface (p. 328)
global interpolation (p. 329)
trends (p. 329)
local interpolation (p. 329)
inverse distance weighted (IDW) (p. 335)
Zonal Statistics (p. 340)
Spline (p. 342)
regularized (p. 342)
tension (p. 342)
deterministic (p. 346)
geostatistical (p. 347)
Kriging (p. 347)
semivariogram (p. 347)
Ordinary Kriging (p. 347)
Universal Kriging (p. 348)
Geostatistical Analyst (p. 348)

15

How to Work with Digital Elevation Models in ArcGIS Pro

ArcGIS Pro Skills

In this module, you will learn how to do the following in ArcGIS Pro:
- Navigate the scene environment.
- Work with the elevation surface in a scene.
- Create 3D symbology for objects in a scene.
- Derive a hillshade from a DEM.
- Calculate slope and aspect raster layers.
- Create and use a profile graph.
- Use line of sight and viewsheds for visibility analysis.
- Make a layout using a scene rather than a map.

Learning Outcomes

After studying this module, you should be able to:
- Define what a digital elevation model (DEM) is.
- Explain what a z-value is.
- Explain why a DEM is a 2.5-dimensional model.
- Describe how to obtain DEM layers from The National Map and 3DEP.
- Describe how slope is calculated.
- Explain what slope aspect is.
- Describe how a hillshade is calculated.
- Explain the difference between a line of sight and a viewshed.

Introduction

For 14 modules now, we've been working with two-dimensional GIS data, in that we've not been concerned with factors such as heights or elevations. We've not taken into account any information about Earth's landscape while examining building footprints, roads, bodies of water, or land cover. However, Earth isn't just a flat, featureless plane, and often you'll need to assess the changing features of the terrain (like hills, valleys, or mountains) as part of your analysis.

For example, an engineer laying down a new road will have to take into account the elevation of the landscape or the steepness of the grade of the land that's being built on. This module introduces you to the use of **digital terrain models (DTMs)** and elevation data in ArcGIS Pro, including the basics of viewing and examining them in a 3D environment. Many applications of GIS have their basis in DTMs, including applications that allow you to examine the slope of the landscape and determine what areas can or cannot be seen due to the intervening terrain. Keep in mind that these types of digital terrain models are measuring the elevation of Earth's surface, not the objects on top of it (such as trees or buildings).

With elevation data, a **z-value** is used to indicate the height or elevation of an x/y coordinate. The z-value indicates the height above some level. So, for

digital terrain model (DTM) A representation of a terrain surface calculated by measuring elevation values at a series of locations.

z-value The elevation assigned to an x/y coordinate.

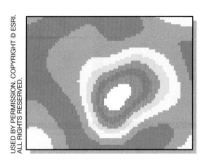
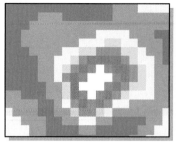

FIGURE 15.1 A finer-resolution raster elevation surface versus a lower-resolution raster elevation surface.

vertical datum A baseline used in measuring elevation values (above or below this level).

NAVD88 The North American Vertical Datum of 1988, a commonly used vertical datum for U.S. digital terrain models.

NAPGD2022 The North American-Pacific Geopotential Datum of 2022, a new vertical datum set to be adopted for U.S. geospatial data in 2022.

digital elevation model (DEM) A representation of a terrain surface created by measuring elevations at evenly spaced sampling points.

functional surface A continuous representation of values that can have a single value assigned to each location.

example, the z-value may indicate the height of the terrain above mean sea level. GIS data with a z-value need to be referenced against this level, referred to as a **vertical datum**, for use in any elevation measurements. As with other coordinate systems, there are many types of vertical datums, but a common one is **NAVD88** (the North American Vertical Datum of 1988), which is often used for U.S. elevation datasets. A new vertical datum for U.S. elevation data, set to replace NAVD88 and take effect in 2022, is **NAPGD2022** (North American-Pacific Geopotential Datum of 2022).

A common way of organizing elevation data within GIS is by using a **digital elevation model (DEM)**. With a DEM, elevation values are evenly spread across a surface at regularly spaced sampling points. DEMs are often used as raster datasets in which the grid cell value is the elevation measurement applied to the entire cell; the cell size indicates the spacing of the elevation points. For instance, a 10-meter raster DEM would be composed of 10-meter-resolution grid cells, and each cell would contain an elevation value. DEM points are always regularly spaced, so the finer the resolution of the DEM, the more elevation values it's able to contain (and the larger the size of the file). In a coarser-resolution DEM, the elevation values are spaced farther apart from one another, indicating larger cells and less detailed elevation information (but a smaller file size); see Figure 15.1 for an example. Numerous DEMs are available for use in GIS at a variety of resolutions. A digital elevation model is considered to be a **functional surface**, in that it is continuous so that you can measure a single elevation value at any location.

 ## Module Scenario and Applications

This module puts you in the role of a planner in Hocking County, Ohio. Your job is to assess the landscape features of the county as part of a study involving the building of a new observation platform in the county. You will be examining the terrain and land-cover characteristics of the county, particularly at the building site for the platform. You want to determine the minimum height at which the platform can be constructed in order to have an unobscured view of the city of Logan's fairgrounds over the terrain. From there, you'll determine what areas of the entire city can or cannot be seen from this projected minimum platform height.

The following are additional examples of real-world applications of this module's skills:

- A developer is designing the layout of a new ski resort. She could use a digital elevation model as the basis for slope and aspect calculations for planning ski runs.

- A forest manager is examining several possible sites for the location of a new lookout tower. He can perform viewshed analysis from each site to determine what areas can and cannot be seen.

- A county engineer in a coastal region wants to examine the visual impact of the development of new high-rise

hotels being built on the coast. She can build viewsheds and lines of sight from a variety of key sites to determine how the addition of tall buildings will impact tourists' views of the beaches and shores.

 ## Study Area

The module data is of Hocking County, located in southeastern Ohio. The county seat (and main urban area) is the town of Logan.

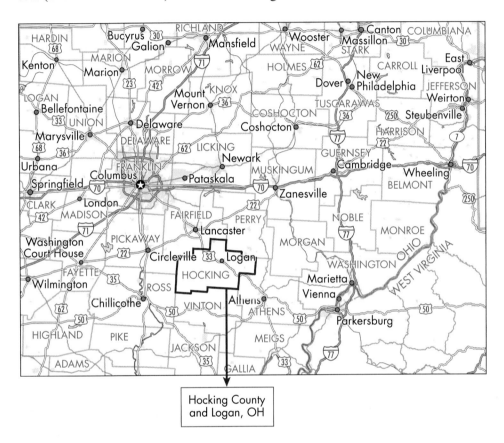

 ## Data Sources and Localizing This Module

The data in this module focus on Hocking County in the state of Ohio. However, you can easily modify this module to use data from your own local area—for instance, Marquette County, Michigan. The DEM, county boundary, and roads layers were all downloaded and extracted for free from The National Map. Use GU_countyorequivalent for the county boundary and trans_roadsegment for the roads layer. The boundary for a city within the county (such as the city of Marquette) can also be downloaded from The National Map (use GU_Incorporated_Place). The four point layers of locations in the county were digitized into their own feature classes. (See Module 6 for how to digitize points.)

The NLCD data were downloaded for free from the Multi-Resolution Land Characteristics (MRLC) website at **https://www.mrlc.gov**. NLCD 2011 data for the entire United States can be downloaded from that website, and the data for Marquette County (or your individual county) can be extracted by using the

Extract By Mask tool (see Module 12) and the county boundary layer. Alternatively, NLCD 2011 data can also be downloaded directly via The National Map.

The Mask layer was created by processing the NLCD layer; map algebra was used to select all grid cells with a value of greater than 0, which resulted in a new grid in which all cells in the county had a value of 1 and cells outside the county had a value of 0. (See Module 20 for more about map algebra.) Then the grid was reclassified so that all cells with a value of 0 were given a value of NoData. (See Module 19 for more about reclassifying cells.)

Step 15.1 Getting Started

- This module's hands-on applications use the data folder called Module15. Your instructor will be able to supply you with this data, or you can download it directly from this book's website at **https://www.macmillanlearning.com/college/us/product/Discovering-GIS-and-ArcGIS-Pro/p/131923075X**. The text in this module assumes that you have this Module15 folder in a computer location referenced as C:\GIS; if you have it somewhere else (for instance, in a flash drive referenced as G:\GISClass), substitute that location and path to the Module15 folder throughout this module.

- The Module15 folder contains the following:
  - Hockingcounty: a file geodatabase that contains the following items:
    - DEMhocking: a 1 arc-second 3DEP DEM resampled to 30-meter resolution (with elevation values for the DEM in meters) for this module
    - NLCDhocking: a 30-meter-resolution NLCD 2011 raster layer
    - Maskhocking: a 30-meter grid that conforms to the boundaries of Hocking County
    - Boundaryhocking: a polygon feature class of the boundary of Hocking County
    - Loganboundary: a polygon feature class of the boundary of the city of Logan
    - RoadSegmenthocking: a line feature class of the county's roads
    - HockingValleyHospital, HockingCoFairgrounds, LoganDam, and Platform: point feature classes with a digitized point representing the location of each place

- Start ArcGIS Pro.

- Sign in with your Esri account username and password.

- Create a new project using the **Local Scene** template. Call this project **Module15** and place it in your **C:\GIS\Module15** folder. Ensure that there is not a checkmark in the box next to **Create a new folder for this project**.

- A **scene** works similarly to a map but is used for viewing and analyzing data in 3D. In the Contents pane, right-click on the scene and choose **Properties**. As you've done with maps in previous modules, change this scene's name to **Hocking County**.

- Several of the tools you will use in this module are part of both the Spatial Analyst and the 3D Analyst extensions. **3D Analyst** is an ArcGIS Pro extension used for viewing and analyzing data that have

scene The ArcGIS Pro environment used for working with 3D data and visualization.

3D Analyst An ArcGIS Pro extension used for 3D visualization and working with data that have z-values.

z-values (as digital elevation models do) in a 3D environment. You need to have both of these extensions licensed to your Esri account and active for use in this module. If you don't have the Spatial Analyst and 3D Analyst extensions licensed and active, some tools will not work: They will have a lock symbol next to them and will be unavailable. See Smartbox 11.1 and Troublebox 11.1 in Module 11 for more about extensions and ensuring that you have both the Spatial Analyst and 3D Analyst extensions available for use.

- Add the **NLCDhocking** layer to the Contents pane. As you did in Module 12, change the symbology of the various land-cover types to better represent each one (for example, shades of green for the forested land-cover types, blue for the water, and so on).
- Also add the **Maskhocking** layer to the Contents pane.
- On the **Analysis** tab, in the **Geoprocessing** group, click the **Environments** button and then set the following geoprocessing environment settings for this module:
  - For Current Workspace, use the **Module15.gdb** geodatabase within your C:\GIS\Module15 folder.
  - For Scratch Workspace, use the **Module15.gdb** geodatabase within your C:\GIS\Module15 folder.
  - For Output Coordinate System, choose **NLCDhocking**. ArcGIS Pro then updates the coordinate system to that of the NLCDhocking layer.
  - For Extent, choose **NLCDhocking**. ArcGIS Pro then updates the x/y coordinates of the lower-left corner and upper-right corner of the grid extent to those of the NLCDhocking layer.
  - For Cell Size, choose **NLCDhocking**. ArcGIS Pro then updates the cell size for raster output in this project to 30 meters, as that is the cell size of the NLCDhocking layer.
  - For Mask, choose **Maskhocking**.
- Leave the other settings alone and click **OK** to put these geoprocessing environment settings into place.
- The NLCDhocking layer was projected from its initial coordinate system to another one for use in this module. The coordinate system being used in this module is **North America Albers Equal Area Conic (NAD 83)**, with **meters** as the map units. Check the Hocking County scene properties (on the Coordinate System tab) to verify that this projection system is being used.
- You can turn off the Maskhocking layer at this point.

Step 15.2 Working with Scenes and Pseudo-3D Visualization

- You'll see that a scene looks very similar to a map in terms of how layers are added and displayed in a Contents pane. However, scenes are designed to work with your data in a 3D environment. For more about working with scenes, see **Smartbox 15.1**; see **Smartbox 15.2** for how to navigate the scene environment.

SMARTBOX 15.1

What is a scene in ArcGIS Pro and how does it operate with 3D visualization?

perspective view A view of GIS data at an oblique angle that causes it to take on a "three-dimensional" appearance.

pseudo-3D The perspective view of a digital terrain model, which is often a 2.5D model rather than a 3D model.

two-and-a-half dimensional (2.5D) A model of the terrain that allows for a single z-value to be assigned to each x/y coordinate.

three-dimensional (3D) A model of the terrain that allows for multiple z-values to be assigned to each x/y coordinate.

elevation surface The source of z-values used for a location from a digital terrain model surface.

Ground The source of z-values used to represent the base heights of the digital terrain model at the surface level.

A scene in ArcGIS Pro is used for visualizing GIS layers in a **perspective view** (such as from an angle) to give the data a three-dimensional appearance. Scenes are either local (used for smaller geographic areas) or global (used for larger geographic areas, where the curvature of Earth is an important viewing consideration). The Contents pane for a scene works the same way as the Contents pane for a map or a layout view, and the corresponding tabs, groups, and buttons work the same way. ArcGIS Pro also enables you to move all items from the Contents pane of a map into the Contents pane of a scene: Just go to the map's **View** tab, and in the **View** group, click the **Convert** button.

Viewing datasets in 3D is the default mode for using a scene. However, although data (such as the digital elevation model of Hocking County) can be viewed and examined in a perspective view and heights can be added to make it look 3D, the resulting model is not necessarily three-dimensional. Surfaces such as DEMs are considered to be **pseudo-3D** or **two-and-a-half-dimensional (2.5D)** models. In a two-dimensional (2D) surface, only x/y coordinates are measured, and no z-value is added. In a 2.5D surface, there can be only one z-value measured for each x/y coordinate. In DEM data, there is only a single value representing the elevation for that location above the vertical datum, so it is considered 2.5D. In a full **three-dimensional (3D)** surface, there can be multiple z-values assigned to each x/y coordinate (Figure 15.2). For instance, if a digital terrain model can simultaneously measure the elevation of the ground and the elevation height of the building on the ground at that same location, it is considered a 3D model. Because most digital terrain models measure a single z-value for each location, they are considered 2.5D.

In a scene, a layer is a feature class or a raster with its height represented using the z-values of the terrain. This **elevation surface** represents the z-value drawn from a digital terrain model surface. By default, a scene contains an Esri digital terrain model of Earth (called the World Elevation 3D/Terrain 3D) that serves as a source of z-values. In the scene's Contents pane, this is listed as the **Ground**, an initial digital terrain model that represents the measured elevation values at the ground level. By default, you can use the elevation surface representing the ground as the starting point for z-values for your layers, but ArcGIS Pro also allows you to specify your own elevation source to use instead.

FIGURE 15.2 A comparison of a (a) 2D surface, (b) 2.5D surface, and (c) 3D surface.

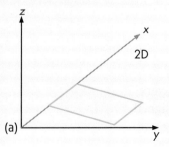
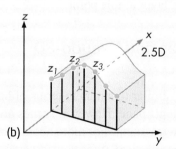
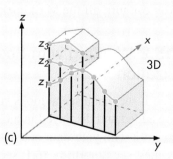

In a scene, the z-value of each x/y location can thus be shown at its proper elevation because ArcGIS Pro extends these values to their proper heights on the elevation surface.

A scene's Contents pane makes the distinction between GIS layers that are handled as 2D layers and also as 3D layers. A 2D layer does not have a z-value of its own; for instance, a feature class of roads doesn't have a z-value assigned to the elevation of each road, and a raster of NLCD data does not have a z-value assigned to each grid cell. We normally handle this data in ArcGIS Pro without taking into account a z-value, but we can match up the locations of the data in these layers with their corresponding places on the elevation surface to display them in pseudo-3D, using the z-values of the elevation surface. This process of showing a 2D layer set properly on the terrain is called **draping**, and it can be thought of as stretching a sheet over a lumpy unmade bed. Wherever the bed has lumps, the sheet will have the same lumps.

You can have ArcGIS Pro show all types of 2D layers (rasters, feature classes, and imagery) in pseudo-3D in a scene by applying the z-values of an elevation surface to the corresponding locations of the other data. As long as their geographic areas are within the boundaries of the DTM supplying the z-values, layers can be draped; satellite images or aerial photographs show an area's hills and valleys, and with roads it is possible to see how they traverse up and down steeply sloped areas. For instance, if a DTM covers the same geographic area as an NLCD raster, the z-values of the DTM may be matched with the x/y coordinates of the NLCD layer. Then, rather than extending the DTM locations up to their z-value heights, the NLCD layer can have its locations extended instead (Figure 15.3).

draping A process in which GIS layers are given z-values to match the corresponding heights in a digital terrain model.

(a)   (b)

FIGURE 15.3 The Hocking County DEM surface (a) and the NLCD layer draped over it (b).

In a scene, a 3D layer is an object (such as a point, a line, or a polygon) that is not only draped on the elevation surface but also displayed using 3D symbology. ArcGIS Pro contains specialized 3D symbology for showing objects in pseudo-3D, such as viewing a point as a three-dimensional box or cone or showing a line as a three-dimensional tube. (See Figure 15.4 for an example of displaying 3D layers in a scene.) Other layers, such as buildings, can have z-values to show their height above the elevation surface and can also be shown as 3D layers. (See Module 18 for more information about working with 3D layers.)

FIGURE 15.4 Points shown as 3D layers (red 3D cone, blue 3D cylinder, and green 3D pyramid) in an ArcGIS Pro scene.

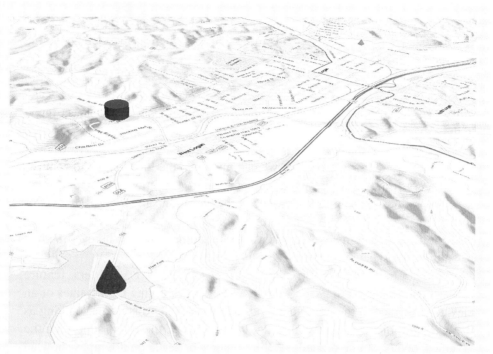

SMARTBOX 15.2

How can I navigate in the scene environment in ArcGIS Pro?

Because a scene works with a third dimension, you use an additional set of controls for navigating a scene (besides the regular Explore tool) using the onscreen navigator tool. To turn the navigator on and off, go to the **View** tab, and in the **Navigation** group, click the **Navigator** button. By default, the navigator appears in the scene in minimized mode (Figure 15.5a), which works as follows:

- The arrow shows the direction of north.
- Clicking and holding on the ring moves the camera view in that direction.
- While clicking and holding on the ring, you can drag the mouse further out from the ring to accelerate your speed of movement; a transparent arrow shows both your movement direction and speed.
- Clicking the button with the four arrows switches to planar navigation, which allows you to pan in 2D mode rather than 3D.
- Clicking the ^ button changes the navigator to the full control state.

When maximized into the full control state (Figure 15.5b), the controls work as follows:

- Clicking on the large arrow adjusts the camera view to the north.
- Clicking and holding on the bar under the large arrow allows you to rotate the direction of the camera view away from north to a different direction.

(a)

(b)

FIGURE 15.5 The scene navigator in (a) minimized mode and (b) full control mode.

- While clicking and holding on the large outer ring, you can drag the mouse further out from the ring to accelerate your speed of movement; a transparent arrow shows both your movement direction and speed.

- Rotating the ball in the center of the tool allows you to pivot the camera around in spherical movement and allows you to readjust the view as you need.

- Clicking and dragging on the inner ring allows you to adjust the camera 360 degrees around the center point of the camera.

- Clicking the up and down arrow buttons moves the camera directly up and down.

- Clicking the plus and minus buttons zooms the camera in and out from the center.

- Clicking the button with the four arrows switches to planar navigation, which allows you to pan in 2D mode rather than 3D. Clicking it again switches you back to 3D navigation mode.

- Turn off the NLCDhocking layer for now and add the DEMhocking layer to the scene's Contents pane. The DEMhocking raster is a DEM of the county, in which each cell has a value for elevation. The scene contains its own default terrain surface, but for this module, you will be using the DEMhocking raster as the terrain surface for all your analysis. So the first step is to get ArcGIS Pro to use this DEM rather than its default as the source for elevations.

- On the **Map** tab, in the **Layer** group, click the **Add Data** button. From the set of pull-down options that appears, choose **Add Elevation Source**. In the dialog box that opens, navigate to your **C:\GIS\Module15** folder, choose the **Hockingcounty.gdb** geodatabase, and choose **DEMhocking**. Click **OK** to close the dialog.

- Next, in the Contents pane under Elevation Surfaces, turn off the option **World Elevation 3D/Terrain 3D**. Your ground terrain elevations are now drawn from the DEMhocking layer. The scene now uses DEMhocking instead of its default terrain. This DEMhocking raster was drawn from the 3D Elevation Program (3DEP). See **Smartbox 15.3** for more about 3DEP.

SMARTBOX 15.3

What is the 3D Elevation Program?

The **3D Elevation Program (3DEP)** is a U.S. government initiative to provide digital elevation data of the United States via the Internet. Through 3DEP, digital elevation models (of equally spaced elevation values) are made available from the USGS. The digital elevation model layers available through 3DEP cover the United States and are compiled from a variety of different sources,

3D Elevation Program (3DEP) A digital elevation model of the entire United States, provided by the USGS.

FIGURE 15.6 3DEP elevation data available from The National Map.

with the goal of providing the best possible resolution elevation data for the United States. 3DEP elevation data provide both "bare Earth" models of the landscape (that is, the elevation layers of a DEM) and lidar source data that can be used for measuring elevations of the ground surface as well as heights of objects on the surface. (See Module 17 for more about lidar.) 3DEP elevation data are available for free download as a layer of The National Map (see Module 5). For ease of delivery, The National Map has prepackaged 3DEP DEM data as a series of tiles. A user can select an area of interest and then download all DEM elevation layers that correspond with the selected area.

3DEP elevation data use the NAVD88 vertical datum and are available as DEMs at multiple resolutions (Figure 15.6):

- 1 arc-second (about 30 meters) DEM: This resolution is available for the entire United States with the exception of Alaska (which is available at 2 arc-second resolution).

- 1/3 arc-second (about 10 meters) DEM: This resolution is also available for the entire United States with the exception of Alaska.

- 5 meter: This dataset is available for Alaska only.

- 1/9 arc-second (about 3 meters) DEM: This resolution is available for only certain areas in the United States.

- 1-meter DEM: This resolution is currently available for only certain areas in the United States.

Step 15.3 Working with Digital Elevation Models and Their Derivations

- To help familiarize yourself with the terrain of Hocking County, you'll next create a hillshade of the DEM. For more about hillshades, see **Smartbox 15.4**.

SMARTBOX 15.4

What is a hillshade?

A **hillshade** is a representation of a surface that depicts how the elevation features would look with a light source (such as the Sun) applied to them. Hillshades are used to better visualize features on a landscape and to examine what areas are illuminated (from the Sun) and what areas are in shadow under different Sun conditions or at different times of day. You can create several hillshades by altering the relative position of the Sun to the surface to model how the area would look or how shadows would fall throughout the day.

Creating a hillshade involves setting two parameters for the illumination source: azimuth and altitude (Figure 15.7). *Azimuth* is set up like a circle, with values from 0 to 360 degrees, and reflects the location of the Sun in relation to the terrain. Assuming 0 degrees to be due north, 90 degrees would be due east, 180 degrees due south, and 270 degrees due west. *Altitude* refers to the elevation of the illumination source above the terrain, measured from 0 degrees (flat on the ground) up to 90 degrees (directly over top of the terrain).

hillshade A shaded relief map of the terrain created by modeling the position of an illumination source (such as the Sun) in relation to the terrain.

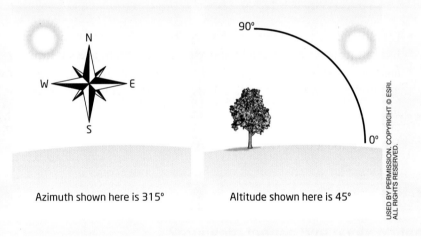

FIGURE 15.7 Settings for azimuth and altitude in a hillshade.

- Go to the **Analysis** tab, and in the **Geoprocessing** group, click the **Tools** button to open the Geoprocessing pane. The tool you will be using is Hillshade. You can either search for it or find it directly from the **Spatial Analyst Tools** toolbox, in the **Surface** toolset. Using either method, open the **Hillshade (Spatial Analyst)** tool.
- For Input raster, use **DEMhocking**.
  - For Output raster, click the **browse** button and navigate to your **C:\GIS\Module15** folder, go to the **Module15.gdb** file geodatabase, and name the output raster **hillshadehock**.
- For Azimuth, use **315** degrees (this is the ArcGIS Pro default).
- For Altitude, use **45** degrees (this is the ArcGIS Pro default).
- Place a checkmark in the **Model shadows** box.
- For Z factor, use **1**.
- **Run** the Hillshade tool.

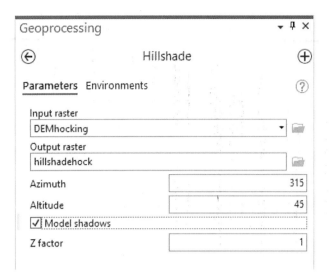

- Next, you will examine the hillshade in relation to the land cover features of the DEM layer. Turn off all the other layers except for hillshadehock and DEMhocking and make sure hillshadehock is placed above DEMhocking in the Contents pane; then click on the hillshadehock layer. On the **Appearance** tab, in the **Effects** group, find the **Layer Transparency** slider. By moving this slider, you can make the chosen layer (in this case hillshadehock) semitransparent so you can see that layer in relation to the one below it (in this case DEMhocking). Examine the hillshade in relation to the DEM layer and answer Question 15.1. Make the hillshade fully visible when you're done.

Question 15.1 How does the hillshade aid in visualizing the landscape of Hocking County? Where are the areas in the county with lower elevations located (as viewed with the hillshade)?

- Elevation doesn't tell the whole story about what the terrain is really like; all it shows is the height above a vertical datum at each location. In answering Question 15.1, you likely saw areas of changes between lower and higher elevations as well as hills and flatter sections of lands. For information about how steep or flat areas are, you have to use the DEM to create new grids, such as slope and slope aspect grids. (See **Smartbox 15.5** for more about these two concepts.)

SMARTBOX 15.5

What do slope and aspect measure?

slope A measurement of the rate of elevation change at a location found by dividing the vertical height (the rise) by the horizontal length (the run).

The slope of the land can be described using terms like "steep" or "flat" when applied to hills, mountains, or cliff faces. **Slope** represents the rate of change of elevations at a particular location and is calculated by measuring the vertical distance (the rise) over the horizontal distance (the run). The slope is computed as an angular degree (Figure 15.8) with values between 0 (completely flat, with no slope) and 90 (a straight vertical drop).

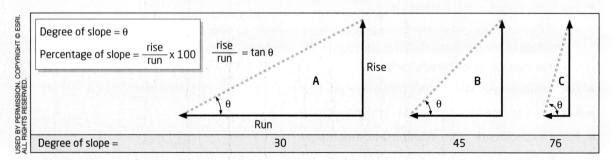

FIGURE 15.8 Calculating slope.

In a raster elevation model, each cell receives a value for slope, calculated as the maximum rate of change for that location in relation to its eight neighboring cells, and a new slope raster is computed. The direction of the steepest slope is referred to as the slope **aspect** and can also be computed as a separate raster layer. In the aspect grid, each cell receives a value that indicates in which of the eight cardinal directions the steepest slope faces for each cell (or it can receive a value indicating it is classified as "flat" and has no slope). Figure 15.9 shows an example of an aspect grid and how to interpret it.

aspect A determination of the direction the steepest slope is facing at a location.

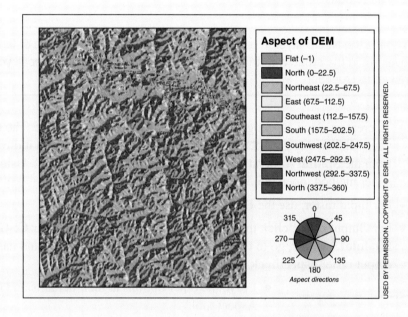

FIGURE 15.9 A raster of slope aspect, along with the corresponding aspect directions.

- To create a slope grid, return to the Geoprocessing pane. The tool you will be using is Slope. You can either search for it or find it directly from the **Spatial Analyst Tools** toolbox, in the **Surface** toolset. Using either method, open the **Slope (Spatial Analyst Tools)** tool.

- For Input raster, use **DEMhocking**.

- For Output raster, click the **browse** button and navigate to your **C:\GIS\Module15** folder, go to the Module15.gdb file geodatabase, and name the output raster **slopehock**.

- For Output measurement, choose **Degree**.

- Leave the other options at their defaults and run the Slope tool.

- Examine the slopehock grid, and you see that each cell contains the slope angle (measured in degrees) at that location. Answer Question 15.2.

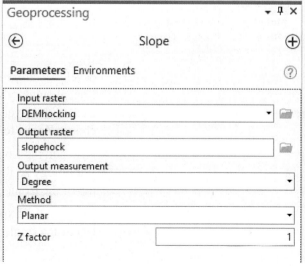

Question 15.2 Where, spatially, are the least-steep slopes in the county located?

- Next, you will examine the slope layer in relation to the land cover features of the NLCD layer. Turn off all the layers except for slopehock and NLCDhocking and make sure that slopehock is placed above NLCDhocking in the Contents pane. Also in the Contents pane, click once on **slopehock** to highlight it. Then, on the **Appearance** tab, in the **Effects** group, use the **Layer Transparency** slider to compare the two layers. Answer Question 15.3. Restore the slopehock layer to full visibility when you're done.

Question 15.3 Examine the NLCD layer in relation to the slope raster. What kinds of land cover are in the areas with the least-steep slopes in the county (recall your answer to Question 15.2)?

- To create a slope aspect grid, return to the Geoprocessing pane. The tool you will be using is Aspect. You can either search for it or find it directly from the **Spatial Analyst Tools** toolbox, in the **Surface** toolset. Using either method, open the **Aspect (Spatial Analyst Tools)** tool.
- For Input raster, use the **DEMhocking** grid.
- For Output raster, click the **browse** button and navigate to your **C:\GIS\Module15 folder**, go to the **Module15.gdb** file geodatabase, and name the output raster **aspecthock**.

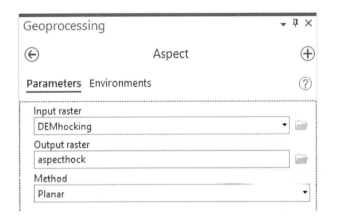

- Leave the other options at their defaults and run the Aspect tool.
- Examine the aspecthock grid. It shows the direction of the steepest slope at that particular location. For instance, all grid cells with the value "south" have their steepest slope as a south-facing slope at that cell.
- Next, you'll use these slope and aspect grids to get information about the potential building site for the observation platform. First, add the Loganboundary and RoadSegmenthocking layers to the Contents pane. These layers will allow you to see other developed areas in Hocking County, as well as the city of Logan. Change Loganboundary's symbology from a solid color to Black Outline (so you can see the layers underneath it). Right-click on **Loganboundary** and choose **Zoom to Layer** to zoom to the extent of the city.
- Also add the LoganDam feature class to ArcGIS Pro, change its symbology so that it stands out from the other layers, and position it at the top of the Contents pane.
- By using the Explore tool, you can get information about multiple visible layers at a single point. (See Module 14 for how to use the Explore tool to

examine all visible layers with one click.) Zoom in closely to the LoganDam location and use the **Explore** tool to answer Questions 15.4 and 15.5.

> What direction is the steepest slope at the Logan Dam location facing? — **Question 15.4**

> What is the degree of slope at the Logan Dam location? — **Question 15.5**

- Turn off all your layers except for DEMhocking, Roadsegmenthocking, and LoganDam.

Step 15.4 Display and Visualization of 3D Layers in a Scene

- Add the HockingCoFairgrounds and HockingValleyHospital layers to the scene.

- Right now, each of your three point layers (LoganDam, HockingCoFairgrounds, HockingValleyHospital) is being shown draped on the DEMhocking surface, and they are likely difficult to see. A better option would be to represent each of these three points as a 3D layer in the scene. You can see that each of these three points is listed in the scene's Contents pane as a 2D layer. To begin, click on the **LoganDam** layer and drag it up into the 3D layer heading in the Contents pane.

- Zoom in closely on the LoganDam point. You can see that it is no longer a flat 2D point draped on the surface but is instead a 3D point; however, you can assign it much better symbology so that it really stands out. Open the Symbology pane for the LoganDam layer.

- In the Symbology pane, click on the point symbol to go to the Format Point Symbol part of the Symbology pane. Close the ArcGIS 2D options and expand the **ArcGIS 3D** options.

- From the available ArcGIS 3D symbol options, choose one to represent LoganDam. You see the point in the scene replaced with the 3D object you chose. To make further changes to this layer (such as increasing its size or changing its color), in the Symbology pane click on **Properties** to access these further options.

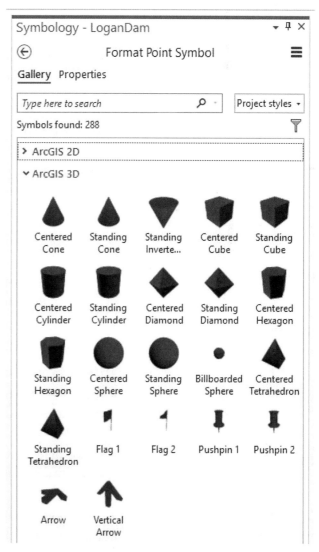

(If you change things under the Properties option, be sure to click the **Apply** button at the bottom of the Symbology pane to save those changes.)

- In the Contents pane, drag the HockingCoFairgrounds layer up into the 3D Layers position and change its symbology to 3D. Then do the same with the HockingValleyHospital layer.

Step 15.5 Working with Profile Graphs

- By now, you should have a fairly good feel for the landscape of Hocking County. Your final goal in this module is to determine how high to build the observation tower at the Logan Dam location. A key point for visualization will be the Hocking County Fairgrounds location, so you need to examine the terrain between these two points. Zoom in to the extent of the town of Logan in the northeastern portion of Hocking County. You should be able to identify it by boundary as well as the pattern of roads in the area.

- As you will be examining the view on a line between these two points, it will be important to have that line anchored at each of the points to ensure that the line of sight goes directly from point to point rather than just ending somewhere near a particular point. On the **Edit** tab, in the **Snapping** group, click the **Snapping** button.

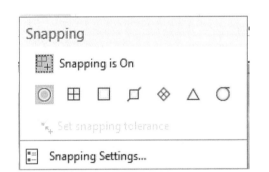

- In the Snapping dialog, make sure that the option **Snapping is On** is activated, and ensure that only the option for **Point Snapping** is highlighted in blue. (See Module 7 for more about snapping.)

- Next, you will digitize a line (as in Module 6) between the Logan Dam and the Hocking County Fairgrounds that will be the basis of the line of sight. First, you will make an empty feature class to hold the line. Return to the Catalog pane and right-click on the **Module15.gdb** geodatabase, choose **New**, and then choose **Feature Class**.

- For Name, type **lineofsight**.
- For Feature Class Type, choose **Line**.
- Under Geometric Properties, place a checkmark in the box for **Z Values**.
- Leave the other options alone and click **Next** at the bottom of the dialog. A second screen of options appears, with information about fields. Just click **Next** again.
- On the third screen of options, you are asked to choose a spatial reference. Under the options for Layers, choose **Albers Conical Equal Area** and then click **Next**.
- The fourth screen of options asks about tolerance. Just click **Next**.
- The fifth screen of options asks about resolution. Just click **Next**.
- The sixth screen of options asks about storage configuration. Now click **Finish**.

- Next, add the new lineofsight layer to the Contents pane as a 2D layer.

- Now that you have an empty feature class, you need to digitize a line into it. On the **Edit** tab, in the **Features** group, click the **Create** button. The Create Features pane opens.

- In the Create Features pane, choose the option for the **lineofsight** feature class.

- A new set of tools appear below the lineofsight option to allow you to digitize a line (as in Module 6). Choose the first option (the line), and then move the cursor into the scene.

- Hover the cursor over the **LoganDam point**, and you see the snapping for that point appear. When it does, click the left mouse button to start the line.

- Next, move the cursor over to the **HockingCoFairgrounds point**, and you see the snapping for that point appear. Double-click the left mouse button to complete the line. Your line is now digitized.

- To finalize it, on the **Edit** tab, in the **Manage Edits** group, click the **Save** button. If ArcGIS Pro prompts you about saving edits, click **Yes**.

- Finally, on the **Edit** tab, in the **Selection** group, click the **Clear** button to clear the selection of the digitized line.

- Next, you will give the digitized line the z-values of the elevation surface that it is on top of rather than just visualizing it as being draped over the surface. You do this by interpolating the elevation values that correspond with the line. To do this, return to the Geoprocessing pane. The tool you will be using is Interpolate Shape. You can either search for it or find it directly from the **3D Analyst Tools** toolbox, in the **Functional Surface** toolset. Using either method, open the **Interpolate Shape** tool.

- For Input Surface, use **DEMhocking**.

- For Input Features, use **lineofsight**.

- For Output Feature Class, click the **browse** button and navigate to your **C:\GIS\Module15** folder, go to the **Module15.gdb** file geodatabase, and name the output feature class **lineofsightz**.

- Use the defaults for the other options and run the Interpolate Shape tool.

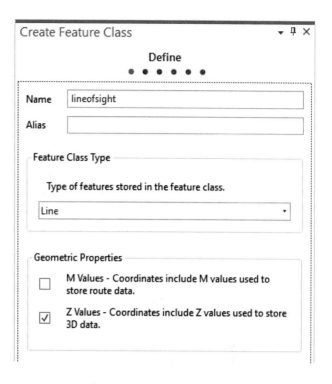

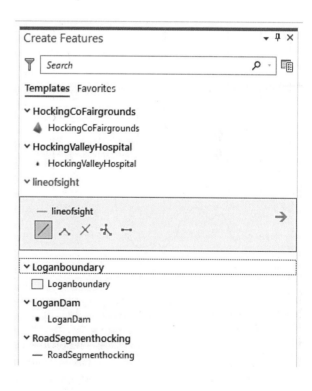

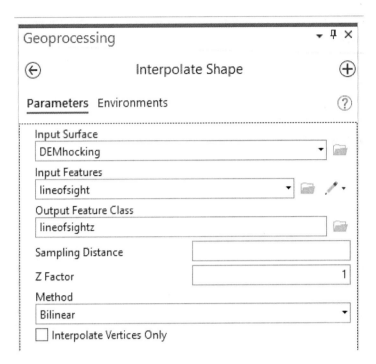

- The lineofsightz feature class is added to the scene's Contents pane. To examine the features along this line as more than just a line drawn on the DEM, click it once to highlight it and then, on the **Data** tab, in the **Visualize** group, click the **Create Chart** button and choose **Profile Graph**.

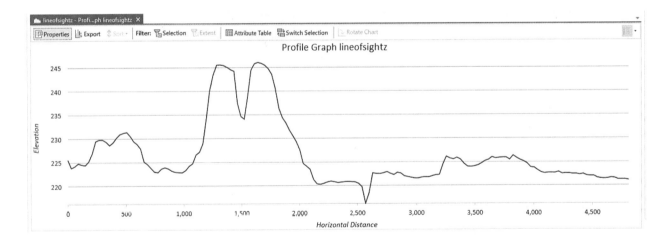

- The profile graph for the line opens in a window below the scene (where tables would normally be opened). It shows a profile view of the terrain along the length of the line from the Logan Dam to the Hocking County Fairgrounds, with the elevation values of the terrain on the y-axis and the horizontal ground distance on the x-axis. Examine the profile graph and answer Questions 15.6 and 15.7. When you're done, you can close the profile graph (for now).

Question 15.6 What is the elevation of the start of the line (that is, the dam), and what is the elevation of the end of the line (that is, the fairgrounds)?

> **Question 15.7**
>
> The profile graph shows a view along a straight line between the Logan Dam and the Hocking County Fairgrounds. What is the elevation of the tallest part of the terrain that is blocking your path between the two points?

Step 15.6 Working with Visibility Analysis

- As you've seen in the profile graph, several terrain features between the dam and the fairgrounds will obstruct your view. The next set of analyses you'll be performing is to determine what areas specifically can and cannot be seen from the proposed platform location, based on the surrounding terrain. This visibility analysis will help you in figuring out how tall the platform should be so that observers can look over the terrain features and into the town of Logan (particularly to see its fairgrounds). For further information about visibility analysis in GIS, see **Smartbox 15.6**.

SMARTBOX 15.6

What is visibility analysis?

In GIS, **visibility analysis** is a technique used to determine what areas on a landscape can be seen and what areas cannot be seen from a particular vantage point. Visibility analysis uses a layer with z-values (typically a digital terrain model) as the basis for analysis because elevations may block an observer's view of other areas of the landscape. This type of information is especially useful for a variety of applications, including planning for coastal developments and gauging views of the water or shore, determining what areas of a forest can be seen from a ranger's observation tower, or checking a witness statement to verify if an area where a crime occurred could have been visible to the witness or not.

There are two types of visibility analysis in ArcGIS Pro. The first is **line of sight (LOS)**, which simulates looking in one direction between two points and examining what can be seen or cannot be seen along that line, based on the elevations between the points. ArcGIS Pro presents the results of this analysis by using red to show areas that are not visible from a particular point and green to show areas that are visible. The second is a **viewshed**, which is a raster showing the areas 360 degrees around the observer and whether or not they can be seen. Keep in mind that both LOS and viewshed results are *binary*—that is, something is either seen or not seen, based on intervening elevations.

visibility analysis Techniques used in GIS to determine what areas can be seen and what areas cannot be seen from a particular vantage point.

line of sight (LOS) The visibility in a straight line between two points.

viewshed The visibility within 360 degrees around a location.

- To judge the best height for the observation tower, you can begin by drawing a line of sight between the LoganDam point location and the HockingCoFairgrounds location. Zoom in to the Logan area so you can see both of these point locations and turn off the lineofsight and lineofsightz layers.

- To create a line of sight, return to the Geoprocessing pane. The tool you will be using is Line of Sight. You can either search for it or find it directly from the **3D Analyst Tools** toolbox, in the **Visibility** toolset. Using either method, open the **Line Of Sight** tool.

- For Input Surface, use **DEMhocking**.

368 CHAPTER 15 How to Work with Digital Elevation Models in ArcGIS Pro

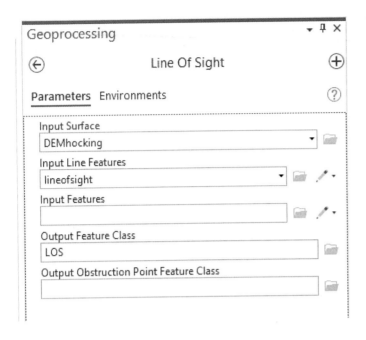

- For the Input Line Features, choose **lineofsight**.
- For Output Feature Class, click the **browse** button and navigate to your **C:\GIS\Module15** folder, go to the **Module15.gdb** file geodatabase, and name the output feature class **LOS**.
- Leave the other options alone and run the Line Of Sight tool. A new feature class called LOS appears as a 2D layer in the Contents pane. This class, which represents looking in a straight line from the LoganDam location to the HockingCoFairgrounds location, has two different lines: a green line showing what parts of that line you can see from the LoganDam location and a red line indicating what parts you cannot see. Most of the line between the two spots is red, with only a little portion in green, indicating that you can see part of the way from the dam to the fairgrounds but that something on the terrain is blocking most of the view.

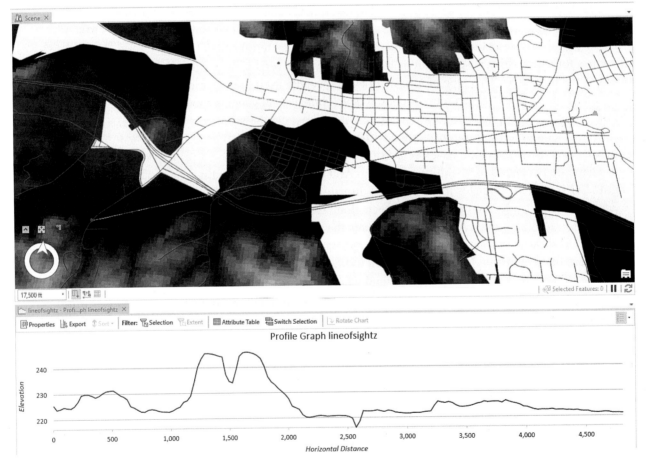

- Adjust what you're looking at in the scene to view the entire line of sight and then reopen the profile graph for lineofsightz. By looking at the graph and the LOS class, you should be able to compare the red portion of the line with what terrain features are obstructing your view. Answer Question 15.8.

Question 15.8
What is the elevation of the terrain feature blocking your immediate view of the fairgrounds (that is, the terrain feature that is turning your line of sight from green to red)?

- To see the fairgrounds from the dam, you're going to have to apply an "observer offset." This offset simulates you standing on a tall object (like the observation platform you're designing) in an effort to see over top of the terrain features in your way. ArcGIS Pro makes this possible by reading a value in the attribute table of the digitized line between the two points. If this value is in a field called OffsetA, ArcGIS Pro applies that offset to the start of the line (that is, the observer); if this value is in a field called OffsetB, ArcGIS Pro applies that offset to the end of the line (that is, the target). Because you want to adjust the height of the observer, you need to add a field called OffsetA to the attribute table and then put a value into that field to represent how high you want to raise the observer. To get started, open the attribute table for the **lineofsight** layer.

- In the lineofsight attribute table, click the **Add** button, and a new field is added. Change the name of this field to **OffsetA** and press the Enter key on the keyboard. Then, on the **Fields** tab, in the **Changes** group, click the **Save** button. Then close the **Fields: lineofsight** tab that is open in the table view.

- You want to determine the minimum value necessary to use as an observer offset to be able to see the fairgrounds from the dam. Keep in mind that you're designing an observation platform at the dam site, and you want it to be the minimum height necessary to look down toward Logan to see the fairgrounds. Recall that the observer offset is the distance you have to go above the base of the terrain on which you're standing. For instance, to try an observer offset of 5 meters, go to the lineofsight attribute table, click once in the **OffsetA** field and type **5**, and press the Enter key on the keyboard. Then, on the **Edit** tab, in the **Manage Edits** group, click the **Save** button. ArcGIS Pro saves the value 5 in the OffsetA field.

- Return to the Geoprocessing pane and the Line Of Sight tool.
- For Input Surface, use **DEMhocking**.
- For Input Line Features, choose **lineofsight**.
- For Output Feature Class, click the **browse** button and navigate to your **C:\GIS\Module15** folder, go to the Module15.gdb file geodatabase, and name the output feature class **LOS5**. Use this name to show what the line of sight is like between the dam and the fairgrounds at a height of 5 meters tall at the dam.
- The new LOS5 layer appears in the Contents pane. Compare what you can see along this new line with the profile graph and answer Question 15.9.

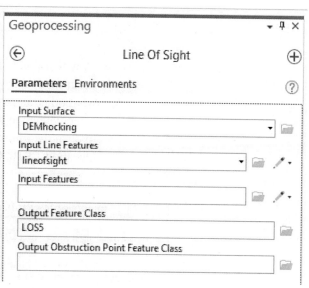

Question 15.9 You still cannot see the fairgrounds from a 5-meter-high observation platform at the dam, but more of the terrain is now visible. Based on the information on the LOS5 layer and its profile graph, what can you now see that you couldn't see before?

- You need to determine the minimum height of the platform (that is, the lowest value for observer offset) needed to see the fairgrounds from the platform location. Look carefully at the profile graph to try to gauge what a reasonable minimum height would be. You need to construct a new green and red LOS line at that height. To do so, follow these steps:
  - In the lineofsight attribute table, type a new number for the observer offset (that is, the height of the tower) in the OffsetA field.
  - On the **Edit** tab, in the **Manage Edits** group, click the **Save** button to save this new value.
  - In the Geoprocessing pane, run the Line Of Sight tool and create a new feature class called LOS*x*, where *x* is the height you are using; for instance, if you are using a height of 20 for the tower, use LOS20 for the name of the feature class.
- Try a few different observer offset values and draw LOS lines and create profile graphs that correspond to them. If the results of the line of sight are not what you need to see the fairgrounds at the minimum observer offset (either the offset is too high or too short), try different OffsetA values until you determine the proper height.
- When you're able to clearly see the fairgrounds from the dam at the minimum height necessary, you know how tall the observer platform has to be. Answer Question 15.10.

Question 15.10 What is the minimum observation platform height required to see the fairgrounds from the Logan Dam? (*Important Note:* This should be the height of the platform itself—for example, a 15-meter-tall platform—not the height the platform will be above the vertical datum.)

- Now that you know the height of the platform necessary to see the fairgrounds from the potential area at the dam, you also need to determine what other areas can and cannot be seen from the platform. To do so, you can create a viewshed from the Logan Dam location, and you will want it to be at the minimum observer offset height that you have determined for Question 15.10. Open the attribute table for the LoganDam feature class.

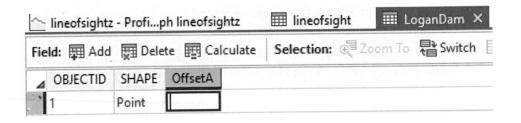

- A field called OffsetA has already been added to this feature class for you. In the OffsetA field, type the value you want to use for the observer offset height and press the Enter key on the keyboard. Then, on the **Edit** tab, in the **Manage Edits** group, click the **Save** button. ArcGIS Pro saves your typed value into the OffsetA field.

- Finally, you need to compute a viewshed from the Logan Dam, and it will be computed at the OffsetA height. To create a viewshed, return to the Geoprocessing pane. The tool you will be using is Viewshed. You can either search for it or find it directly from the **3D Analyst Tools** toolbox, in the **Visibility** toolset. Using either method, open the **Viewshed (3D Analyst Tools)** tool.

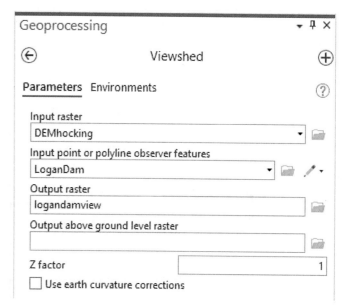

- For Input raster, use **DEMhocking**.

- For Input point or polyline observer features, use **LoganDam**.

- For Output raster, click the **browse** button and navigate to your **C:\GIS\Module15** folder, go to the **Module15.gdb** file geodatabase, and name the output raster **logandamview**.

- Use the defaults for the other options and run the Viewshed tool. Answer Question 15.11.

> Can Hocking Valley Hospital be seen from Logan Dam at the minimum observer height of the proposed tower? **Question 15.11**

Step 15.7 Presenting Results of 3D Visualization Analysis

- Close any open attribute tables and the profile graph.

- Your new logandamview layer shows you what areas can be seen and not seen (due to intervening terrain) from the platform location at the proposed height. The last analysis you need to perform is to determine how much of the entire city can be seen (or not seen) from the platform at that height. However, because your logandamview viewshed covers the entire county, you need to examine only the parts of the viewshed that are within the boundaries of Logan. To do so, you have to create a subset of the viewshed raster (as in Module 12) that conforms to the Logan boundaries. To begin, turn on the Loganboundary feature class, and if it is still a solid color, change its symbology to Black Outline (so you can see the layers underneath it) and zoom in to its boundaries.

- To create a subset, return to the Geoprocessing pane. The tool you will be using is Extract By Mask. You can either search for it or find it directly from the **Spatial Analyst Tools** toolbox, in the **Extraction** toolset. Using either method, open the **Extract By Mask** tool.

- For Input raster, select the **logandamview** raster. This is the raster from which you want to extract cells.

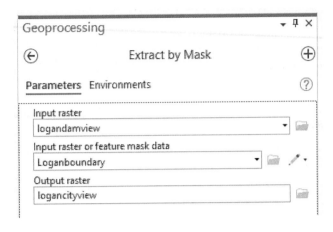

- For Input raster or feature mask data, select the **Loganboundary** feature class. This is the shape (or mask) that will be used to extract cells.
- For Output raster, click the **browse** button and navigate to your **C:\GIS\Module15** folder, go to the **Module15.gdb** file geodatabase, and name the output raster **logancityview**.
- Run the Extract By Mask tool. The new Logancityview raster is added to the Contents pane.
- Next, to analyze the Logancityview raster, you want to know how many of its cells have a value of 0 (that is, cannot be seen from the platform) and a value of 1 (that is, can be seen from the platform). Open the attribute table of the Logancityview raster to obtain these numbers and then answer Question 15.12.

Question 15.12 What is the percentage of the area within the boundaries of Logan that can be seen from the height of the platform? (*Hint:* Use the attribute table to find the total number of cells that have a 0 or a 1 and then determine the percentage of the whole that can be seen.)

- Make sure you have only the Logancityview, Loganboundary, HockingCoFairgrounds, LoganDam, and RoadSegmenthocking layers displayed, along with your final LOS. Use the navigator to change the camera view so it looks as if you are hovering beside the observer offset height (at the Logan Dam) and looking at an angle down toward the fairgrounds (Figure 15.10).

FIGURE 15.10 A 3D view of Logan and Hocking County at the proposed platform height.

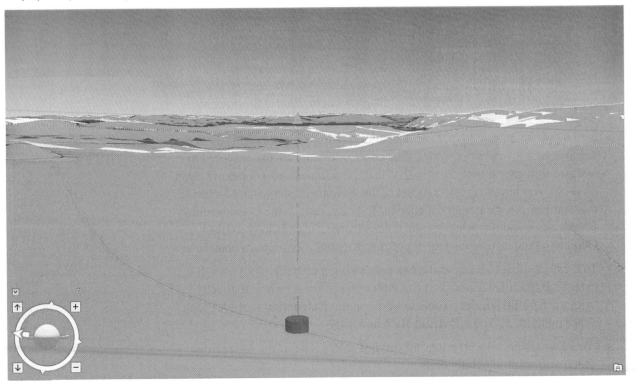

Step 15.8 Saving and Printing Your Results

- Save your Module15 project.

- Print a layout (see Module 3) showing the final 3D camera view from the Logan Dam looking toward the Hocking County Fairgrounds. When you insert a Map frame into a layout, you can have the map frame show the contents of a scene the same way you do with a map. Use the Hocking County scene as the source for the map frame when you are creating your layout. Then you can add the usual elements, such as a legend, title, and scale bar, to the layout.

Closing Time

This module got you started working with elevation data and several derivations and applications of it. Elevation data are commonly used in many different types of GIS analysis, and you'll continue to use elevation data from 3DEP in forthcoming modules. The ability to visualize and interact with data in a 3D-style environment is a big advantage of working in ArcGIS Pro, and several upcoming modules will continue to utilize scenes for 3D elevation-based data or analysis. ArcGIS Pro also contains several other interactive tools for working with visibility analysis. See *Related Concepts for Module 15* for more information.

Digital elevation models are particularly useful for modeling terrain and landscapes, although they're not the only method of doing so. DEMs are limited because they always use evenly spaced sampling points, and sometimes a more versatile method of working with elevation data is needed (in which you would focus on the most important points on the landscape's surface, not just evenly spaced ones). Module 16 examines other elevation models and applications in ArcGIS Pro, including the TIN data structure.

RELATED CONCEPTS FOR MODULE 15

Interactive 3D Visibility Analysis

Visibility analysis in GIS is more than just determining what can be seen or not seen based on intervening terrain; lots of other things can block what you're looking for, such as trees, buildings, signs, or billboards. ArcGIS Pro gives you the ability to include these types of objects in visibility analysis by using the geoprocessing tools to model each individual element and include them as feature classes. However, often you might want to do analysis on the fly. For example, after setting up 3D objects as trees or houses, you might want to perform visibility analysis from multiple different vantage points or heights. This would be useful for examining what could be seen or not seen from police positions for security at an event, crowd viewing of a parade, or a security camera's field of vision.

ArcGIS Pro contains several other features that very useful for interactive visibility analysis when working in a scene environment. These interactive tools, which are available in the 3D Exploratory Analysis group on the Analysis tab, enable you to generate lines of sight and viewsheds, as well as **view domes** (volumetric versions of viewsheds) interactively with a few mouse clicks rather than by using geoprocessing tools. The interactive tools allow for more specialized analysis, such as a viewshed showing what can and can't be seen along an angular field

view dome A volumetric version of a viewshed, which shows what can or can't be seen in a spherical area around an observer point.

of vision (which would simulate what a person could see by looking across an area rather than down one narrow line of sight).

For example, in Figure 15.11, a vantage point for the viewshed has been chosen, and an observer offset height has been interactively selected. ArcGIS Pro automatically calculates the viewshed within the angular field of view. (Locations that can be seen are shown in green, and locations that are hidden from view are shown in red.) For instance, an observer set at a particular height looking toward the northwest would be able to see the ground leading up to the lake and part of the lake itself, but the hill leading down to the lake would be hidden from view. This tool is interactive, and the user can move the field of view, the viewing depth, the observer position, and the observer height by grabbing the observer location and moving it with the mouse. As the observer location is moved, the viewshed is recalculated.

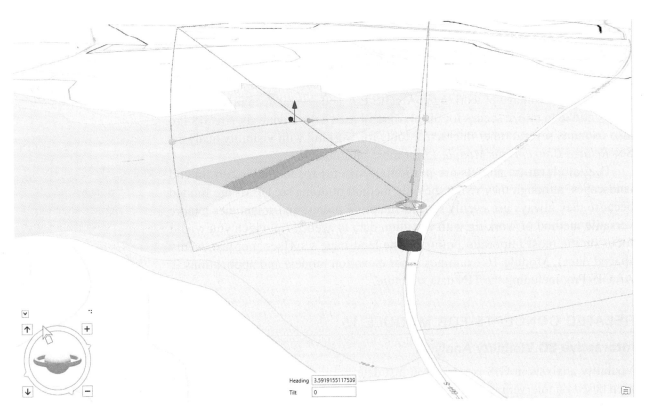

FIGURE 15.11 An interactive viewshed created using the 3D Exploratory Analysis tools.

Key Terms

digital terrain models (DTMs) (p. 349)
z-value (p. 349)
vertical datum (p. 350)
NAVD88 (p. 350)
NAPGD2022 (p. 350)
digital elevation model (DEM) (p. 350)
functional surface (p. 350)
scene (p. 352)

3D Analyst (p. 352)
perspective view (p. 354)
pseudo-3D (p. 354)
two-and-a-half-dimensional (2.5D) (p. 354)
three-dimensional (3D) (p. 354)
elevation surface (p. 354)
Ground (p. 354)
draping (p. 355)

3D Elevation Program (3DEP) (p. 357)
hillshade (p. 359)
slope (p. 360)
aspect (p. 361)
visibility analysis (p. 367)
line of sight (LOS) (p. 367)
viewshed (p. 367)
view dome (p. 373)

16

How to Work with Contours, TINs, and 3D Imagery in ArcGIS Pro

ArcGIS Pro Skills

In this module, you will learn how to do the following in ArcGIS Pro:
- Examine contour lines in ArcGIS Pro.
- Construct a TIN from a feature class.
- Generate slope and slope aspect from a TIN.
- Create a pseudo-3D visualization of a TIN landscape.
- Drape high-resolution imagery on a TIN.
- Apply vertical exaggeration to a scene.
- Create an animation using keyframes.
- Export an animation to MP4 format.

Learning Outcomes

After studying this module, you should be able to:
- Define what a contour line is.
- Define what a contour interval is.
- Describe what the features of a TIN are.
- Explain the steps in creating a TIN.
- Explain the difference between a hard breakline and a soft breakline.
- Describe what a keyframe is and how it is used for animations.
- Explain what a US Topo is and how it is different from a traditional topographic map.

Introduction

Beyond using the evenly spaced sample points of a digital elevation model, there are other ways of representing a surface as a digital terrain model. The data sources used to produce an elevation surface may not be evenly distributed, or there may be certain areas (or smaller sections of the terrain) where the landscape undergoes rapid elevation changes and many more elevation points are necessary to accurately capture these features in GIS. In such cases, you need a lot of points in the most important areas and fewer points for the less variable terrain areas. A data structure that accommodates irregularly spaced points would help to better represent the terrain.

A **TIN** (triangulated irregular network) represents this kind of a data structure in ArcGIS Pro. TINs do not need to have evenly spaced sample points; they can be built from selected points (or points deemed the most important for representing the landscape). Unlike digital elevation models, TINs use points, lines, and a series of non-overlapping triangular polygons to represent the surface (Figure 16.1). TINs can often provide a better display of the terrain than digital elevation models (DEMs) can. They are also more versatile in representing landscape features, but they are more complicated to work with than evenly spaced sample points. Like the surfaces described in Module 15, a TIN acts as a functional surface, and it is

TIN Triangulated irregular network, a terrain model formed from non-overlapping triangles that allows for unevenly spaced elevation points.

also considered a 2.5D model of terrain. Also, as a TIN is a digital terrain model, all of the terrain analysis functions used in Module 15 (hillshades, slope, aspect, viewsheds, line of sight (LOS), and draping) can also be performed using a TIN.

FIGURE 16.1 A TIN representation of a surface.

TINs can be constructed from a variety of different data sources, including points drawn from DEMs (see Module 15), lidar (see Module 17), or contour lines from a map. Contours have long been a means of representing terrain features such as hills, peaks, or valleys on maps, and they can also be used in ArcGIS Pro. Digitized versions of contours can be used as a basis for constructing digital terrain models. Using ArcGIS Pro, you can create detailed 3D representations of surfaces for analysis.

Module Scenario and Applications

This module puts you in the role of a member of the local chamber of commerce in Youngstown, Ohio. As part of a promotional effort to showcase the city to potential business investors and property developers, you will be creating a virtual 3D tour of the city and its nearby areas. To this end, you'll be showing high-resolution imagery draped over a 3D TIN of the city's landscape and creating a video animation of your results rather than a layout or web map.

The following are additional examples of real-world applications of this module's skills:

- An archaeologist is beginning to reconstruct the original appearance and layout of a Mayan city in 3D format. She could begin by building a detailed TIN of the surface and draping imagery of the environment over the TIN.

- An environmental scientist is performing analysis of the aftermath of a large-scale forest fire. She could use contours and a TIN to model the landscape and use draped imagery from before and after the fire to examine the extent and geographic spread of the burn.

- An urban planner wants to examine the potential changes in an area and present the results to a group of stakeholders. He can use elevation models and imagery to visualize the city layout in 3D and present the results using an animation.

 ## Study Area

For this module, you will be working with data from Youngstown, Ohio.

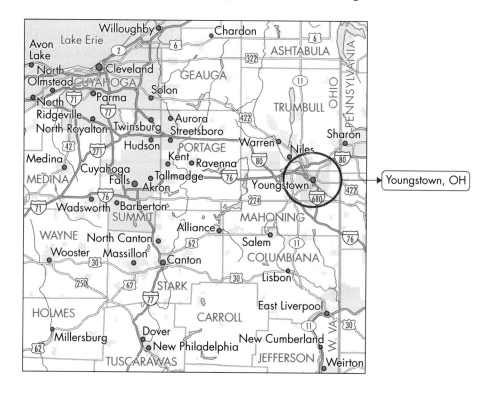

 ## Data Sources and Localizing This Module

The data in this module focus on features and locations within Youngstown, Ohio. However, you can easily modify this module to use data from your own city or local area instead. The contours and NAIP imagery used in this module were downloaded for free from The National Map, and similar datasets are available for the entire country. For instance, if you were doing this same analysis for Blacksburg, Virginia, you could find contours and NAIP imagery for that city via The National Map.

Important Note: When obtaining data for your own location, keep in mind that The National Map contours and NAIP imagery are projected into UTM using meters as their linear units. The contours, however, have their elevations (z-values) measured in feet. By default, ArcGIS Pro uses the units of measurement of the horizontal projection for the vertical measurements as well. Thus, even though the contour elevations are in feet, ArcGIS Pro treats these same z-values as meters (because the contour layer was in the UTM projection using meters). When working with such data, however, ArcGIS Pro can control these types of conversions for each layer without requiring you to project them. In the layer's properties under the Elevation tab, you need to choose what elevation units to use. For ease of use in this module, both the contours and the NAIP imagery have been projected for you into a projection using feet as linear units (State Plane) to keep both x/y values and z-values consistent.

Step 16.1 Getting Started

- This module's hands-on applications use the data folder called Module16. Your instructor will be able to supply you with this data, or you can download it directly from this book's website, at **https://www.macmillanlearning.com/college/us/product/Discovering-GIS-and-ArcGIS-Pro/p/131923075X**. The text in this module assumes that you have this Module16 folder in a computer location referenced as C:\GIS; if you have it somewhere else (for instance, in a flash drive referenced as G:\GISClass), substitute that location and path to the Module16 folder throughout this module.

- The Module16 folder contains the file geodatabase Ytowncontours, which contains the following items:

  - Ytownconsstateplane: This line feature class contains contours for the Youngstown area. The feature class has been projected into the NAD 83 State Plane Ohio North coordinate system (using U.S. feet as the linear units) and clipped to the boundaries of the orthoimage.

  - Ytownimage: a 2011 NAIP orthoimage of the Youngstown area. The raster image has been projected into the NAD 83 State Plane Ohio North coordinate system (using U.S. feet as the linear units).

- Start ArcGIS Pro.

- Sign in with your Esri account username and password.

- Create a new project using the **Map** template. Call this project **Module16** and place it in your **C:\GIS\Module16** folder. Ensure that there is not a checkmark in the box next to **Create a new folder for this project**.

- When ArcGIS Pro opens, change the map's name to **Youngstown**.

- Several of the tools you will use in this module are part of both the Spatial Analyst and the 3D Analyst extensions. You need to have both of these extensions licensed to your Esri account and active for use in this module. If you don't have the Spatial Analyst and 3D Analyst extensions licensed and active, some tools will not work: They will have a lock symbol next to them and will be unavailable. See Smartbox 11.1 and Troublebox 11.1 in Module 11 for more about extensions and ensuring that you have both the Spatial Analyst and 3D Analyst extensions available for use.

- Add the **Ytownconsstateplane** feature class to the map.

- The coordinate system being used in this module is **State Plane Ohio North (NAD 83)**, with **feet** as the map units. Check the Youngstown map Properties (on the Coordinate System tab) to verify that this projection system is being used.

Step 16.2 Examining Contour Lines

- Change the symbology of the Ytownconsstateplane feature class to a graduated color (see Module 3). Use the **ContourElevation** attribute for display and use the **Natural Breaks** method with five classes. Choose an appropriate and distinctive color ramp for the contours. Also, turn on the labels for the layer by using the **ContourElevation** field (see Module 2). For more about contour lines and contour intervals, see Smartbox 16.1.

SMARTBOX 16.1

How can contour lines be used in ArcGIS Pro?

A **contour** is a line drawn on a map that joins points of equal elevation. For instance, if a contour line is labeled 800, then all points on that line are assumed to have an elevation value of 800 units above the vertical datum being used (such as 800 feet above the NAVD88 datum). If the contour line adjacent to it is labeled 810, then the points along those elevations are assumed to be 10 units higher than the points on the 800 contour (and there is a 10-Unit elevation change between the two lines). Contour lines are imaginary in that they don't exist on real landscapes; they're just a device used to represent elevations on a map.

In theory, a map could have an infinite number of contours, but the map would quickly become unreadable because it would be filled to the maximum with lines detailing every tiny change in elevation. Thus, contours are drawn only at a certain distance apart, such as every 10 feet or every 50 feet of elevation change. This vertical distance between contours on a map is called the **contour interval**. For instance, if a contour is drawn at 800 feet and the next one is drawn at 850 feet, the contour interval is 50 feet. While any contour interval could be selected, in general maps of more mountainous terrain tend to use a wider contour interval, while maps of flatter terrain tend to use narrower contour intervals. Likewise, smaller-scale maps tend to use a wider contour interval, while larger-scale maps tend to use narrower contour intervals.

When displaying contours on a map or in ArcGIS Pro, different types of contours are available. The thicker lines (which are often labeled on topographic maps) are called **index contours**, and the thinner (often unlabeled) lines drawn at the contour interval between the index contours are called **intermediate contours** (Figure 16.2). In ArcGIS Pro, a digital terrain model (such as the DEM layers from 3DEP) can be used as a source from which to derive contours, using tools available from the Surface toolset in the Spatial Analyst toolbox, as well as the Create Contours tool on the 3D Analyst toolbar. The Youngstown contours used in this module were derived from 3DEP elevation data.

contour An imaginary line drawn on a map that connects points of common elevation.

contour interval The vertical distance between contour lines.

index contour The main contour lines on a map.

intermediate contour The contour lines drawn in between index contours at the contour interval.

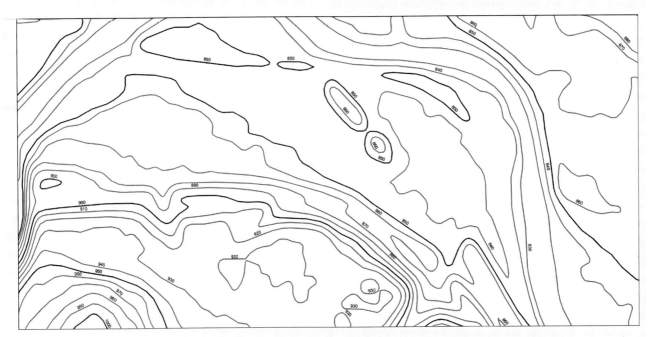

FIGURE 16.2 Contour lines at a 10-foot interval; index contours are in black, and intermediate contours are in red.

- Pan and zoom around the Youngstown area to familiarize yourself with the elevation and landscape of the area. Answer Questions 16.1 and 16.2.

Question 16.1 What is the contour interval of the Ytownconsstateplane layer?

Question 16.2 Where are the highest (1060 feet) and lowest (820 feet) elevations in the Youngstown area?

Step 16.3 Creating a TIN

- To construct a 3D version of the area, you can construct a TIN from the contours. To do so, open the Geoprocessing pane (on **Analysis** tab, go to the **Geoprocessing** group and click the **Tools** button). The tool you will be using is Create TIN. You can either search for it or find it directly from the **3D Analyst Tools** toolbox, in the **Data Management** toolset and then under **TIN**. Using either method, open the **Create TIN** tool.

- For Output TIN, click the **browse** button and navigate to your **C:\GIS\Module16** folder and name the output TIN **YTIN**.

- For Coordinate System, choose **Ytownconsstateplane**. ArcGIS Pro adjusts the coordinate system being used to match that layer (that is, NAD 1983 State Plane Ohio North).

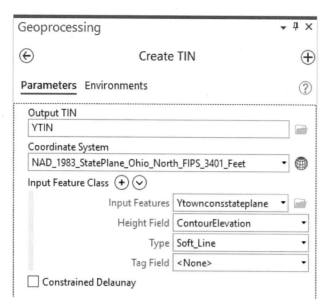

- Under Input Feature Class, make the following selections:
  - For Input Features, choose **Ytownconsstateplane**.
  - For Height Field, choose **ContourElevation**. This field contains the elevation values (above the vertical datum) assigned to each contour. (This is why you labeled the contours using this field in Step 16.2.)
  - For Type, choose **Soft_Line**. This option allows you to specify the type of features (mass points, soft lines, or hard lines) from which to construct the TIN.
- Run the Create TIN tool. For more information about how ArcGIS Pro constructs a TIN from the contours and these settings, see **Smartbox 16.2**.

SMARTBOX 16.2

How is a TIN created in ArcGIS Pro?

When you use the Create TIN tool in ArcGIS Pro, a number of processes happen behind the scenes to build the TIN, and several options affect the end result. First, the TIN needs a set of points with z-values; these points are often called **mass points**. A variety of sources can be used for the mass points: field point measurements or, as you'll see in Module 17, lidar point cloud data. Elevation measurements from other features, such as DEMs or contours, can be used as well.

Because a TIN is constructed from a series of non-overlapping triangles, the selected points must be connected to form these triangles. However, the points can't be connected in just any fashion; a method for building the triangles called **Delaunay triangulation** is used. A **Delaunay triangle** is created when you can draw a circle through the three points of the triangle without having the circle pass through any other points. Creating these types of "fat" triangles (as close to 60-degree triangles as possible) is the goal of the Delaunay triangulation method.

mass points The selection of points used in creating a TIN.

Delaunay triangulation The process of connecting sets of points to build the triangles and edges of a TIN.

Delaunay triangle A triangle that can have a circle drawn through its three points that does not pass through any additional points.

FIGURE 16.3 (a) A sample set of mass points, (b) Thiessen polygons formed equidistant around the points, (c) Delaunay triangles constructed by drawing edges perpendicular to the Thiessen polygons.

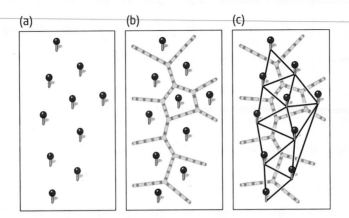

In Delaunay triangulation, a series of polygons (referred to as *Thiessen polygons* or *Voronoi regions*) are created around the points. These polygons are created equidistant between pairs of points with no overlapping polygons. Next, lines are drawn to connect the points so that the lines are perpendicular to the Thiessen polygon boundaries. These connected lines form the three **edges** of each of the Delaunay triangles of the TIN (Figure 16.3). These triangles make up the faces of the TIN. Note, however, that when contours are used as a source for creating TINs, sometimes points of equal elevation are connected, resulting in the faces being "flat triangles" without any slope.

edges The lines that connect the points of the triangles of a TIN.

hard breaklines Lines used in a TIN to enforce surface discontinuities.

soft breaklines Enforced lines in a TIN that do not represent discontinuities.

hull The polygon boundary around the area of a TIN.

To add more realism to a TIN, you can add breaklines. **Hard breaklines** represent areas on the landscape where the edges of the TIN should not cross. For instance, the faces of the triangles shouldn't cross areas such as stream boundaries, building footprints, or flat, paved roads. Incorporating hard breaklines to ensure that features like these do not have changing elevation values (when they should be flat, such as the surface of a road or stream) makes the TIN surface more realistic (Figure 16.4). **Soft breaklines** represent areas that should also be kept as edges in the TIN, but they do not represent the types of surface discontinuities that hard breaklines do. The last feature that is included in a TIN is a **hull**, or a polygon that defines the area of the TIN. After a TIN is created, you can add more points or breaklines to it in order to create more realistic representations of the terrain.

FIGURE 16.4 A comparison of two TINs—(a) one formed from only mass points and (b) one formed from both mass points and breaklines.

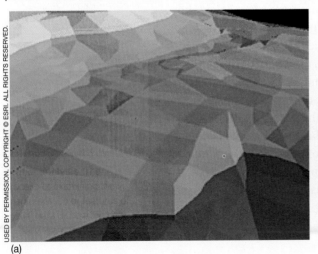

(a)   (b)

- After the Create TIN tool finishes, the YTIN is added to your Contents pane. Turn off the Ytownconsstateplane layer. (You won't be using the contours anymore now that you have a TIN.)

- Also add the **Ytownimage** raster layer to the map. Be sure to add the composite image so that it appears in color, not just as a single band (see Module 13).

- Note that Questions 16.3 through 16.7 will have you making reference to the TIN in relation to the city area. To do this, you could place the orthoimage on top of the TIN (by selecting the **List By Drawing Order** button in the Contents pane; see Module 1) in the Contents pane and turn the orthoimage on and off, or you could use the Swipe tool (see Modules 13 and 14) to look at both the TIN and the image. However, it might be easier to place the orthoimage on top of the TIN in the Contents pane and then make it semi-transparent; this is similar to what you did with a polygon layer in Module 7 and raster layers in Module 15. To do so, do the following:

  - In the Contents pane, click once on the **Ytownimage** layer to highlight it and then, in the **Appearance** tab, in the **Effects** group, you see the **Transparency** slider.

- At the top of the **Effects** group, with the Transparency options, type a value higher than 0 (0% means the orthoimage will be completely solid, and 100% means the ortho will be completely transparent or invisible) or use the **Transparency** slider to adjust the transparency value. Try values like 50%, 60%, or 75% until you get a good "ghost" of the orthoimage positioned on top of the TIN so that you can see both the features of the TIN and the high-resolution orthoimage at the same time. Adjust the value until you get something useful to you and then answer Question 16.3.

> **Question 16.3** Locate the highest and lowest elevation areas again (your answer to Question 16.2) by using the TIN. From your examination of these same areas on the orthoimage, what is actually at these areas in the Youngstown region?

Step 16.4 Performing Surface Analysis of a TIN

- Next, you will be examining the slope and slope aspect of the TIN you've created. Slope and slope aspect function the same as in Module 15, but now they are computed for each triangular face of the TIN rather than for each grid cell of a DEM. To compute slope, return to the Geoprocessing pane. The tool you will be using is Surface Slope. You can either search for it or find it directly from the **3D Analyst Tools** toolbox, within the **Triangulated Surface** toolset. Using either method, open the **Surface Slope** tool.

- For Input Surface, choose **YTIN**.

- For Output Feature Class, click the **browse** button and navigate to your **C:\GIS\Module16** folder, go to the **Module16.gdb** file geodatabase, and name the output feature class **Yslope**.

- For Slope Units, choose **Degree**.
- Leave the rest of the options at their default values and run the Surface Slope tool. The Yslope layer is added to the Contents pane. Each of the SlopeCode values represents a different value for the degree of the slope for that face.

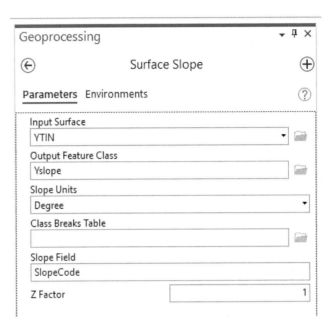

- Examine your slope map carefully in relation to your semi-transparent orthoimage of the city; you may have to turn off the TIN and rearrange the layers to examine the Yslope layer under the orthoimage. Answer Questions 16.4 and 16.5.

Question 16.4 How is the slope of the terrain distributed around the area? (That is, what areas have very flat slopes, and what areas have steeper slopes?)

Question 16.5 Where are the very steepest slopes on your slope map? What accounts for these very high slope values and their spatial distributions?

- Next, you will calculate slope aspect. To do so, return to the Geoprocessing pane. The tool you will be using is Surface Aspect. You can either search for it or find it directly from the **3D Analyst Tools** toolbox, within the **Triangulated Surface** toolset. Using either method, open the **Surface Aspect** tool.
- For Input Surface, choose **YTIN**.
- For Output Feature Class, click the **browse** button and navigate to your **C:\GIS\Module16** folder, go to the **Module16.gdb** file geodatabase, and name the output feature class **Yaspect**.

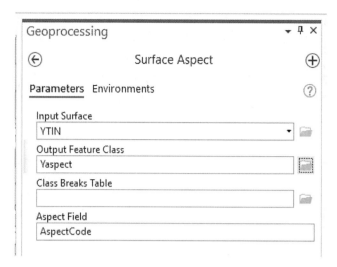

- Leave the rest of the options at their default values and run the Surface Aspect tool. The Yaspect layer is added to the Contents pane. Each of the aspect values represents a different value for the direction of the slope for that face. They are also separated out by color. Examine your aspect map carefully in relation to your semi-transparent orthoimage of the city; you may have to turn off the TIN and rearrange the layers to examine the Yaspect layer under the orthoimage. Keep in mind that there are two values that represent north (1 and 9) as well as a value that represents a flat area without an aspect value (−1). Answer Questions 16.6 and 16.7.

Question 16.6 How are your values for aspect distributed around the city? (That is, in what directions are the majority of the steepest slopes facing in the campus areas versus in some surrounding areas?)

Question 16.7 Based on your examination of the orthoimage, what is physically present at the areas around the city that are considered flat with respect to aspect?

Step 16.5 Viewing the TIN and Imagery in 3D

- For the remainder of the module, you'll be examining the TIN in 3D and draping the orthoimagery on it, and then you'll use this 3D visualization of the area to complete your virtual tour. On the **Insert** tab, in the **Project** group, click on the **New Map** button and select **New Local Scene** to create a new scene to work with. In the scene properties, change its name to **Ytown3D**.
- Add the **YTIN** layer to the scene.
- Next, you need to tell the scene to get its z-values from the TIN rather than from the default ground elevation surface. On the **Map** tab, in the **Layer** group, click the down arrow under the **Add Data** button and choose **Elevation Source**.
- In the Add Elevation Source dialog box that opens, navigate to your **C:\GIS\Module16** folder and choose **YTIN**. Click **OK** to close the dialog.

- Next, in the Contents pane, under Elevation Surfaces, turn off the option **World Elevation 3D/Terrain 3D**. Your ground terrain elevations are now drawn from the YTIN layer.

- Now add the **Ytownimage** layer to the scene as a 2D layer.

- Under 2D Layers, turn off (but do not remove) the YTIN layer and the Topographic layer to view only the Ytownimage layer. Under Elevation Surfaces, keep a checkmark next to **YTIN**.

- Because YTIN is the first source of the elevations for the scene, the Ytownimage layer will be draped over the YTIN. (See Module 15 for more about draping.) Use the scene navigator to examine the image in pseudo-3D. (See Module 15 for how to use this tool.) Answer Question 16.8.

> **Question 16.8** You can see several bridges crossing the river on the southern side of the city, as well as several freeway overpasses in the surrounding areas. Zoom in closely on them and focus your attention on each of them. How did the orthoimage drape over the bridges and overpasses, and why did this happen?

- Despite the changes in elevations, the Youngstown area is relatively flat; however, if you zoom in, you can see some terrain relief. For visualization purposes, you will add some vertical exaggeration to the scene. (See **Smartbox 16.3** for more about vertical exaggeration.) To begin, in the Contents pane, click on the **Ground** option.

SMARTBOX 16.3

What is vertical exaggeration?

vertical exaggeration
A 3D visualization technique that alters the vertical scale but keeps the horizontal scale the same.

With **vertical exaggeration**, a technique used in 3D visualization, the horizontal (x/y) scale remains the same, but the vertical (z) scale is altered by some amount. This method makes the differences in heights of features really stand out, such as making mountains look higher or valleys look deeper. Vertical exaggeration multiplies the *z*-values by a number. For instance, with a vertical exaggeration of 3, all elevation heights are shown at three times what they should be. (See Figure 16.5 for two examples of vertically exaggerated terrain.) Because vertical exaggeration alters the scale of the data, it's used only for visualization and should not be used for measurements.

FIGURE 16.5 Draped imagery on a TIN shown at vertical exaggerations of (a) 1.5 and (b) 7.

(a) (b)

- On the ribbon, you now see a new contextual tab called Appearance. Click on the **Appearance** tab. In the **Drawing** group, you see the heading Vertical Exaggeration with its default value set to 1.0. Use the arrows next to this to set Vertical Exaggeration to **3.0**. In a few seconds, ArcGIS Pro change the display to add a vertical exaggeration of 3.0 to the scene. Answer Question 16.9. (For visual purposes of this module, return Vertical Exaggeration to **3.0** after you've answered Question 16.9.)

> **Question 16.9** Test several choices for vertical exaggeration: 1.5, 2, 3, 7, and 10. Given that the area is actually very flat, which of these exaggeration levels is best for visualizing the landscape features of Youngstown for this project (from what you've seen in examining the contours, slope, and slope aspect)?

Step 16.6 Creating an Animation in ArcGIS Pro

- The next step in creating the virtual tour of the city is to create a video file simulating flying around the area by moving the camera to various locations and creating a full animation of the scene. For more information about animations in ArcGIS Pro, see **Smartbox 16.4**.

SMARTBOX 16.4

How can animations be used in ArcGIS Pro?

While 3D visualization is very cool and can effectively communicate concepts such as elevations, heights, and depths, an exported view of a scene is still just a graphic. To capture the interactivity allowed by the 3D visualization tools of ArcGIS Pro (or ArcGIS Pro), you can create an animation of actions occurring in the scene. An **animation** captures movements, actions, or changes to objects or layers in a video format. Animations also allow you to show the changes to layers of data, permitting you to look at population changes over time, watch layers slowly becoming transparent and showing other layers underneath them, or follow a predetermined camera path (such as one simulating the movement of a car down a road). ArcGIS Pro contains numerous types of animation tools that allow you to create and edit animations and then export them to video formats such as MP4 or to play them on YouTube or Instagram.

In ArcGIS Pro, animations are made by creating **keyframes**, or "snapshots" of single views, and then ArcGIS Pro fills in the necessary movements between the keyframes. For instance, if you wanted to create a smooth flyby of a mountain, you could capture keyframes of the mountain from three different angles (one at the base of the mountain, one at the top, and one around the side). When the animation is created, ArcGIS Pro shows the view of the first keyframe at the mountain base, then the view moves to the second keyframe to show the top of the mountain, and then it moves around the mountain to the third keyframe.

animation The capture of movements, actions, or changes to layers in a video format.

keyframe A snapshot of a frame used in an animation.

- You can begin the video tour by starting at the large lake and traversing around the region from there. During the tour, you'll likely want to highlight features such as the city, the valley, the path along the river, the football stadium, and other areas. To start, use the navigator to create a good perspective view, beginning your virtual tour at the lake. This will become the starting point and appearance for your animation.

- To begin recording an animation, on the **View** tab, in the **Animation** group, click the **Add** button.

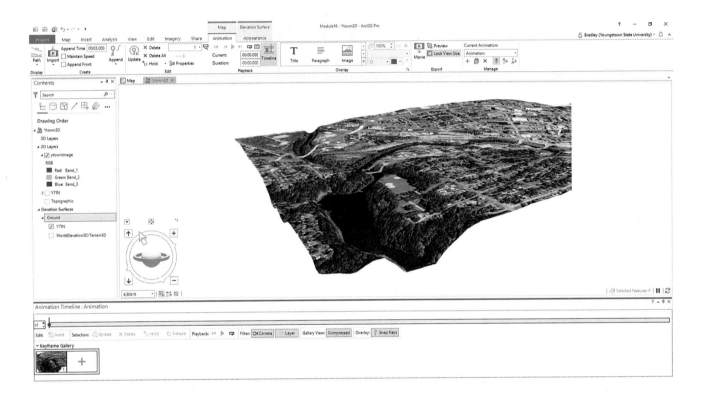

- The ribbon changes to an Animation tab, with many new options available. In addition, an Animation Timeline pane opens at the bottom of the screen. In the Animation Timeline pane, click on the **Create first keyframe** button. ArcGIS Pro takes a snapshot of the camera view and saves it as the first keyframe of the animation.

- Use the navigator to move the camera to highlight another prominent place on campus. When you have the view set up as you want it, click the large **+** button in the Animation Timeline pane to add a second keyframe. You should also see a time bar stretching across the top of the Animation Timeline pane, indicating how many seconds pass in the animation to move from the first keyframe to the second.

- You can control how the camera moves from keyframe to keyframe. On the **Animation** tab, in the **Create** group, you can click the arrow under the **Append** button to access different choices for how the camera should transition between keyframes. By default, the Fixed option just moves the camera from one position to another, but Linear, Hop, and Stepped give you different ways to change the transition between keyframes. On the Animation Timeline pane, a symbol appears between keyframes to show you which transition method was used.

- Set up a variety of keyframes so that your video is about 30 seconds long. (You will want to experiment with different transitions between the keyframes.) Consider these tips for choosing keyframes for the tour:

  - Choose your flight path carefully. Because your goal is to show off the city and surrounding areas, you might want to use the navigator for touring various locations. With the campus buildings, try for something creative instead of just a generic flyover of campus. (Try to use a few keyframes to follow the river's path if you feel adventurous.)

  - To see the path that the animation will take from the keyframes, on the **Animation** tab, in the **Display** group, click the **Path** button and then zoom back out. ArcGIS Pro labels the location of each keyframe and shows you with a line the path that the camera will take.

  - If you want to remove a keyframe, simply choose it in the Animation Timeline pane and then, on the **Animation** tab, in the **Edit** group, click the **Delete** button.

  - Whatever you show in the view will be captured by the animation, so you might want to add a keyframe or two to the end to finish out the animation smoothly. For instance, if you use the Hop option to move a large distance, you might want to use some additional keyframes after that to be able to spend a couple of seconds viewing the location you just hopped to.

  - To expand the area seen in the view for the recording, you might want to close extra windows (such as the Contents pane or the Catalog pane) when working with the animation tools.

- After you have set up a series of keyframes, you need to export the animation to a video file. To do so, on the **Animation** tab, in the **Export** group, click the **Movie** button.

- In the Export Movie pane that opens, for this first video, choose **Draft** to minimize the time and file size.

- For File Name, click the **browse** button and navigate to your **C:\GIS\Module16** folder as the location to which to save your video and give it an appropriate name.

- Click **Export** when you're ready.

- ArcGIS Pro starts the export process, and you see a counter in the pane moving as the file is exported. Depending on the file size and export type, this may take a couple of minutes. Don't touch any controls or switch windows; just let the process run.

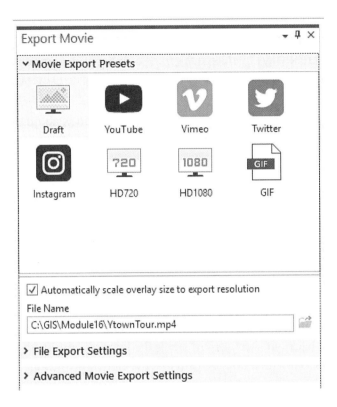

- When the countdown is finished in ArcGIS Pro and the exporting is complete, in the Export Movie pane, click **Play the video** to open the MP4 file using a video player on your computer (such as Windows Media Player) and view your animation. If you're unhappy with it (for example, if there is too much hesitation or if the flight is not smooth enough), return to ArcGIS Pro and record a new animation and then export it to a new file.

- If you want to create a better-quality video for viewing, choose **HD720** for Movie Export Preset and re-export the video to a file called **YtownTour720**. It will take a couple of minutes to export, but you'll have a much better-quality video file than the Draft version.

- When you have the video set up the way you want it, answer Question 16.10.

Question 16.10 Submit (or show) your final MP4 file to your instructor, who will check over your work for its originality and the quality of the video tour to make sure you get credit for this question.

Step 16.7 Saving and Sharing Your Results

- Close the animation and remove the keyframes by selecting the **View** tab, going to the **Animation** group, and clicking the **Remove** button.
- Save your Module16 project. The video file you created can be submitted to your instructor to complete this module.

Closing Time

TINs are exceptionally useful and versatile data structures to use in modeling and in 3D landscape visualization. Being able to pick and choose the most important points on the landscape helps create a more detailed landscape where necessary, and breaklines can help in fashioning more realistic landscapes. TINs can be created from a variety of different sources, including digital elevation models or contour lines, and also from point measurements of elevation. Module 17 looks at another technique called lidar, which can also be used as a source for a TIN. Lidar allows for the collection of millions of elevation points on a landscape, and you will use it to do more work with TINs.

The contour lines used in this module were drawn from The National Map, and for a long time, contour lines were mainstays of USGS topographic maps. Although standard topographic maps are no longer being produced, they can still be used as data sources within ArcGIS Pro. In addition, the next generation of topographic maps (the US Topos) can be very valuable tools for geospatial research. See *Related Concepts for Module 16* for more about the use of topographic mapping products with ArcGIS Pro.

RELATED CONCEPTS FOR MODULE 16

USGS Topographic Maps and US Topos

The traditional source for examining contour information was the USGS **topographic map** series. The function of topographic maps was to examine the features on Earth's surface with information about land cover and especially topography, as represented by contours on the map. Topographic maps were produced as large paper maps in a variety of formats. The 1:24000 scale topographic maps (sometimes referred to as "quadrangles" or "topoquads") each covered a geographic area of 7.5 minutes of latitude by 7.5 minutes of longitude. Topographic maps were also produced as smaller-scale maps, such as 1:100000 and 1:250000.

The USGS no longer produces topographic maps, but digital versions of them remain available for use in GIS. Examination of landforms, contours, and historical land-cover patterns can still be done with topographic maps. In ArcGIS Pro, topographic maps (at a variety of scales) are available through a basemap called USA Topo Maps, available in the standard basemap options. Instead of showing individual quadrangles, the basemap combines many quads of the same scale into a "seamless" map (Figure 16.6).

Scanned and georeferenced versions of topographic maps, called **digital raster graphics (DRGs)** are available and can be used as separate layers in ArcGIS Pro. (See Module 5 for sources for obtaining DRGs.) Because they've been georeferenced, DRGs align with other layers in ArcGIS Pro. DRGs may be useful for draping over a terrain model in ArcGIS Pro to see how landforms match up with the features or contours of a topographic map. DRGs are considered "legacy" data as topographic maps are no longer being produced—and so neither are the corresponding DRGs.

The current generation of topographic maps is the **US Topo** map series being produced by the USGS. US Topos are standalone products available for free digital download via The National Map and are being produced for areas on a

topographic map A map created by the USGS to show landscape and terrain as well as the locations of features on the land.

digital raster graphic (DRG) A scanned version of a USGS topographic map.

US Topo A digital topographic map series created by the USGS to allow multiple layers of data to be used on a map in GeoPDF file format.

FIGURE 16.6 A USA Topo Maps basemap shown in ArcGIS Pro.

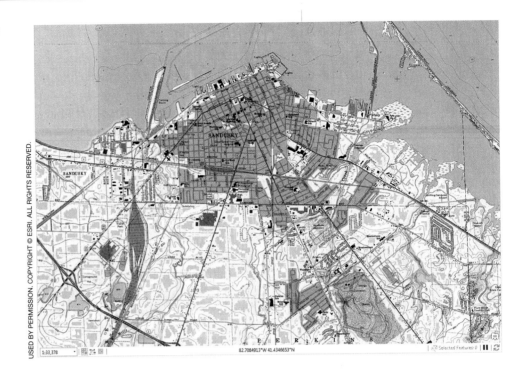

GeoPDF A format for maps to be used as PDFs that can contain geographic information and multiple layers.

three-year cycle. US Topos contain multiple layers, such as contours, transportation labels, and orthoimagery, all of which can be turned on and off like layers in ArcGIS Pro's Contents pane.

US Topos are available in **GeoPDF** format, in which a PDF document has interactive capabilities and can return coordinate information (in GCS or UTM) when a user selects a location on the map (Figure 16.7); a GeoPDF also allows a user to make measurements of lines or areas directly on the US Topos. The National Map also makes historic topographic maps available as GeoPDFs. Keep in mind, however, that a GeoPDF is a standalone document and cannot be used as a layer in ArcGIS Pro. You can access the necessary GeoPDF plug-in and toolbar to work with US Topos from the TerraGo website, at **https://terragotech.com/products/geopdf/toolbar/**.

FIGURE 16.7 Finding coordinates on a US Topo map of Youngstown, Ohio.

Key Terms

TIN (p. 375)
contour (p. 379)
contour interval (p. 379)
index contour (p. 379)
intermediate contour (p. 379)
mass points (p. 381)
Delaunay triangulation (p. 381)
Delaunay triangle (p. 381)
edges (p. 382)
hard breaklines (p. 382)
soft breaklines (p. 382)
hull (p. 382)
vertical exaggeration (p. 386)
animation (p. 387)
keyframe (p. 387)
topographic map (p. 391)
digital raster graphic (DRG) (p. 391)
US Topo (p. 391)
GeoPDF (p. 392)

17

How to Work with Lidar Data in ArcGIS Pro

Learning Outcomes

After studying this module, you should be able to:

- Explain the concept of active remote sensing.
- Define the term *lidar*.
- Describe how a height measurement is made using lidar.
- Explain what a digital surface model is.
- Describe how LAS files and a LAS dataset are used for storing lidar data.

lidar A method for measuring elevation values using laser pulses sent to the ground from an aircraft.

active remote sensing A process in which a sensor generates its own energy source and measures the reflection or backscatter of that energy.

ArcGIS Pro Skills

In this module, you will learn how to do the following in ArcGIS Pro:

- How to create a LAS dataset and view lidar data (such as LAS files) in ArcGIS Pro.
- How to examine elevation and class features of lidar data.
- How to edit lidar class codes.
- How to obtain information and measurements about lidar points.
- How to work with lidar data in a scene environment.

Introduction

Obtaining accurate information about elevation and heights is critical for numerous GIS applications. Landscape elevations, terrain profiles, and height measurements of trees or buildings are necessary for forest managers, urban planners, archeologists, and environmental scientists (among others). While digital elevation models or contours can provide some of this information, a geospatial technique called lidar is used often to determine the elevation and height of the ground or objects on the surface. With **lidar**, an aircraft uses a laser beam to determine the heights of the terrain and objects over which it flies. The term *lidar* is a combination of the words *light* and *radar*, although it has also been used as an acronym for *light detection and ranging*.

Lidar is a form of **active remote sensing** (see Module 13) in which the sensor generates its own energy source, fires it at a target, and measures the reflection of that particular type of energy. In the case of lidar, a laser beam on an aircraft is fired at the ground below, the laser bounces off a target, and the reflection is received back at the aircraft (Figure 17.1). The times the laser is sent and received are precisely measured. With information regarding the time that it took for the beam to travel from the plane to a target and back again as well as the speed of the beam (that is, the speed of light), the distance from the plane to the target can be calculated. The plane's GPS and inertial navigation system (INS) are used to accurately determine the plane's position and altitude. With the known altitude of the plane and the distance from the plane to the target (and information about the angle of the beam), the elevation of the target can be measured.

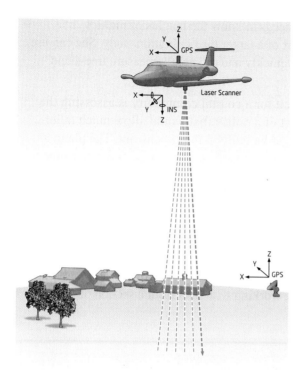

FIGURE 17.1 The operation of airborne lidar.

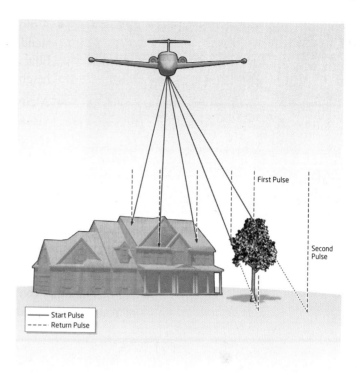

FIGURE 17.2 The different types of measurements made by lidar.

Millions of lidar pulse measurements can be quickly taken over a small area. These lidar data are then processed into a series of points, where each point contains x and y measurements as well as a z-value for the elevation at that point. Because they have so many processed points, the lidar data comprise a **point cloud**, and this large dataset can be used in GIS. The point cloud represents the measured elevations not just of the ground surface but also of all the objects from which the laser beams were reflecting, such as buildings, bridges, or trees (Figure 17.2). Note that a laser beam with a different wavelength is used for studies involving water or measuring the depths of bodies of water, in a method called *bathymetric lidar*.

point cloud The processed elevation points of a lidar dataset.

Module Scenario and Applications

This module again puts you in the role of a member of the Youngstown, Ohio, chamber of commerce. As part of a marketing effort for the city, a 3D model of the urban landscape is being developed, and as part of the process, accurate heights of the city's buildings are needed. Your job is to investigate the use of a set of lidar data in measuring building heights as well as to create a simplified 3D representation of the city to use as a basis for the model.

The following are additional examples of other real-world applications of this module's skills:

- An urban planner needs to know the height of the city's bridges as part of a redevelopment effort. He can use lidar data to quickly obtain the bridge height information.

- A biologist needs to know the heights of multiple tree stands as part of a larger bird migration study. She can use lidar data to quickly and efficiently measure tree-stand heights.
- A city council for a coastal community is assessing the visual impact of a policy that would allow much taller resort hotels to be built on the beachfront. The planners and developers can use lidar data to quickly determine the heights of existing structures as part of a viewshed analysis.

Study Area

- For this module, you will be working with data from a section of the city of Youngstown, Ohio.

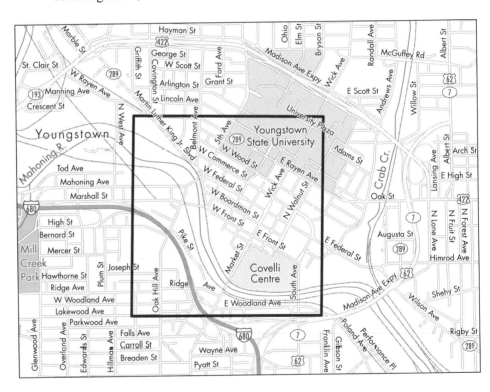

Data Sources and Localizing This Module

The data for this module focus on a section of the city of Youngstown, Ohio. However, you can easily modify this lab to use data from your own local area instead if lidar data are available for free download for areas near you. The LAS dataset used in this lab was downloaded from EarthExplorer (**https://earthexplorer.usgs.gov**; see Module 5). Free lidar datasets are also available via The National Map website, under the heading Elevation Source Data. Beyond what is available through EarthExplorer or The National Map, there may also be free lidar data available for specific states or counties. For example, the U.S. Interagency Elevation Inventory (available at **https://coast.noaa.gov/inventory/**) lists which states have available downloads of lidar data.

Step 17.1 Getting Started

- This module's hands-on applications use the data folder called Module17. Your instructor will be able to supply you with this data, or you can download it directly from this book's website at **https://www.macmillanlearning.com/college/us/product/Discovering-GIS-and-ArcGIS-Pro/p/131923075X**. The text in this module assumes that you have this Module17 folder in a computer location referenced as C:\GIS; if you have it somewhere else (for instance, in a flash drive referenced as G:\GISClass), substitute that location and path to the Module17 folder throughout this module.
- The Module17 folder contains the following:
  - OH_North_2006_015328.las: This is a LAS file of lidar points of a section of Youngstown, Ohio, from 2006.
  - OH_North_2006_015328.xml: This is a metadata document for the LAS file.
- Start ArcGIS Pro.
- Sign in with your Esri account username and password.
- Create a new project using the **Map** template. Call this project **Module17** and place it in your **C:\GIS\Module17** folder. Ensure that there is not a checkmark in the box next to **Create a new folder for this project**.
- When ArcGIS Pro opens, change the map's name to **Ytown Lidar**.
- Make sure the **3D Analyst** extension is licensed and activated. See Smartbox 11.1 and Troublebox 11.1 in Module 11 for more information about this.

Step 17.2 Working with LAS Files in ArcGIS Pro

- Your lidar data have the extension .las, which indicates a LAS file. For more about LAS and working with it in ArcGIS Pro, see **Smartbox 17.1**.

SMARTBOX 17.1

What is a LAS file, and how is it used in ArcGIS Pro?

When lidar data are being processed, these points are classified according to what they represent, such as the ground or a building. Lidar point cloud data are accessed in the **LAS file** format, a standard that is commonly used with GIS. The LAS file stores all the information related to the point cloud data (including x, y, and z coordinates, as well as the classification data). They are also commonly accessed as **LAZ files** (which are compressed versions of lidar files).

LAS file A standard file format for holding lidar data.

LAZ file A compressed file format for holding lidar data.

LAS dataset An ArcGIS Pro file structure used for storing, viewing, and analyzing one or more LAS files.

A **LAS dataset** is a specific file structure that allows for the storage of one or more LAS files for use in ArcGIS Pro. A LAS dataset also allows users to easily view, analyze, query, edit, and visualize the point cloud data within LAS files. A LAS dataset can also store any types of features representing surface constraints (such as hard breaklines or polygons representing the boundaries of bodies of water) that affect how a surface is constructed.

When a LAS dataset is created, it's empty (just as a new geodatabase is empty when it's created), and LAS files can be added to it. Typically, when LAS files are obtained, they are tiled—like the LAS file of Youngstown you're using in this module. Tiles of data are small rectangular sections of point cloud data adjacent to one another. To look at the point cloud data for just north of the area you're working with in this module, you would need to obtain the LAS file representing the next tile to the north. A LAS dataset can contain multiple LAS files, and you can work with the data from all the files together instead of using only individual data tiles (Figure 17.3). In ArcGIS Pro, the geoprocessing tools Add Files to LAS Dataset and Remove Files from LAS Dataset allow to add and subtract individual LAS files from a LAS dataset.

FIGURE 17.3 The relationship between LAS files, a LAS dataset, and the geospatial data they contain.

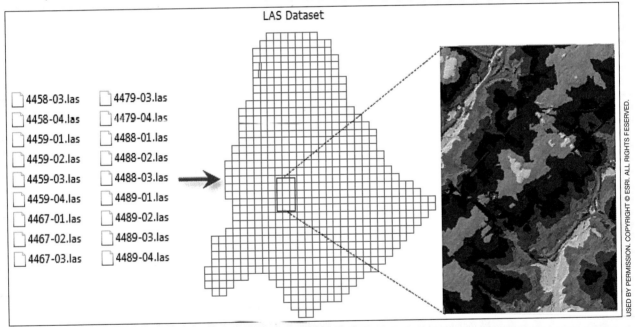

- To obtain some basic information about your LAS file, use the Catalog pane to navigate to the Module17 folder and find the **OH_North_2006_015328.las** file. Right-click on it and choose **Properties**.

- In the LAS File Properties dialog, choose the **General** tab and Answer Question 17.1. Then choose the **LAS Files** tab and answer Question 17.2. Then click **OK** to close the dialog box.

Question 17.1 How many points does this LAS dataset (which is composed of a single LAS file) contain?

> What is the point spacing in this dataset? That is, how many feet apart are the lidar points?
>
> **Question 17.2**

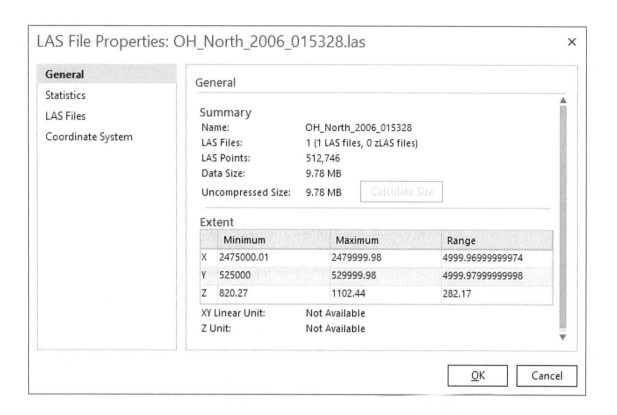

- To create a LAS dataset to work with in this module, open the Geoprocessing pane. The tool you will use is Create LAS Dataset. You can either search for it or find it directly in the Data Management toolbox in the LAS Dataset toolset. Using either method, open the **Create LAS Dataset** tool.

- For Input Files, click the **browse** button and navigate to your **C:\GIS\Module17** folder, select the **OH_North_2006_015328.las** file, and click **OK**. Your OH_North_2006_015328.las file is added as the first line. If you had multiple LAS files, you could add them here as well to combine several LAS files into a single LAS dataset.

- For Output LAS Dataset, click the **browse** button, navigate to your **C:\GIS\Module17** folder, and name it the dataset **YLASD**.

- For Coordinate System, click the spherical globe button to open a new set of choices.

According to the metadata, the LAS file is State Plane, Ohio North, and it uses U.S. feet as linear units, and you can define the LAS dataset using this information. Expand the **Projected Coordinate Systems** options, select **State Plane**, select **NAD 1983 (US Feet)**, and select **NAD 1983 StatePlane Ohio North FIPS 3401 (US Feet)**. Click **OK** to set the coordinate system.

- Leave the other options alone and run the Create LAS Dataset tool.

Step 17.3 Examining Lidar Data in ArcGIS Pro

- If the YLASD LAS dataset is not automatically added to the Contents pane, add it manually from the Catalog pane. The point cloud of the area is displayed by default as points with elevation values (in feet).

- In order to give your lidar points some context, on the **Map** tab, in the **Layer** group, click the **Basemap** button, and select the **Imagery Hybrid** option. You now see your lidar points sitting on top of the imagery.

- To get a better sense of how your lidar points match up with the corresponding locations on the imagery, you can change your lidar points to be semi-transparent. In the Contents pane, choose the **YLASD** layer and then, on the **Appearance** tab, in the **Effects** group, use the **Layer Transparency** slider bar to adjust the transparency of the points. Set the transparency at a level where you can still see not only the points distinctly but also some of the corresponding imagery the points are sitting on.

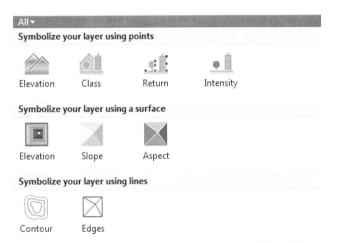

- There are hundreds of thousands of points in this one point cloud dataset, and the LAS Dataset toolbar gives you several options for which points to examine and how to do so. By default, all the lidar points are shown, and they are displayed according to their elevation, but different visualization options are available. In the Contents pane, choose the **YLASD** layer and then, on the **Appearance** tab, in the **Drawing** group, click the **Symbology** button and then under the **Symbolize your layer using points** header, choose **Class**.

- Each point has a classification assigned to it for what that point represents; you can see the value and the classification displayed in the Contents pane and also in the Symbology pane. For more about LAS classifications, see **Smartbox 17.2**.

- Zoom in on some of the points and carefully examine them according to their classification in relation to the basemap imagery (keeping in mind the difference between the 2006 lidar data collection and the more recent imagery). Some points don't exactly line up with the imagery; for example, some ground points show up on top of the imagery of a building. This is due to the fact that the locations in the imagery were not all captured looking straight down at the nadir point, and the lidar points were collected differently. Answer Question 17.3.

SMARTBOX 17.2

How are LAS classification values used in ArcGIS Pro?

The American Society for Photogrammetry and Remote Sensing (ASPRS) created the standardized classification system for LAS 1.1 and later versions. Table 17.1 lists each classification value and the class to which it should correspond for the most recent LAS version (1.4).

ArcGIS Pro gives you the ability to edit the classification values (referred to as class codes) and change all points assigned to one value to a different value. You do this with the Change LAS Class Codes geoprocessing tool, but as you'll see in this module, a total attribute switch may introduce errors into the data. A different strategy would be to use the Set LAS Class Codes Using Features tool, in which feature classes are used when classifying points. For instance, you could digitize the footprints of buildings as a polygon feature class, and any lidar points that fall within these boundaries could be reclassified as the Buildings class code.

Classification Value	Class
0	Never classified
1	Unclassified
2	Ground
3	Low vegetation
4	Medium vegetation
5	High vegetation
6	Building
7	Low point (Noise)
8	Reserved
9	Water
10	Rail
11	Road surface
12	Reserved
13	Wire-guard (Shield)
14	Wire-conductor (Phase)
15	Transmission tower
16	Wire-structure connector
17	Bridge deck
18	High noise
19	Overhead structure
20	Ignored ground
21	Snow
22	Temporal exclusion
23–63	Reserved
64–255	User-definable

Table 17.1 The ASPRS LAS classification system

Question 17.3 The points classified with the value 1 (Unassigned) should likely have been classified according to which of the LAS classifications? How can you tell?

- ArcGIS Pro allows you to edit the classification values assigned to a LAS file to update data or make necessary corrections. In this case, you'll want to change the value of the points classified as 1. Return to the Geoprocessing pane. The tool you will use is Change LAS Class Codes. You can either search for it or find it directly in the **3D Analyst Tools** toolbox, within the **Data Management** toolset, in the **LAS Dataset** toolset. Using either method, open the **Change LAS Class Codes** tool.
- For Input LAS Dataset, choose **YLASD.lasd**.
- Under Class Codes, for Current Class, type the value you want to change—in this case, type **1**.
- For New Class, type the value to which you want the current class to be changed—in this case, type **6**.
- Leave the other settings alone and then run the Change LAS Class Codes tool.
- Re-examine your points and their new classifications. Answer Question 17.4. If the newly reclassified points are not appearing on the map, see **Troublebox 17.1** for an explanation and a way to retrieve them.

Question 17.4 Switching classifications from 1 to 6 did not solve all the misclassifications. What features are now incorrectly classified as 6? (*Hint:* Zoom in on some areas and compare the lidar points with the imagery on the basemap.)

TROUBLEBOX 17.1

What happened to the reclassified lidar points?

After you ran the Change LAS Class Codes tool, you may have seen that the set of points that were previously classified as 1 and are now classified as 6 disappeared from the map and that a value of 6 is not shown in the YASD.lasd legend in the Contents pane. If so, don't worry: Those points are still there, but their new value of 6 isn't being displayed in the layer's symbology.

To correct this, open the Symbology pane for the YASD.lasd layer and in the section labeled Values, click the **More** button. Then click **Add all values**. This forces ArcGIS Pro to add the value 6 to the symbology being displayed for the layer. Thus, you should see the points with the value 6 shown in the Contents pane and on the map.

- ArcGIS Pro gives you the option to examine whether a point is measuring a ground feature or a non-ground feature. In the Contents pane, choose the **YLASD** layer and then, on the **Appearance** tab, in the **Filters** group, click the **Filters** button down arrow and choose the option for **Ground**. Now, the only points displayed are those identified as measuring elevations on the ground. Click the **Filters** button down arrow and choose **Non Ground**. Now the only points displayed are those measuring points not on the "ground."
- Answer Question 17.5. For more information about what lidar measurements of ground and non-ground features represent, see **Smartbox 17.3**.

SMARTBOX 17.3

What do lidar measurements represent?

When the laser pulses from the lidar system strike areas on the ground, they do so in "laser footprints" that are circular areas 30 centimeters in diameter. The strength of the laser pulse returned after striking an object is called the **intensity** of that pulse. The **return** (or reflection) of this pulse is what is measured back at the airborne sensor. However, a pulse may have several returns, depending on the object that is struck with the pulse. For instance, Figure 17.4 shows a large tree illuminated by the laser pulse; different places (at different heights) on the tree will generate multiple returns. The **first return** is in many ways the most important return, as it represents the top of the object (or the maximum elevation) and is used in determining the height of the tree. The intermediate returns can be used for other measurements or the profile of the tree, and the **last return** is often the one from the ground. All these returns are used in constructing a **digital surface model (DSM)** to show the heights of the terrain and the objects on top of it (including buildings, tree canopies, and vegetation).

intensity The strength of the reflected return of the laser pulse from an object.

return The reflection of energy from a lidar laser pulse.

first return The initial reflection of the lidar laser pulse, which usually indicates the height of an object.

last return The final reflection of the lidar laser pulse, which usually indicates the elevation height of the ground.

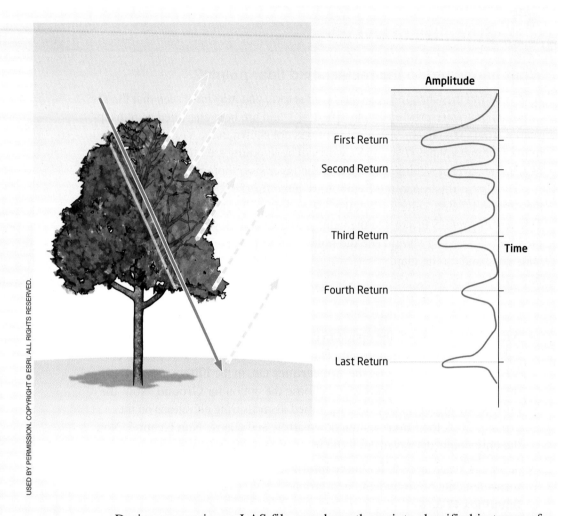

FIGURE 17.4 The returns associated with lidar pulses.

digital surface model (DSM) A representation of the surface and features generated from a set of lidar points.

During processing, a LAS file may have the points classified in terms of whether they are first return or not. (Unfortunately, the data used in this module do not contain this information, though they do contain classifications for points to be separated as ground and non-ground.) One of the options under Filters on the LAS dataset toolbar allows you to examine only first-return points. Last-return data are utilized to create a bare-Earth digital elevation model, which shows only the ground elevations. Algorithms are used to remove return data (including last-return data) that contain vegetation, trees, or structures from the dataset to use only those points that represent the bare surface of the ground. Some of the data within the 3DEP program (see Module 15) consist of lidar-derived bare-Earth digital elevation models.

Question 17.5 Aside from buildings, what other types of specific features on the landscape are considered non-ground in terms of lidar measurements?

- The elevation value of an individual lidar point can be accessed with the Explore tool. For instance, zoom in closely on the area shown in Figure 17.5. This is the Lincoln Building on the Youngstown State University campus; it is located at the corner of Lincoln Avenue and North Phelps Street.

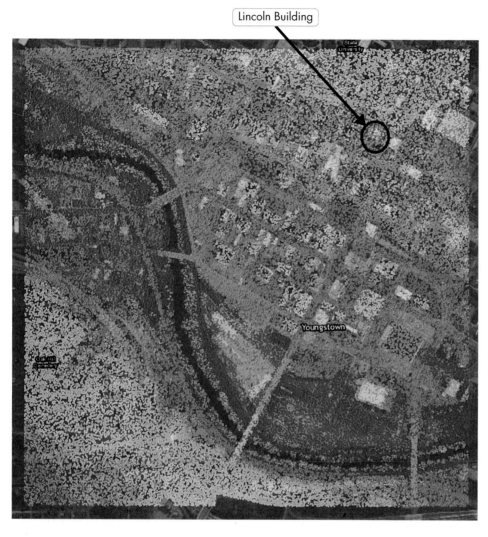

FIGURE 17.5 The point cloud of the study area and the highlighted Lincoln Building.

- Zoom in closely, click the **Filters** button down arrow, and choose the **All Points** option to show all the lidar points once more.
- Lincoln is a flat building with an air-conditioning unit on the roof. (See the close-up in Figure 17.6 and note that the top of the building in the image doesn't match up directly with the lidar points due to relief displacement.)

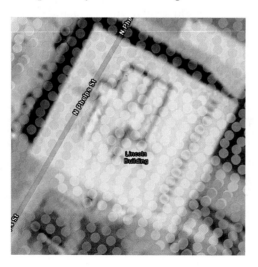

FIGURE 17.6 The lidar points obtained from the Lincoln Building and its surrounding areas.

Use the Explore tool on one of the points on the flat part of the building's roof and answer Question 17.6.

> **Question 17.6** What is the elevation height of the flat roof of the Lincoln Building (in feet)?

- Keep in mind that the elevation value for this point is the height above the vertical datum, not the height of the building itself. To obtain a relative value for the building's height, use the Explore tool on one of the lidar points on the flat sidewalk in front of the building. Compare the elevation value of the sidewalk and the elevation value of the flat roof of the building and answer Question 17.7.

> **Question 17.7** Based on these two measurements (the elevation value of the sidewalk and the elevation value of the flat roof of the building), what is the approximate height of the Lincoln Building?

Step 17.4 Examining Lidar Data as a Surface

- As you've seen in this module, you can look at points on the screen to get a sense of their elevations. However, ArcGIS Pro provides tools for better visualization of lidar data. ArcGIS Pro gives you the capability to use the points and interpolate a surface from them to work with. In the Contents pane, choose the **YLASD** layer and then on the **Appearance** tab, in the **Drawing** group, click the **Symbology** button. Under Symbolize your layer using a surface, choose **Elevation**.

- Your semi-transparent points are replaced with a semi-transparent surface, where each location contains a value for elevation. Zoom in to the section of downtown Youngstown shown in Figure 17.7. This section of downtown contains the Home Savings and Loan Building (which is the building with the pointed spire on the roof and is the tallest structure on that block). See Figure 17.8 for an aerial image of what this section of downtown really looks like. You should be able to identify this area based on the elevation surface in relation to the imagery (it should have some of the highest elevation values); you should also be able to identify it from the labels on the imagery (at the corner of Vindicator Square and W Federal St).

- Zoom in closely on this area and use the lidar surface and the imagery to find the highest elevation of the Home Savings and Loan Building (along the spire of the building). (Keep in mind that the imagery does not have an exact one-to-one correspondence with the locations of the lidar points.) Use the Explore tool to obtain the lidar elevation value at that point. Next, use the Explore tool to obtain the lidar elevation value of a point on the flat sidewalk adjacent to the building. With these values in mind, answer Question 17.8.

> **Question 17.8** What is the height of the Home Savings and Loan Building?

Step 17.4 Examining Lidar Data as a Surface   407

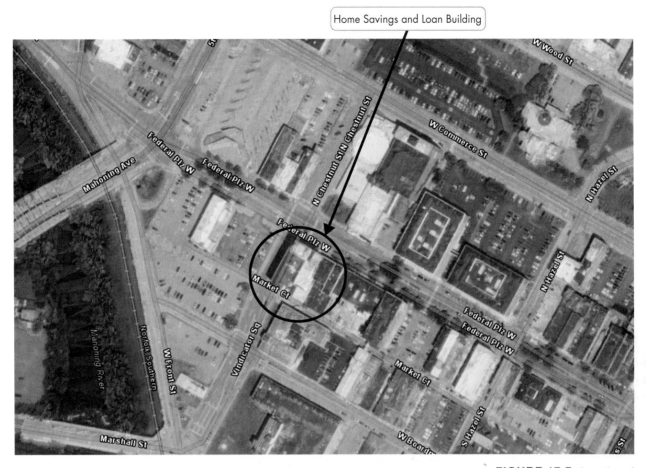

FIGURE 17.7 A section of downtown Youngstown as shown in imagery and lidar data shown as a surface.

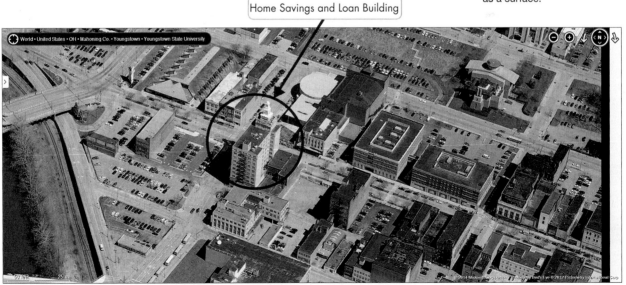

- In the Contents pane, choose the **YLASD** layer and then, on the **Appearance** tab, in the **Drawing** group, click the **Symbology** button. Under Symbolize your layer using a surface, choose **Slope**. Examine the calculated slope TIN surface and answer Question 17.9.

FIGURE 17.8 The same section of downtown Youngstown as in Figure 17.7, here shown in oblique aerial imagery.

Question 17.9 Why do all the buildings have their sides shown in a solid red color? What is being measured and shown in this display?

Step 17.5 Examining Lidar Data in a 3D Scene

- Next, you will represent the lidar data in 3D. After all, these are points that contain z-values and would be best represented in a 3D environment. The easiest way to do this is to convert the data shown on the 2D map into a 3D scene. To do so, on the **View** tab, in the **View** group, click the **Convert** button and choose **To Local Scene**.

- ArcGIS Pro adds a new scene (called Ytown Lidar_3D) with your YLASD layer added as a 3D layer. Use the navigator (see Modules 15 and 16) to examine this scene in perspective view. If your lidar points are not displayed as 3D points, see **Troublebox 17.2** for how to fix this.

TROUBLEBOX 17.2

How do I restore my lidar measurements to points?

If your lidar points are being displayed as a surface rather than as points, in the Contents pane, choose the **YLASD** layer and then, on the **Appearance** tab, in the **Drawing** group, click the **Symbology** button. Under Symbolize your layer using points, choose **Elevation**. Note that in the Symbology options, there are two choices using Elevation: Symbolize your layer using points and Symbolize your layer using a surface.

- Use the navigator to examine the full scene as the point cloud is now being shown in 3D. The ground, the tops of buildings, and the surrounding areas outside the downtown are now shown with points set to their calculated heights. Some of the buildings do not have sides due to the relatively sparse nature of the point cloud you're working with in this module.

- Return to the Home Savings and Loan Building. You should now see a 3D representation of points creating the shape of the building, its features, and the spire. Even though the free LAS file you're using in this module is relatively sparse, you should be able to determine the building's profile and how it stands out among the rest of the downtown. Zoom in closely on the building and use the Explore tool to obtain the value of the point with the highest elevation. Answer Question 17.10.

> **Question 17.10**
> What is the elevation value of this highest measured point of the Home Savings and Loan Building? Assuming that the ground has an elevation height of 859 feet, what is the height of the highest part of the spire?

- Return to the Lincoln Building and use the Explore tool to get a good measurement of the height of the building and also the height of the air conditioner unit in the center of the roof. Answer Question 17.11.

> **Question 17.11**
> Assuming that the ground has an elevation height of 893 feet, what is the height of the rooftop of the Lincoln Building? Using this rooftop height, what is the additional height of the rooftop air conditioner (that is, the height of the air conditioner above the actual rooftop)?

Step 17.6 Exporting and Saving Your Results

- Save your Module17 project.
- Print a layout (see Module 3) showing the scene representation of the 3D point cloud view of downtown Youngstown and the campus area you've been working with in this module. When you create a map frame in the layout, you should draw from the Ytown Lidar_3D scene. Then you can add the usual elements, such as a legend, a title, and a scale bar, to the layout.

- If you want to share your 3D scene with the point cloud data through ArcGIS Online, you can do this by using the Create Scene Layer Package geoprocessing tool to create a 3D web scene. (See Module 18 for more about web scenes.)

Closing Time

Lidar data provide a rich and versatile dataset for use in measuring biomass, measuring building heights, and creating digital elevation models of an area. A LAS dataset allows you to easily work with several LAS files of point cloud data to work with height measurements and digital surface models. As lidar data are becoming more accessible, it's important to have new tools to take advantage of this type of geospatial data. For instance, ArcGIS Pro has a separate type of dataset for use in visualizing and working with large amounts of lidar data, called a *terrain dataset* (see *Related Concepts for Module 17*). Module 18 examines further techniques for creating other types of three-dimensional visualizations of objects, using more of the tools that are available when working with 3D scenes.

RELATED CONCEPTS FOR MODULE 17

Terrain Datasets in ArcGIS Pro

Lidar point cloud data can be used to create a TIN or DEM. In addition, a special file structure exists in ArcGIS Pro specifically for containing, visualizing, and analyzing large quantities of elevation data. A **terrain dataset** is a TIN-based file structure that can "consume" extremely large amounts of elevation values to produce an elevation model. An advantage of a terrain dataset is that it can have pyramids generated for it, enabling quick visualization of the large amount of data at different scales.

A terrain dataset is created from a feature dataset within a geodatabase (see Module 6). This feature dataset must contain the feature classes that will be used to create the terrain dataset. One caveat is that, while several LAS files are often used as sources of elevation data to construct a terrain dataset, LAS files cannot be stored as a feature class. The LAS files must first be converted into a **multipoint** feature class in order to be used. Multipoint is a special type of feature class designed to hold arrays of millions of points (such as lidar point cloud data). Ordinarily, a point feature class has a separate record for each individual point—which would not be feasible with millions (or billions) of points—but the multipoint feature class gathers these points into rows of data, thus allowing their use in a geodatabase.

A terrain dataset can be constructed from lidar points stored as a multipoint feature class along with any other feature classes that define surface considerations (such as polygon feature classes to show the boundaries of water bodies or line feature classes to show hard breaklines). The resultant terrain dataset enables you to quickly visualize the elevation values and surface features because it does not need to create and use a TIN but rather references the feature class data and creates a TIN representation "on the fly." With the terrain dataset's pyramids, different scales of massive datasets can be rendered quickly (see Figure 17.9).

terrain dataset An ArcGIS Pro file structure that can hold large amounts of elevation data and render them quickly at a variety of scales and resolutions.

multipoint A geodatabase feature class that can hold large amounts of point data.

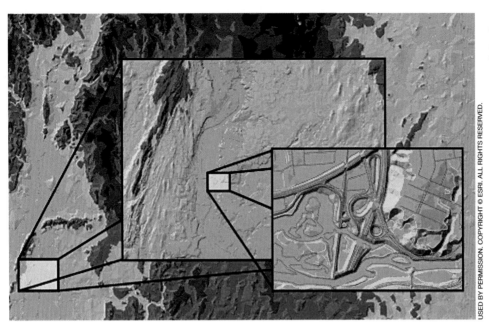

FIGURE 17.9 The use of a terrain dataset for quickly displaying a very large GIS dataset at various scales.

Key Terms

lidar (p. 394)
active remote sensing (p. 394)
point cloud (p. 395)
LAS file (p. 397)
LAZ files (p. 397)

LAS dataset (p. 398)
intensity (p. 403)
return (p. 403)
first return (p. 403)
last return (p. 403)

digital surface model (DSM) (p. 403)
terrain dataset (p. 410)
multipoint (p. 410)

18

How to Represent Geospatial Data in 3D with ArcGIS Pro

Learning Outcomes

After studying this module, you should be able to:

- Describe how extrusion is used to create objects with a 3D appearance.
- Explain how an offset is used in 3D visualization.
- Define what a multipatch is.
- Describe what a COLLADA file is and how it is used in 3D visualization.
- Explain how digitized points can be symbolized as 3D objects.

ArcGIS Pro Skills

In this module, you will learn how to do the following in ArcGIS Pro:

- Extrude polygon layers.
- Publish a 3D web scene.
- Use a web scene in ArcGIS Online.
- Apply an offset to polygon layers.
- Create multipatch versions of extruded polygon layers.
- Replace a multipatch object with a COLLADA file.
- Create and visualize representations of 3D objects.

Introduction

The last three modules have focused on three-dimensional representations of Earth-based concepts. Land cover, terrain, and imagery and lidar provide means of assessing heights of non-landscape objects as well as rough visualizations of them. Now it's time to focus on 3D visualizations of objects and items on the surface of Earth—buildings, structures, trees, cars, and pretty much anything else you can envision (beyond what you saw with lidar in Module 17). It's one thing to represent a neighborhood as a series of two-dimensional polygons of houses' footprints; it's another thing entirely to view the same neighborhood as three-dimensional polygons that show the height and dimensions of each house.

3D visualization communicates some geospatial concepts quickly and intuitively and can help present ideas effectively. For instance, back in Module 6, you created a map of the Youngstown State University campus by digitizing a representation of its buildings. A newcomer to the campus using that map would have an idea of the relative sizes of the structures but would have no feel for what the buildings actually look like or how tall they are. However, if you could use those digitized polygons to create three-dimensional blocks showing the heights of the buildings and use different textures and colors on the sides of those blocks to represent the physical appearance of each building, you would have a GIS dataset that would provide a more useful map. If you then added to the map some

3D objects, such as trees, cars, statues, or building signs, you would have an even better visual representation of the campus.

This module provides an introduction to creating three-dimensional representations from two-dimensional GIS data. It's becoming more common to view and analyze geospatial data in a 3D format, and Esri has produced numerous tools for visualizing and working with 3D data. For instance, Esri's **CityEngine** program is designed to generate realistic-looking 3D visualizations of entire cities quickly and efficiently (Figure 18.1). In this module, you won't be using anything as complex as CityEngine. Rather, you'll be working with a global scene in ArcGIS Pro for creating a 3D representation of GIS data.

FIGURE 18.1 An example of 3D design in GIS, using Esri CityEngine.

Throughout this module, keep in mind the previous modules' distinction between 2.5D and 3D data: A layer or an object can be considered 3D only if it can hold volume or contain multiple z-values for each x/y coordinate. For ease of use, this module refers to any objects that contain a third dimension as 3D, but keep in mind that many of them (such as the buildings you will be examining) are actually "pseudo-3D," or 2.5D.

CityEngine Powerful 3D design software that can quickly create large-scale 3D models (such as models of cities).

 ## Module Scenario and Applications

This module puts you in the role of a campus marketing manager at Youngstown State University. As part of a marketing program, your task is to create a new 3D version of the campus, as well as short videos to show off the 3D campus. Your goal is to start with a two-dimensional representation of the campus buildings and create a simple three-dimensional visualization of the data. You can then use your simple 3D campus model as a starting point for creating other campus maps (or creating more detailed 3D visualizations).

The following are additional examples of other real-world applications of this module's skills:

- The city council for a coastal tourist town has to make decisions regarding changes to the size and height of new resort developments on the beach. Rather than looking at a two-dimensional map or architectural diagrams, members of the council can use GIS to create three-dimensional views of the current building arrangement and of the appearance of the proposed changes. They can then assess the visual impacts of the decisions on the area.

- Environmental planners are examining potential locations for the construction of wind turbines for a region's alternative energy sources. They want to

construct a 3D visualization of the proposed sites to show a representation of the windmills (at their proper sizes and heights) in relation to the structures and land use of the surrounding areas.

- An archaeologist wants to create a 3D visualization of a Mayan village that is the subject of a dig site. Her goal is to make a 3D version of the site being excavated to show what the area looked like centuries ago. She will then use the visualization to present the findings from the site.

 ## Study Area

For this module, you will be examining a portion of the Youngstown State University (YSU) campus in Youngstown, Ohio.

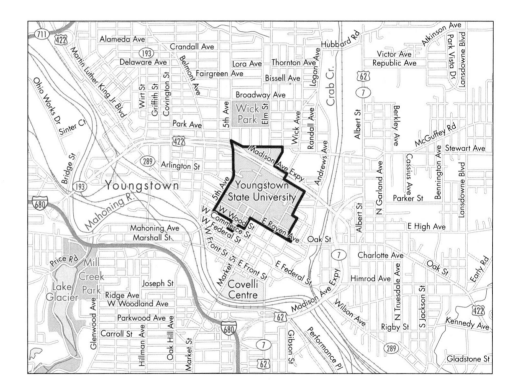

 ## Data Sources and Localizing This Module

The data for this module focus on features and locations on the YSU campus. However, you can easily modify this module to use data from your own campus or local area instead. By using the Imagery basemap, you can digitize a set of building footprints of your own campus (see Module 6), as well as points representing cars and trees. In the building footprints layer, you can create a new field called Height (using the Add Field techniques from Module 2) and add a value representing the heights of the buildings. Note that, for the purposes of this module, the heights of some buildings and features have been exaggerated or changed from their actual sizes to aid in visualization. To use exact values for building heights, you should consult blueprints or lidar height data or make field measurements of the buildings.

Step 18.1 Getting Started

- This module's hands-on applications use the data folder called Module18. Your instructor will be able to supply you with this data, or you can download it directly from this book's website at **https://www.macmillanlearning.com/college/us/product/Discovering-GIS-and-ArcGIS-Pro/p/131923075X**. The text in this module assumes that you have this Module18 folder in a computer location referenced as C:\GIS; if you have it somewhere else (for instance, in a flash drive referenced as G:\GISClass), substitute that location and path to the Module18 folder throughout this module.

- The Module18 folder contains a COLLADA file called Butler (no 4).dae. This is a 3D model of the Butler Institute of American Art created using SketchUp and converted to COLLADA format. It was created by members of the YSU 3D Campus Model team. (See the Preface for more information about this team and the names of the students involved.)

- The Module18 folder contains a file geodatabase called Campus3D that contains the following items:
  - bridge: a feature class containing a simplified digitized polygon of the bridge crossing Wick Avenue from the M1 parking deck.
  - cars1, cars2, cars3, and trees: four point feature classes, each representing digitized points.
  - church: a polygon feature class representing a simplified outline of St. John's Church.
  - campusbldgutm: a polygon feature class with digitized polygons of YSU campus buildings that has been projected into UTM Zone 17 for use in this module. (The original version of these campus data was created by YSU students Rob Carter, Paul Gromen, Sam Mancino, and Jaime Webber.)
  - museum: a polygon feature class containing a simplified version of the Butler Institute of American Art.

- Start ArcGIS Pro.
- Sign in with your Esri account username and password.
- Create a new project using the **Local Scene** template. Call this project **Module18** and place it in your **C:\GIS\Module18** folder. Ensure that there is not a checkmark in the box next to **Create a new folder for this project**.
- When ArcGIS Pro opens, change this scene's name to **YSU**.
- Ensure that you have access to the **3D Analyst** extension in ArcGIS Pro. See Smartbox 11.1 and Troublebox 11.1 in Module 11 for more information on how to set up this extension.

Step 18.2 Extruding Polygon Layers

- Add the campusbldgutm feature class to ArcGIS Pro. It should be shown in the Contents pane as a 2D layer. You can see the building footprints on the surface of the global imagery.
- Use the navigator to tilt the scene so you're looking at a perspective view of the campus. Use the Navigate and Zoom controls to look around the campus area. You can see the polygons representing the

buildings draped over the default ArcGIS Pro terrain, but they are shown flat. To show them as pseudo-3D buildings, you need to extrude the building footprints into blocks. For more information about extrusion, see **Smartbox 18.1**.

SMARTBOX 18.1

What is extrusion and how does it work in ArcGIS Pro?

extrusion A GIS technique used to give an object height.

Extrusion allows you to give a third dimension to objects by extending a two-dimensional item to a height. For example, a polygon might show a 2D view of a house, and an extruded polygon might show the polygon at 12 feet high. By assigning an attribute value for a height to each of a number of polygons and then extruding the polygons to that height, you can visualize the polygons in 3D (Figure 18.2).

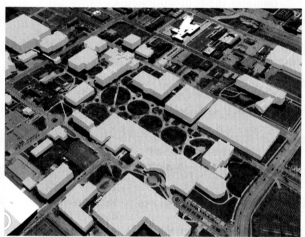

FIGURE 18.2 Flat 2D building footprint polygons (a) and their pseudo-3D extruded versions (b).

All three vector objects can be extruded. The extrusion process turns points into poles, lines into walls, and polygons into blocks. ArcGIS Pro offers several different types of extrusion methods:

- Base height: This method adds the extrusion value to the base height of the underlying terrain. Note that, as in Module 16, the height determination used here is based on the horizontal coordinate system; for example, if the layer's coordinate system is measured in feet, feet will be used for the extrusion height, and if the layer's coordinate system is measured in meters, meters will be used for the extrusion height.

- Minimum height: This method adds the extrusion value to the smallest available height value and extrudes up from there.

- Maximum height: This method adds the extrusion value to the highest available height value and extrudes up from there.

- Absolute height: This method uses the extrusion value as the height to extrude the objects to but does not incorporate base heights into the equation (and thus gives objects flat tops).

- On the **Appearance** tab, in the **Extrusion** group, click the **Type** button and choose **Base Height**.
- Also in the Extrusion group, make sure the Unit pull-down menu is set to **Meters** and then, from the **Field** pull-down menu, choose **HMeters** (which is the field containing the z-value the polygons will be extruded to).
- The campus building polygons are now extruded, and as a result, campusbldgutm is now displayed in the Contents pane as a 3D layer. Use the navigator to look around campus and examine the appearance of the extruded buildings. Answer Question 18.1.

> **Question 18.1** Why do many of the campus buildings look so strange, as if they're crooked or distorted?

- On the **Appearance** tab, in the **Extrusion** group, click the **Type** button and choose **Min Height**. The extruded appearance of the buildings changes. Use the navigator tool to focus on the buildings in the southeastern portion of campus and answer Question 18.2.

> **Question 18.2** What appearance do the buildings in the southeastern portion of campus have, and why do they appear this way?

- On the **Appearance** tab, in the **Extrusion** group, click the **Type** button and choose **Max Height**. The extruded appearance of the buildings changes again.
- This 3D model is a basic, simplified representation of YSU's campus shown in ArcGIS Pro. The heights of buildings are approximated, and the polygons are extruded to those heights. Use the navigator to move around the 3D campus and get familiar with the polygons that represent various campus features. Answer Question 18.3.

> **Question 18.3** Focus on the bleacher seating for Stambaugh Stadium (the block on the west side of the football field). Why does the stadium look like this rather than the way stadium seating should look? How would you correct this in ArcGIS Pro?

Step 18.3 Working with Multipatches in ArcGIS Pro

- Right now you have a set of extruded buildings. However, as noted at the beginning of this module, your extruded buildings are really only pseudo-3D, or 2.5D. In order to work with full 3D models, your buildings also need to be 3D. To work with true 3D objects and structures in ArcGIS Pro, you use a different storage format, referred to as a *multipatch*. (See **Smartbox 18.2** for more about working with multipatch data in ArcGIS Pro.) In order to progress through the more advanced steps of the remainder of this module, you'll convert your buildings and bridge into multipatches. Open the Geoprocessing pane. The tool you will be using is Layer 3D To Feature Class. You can either search for it or find it directly from the **3D Analyst Tools** toolbox, in the **Conversion** toolset, and then within the **From TIN** toolset. Using either method, open the **Layer 3D To Feature Class (3D Analyst)** tool.

SMARTBOX 18.2

What is a multipatch and how is it used in ArcGIS Pro?

multipatch An ArcGIS Pro data format that allows for fully 3D objects and structures to be represented.

A **multipatch** is an ArcGIS Pro data format that allows for a true 3D representation of an object (such as a polygon representing a building), as opposed to 2.5D. A multipatch object can hold volume and can contain multiple z-values for each x and y coordinate. Its name is derived from how the boundaries of a 3D object are saved in one row of the GIS database: as a collection of patches comprising the object.

In ArcGIS Pro, multipatches are used to make full 3D representations of objects, and they are also used for publishing 3D objects to the Web, creating more detailed or textured objects, and 3D editing and design. As a result, when you want to go beyond basic pseudo-3D visualization of data in ArcGIS Pro, you often end up converting your data into a multipatch version.

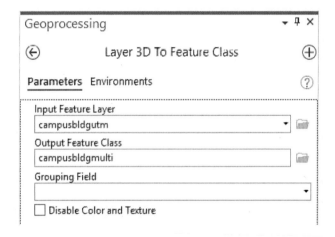

- For Input Feature Layer, choose **campusbldgutm**.
- For Output Feature Class, click the **browse** button and navigate to your **C:\GIS\Module18** folder, go into the **Module18.gdb** geodatabase, and name the output feature class **campusbldgmulti**.
- Run the tool. You now see a new set of buildings added to the scene. They should look very similar to the extruded campus buildings but a bit smoother in appearance. These are multipatch versions of the campus buildings, and you'll be working with them for the remainder of the module, so you can remove the original campusbldgutm from the Contents pane.
- Notice that campusbldgmulti is now listed under 3D Layers in the Contents pane. This multipatch version is listed here because it is a full 3D representation rather than extruded 2D layers.

Step 18.4 Creating an Interactive 3D Web Scene

- Now that you have a set of multipatches of the buildings, you can place them in an environment that can be used by everyone as part of the presentation you'll be making in this module's scenario. To do so, you'll be creating a virtual tour around the city and creating a web scene for delivery. For more information about web scenes, see **Smartbox 18.3**.
- The next thing to do is to set up the places you want to highlight during the virtual tour. Think of the ArcGIS Pro view as a camera: Whatever you

SMARTBOX 18.3

What is a web scene in ArcGIS Pro?

One of the big advantages of sharing your results through a cloud application like ArcGIS Online is that they are accessible by anyone who has a web browser. End users don't need to be running ArcGIS Pro on their computer, and they don't even have to have experience working with ArcGIS Pro; they only need a browser and an Internet connection. For multiple modules now, you've been able to share only the results of two-dimensional analysis through ArcGIS Online as web maps, but with a web scene, you can share three-dimensional results. A web scene is a format that allows results shown in ArcGIS Pro to be uploaded to ArcGIS Online. By converting your results to a web scene, you can allow anyone to interactively view your ArcGIS Pro data by using only a web browser.

A **web scene** is an item uploaded to ArcGIS Online that can be shared with the members of your organization or anyone using ArcGIS Online. A web scene can be viewed in any web browser that is WebGL compatible. **WebGL** is a standard for viewing 3D graphics in a browser without having to install a plug-in. (Compatible WebGL browsers include Google Chrome, Mozilla Firefox, and Safari, as well as many mobile device browsers.) The properties of the scene document in ArcGIS Pro (such as the viewing extent or vertical exaggeration) are saved to the web scene for viewing. By setting up bookmarks in ArcGIS Pro, you can create a series of preset views of key areas that will also be carried over to the web scene viewing format.

web scene An interactive version of an ArcGIS Pro scene that can be viewed and used through a web browser.

WebGL Web Graphics Library, a web standard for viewing 3D graphics.

focus the view on will be one "snapshot" of the camera. By taking several of these snapshots, you can set up the places you want to automatically visit on the virtual tour without having to manually navigate to each one. You do this by setting up a series of bookmarks in ArcGIS Pro, and these bookmarks will act as the snapshots of the camera. To begin, use the navigator to change the camera view to a part of the city you want to highlight on your virtual tour. Some potential suggestions for the Youngstown area include the YSU football stadium, the buildings on the edges of campus, a view of the entirety of campus, and a view of the greenery in the center of campus.

- When you have the camera set up the way you want it, on the **Map** tab, in the **Navigate** group, click the **Bookmarks** button and then choose **New Bookmark**. In the dialog box that appears, give the new bookmark an appropriate descriptive name and click **OK**. Back on the **Map** tab, in the **Bookmarks** pull-down menu, you now see a new entry for the bookmark you just created. Selecting this option causes the camera to switch to that particular view.

- Set up four additional bookmarks highlighting interesting places to showcase during the virtual tour and give each bookmark an appropriate descriptive name. These bookmarks will be all the places that will also carry over to your web scene.

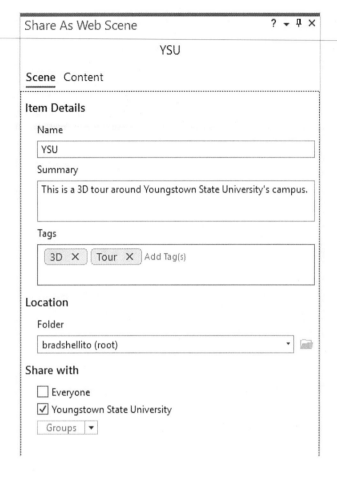

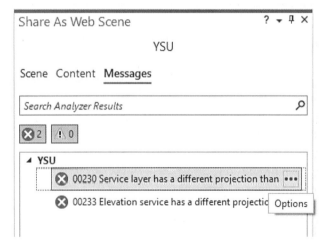

- Save your Module18 project.
- To create the web scene, on the **Share** tab, in the **Share As** group, click the **Web Scene** button. The Share As Web Scene pane opens.
- For Name, type in **YSU** (as that is the name of the ArcGIS Pro scene being converted to a web scene).
- For Summary, type an appropriate description of the purpose of the web scene.
- For Tags, type **3D** and **Tour**.
- For Location, choose your root folder from the pull-down menu. (This is the default folder associated with your Esri account from when you signed in back in Step 18.1.)
- Under Share with, place a checkmark in the box corresponding to your organization. (In the graphic to the left, it is Youngstown State University, but your own organization will be something else.)
- Click **Analyze** at the bottom of the pane, and ArcGIS Pro scans for errors.
- ArcGIS Pro does not allow you to publish anything that has errors in it (indicated under Messages with red-and-white X icons). The errors need to be fixed before you can publish the scene and its contents as a web scene. To start fixing them, hover your cursor over the right side of the first error, and you see three dots with the word Options. This is ArcGIS Pro's way of trying to find a solution to the problem. Click on the **three dots**.
- You now see a pop-up in which ArcGIS Pro suggests a means of fixing this error. Click on **Update Map to Use Basemap's Coordinate System**.
- Although there are still errors, click **Analyze** again in the Share As Web Scene pane. You now see that repairing the first error also fixed the remainder of the errors. Click **Share** at the bottom of the pane. ArcGIS Pro converts the scene to a web scene and uploads it to ArcGIS Online.
- It may take a few minutes for ArcGIS Pro to upload the web scene. After you receive the message that the scene has been successfully shared, click on **Jobs** at the bottom of the Share As Web Scene pane to make sure the layers have been successfully published and cached before proceeding (because the web layers continue to be prepared for proper display and usage in the cloud after being uploaded). Under the Jobs option, you see two new items appear

in the pane. When they both have green checkmarks next to them, your web scene has been successfully published and is ready to go. If you don't see green checkmarks but see other symbols instead, see **Troublebox 18.1** for suggestions on how to finish publishing your web scene.

TROUBLEBOX 18.1

How can I successfully publish a web scene?

Publishing a web scene involves transferring your files from ArcGIS Pro to ArcGIS Online and then publishing those layers as web layers. If you do not see green checkmarks but instead see different symbols, such as yellow triangles or red-and-white X icons, be sure that you are logged in to your Esri account on ArcGIS Online. Next, try running the Share As Web Scene tool again but this time using a different name (such as YSUa or YSUb). If the errors persist, save your project, close ArcGIS Pro, and reopen the project and again attempt to use the Share As Web Scene tool.

Step 18.5 Working with Web Scenes in ArcGIS Online

- Now that you have a web scene created, you can access it via ArcGIS Online. Open a web browser, go to the ArcGIS Online website (**https://www.arcgis.com**), and log in with your Esri account. (See Module 4 for more about logging into ArcGIS Online.) Note that you can work with a web scene by using either the free version of ArcGIS Pro Online or the organization-level version.

- When you are logged in, click **Content**.

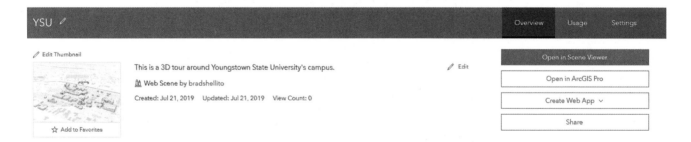

- You now see an option for the YSU file listed as a web scene. Click on it to open the main page for the YSU web scene. Click on **Open in Scene Viewer** to view your web scene. The YSU web scene opens in the browser.

- Under the Layers option, make sure there is a checkmark in the campusbldgmulti box to display the multipatch representation of campus. For more about the controls for using the web scene, navigating it, and using the bookmarks, see **Smartbox 18.4**.

SMARTBOX 18.4

How do I work with a web scene in ArcGIS Online?

There are several controls for working with web scenes in ArcGIS Online (Figure 18.3).

FIGURE 18.3 The basic layout of a web scene in ArcGIS Online.

The buttons on the left side of the screen operate as follows:

- The **Modify scene** button opens a side pane in the web scene that allows you to make changes to the scene, such as editing the scene's properties and adding new layers to the scene.

- The **Initial view** button resets the view of the scene to its starting point.

- The **Zoom in (plus)** and **Zoom out (minus)** buttons allow you to zoom in and out from the center of the view.

- The **Pan** button allows you to use the mouse to shift the view backward and forward.

- The **Rotate** button allows you to tilt the scene to examine it in 3D or perspective view.

- The **Compass** button shows you what direction you are facing. The red arrow indicates north, and you can click it to reset the direction to north.

The large icons at the bottom of the screen operate as follows:

- Clicking each of the **Bookmark** buttons moves the view to center on that spot. Bookmarks are premade items created before you export to a web scene.

The buttons on the right side of the screen operate as follows:

- The **Search** button allows you to locate a place or an address in the scene.

- The **Layers/Legend** button activates a pop-up window that allows you to turn scene layers on and off and also view the legend for each of the layers.

- The **Basemap** button allows you to add a basemap (see Module 4) to the scene as well as change the transparency of the ground layer.

- The **Daylight** button allows you to model Sun conditions for the scene.

- The **Measure** button allows you to work with the Measure tool (see Module 8) in the scene.

- The **Share** button allows you to share the scene with others.

- The **Settings** button allows you to adjust the mouse navigation and optimize the graphical display.

- Navigate around the web scene and click on the **Bookmarks** buttons to move about. After trying out your web scene, answer Question 18.4.

> **Question 18.4**
> Submit the URL of your final web scene to your instructor, who will check over your work for completeness and accuracy to make sure you get credit for this question.

Step 18.6 Working with 3D Editing in ArcGIS Pro

- Minimize your web browser and return to ArcGIS Pro. Next, you'll be making your basic version of campus look more realistic by using several GIS techniques. First, as you'll be creating different 3D visualizations of campus to work with, you'll want to obtain the best-quality imagery available, so you will want to use World Imagery Clarity (as in Modules 6 and 7). To do so, on the **Map** tab, in the **Layer** group, click the **Add Data** button.

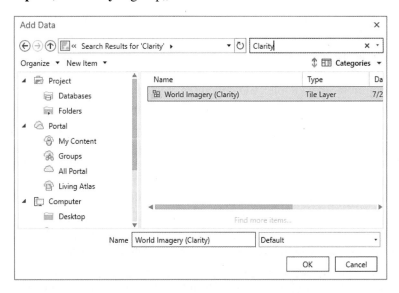

- In the dialog that appears, click on Living Atlas and then, in the search box in the upper-right corner, type **Clarity** and press **Enter**. Choose the option **World Imagery (Clarity)** and click **OK**. The Clarity imagery basemap is added to the view.

- You still have a campus composed of solid-colored buildings. While you could change the colors of the different buildings, as a campus marketing person, you'd probably want to have a campus visualization of something that really looks like the campus. For instance, buildings should have doors and windows; you might also want to give them brick textures and make them look like real buildings in other ways. To make buildings look more realistic in ArcGIS Pro, you can create the buildings with another 3D design software package and import the buildings into ArcGIS Pro. *Related Concepts for Module 18* discusses different ways of integrating other 3D design software into ArcGIS Pro, but for now, you can use the COLLADA file provided in the Module 18 folder, which provides a realistic-looking version of a campus building. For more information about COLLADA, see **Smartbox 18.5**.

SMARTBOX 18.5

What is COLLADA and how is it used with 3D visualization in ArcGIS Pro?

While ArcGIS Pro can do a lot with 3D visualization, many different types of specialized 3D design software are much better suited to crafting detailed, realistic 3D structures. Unfortunately, many types of specialized 3D models are not directly compatible with ArcGIS Pro. However, you can easily integrate 3D designs into your ArcGIS Pro visualizations by exporting the models to the **COLLADA** format.

COLLADA Collaborative Design Activity. An XML-based file type (with the extension .dae) that works as an interchange format from one kind of 3D modeling file type to another.

COLLADA (Collaborative Design Activity) acts as an intermediate file format that allows 3D models to move from one platform to another. For example, you could design a house as a 3D model using a 3D modeling program, convert the model to COLLADA format, and then use the COLLADA version of the model in ArcGIS Pro. COLLADA is a common format that many 3D design software packages can export to, and it allows you to move easily from one software package to another.

ArcGIS Pro uses COLLADA files in conjunction with the multipatch file format. Therefore, the complex 3D structures created as COLLADA files can be represented in ArcGIS Pro as multipatches. For instance, you would first convert your extruded buildings to multipatches. Then, using the editing tools of ArcGIS Pro, you could select one of the multipatch buildings and exchange the multipatch blocks that comprise a building with the COLLADA version of that building created in another 3D design software. ArcGIS Pro then has the fully visualized version of the building stored as a multipatch. By going through this process for all of the campus buildings, you can create a rich, detailed, 3D version of the campus.

- Add the museum feature class from the campus3D.gdb geodatabase to the Contents pane of the scene. Extrude its polygons by using the **Max Height** option and the **Height** attribute and using **US Feet** for the units. Set the elevation so that the museum sits on the ground. Change the symbology so that the museum polygon is a different color from the rest of campus. You should now have a simplistic digitized boundary of the Butler Art Museum on campus.

- You are going to replace your basic extruded polygon with a detailed 3D COLLADA model. To do so, you need to convert this polygon to a multipatch. Open the Geoprocessing pane. The tool you will be using is Layer 3D To Feature Class. You can either search for it or find it directly from the **3D Analyst Tools** toolbox, in the **Conversion** toolset, then within the **From TIN** toolset. Using either method, open the **Layer 3D To Feature Class (3D Analyst)** tool.

- For Input Feature Layer choose **museum**.

- For Output Feature Class, click the **browse** button and navigate to your **Module18** folder, select the **Module18.gdb** file geodatabase, and name the file **museummulti**.

- Leave the other options alone and run the tool.

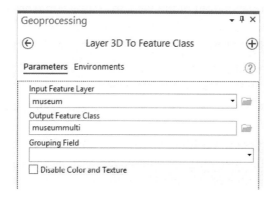

- When the museummulti layer is added to the Contents pane, turn off the old museum layer. (Although the two layers look similar, remember that museummulti is the multipatch version of the extruded polygon museum layer.)

- Now you can replace the museummulti layer with a COLLADA file containing a detailed 3D model of the museum. On the **Edit** tab, in the **Features** group, click the **Modify** button. The Modify Features pane opens. In this pane, under the **Reshape** heading, click the **Replace Multipatch** button.

  - Your cursor switches to a four-pointed symbol cursor (see Module 7). In the scene, click on the **museummulti** multipatch object. The multipatch is selected, and a new Browse 3D Models dialog box opens. Navigate to your Module18 folder and choose the **Butler.dae** COLLADA file and click **OK**. The museummulti layer disappears and is replaced with the 3D COLLADA model version of the museum. (Note that this process may take a few minutes.)

- The COLLADA file may not have the same scale as the museum. (You can see onscreen whether the dimensions of the COLLADA file conform or do not conform to the boundaries shown on the imagery basemap.) If you need to adjust the location of the COLLADA file, in the Modify Features pane, under the **Alignment** heading, click the **Move** button. The cursor

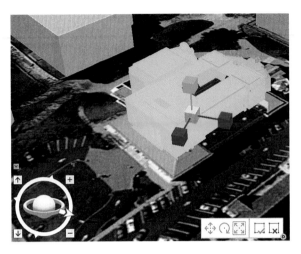

changes into a four-pointed symbol cursor, and you can grab and drag the COLLADA model in the scene.

- If you want to change the size and dimension of the COLLADA file, in the Modify Features pane, under the **Alignment** heading, click the **Scale** button.

- The cursor again turns into the four-pointed symbol cursor. Click on the COLLADA file of the museum, and new items appear. The colored boxes allow you to change the scale; use the mouse to grab a blue or red box to change the scale along the x-axis and use the mouse to grab a green box to change the vertical scale. Carefully adjust the scale of the museum. As you do, be sure to change it equally along both directions on the horizontal axis to avoid skewing the model. You may need to adjust the vertical axis as well. When you have adjusted the model until you like it, click the box with the green checkmark at the bottom of the view to complete the process; if you're unhappy with your result, click the box with the black X to cancel the process and start over.

- When you are satisfied with the appearance of the COLLADA file, on the **Edit** tab, in the **Manage Edits** group, click the **Save** button to save your edits and changes. If ArcGIS Pro prompts you to save your edits, click **Yes**.

- Use the navigator to position the camera so that the viewer can see your COLLADA file matching up with the dimensions shown on the imagery. To create a graphic of this, on the **Share** tab, in the **Export** group, click on the **Map** button. In the Export Map dialog box, navigate to your **C:\GIS\Module18** folder, give the exported graphic and name, and save it as type **PNG**. Click **OK** to export the scene as a graphic and save it on your computer's hard drive. Answer Question 18.5.

Question 18.5 Submit to your instructor the graphic of your final ArcGIS Pro 3D rendering of the COLLADA file of the museum matching up with the imagery. Your instructor will check over your work for completeness and accuracy and make sure you get credit for this question.

Step 18.7 Using Offsets with Extrusion

- This section introduces the concepts of offsets and extrusion. The bridge feature class represents a simplified version of the bridge crossing Wick Avenue from the M1 parking deck to the main part of campus near Maag Library and the Butler Art Institute. Add this feature class from your campus3D geodatabase to the Contents pane. Focus your view in ArcGIS Pro on the area where the bridge crosses Wick Ave. By default, the bridge is draped on the terrain surface as a 2D layer.

- You need to extrude the bridge, so in the Contents pane, click once on the **bridge** feature class and then, on the **Appearance** tab, in the **Extrusion** group, click the **Type** button and choose **Max Height**. Also in the **Extrusion** group, make sure the Unit pull-down menu is set to **Meters**, and then from the Field pull-down menu, choose **Height** (which is the field containing the z-value the bridge polygon will be extruded to). The bridge polygon is extruded and displayed as a 3D layer. Answer Question 18.6.

> **Question 18.6** Why does the bridge look like a wall stretching from the parking deck through Wick Avenue?

- Next, you need to assign the bridge an offset value to make it "float" above the terrain. For more information about offsets, see **Smartbox 18.6**.

SMARTBOX 18.6

How does an offset function in ArcGIS Pro?

When visualizing geospatial data in 3D, there are often features that you don't want to extrude from the ground level up. Instead, you want those features to "float" above the ground's surface. For instance, an elevated train track would first need to be raised off the ground surface before more modeling could be done with it; the same would be needed with a glass skywalk.

You can make an object "float" in ArcGIS Pro by assigning an **offset** value to the feature. The feature is then placed in a floating position above the ground level at a height (in real-world units) equal to the value of the offset. If a monorail track should be 10 meters above the ground, assign the object representing the track an offset value of 10 with meters as the elevation units. ArcGIS Pro positions the object above the ground surface (Figure 18.4). You can then extrude the object or otherwise change its appearance.

offset A value that is used in ArcGIS Pro to raise an object above the ground surface.

FIGURE 18.4 An offset applied to the polygon representing the bridge "lifts" it above the terrain.

- Right-click on the **bridge** layer, choose **Properties**, and then choose the **Elevation** tab.

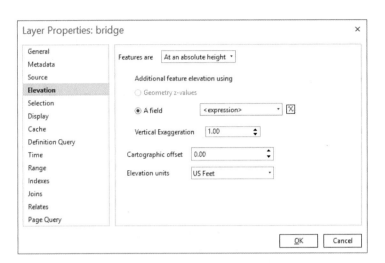

- For Features are, choose **At an absolute height** to position the bridge at a fixed height rather than extending the draped version from the surface.

- For Additional feature elevation using, choose the **A field** radio button.

- For Elevation units, choose **US Feet**.

- Click the green X button next to <expression> and type an expression to use in calculating the offset.

- In the Expression Builder dialog box that appears, delete the **0** that is in the Expression box. Next, double-click on the **Height** option under Fields to add [Height] to the Expression box. Next, click in the Expression box to the right of [Height] and type **+930**. This tells ArcGIS Pro to raise the bridge to a height of 930 units above the vertical datum. This means that the bridge will be offset at 930 feet from the zero point of the vertical datum, not 930 feet above the surface of the ground.

- Click on the green checkmark icon to make sure your expression is valid. When you see the message "Expression is valid," click **OK** in the dialog box.

- In the Layer Properties: bridge dialog box, click OK. The bridge is now set to the proper offset value and floats above the terrain surface. Answer Question 18.7.

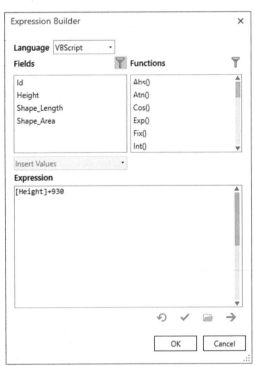

Question 18.7 One of the elements missing from this representation of YSU's campus is an elevated pedestrian bridge that connects two of the buildings (Moser Hall and Cushwa Hall). What steps would be involved in adding this object to the 3D scene?

Step 18.8 Using 3D Objects in ArcGIS Pro

- A university campus is more than just a bunch of buildings. As a campus marketing person, you would want to showcase some unique items around the campus, as well as provide a better representation of the campus's greenery and parking areas. For additional context, add the **church** feature class to the Contents pane. Extrude its polygons by using the **Max Height** option and the **Height** attribute, with feet for the units. Set the elevation so that the church sits on the ground. Change the symbology so that the church polygon is a different color from the rest of campus.

- Add the trees feature class to ArcGIS Pro as a 2D layer. You should see two points draped on the terrain to the northeast of the church.

- To change these 2D points in the trees layer to 3D objects, in the Contents pane use the mouse to move them under the heading **3D Layers**. ArcGIS Pro now interprets these points as a 3D layer. (See Module 15 for more information.)

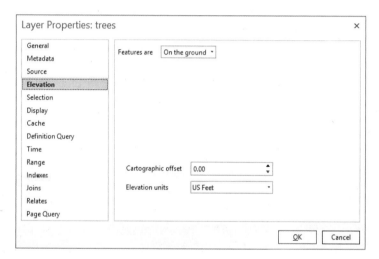

- In the Contents pane, right-click on the **trees** layer and choose **Properties**. On the **Elevation** tab, next to the Features are pull-down menu, choose **On the ground** and click **OK**. The 3D points appear on top of the basemap in the scene. Here you have essentially draped the 3D points on the default surface.

- Next, because you'll be displaying 3D objects that will resemble real-world trees, you'll want them to be displayed in real-world units and at a real-world scale. In the Contents pane, right-click on the **trees** layer and choose **Properties**. On the Display tab, place a checkmark in the box next to **Display 3D symbols in real-world units**.

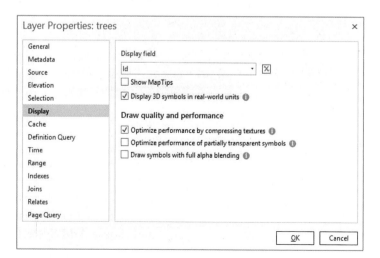

- Rather than displaying simple 3D points, you can now switch them over to 3D objects that resemble trees. (See **Smartbox 18.7** for more information.) To get a feeling for what type of trees these are, examine the shadows being cast by the trees onto the Clarity basemap imagery.

SMARTBOX 18.7

How can 3D objects be represented in ArcGIS Pro?

In ArcGIS Pro, when you have a point object on a map, you can change its symbology to take on a new appearance; for instance, you could represent a point object on a map as a 2D symbol such as a large red triangle, a green square, or a blue circle. In a scene environment, you can instead represent these points as 3D objects—for instance, a red pyramid, a green cube, or a blue sphere—by using different 3D system styles.

ArcGIS Pro contains numerous premade system styles that can be used to represent all manner of features, including different types of cars and other vehicles, trees, road signs, construction equipment, and a multitude of other objects. You could use 3D objects to represent an entire layer of points. For instance, Figure 18.5 shows how a set of digitized points that started as small 3D spheres can have this symbology replaced with realistic-looking 3D trees. You can change the symbology of points to any of the premade system styles.

FIGURE 18.5 You can create 3D objects in ArcGIS Pro by digitizing a set of points (a) and then replacing those points with 3D system styles (b).

- If you want the points to resemble trees, you have to add 3D symbology for the trees. On the **Insert** tab, in the **Styles** group, click the **Add** button and choose **Add System Style**.
- In the System Styles dialog box, in the left-hand pane, expand the **3D** option. The System Styles dialog box shows you the many different preset

symbologies you can use for your 3D objects. For now, place a checkmark in the box for **3D Vegetation - Realistic**. This option is added to the right-hand pane as one of the system styles in the project. If a system style is listed in the right-hand pane, you will have access to it in the current project. Click **OK** to close the dialog box.

- To change the symbology of the 3D points to resemble 3D trees, back in the Contents pane, choose the **trees** layer and then, on the **Appearance** tab, in the **Drawing** group, click the **Symbology** button and choose **Single Symbol**.

- The Symbology pane opens. Double-click on the point symbol that represents the current symbology of your 3D points and choose the **Gallery** option. Scroll through the options and expand the choices for **3D Vegetation - Realistic**. You see that you have multiple options for trees. Choose a tree symbol that corresponds to the shadows of the trees being cast on the Clarity basemap. After you choose a style, the points in the scene change to resemble 3D versions of the trees you chose. Consider your available options and answer Question 18.8.

Question 18.8: Which tree type did you choose to best fit the scene and the imagery?

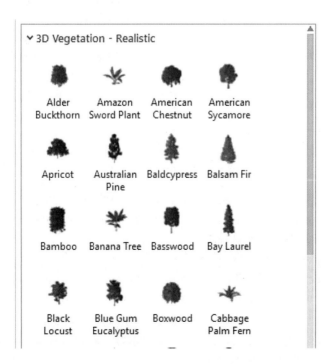

- Next, you need to change the size of the 3D trees so that they are properly scaled in relation to the rest of the scene. In the Symbology pane, click on

Properties. The pane changes to display a new set of options that allow you to alter the size of the 3D trees, the angle of their rotation, and more. For instance, to change the height, type a number into the Size box and press the **Enter** key and then click **Apply**. ArcGIS Pro alters the size of the 3D tree objects. Try some options for the size until you get your chosen trees to suit the scene. Answer Question 18.9.

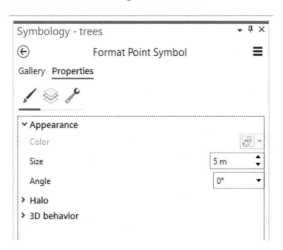

Question 18.9 Which real-world height did you choose for the 3D trees to best suit the scene and the imagery?

- To add more 3D objects, add the following feature classes to ArcGIS Pro from the campus3D geodatabase: **cars1**, **cars2**, and **cars3**. The points in these classes represent the locations of three different rows of cars in the St. John's parking lot. On the imagery, you want each of these sets of points to be represented as cars that will "park" in the striped parking places and be displayed as similar car types, colors, and sizes. To do so, for each of the three layers:
  - Move the layer in the Contents pane into the 3D Layers section.
  - Change the layer's elevation so that the objects are on the ground, as you did with the trees layer.
  - Change the layer's display properties to use real-world units.
- You should now see sets of 3D points at various locations draped on the imagery in the parking lot.
- From the Add System Styles menu, use the option **3D Transportation** to get access to available symbologies for appropriate-looking cars. Choose an appropriate type of car for each of the three rows in the church parking lot.
- To have each car "cover" its parking space, you need to adjust the angle at which the 3D cars are parked. To do this, return to the Symbology pane and the Properties option for each car type and adjust the values for **Angle** to change the angle of the cars in the parking spaces. Also adjust the **Size**

setting for the cars so that the 3D objects "cover" their corresponding cars on the image of the parking lot. Keep playing with the options until you get the cars parked correctly in their spaces. Answer Question 18.10 (which will require some explanation on your part).

> **Question 18.10** What steps would you take to create a more realistic 3D representation with respect to (1) the church, (2) cars, and (3) trees for this area?

- When you have your view of the church set up the way you like it in terms of the representation of the trees, the three sets of cars, and the church itself, you can create a graphic of the scene. Use the navigator to position the camera to get a good view of these three elements in the scene.
- To create the graphic, on the **Share** tab, in the **Export** group, click on the **Map** button. In the Export Map dialog box, navigate to your **C:\GIS\Module18** folder, give the exported graphic a name, and save it as type **PNG**. Click **OK**, and the scene is exported as a graphic and saved on your computer's hard drive. Answer Question 18.11.

> **Question 18.11** Submit the graphic of your final ArcGIS Pro 3D rendering of the church, trees, and cars to your instructor, who will check over your work for completeness and accuracy and make sure you get credit for this question.

Step 18.9 Saving and Sharing Your Results

- Save your Module18 project. You can submit the graphics and web scene URL you create in this process to your instructor to complete this module.

Closing Time

This module examined how geospatial data (such as trees, cars, bridges, and buildings) can be represented in 3D by using the ArcGIS Pro environment. Three-dimensional GIS visualization is an excellent method for communicating and presenting geospatial concepts. Much more can be done with 3D visualization in ArcGIS Pro. See *Related Concepts for Module 18* for information about creating more realistic 3D data to use in ArcGIS Pro.

Starting in the next module, you'll be working with measuring distances as a surface, which requires understanding and using many concepts for modeling and problem solving, particularly shortest paths over a landscape. You'll also be able to examine the results of your analysis in ArcGIS Pro as 3D visualizations, which are ideal for presenting the results of analyses. For example, you could create an interactive 3D version of the wetlands in an area slated for construction of a new shopping center or a 3D visualization of a shortest path over mountainous terrain; these 3D visualizations would have greater visual impact than simple printed maps.

RELATED CONCEPTS FOR MODULE 18

Working with More Detailed 3D Models in ArcGIS Pro

SketchUp 3D design software owned and distributed by Trimble.

KMZ A compressed version of a KML file.

KML Keyhole Markup Language, an XML-based file format that can hold georeferenced information and that is the file format used by Google Earth.

As noted in Smartbox 18.5, there are several 3D design programs available for creating more detailed models to use (in COLLADA format) in ArcGIS Pro. An easy-to-use design program is **SketchUp**. (To download the free version called SketchUp Make, visit **https://help.sketchup.com/en/downloading-older-versions**.) With SketchUp, you can start with georeferenced imagery, digitize building footprints, extrude and shape polygons, and add realistic textures and other features to create rich 3D visualizations of buildings. SketchUp enables you to create much more detailed and textured 3D representations of buildings and objects than you can create with simple extrusion or symbology changes. SketchUp also gives you the option to start with a photo or map outline of a building and use the photo's dimensions to create a 3D model. SketchUp models can be as basic or complex as you want them to be. You can also create standalone 3D objects such as trees, cars, sculptures, and statues. SketchUp files cannot be directly read into ArcGIS Pro but can easily be exported into COLLADA format and then used in ArcGIS Pro, as shown earlier in this module.

Another format that can be used for working with detailed 3D models in ArcGIS Pro is **KMZ**. Files in KMZ format are compressed versions of the **KML** (Keyhole Markup Language) file format that is used with data in Google Earth. KML is an XML-based format that can contain georeferenced information. KMZ (or KML) files can be added directly to Google Earth to show additional overlays or layers or to add 3D representations of objects to Google Earth. Three-dimensional buildings or structures (everything from baseball fields to the Eiffel Tower) can be constructed in 3D design software, converted to KML or KMZ, and then placed into Google Earth. ArcGIS Pro can also easily read KML or KMZ files and display them in a scene (Figure 18.6).

FIGURE 18.6 A KMZ version of a 3D model of YSU's Stambaugh Stadium in ArcGIS Pro.

Key Terms

CityEngine (p. 413)
extrusion (p. 416)
multipatch (p. 418)
web scene (p. 419)

WebGL (p. 419)
COLLADA (p. 424)
offset (p. 427)
SketchUp (p. 434)

KMZ (p. 434)
KML (p. 434)

How to Use Distance Calculations and Cost Distance in ArcGIS Pro

ArcGIS Pro Skills

In this module, you will learn how to do the following in ArcGIS Pro:
- Calculate a Euclidean distance surface.
- Calculate an allocation grid.
- Create a cost-distance surface (using different rasters for the various costs as different constraints on the problem).
- Calculate a least-cost path based on the different costs for traversing each cell.
- Reclassify the values of a raster to create a new raster with new values.
- Use the Slice tool to create a new raster based on a classification scheme.
- Examine the results of cost analysis in 3D.

Learning Outcomes

After studying this module, you should be able to:
- Describe how lateral and diagonal movements are made across a raster surface.
- Explain how the allocation procedure operates.
- Define the term *Euclidean distance*.
- Explain how the concept of distance can be measured by using a raster surface.
- Explain what a cost-weighted grid is.
- Describe the steps in the cost-weighted distance algorithm.
- Explain what a backlink grid is and what it consists of.
- Explain how a least-cost path over a raster surface is generated.

Introduction

So far, you've seen many types of spatial analysis that relate to proximity (how far certain things are away from other things). You've used spatial joins, selection by location, network analysis, and buffers as measures of proximity, but you haven't yet seen how to use ArcGIS Pro to find the distance between locations that are not on a transportation network (aside from small uses of the Measure tool and one type of spatial join).

Let's look at an example. Say that you want to determine the distance from each location in a park to the closest first-aid station. The tools you've used so far won't help you do that. What you need instead is some sort of a park map, with each location on the map marked to show the distance from that spot to the nearest first-aid station(s). However, a park has a nearly infinite number of locations—for instance, the park entrance, a spot 1 foot away from the park entrance, another spot 2 feet away from the entrance, and so on—and measurements need to be made from all of them to the closest first-aid station. You can think of a distance map as a continuous surface, where each location has a value denoting how far that location is from the nearest first-aid station. From previous modules, you know you can represent this kind of continuous surface as a raster, and that's how ArcGIS Pro represents distances—as raster surfaces, with each grid cell containing a value for the distance to the feature(s) you're trying to represent.

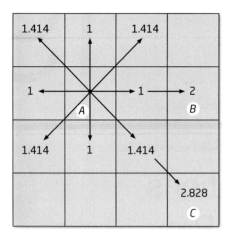

FIGURE 19.1 Lateral and diagonal movement (and accumulation of distance) across a raster surface.

You create a distance raster by determining how far each cell is from a feature (such as the first-aid stations). Distance in a raster can be calculated as movement across a surface of grid cells from an origin cell (or several origin cells) to all other cells in the raster. Because this movement is measured in square blocks, these movements can be either lateral or diagonal.

Figure 19.1 shows movement across a raster from an origin point (the cell marked *A*). Movements are made from the center of the origin cell to the center of an adjacent cell. Moving in a lateral direction (up, down, left, or right) to an adjacent cell costs 1 unit, and moving in a diagonal direction costs 1.414 units (because of the geometry measurements of the Pythagorean theorem). Moving to another cell adds an accumulative cost of either another 1 unit (for a lateral move) or another 1.414 units (for a diagonal move). In Figure 19.1, the distance from the origin to the grid cell marked *B* is 2 units, and the distance from the origin to the grid cell marked *C* is 2.828 units.

Each one of these grid cells represents geospatial data. That is, each grid cell represents a real-world size based on the raster resolution (see Module 12). For instance, if the raster in Figure 19.1 were a 30-meter resolution raster, the measured distance from the center of origin cell *A* to the center of destination cell *B* would be 60 meters (2 × 30 meters). Similarly, the measured distance from the center of origin cell *A* to the center of destination cell *C* would be 84.84 meters (2.828 × 30 meters). This module expands on this basic concept and shows how to use these kinds of distance measurements to determine the shortest path along a raster surface (such as finding the shortest distance from every grid cell location in a park to the grid cells representing the locations of the first-aid stations).

Module Scenario and Applications

In this module, you will be taking the role of a ranger in the Hocking Hills region of Hocking County, Ohio, trying to find the shortest path from the ranger station to a group of stranded hikers and also to a separate campsite emergency. In the analysis, you'll need to take into account certain conditions in the park environment that will influence the overland path you'll have to take to rescue the hikers and get to the emergency site. These conditions include the park's terrain and land cover.

The following are additional examples of other real-world applications of this module's skills:

- An engineer is attempting to determine the best location for a new road through unpaved terrain. He needs to take into account factors such as the slope of the landscape when figuring out the least-cost path of the new road.
- A park manager is laying out potential new trails for land acquired by a park to provide new routes for tourists and hikers. She can include factors such as the steepness of slopes and land cover when determining the locations (and difficulty ratings) of these new paths.
- A game-control warden at a national park wants to identify likely areas for animal migrations and movements within the

park's boundaries to aid in closing certain areas to tourists at certain times of the year. He can incorporate various landscape-based features of the park in determining the likely paths of animal movements.

 ## Study Area

- For this module, you will be working with data from the Hocking Hills State Park region, located in Hocking County, Ohio.

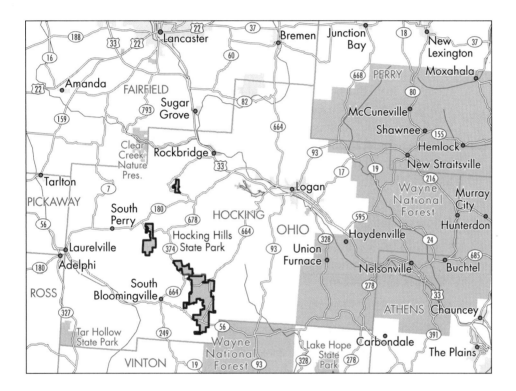

 ## Data Sources and Localizing This Module

The data for this module focus on features and locations within the Hocking Hills State Park region. However, you can easily modify this module to use data from your local area instead (for instance, Petoskey State Park in Michigan).

The 3DEP DEM was downloaded from the USGS National Map. The NLCD 2011 dataset was downloaded (for free) from the Multi-Resolution Land Characteristics Consortium (MRLC), which is online at **https://www.mrlc.gov**. You can download NLCD 2011 data for the entire United States from this website and then extract the data for your individual county (such as Emmet County, Michigan, where Petoskey State Park is located) by using ArcGIS Pro. Alternatively, you can download NLCD 2011 data for a county directly via The National Map. Both the DEM and NLCD grids for this module were extracted to fit the approximate boundaries of the study area.

The other layers of areas of the park were digitized using ArcGIS Pro and the maps and information available from the Ohio Department of Natural Resources at **http://parks.ohiodnr.gov/hockinghills**. Other state parks likely have similar maps and websites available through your state's DNR. For instance, maps and information about Michigan's Petoskey State Park are available through the Michigan Department of Natural Resources at **https://www2.dnr.state.mi.us/parksandtrails/details.aspx?id=483&type=SPRK**.

Step 19.1 Getting Started

- *Important note:* Starting with ArcGIS 2.5, there are many new distance tools and methods available for use. In ArcGIS 2.5, the tools used in this module are located in the **Distance** toolset, then within the **Legacy** toolset. For more information on using the new distance tools with this module, see the book's Website at: **https://www.macmillanlearning.com/college/us/product/Discovering-GIS-and-ArcGIS-Pro/p/131923075X**.

- This module's hands-on applications use the data folder called Module19. Your instructor will be able to supply you with this data, or you can download it directly from this book's website at **https://www.macmillanlearning.com/college/us/product/Discovering-GIS-and-ArcGIS-Pro/p/131923075X**. The text in this module assumes that you have this Module19 folder in a computer location referenced as C:\GIS; if you have it somewhere else (for instance, in a flash drive referenced as G:\GISClass), substitute that location and path to the Module19 folder throughout this module.

- Module19 contains a file geodatabase called hockinghills that contains the following items:
  - DEMhills: a 1 arc-second DEM 3DEP raster layer (resampled to 30-meter resolution for purposes of this module). The DEM elevation values are in meters.
  - NLCDhills: a 30-meter resolution NLCD 2011 raster layer.
  - Maskhills: a 30-meter resolution grid of the boundaries of the study area.
  - ranger: a point feature class of a digitized point showing the location of a ranger station (the state forest headquarters) in the Hocking Hills State Park.
  - hikers: a point feature class of a digitized point signifying the location of a group of stranded hikers in the region.
  - campers: a point feature class of a digitized point signifying the location of an emergency occurring at a campsite in the park.
  - bounds: a polygon feature class of a digitized polygon showing the approximate boundary of the Hocking Hills State Park areas.

- Start ArcGIS Pro.
- Sign in with your Esri account username and password.
- Create a new project using the **Map** template. Call this project **Module19** and place it in your **C:\GIS\Module19** folder. Ensure that there is not a checkmark in the box next to **Create a new folder for this project**.
- When ArcGIS Pro opens, change the map's name to **Hocking Hills**.
- You will be using the ArcGIS Pro Spatial Analyst and 3D Analyst extensions in this module. If you don't have access to these two

extensions, see Smartbox 11.1 and Troublebox 11.1 in Module 11 for more information on how to get them set up in ArcGIS Pro.

- Add the **NLCDhills** layer to the Contents pane. Change the color scheme of the zones in NLCDhills to be more intuitive and to use more distinguishable colors than the defaults (so that each class can be seen distinctively, but forests are in shades of green, water is in shades of blue, and so on). Refer to Module 12 for information about the land-cover classes the NLCD grid-cell values represent.

- Use the following geoprocessing environment settings for this module (on the **Analysis** tab, in the **Geoprocessing** group, click the **Environments** button):
  - For Current Workspace, use the **Module19.gdb** geodatabase from your **C:\GIS\Module19** folder.
  - For Scratch Workspace, use the **Module19.gdb** geodatabase from your **C:\GIS\Module19** folder.
  - For Output Coordinate System, choose **NLCDhills**. ArcGIS Pro updates the coordinate system to that of the NLCDhills layer.
  - For Extent, choose **NLCDhills**. ArcGIS Pro updates the x and y coordinates of the lower-left corner and upper-right corner of the grid extent to those of the NLCDhills layer.
  - For Cell Size, choose **NLCDhills**. ArcGIS Pro updates the cell size for raster output in this project to 30 meters, as that is the cell size of the NLCDhills layer.

- Leave the other settings alone and click **OK** to put these geoprocessing environment settings into place.

- The NLCDhills layer was projected from its initial coordinate system to another one for use in this module. The coordinate system being used in this module is **NAD 1983 (Albers)**, using meters as the map units. Check the Hocking Hills map properties (under the Coordinate System tab) to verify that this projection system is being used.

Step 19.2 Calculating a Euclidean Distance Surface

- In this module you are working at a ranger station in the Hocking Hills region, and you receive a call that hikers are stranded somewhere in the region. You want to calculate the shortest overland path to reach the hikers from the ranger station.

- To get started, add the **DEMhills** layer to the Contents pane.

- Add the **ranger** point feature class and the **hikers** point feature class layers to ArcGIS Pro. Each of these layers consists of a single point, and these points may be difficult to see at a wide scale. Adjust the colors and symbology of each layer and zoom in to see where the two points (the ranger station and the stranded hikers) are located.

- The first calculation you'll make is a Euclidean distance calculation to determine the accumulated cost distance between the ranger station and the hikers. See **Smartbox 19.1** for more about using Euclidean distance in ArcGIS Pro.

SMARTBOX 19.1

How is Euclidean distance calculated and used in ArcGIS Pro?

Euclidean distance The straight-line distance between two points.

A **Euclidean distance** measurement represents the straight-line distance between two points. In Module 11, you looked at accumulating the driving distance along many roads/edges in a network between two destinations. Euclidean distance is more like taking a helicopter directly from one point to another. The old saying about "the shortest distance between two points is a straight line" is what Euclidean distance measurements are all about.

In ArcGIS Pro, Euclidean distance is applied as the measurement of a straight line between the centers of two grid cells. The Euclidean Distance tool calculates this straight-line distance from each source cell to all other cells in the raster, and each cell is assigned the shortest distance from a source. However, if the source from which you're trying to find Euclidean distance is a point, line, or polygon feature class, those features are first converted to raster format (at whatever cell resolution is being used in the distance calculation). For instance, if you have a layer of all the lakes in the county and want to find how far each location in the county is from a lake, you can use the Euclidean Distance tool to convert the lake polygons to raster format and then compute the straight-line distance from the cells that comprise the lakes to all other cells within the extent. The output of the Euclidean Distance tool is a new surface in which every cell has a real-world value for the shortest straight-line distance from the source features/cells to each other cell in the raster.

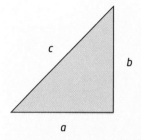

FIGURE 19.2 The basic setup of a triangle using the Pythagorean theorem.

As mentioned earlier in this module, because grid cells are square and distances are calculated from the centers of cells, the Pythagorean theorem is used to determine the Euclidean distance between cells. The Pythagorean theorem measures the sides and the hypotenuse (the diagonal) of a right-angle triangle as $c^2 = a^2 + b^2$. Consider Figure 19.2. If the two sides of the triangle (a and b) are both equal to 1, the hypotenuse of the triangle (c) has a value of 1.414 ($c^2 = 1^2 + 1^2$, or $c = \sqrt{2}$, or 1.414). The application of the Pythagorean theorem makes it clear why lateral movement on a raster has a value of 1, while a diagonal movement has a value of 1.414.

This same basic measurement is applied when calculating Euclidean distances between grid cells. For example, say that in Figure 19.3 you want to determine the shortest distance between the source cell (X) and every other cell in the raster. To find the distance between cell X and cell Y, you use the Pythagorean theorem as follows: a (the bottom of the triangle) has a value of 3 (because it is three units between the centers of the grid cells on the bottom of the triangle), and b has a value of 2 (because it is two units between the centers of the grid cells on the side of the triangle). Thus, the value of c would be found as $(c^2 = a^2 + b^2)$, or $(c^2 = 3^2 + 2^2)$, or $(c^2 = 13)$.

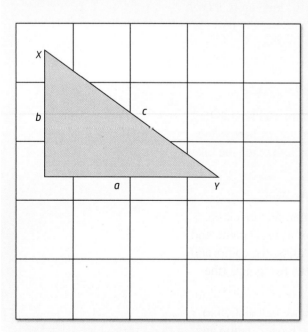

FIGURE 19.3 An example of Euclidean distance calculation of raster distance.

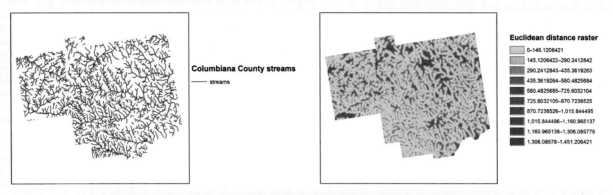

FIGURE 19.4 A feature class of Columbiana County, Ohio, streams and a raster surface of Euclidean distance from each stream.

Thus, *c* equals the square root of 13, or 3.61. The shortest distance between cell *X* and cell *Y* would thus be 3.61 units. As noted previously, the raster resolution determines the real-world size of the grid cells, so if the raster in Figure 19.3 were 30-meter resolution, the real-world Euclidean distance between cell *X* and cell *Y* would be 3.61 × 30 meters, or 108.3 meters.

For an example of calculating a Euclidean distance surface in ArcGIS Pro, see Figure 19.4. An initial feature class representing streams in Columbiana County, Ohio, is used as the input to the Euclidean Distance tool. The resulting output is a raster surface in which every grid cell has a value for the straight-line distance from any stream in the county. The raster surface's values reflect the real-world measurements of distance, based on the specified grid-cell size used by the tool.

- Open the Geoprocessing pane (**Analysis** tab, **Geoprocessing** group, **Tools** button). The tool you will be using is Euclidean Distance. You can either search for it or find it directly from the **Spatial Analyst Tools** toolbox, in the **Distance** toolset. Using either method, open the **Euclidean Distance** tool.

- For Input raster or feature source data, choose **ranger**.

- For Output distance raster, click the **browse** button, navigate to your **C:\GIS\Module19** folder, and name the output raster **Disttoranger**.

- Leave the other options alone (but make sure Output cell size is set to **NLCDhills**, which will be 30 meters, as you set under Environments).

- Run the Euclidean Distance tool.

- In the new Disttoranger grid, each grid cell has a value that represents the distance from the state forest headquarters (HQ) to every cell in the grid. Answer Question 19.1.

> **Question 19.1** Zoom in closely on the grid cell on which the hikers' point lies. Use the Explore tool to obtain the grid-cell value of the Disttoranger grid at that location. How far (using the value from the Euclidean distance grid) are the hikers from the state forest HQ?

- Next, you will calculate a Euclidean allocation grid, which will show each cell's value representing the closest source (the state forest HQ) to which it is allocated. (See **Smartbox 19.2** for more about allocation.)

SMARTBOX 19.2

What is allocation, and how is it used in ArcGIS Pro?

The Euclidean Distance tool allows you to find the shortest straight-line distance to each cell from a source. However, if you have multiple sources, that Euclidean distance surface will show the shortest distance from any of the sources. By looking at the value of an individual cell, you can't tell to which of the sources it is closest. It may be more useful to know to which source a particular cell is nearest. For instance, if an accident occurs in a park, knowing which of several different first-aid stations is closest to the incident location is going to be critical.

allocation The determination of which source is closest in distance to each grid cell.

An **allocation** process involves creating a new grid that shows to which of the sources a cell is closest. For example, in Figure 19.5, the Source_Ras grid has two different sources (which could be the first-aid stations in a park), and each source consists of one or more cells. The Euclidean Allocation tool calculates the Euclidean distance to each cell (as described in Smartbox 19.1), but rather than assigning the distance value to the grid cell, it assigns a value indicating to which source that cell is closest (based on the Euclidean distance to that cell). The EucAllocation grid in Figure 19.5 shows the output of the allocation process. Each cell has a value of 1 if it is closest to source #1 and a value of 2 if it is closest to source #2.

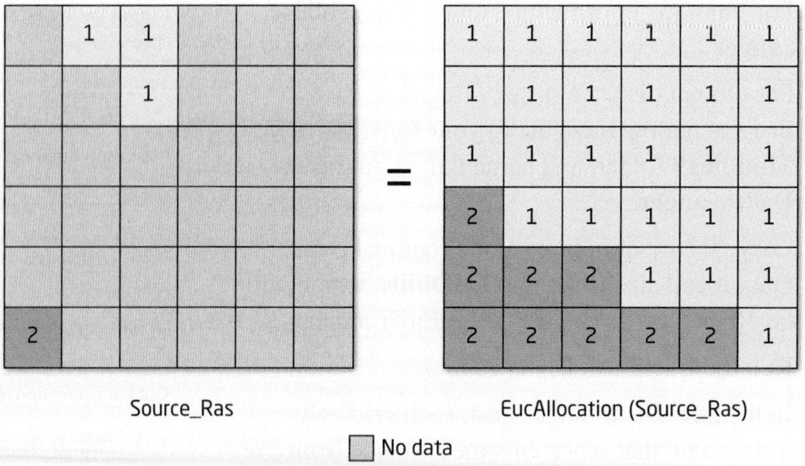

FIGURE 19.5 The Euclidean Allocation calculation in ArcGIS Pro, used for assigning grid cells to their nearest source cell.

- Return to the Geoprocessing pane. The tool you will be using is Euclidean Allocation. You can either search for it or find it directly from the **Spatial Analyst Tools** toolbox, in the **Distance** toolset. Using either method, open the **Euclidean Allocation** tool.
- For Input raster or feature source data, choose **ranger**.
- For the Output raster, click the **browse** button, navigate to your **C:\GIS\ Module19 folder**, and name the output raster **allocaranger**.
- Leave the other options alone (but make sure that Output cell size is set to nlcdhills, which will be 30 meters, as you set under Environments).
- Run the Euclidean Allocation tool.
- Examine the new allocaranger grid. Answer Question 19.2.

> **Question 19.2**
> Why does every cell of the Allocation to ranger grid (allocaranger) have the same value?

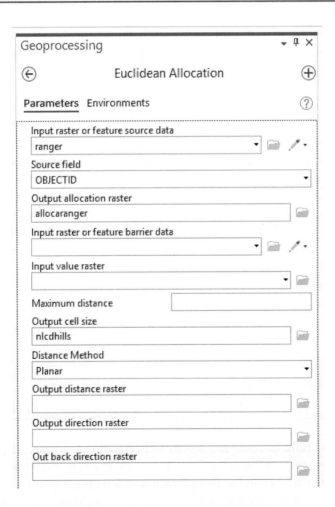

Step 19.3 Shortest Paths over Raster Surfaces

- Now you will set up the constraints on the problem in terms of the *cost* of moving across the surface. (See **Smartbox 19.3** for more about cost distances.) In this case, the accumulated distance to travel will be the cost of traveling from the state forest HQ to the stranded hikers.

SMARTBOX 19.3

How is cost distance used in ArcGIS Pro?

cost The number of units it takes to move from one grid cell to another.

accumulated cost The total cost of traveling from a source to a grid cell.

In ArcGIS Pro, the **cost** of moving from cell to cell is the number of units it takes to do so. For instance, a basic lateral movement costs 1 unit, and a diagonal movement costs 1.414 units. If you move three grid cells in a lateral direction, the total amount of movement—or the **accumulated cost**—is 3 units. If you travel two grid cells laterally and then one grid cell diagonally, the accumulated cost is 3.414 units. However, these basic costs assume that nothing is impeding the progress of movement from one cell to another, and the cost of moving from one cell to another will always remain at constant values of 1 or 1.414. This assumption is not always true in reality.

For example, if you were hiking in a park, crossing a flat section of land would take less exertion (and time) than hiking up a steep hill. Similarly, walking across a paved section of road would take less exertion and time than walking through dense forest or over rocky, uneven terrain. If you were simulating movement through a park to a first-aid station, many things could slow your progress; navigating through a forest, climbing steep terrain, and fording streams would all add to the time it would take to reach the first-aid station. Thus, if you were trying to find the shortest way to the first-aid station, you'd want to take these factors into account and travel along the path that impedes your progress the least.

In ArcGIS Pro, you can simulate these kinds of factors by assigning additional costs (or impedances) to traveling from cell to cell. For instance, moving across rougher terrain or swampy land might have a higher cost value. Grid cells with steeper slopes can be assigned a higher cost value. Similarly, different types of land cover can be assigned higher cost values, depending on the difficulty of traversing them. If you have multiple grids of different costs (for instance, one grid of higher costs based on slope conditions and another grid of higher costs based on land-cover conditions), they need to be combined into a single cost grid for use in ArcGIS Pro.

The number of units (x) for laterally moving from one grid cell to an adjacent one is calculated as:

$$x = 1 \times \frac{(\text{cost of starting grid cell}) + (\text{cost of ending grid cell})}{2}$$

The number of units (x) for diagonally moving from one grid cell to an adjacent one is calculated as:

$$x = 1.414 \times \frac{(\text{cost of starting grid cell}) + (\text{cost of ending grid cell})}{2}$$

For an example of how this works in practice, see Figure 19.6, which shows a sample raster with an additional cost assigned to each grid cell. The cost to move laterally from the cell in the upper-left corner (with a value of 5) to the adjacent cell on its right (with a value of 1) would be $1 \times (5+1)/2$, or 3. The cost of diagonally moving from the cell in the upper-left corner (with a value of 5) to its diagonally lower-right cell (with a value of 6) would be $1.414 \times (5+6)/2$, or 7.77. (Note that if values of 1 are used in the starting and ending cells, the calculated costs will be values of 1 for lateral movements and 1.414 for diagonal movements.)

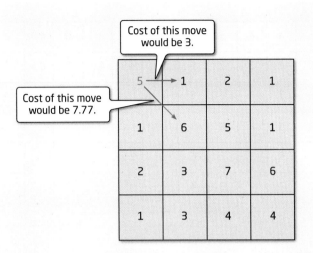

FIGURE 19.6 A sample raster of various cost values and the corresponding computed lateral and diagonal movement costs.

The Cost Distance tool in ArcGIS uses this method to calculate a new grid of output values, with each one containing the lowest accumulative cost value of movement from one cell to another. The algorithm used by the Cost Distance tool requires an initial set of sources (and if these sources are feature classes, they are internally converted to raster format for use); the algorithm also requires a raster of the cost values associated with each grid cell (Figure 19.7).

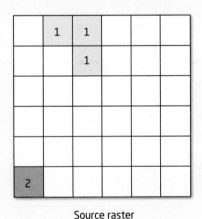
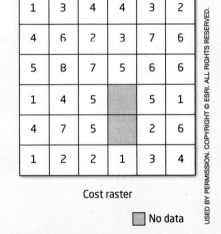

FIGURE 19.7 The two layers necessary for computing cost distance: a layer of source cells from which to find the distance and a second layer of the costs associated with each cell.

The first step of the cost-distance algorithm is to calculate the cost of moving from each source cell to its adjacent cells. The cost formulas described earlier are used (Figure 19.8). Thus, the cost of moving from the source cell labeled with a 2 to the adjacent cell on its right would be a lateral move with a cost of 1.5 (that is, $1 \times (1+2)/2$). The cost of moving from that same source cell to the cell diagonally adjacent to it would be 5.7 (that is, $1.414 \times (1+7)/2$). These new values are the total accumulation costs of moving from a source cell to adjacent cells. ArcGIS Pro keeps track of the cells that now have accumulated costs computed for them.

The source cells are now considered "exhausted," as there is no need to perform any further cost-distance calculation from them. Now the algorithm needs to select a new grid cell from which to begin calculating cost distance. It selects the cell that corresponds with the lowest value on the active accumulated cost

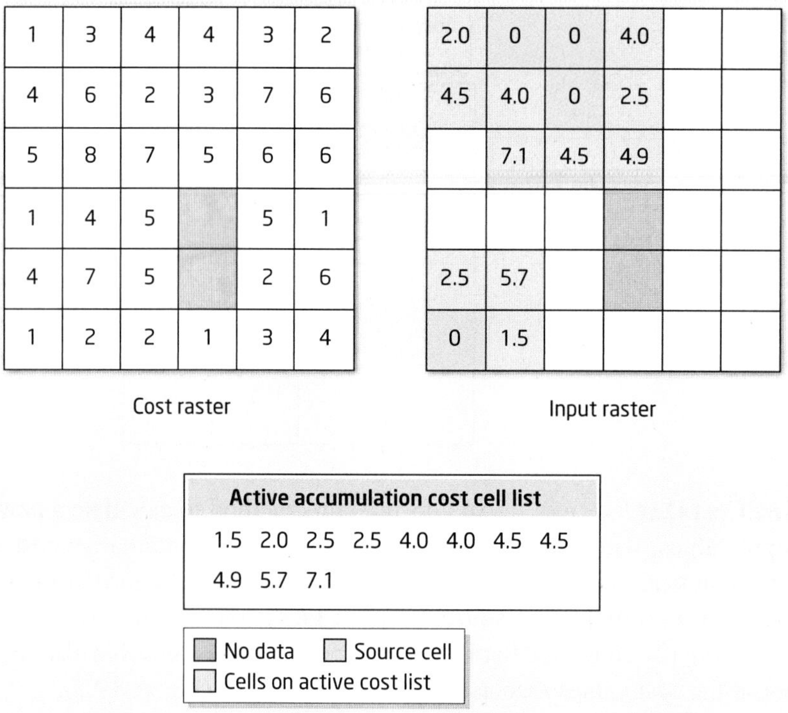

FIGURE 19.8 The movement costs computed for cells adjacent to all source cells and the initial active accumulated cost list that is generated.

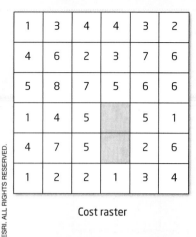

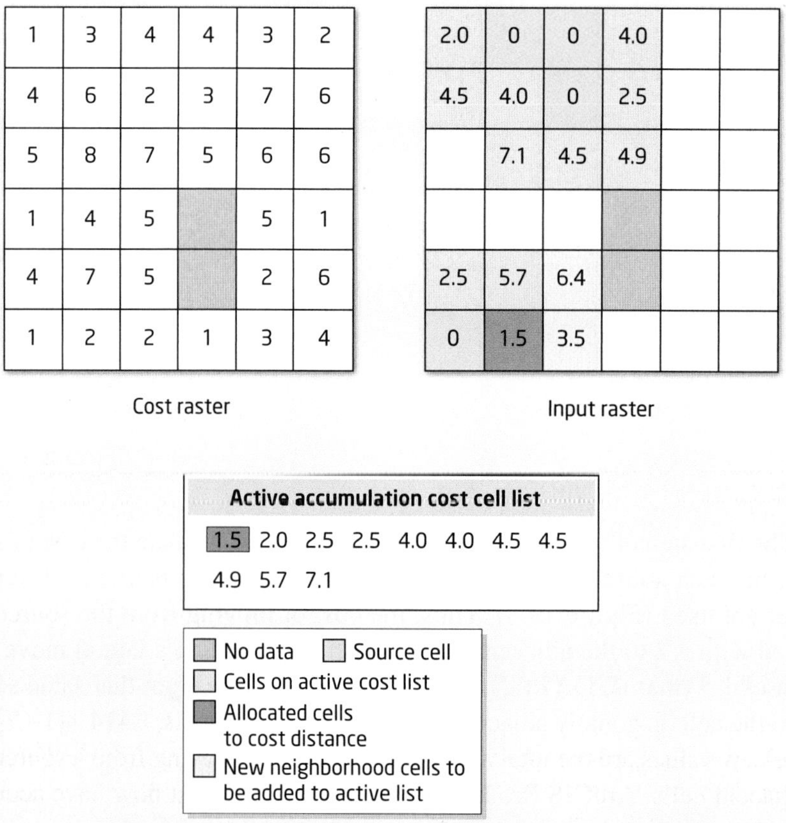

FIGURE 19.9 The computed accumulative cost values for the cells connected to the new searching cell.

list and uses that as the new starting point cell for calculations (Figure 19.9). In this case, the grid cell that corresponds with the lowest active accumulative cost value of 1.5 is chosen (shown in purple).

Cost distances from this cell to its adjacent cells are calculated using the values from the initial cost raster. For instance, the lateral movement from the purple cell (with a value of 1.5) would be 2 (that is, $1 \times (2+2)/2$). The accumulative cost value of the purple cell that's being used for the calculations (1.5) is added to this value of 2 (because it has already cost 1.5 units to travel from a source cell to the purple cell). Thus, the final accumulative cost of moving laterally to the cell to the right of the purple cell would be 3.5 (movement cost of 2 plus the already accumulated cost of 1.5).

The cost distance to move from the purple cell to the one diagonally adjacent to its right would be 4.9 (that is, $1.414 \times (2+5)/2$). The total accumulative cost to move from a source cell to this new cell would be 6.4 (4.9 to move diagonally from the purple cell to the new one plus the already accumulated cost of 1.5).

If a cost-distance calculation results in a lower value than one that's already been calculated for a cell, this new lower value is placed into the cell instead. The higher value is removed from the active accumulative cost-cell list, and this new lower value is placed there instead. Because you're trying to find the least-cost value to reach each cell from a source, these newly calculated lower values represent movement options that are shorter (or "cost less") than previously computed options.

When you've looked at all possible connections from the purple cell, it is removed from the active accumulation cost-cell list and is considered "exhausted." The newly calculated accumulative cost values of 3.5 and 6.4 are added to the accumulative cost-cell list in its place.

This process continues until all cells on the active list are searched and calculations are made from them. The end result of this process is a new raster in which each cell contains a value representing the shortest calculated cost distance (or the "least cost" value) of traveling from a source location to that cell (Figure 19.10). The Cost Distance tool also generates another raster, called a *backlink*, which is used in determining the shortest path from a source to each cell.

1	3	4	4	3	2
4	6	2	3	7	6
5	8	7	5	6	6
1	4	5		5	1
4	7	5		2	6
1	2	2	1	3	4

Cost raster

2.0	0	0	4.0	6.7	9.2
4.5	4.0	0	2.5	7.5	13.1
8.0	7.1	4.5	4.9	8.9	12.7
5.0	7.5	10.5		10.6	9.2
2.5	5.7	6.4		7.1	11.1
0	1.5	3.5	5.0	7.0	10.5

Cost-distance output values

FIGURE 19.10 The initial cost raster and the final computed cost-distance raster, showing the least cost of moving from a source cell to each other cell.

- Return to the Geoprocessing pane. The tool you will be using is Cost Distance. You can either search for it or find it directly from the **Spatial Analyst Tools** toolbox, in the **Distance** toolset. Using either method, open the **Cost Distance** tool.

- For Input raster or feature source data, choose **ranger**.

- For Input cost raster, choose **allocaranger**.

- For Output distance raster, click the **browse** button, navigate to your C:\GIS\Module19 folder, and name the output raster **Eucliddist**.

- For Output backlink raster, click the **browse** button, navigate to your C:\GIS\Module19 folder, and name the output raster **Euclidlink**.

- Leave the other options alone and run the Cost Distance tool to calculate the cost distance based on these parameters.

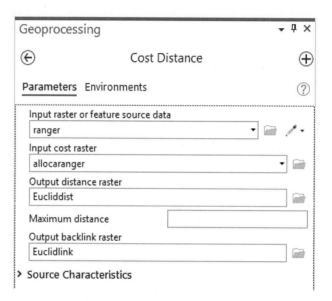

- Examine the Eucliddist grid. Each cell should contain a value that represents the cost distance from the state forest HQ to every cell in the grid. Answer Questions 19.3 and 19.4.

Question 19.3 The allocaranger grid has a value of 1 for each grid cell. Why was this value used for the input cost raster in this calculation? What is the cost of traversing each grid cell in this step?

Question 19.4 What (specifically) were the constraints (in terms of cost-distance factors) placed on this shortest-path calculation?

- The last step is to calculate the shortest overland path from the state forest HQ to the hikers. In ArcGIS Pro, this is done by calculating a least-cost path. (See **Smartbox 19.4** for more information on how a least-cost path is calculated.)

SMARTBOX 19.4

How is a cost path calculated in ArcGIS Pro?

The raster created by the Cost Distance tool has grid-cell values that represent the lowest accumulative cost distance from a source to that cell. However, this information does not tell you which cell path to take to get from a source to that location. If you're trying to find the shortest path from an accident to a first-aid station, you need to take one more step beyond generating the cost distance: Specifically, you need to identify the actual path you should take. In ArcGIS Pro, you can use the Cost Path tool for exactly this purpose.

Because you already have a raster that shows the lowest cost to reach a cell from a source, the Cost Path tool essentially "backtracks" from a location on the surface to the nearest source. Thus, after you've selected a destination, the Cost Path tool determines the **least-cost path** (or the grid cells to move through for the lowest accumulated cost) from that destination backward to a source. Thus, you have to determine the destination. (As usual, if this is a feature class, it is internally converted to a raster for use by the tool.) If the cost-distance raster showed the distance from the aid stations (the sources), the destination point would be the location of an accident, and you would determine the path backward from the accident destination to the first-aid station source.

This process is enabled by the use of the backlink grid, which is also part of the output from the Cost Distance tool. For each cell in the grid, ArcGIS Pro computes the cost-weighted distance from that cell toward its nearby cells. It then chooses the least-cost value, which indicates the least-cost path from that cell back to a source. ArcGIS Pro then records the direction of movement along this least-cost path in a **backlink** grid.

Each cell of the backlink grid is then coded with a number from 1 through 8, with each value indicating the direction the movement should take. Figure 19.11 shows the assigned values and their backlink direction positions. In this case, the least-cost path is the value 8.75, which lies to the west of the destination cell. The value 5 is written into the backlink grid, signifying that if you were at that particular cell, your choice for least-cost movement to get from there to a source would be to the west. If that value had indicated that you should move north, a value of 7 would be saved for that cell in the backlink grid; if the movement should go to the southeast, a value of 2 would be saved for that cell in the backlink grid. A value of 0 indicates a source cell (and thus no movement is necessary).

The Cost Path tool uses the directions from the backlink grid to "trace" the least-cost path from the destination back to the source. For instance, if the destination is at a cell with a backlink value of 5, movement to the west is indicated. That cell to the west becomes part of the least-cost path. The cell to the immediate west of that one has the value 3, indicating movement to the south. Thus, the cell to the south becomes the next grid cell in the least-cost path. This process continues until a source is reached. All the adjacent cells found in this way become the set of cells extracted to form the tool's output, which represents the least-cost path from a destination to a source.

least-cost path The set of grid cells that represents the lowest value for accumulated cost to move from a source to a destination.

backlink The grid that shows the direction to be traveled for the least-cost amount of movement from cell to cell.

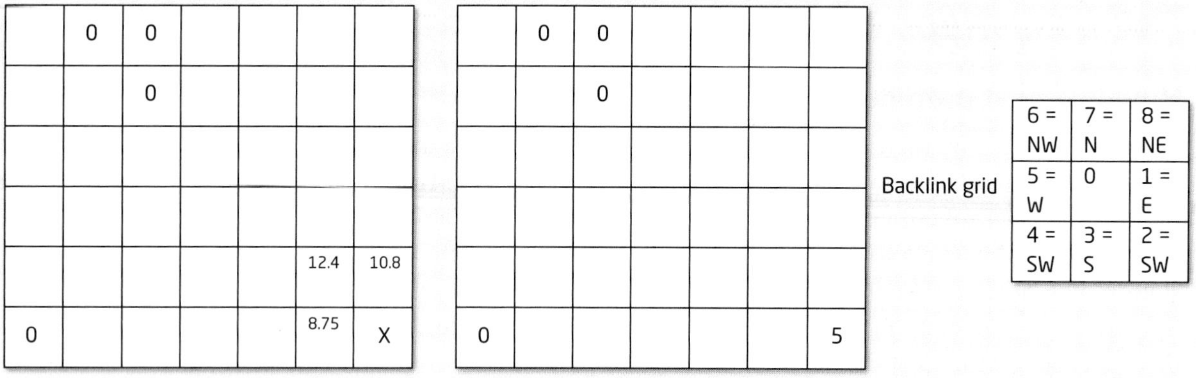

FIGURE 19.11 Assigning values to the backlink grid, based on the shortest distance.

Note that the least-cost path generated in this way by ArcGIS Pro is a new raster surface that consists of the cells making up the least-cost path (and assigned the value 3) and another cell that denotes the location of the destination (assigned the value 1). All other cells on the surface receive the value NODATA (because they are not cells that are part of the least-cost path).

- Return to the Geoprocessing pane. There are two tools for generating the least-cost path in ArcGIS Pro: Cost Path (which creates a new raster as the output showing the path) and Cost Path As Polyline (which creates a new line feature class showing the path). The tool you will be using here is Cost Path As Polyline. You can either search for it or find it directly from the **Spatial Analyst Tools** toolbox, in the **Distance** toolset. Using either method, open the **Cost Path As Polyline** tool.
- For Input raster or feature destination data, choose **hikers**.
- For Input cost distance or euclidean distance raster, choose **Eucliddist**.
- For Input cost backlink, back direction or flow direction raster, choose **Euclidlink**.
- For Output polyline features, click the **browse** button, navigate to your **C:\GIS\Module19** folder and go to the **Module19.gdb** geodatabase, and name the output feature class **Euclidpath**.
- For Path type, choose **Each cell**.
- Leave the other options alone and run the Cost Path As Polyline tool to calculate the least-cost path, based on these parameters. The least-cost path between the hikers and the ranger station will be displayed by a new line feature class.

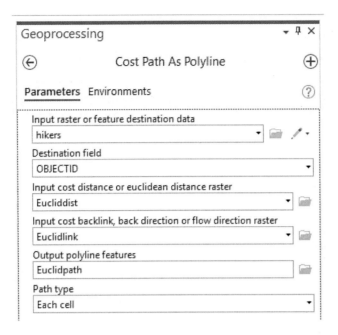

Step 19.4 Visualizing the Cost-Path Results

- To better visualize the least-cost path ArcGIS has just computed in relation to the terrain and land cover of the park, you can display it in 3D. You could just create a new scene and add all the pertinent layers into it, or you could instead simply convert the contents of the Hocking Hills map to a scene with the same content. On the **View** tab, in the **View** group, click the **Convert** button and then choose **To Local Scene**.

- ArcGIS Pro adds a new scene called Hocking Hills_3D that contains all the layers from the Hocking Hills map but now as 2D layers draped on the terrain. Use the navigator to adjust to a perspective view of the area, and you see your raster and vector layers properly draped.

- For now, turn off all the 2D layers you've added or created except for hikers, ranger, and Euclidpath. Make sure to leave the ArcGIS Pro layers Topographic (which is the basemap) and World Elevation 3D (which is the default DEM elevation surface) turned on.

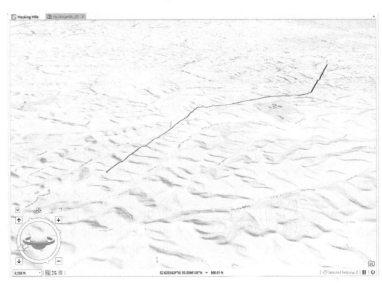

- You may want to adjust the 2D symbology of hikers, ranger, and Euclidpath to make them stand out when they are displayed draped on the Topographic basemap. For instance, you might want to use a thicker or darker line for Euclidpath so that it stands out on the surface.

- Zoom in and examine the draped Euclidpath shortest-path layer and answer Questions 19.5 and 19.6.

Question 19.5 What factors were unaccounted for in calculating this shortest path?

Question 19.6 Is this a path that you would use to rescue the hikers? Why or why not?

Step 19.5 Further Cost-Distance Calculations

- Return to the Hocking Hills map and turn off all layers except for the DEMhills layer. In this step, you'll add more complexity to the problem—specifically, the steepness of the terrain. (See Module 15 for more about slope calculation from an elevation surface.) Return to the Geoprocessing pane. The tool you will be using here is Slope. You can either search for it or find it directly from the **Spatial Analyst Tools** toolbox, in the **Surface** toolset. Using either method, open the **Slope** tool.
- For Input raster, choose **demhills**.
- For Output raster, click the **browse** button, navigate to your **C:\GIS\Module19** folder, and name the output raster **slopehills**.
- For Output measurement, choose **Degree**.
- Leave the other options alone and run the Slope tool. ArcGIS Pro creates the new slope grid. Each grid cell has a value for degree of slope at that cell.

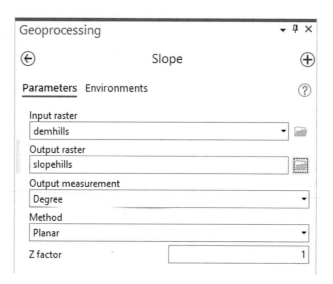

- Next, you need to break the slope grid into several classes to separate out relatively flat slopes from moderate slopes from very steep slopes as a constraint for calculating the shortest path. To do this, choose **slopehills** in the Contents pane and then, on the **Appearance** tab, in the **Rendering** group, click the **Symbology** button and choose **Classify**. The Symbology pane opens, allowing you to alter how the slopehills layer is displayed.
- For Primary symbology, be sure **Classify** is chosen.

- For Method, choose **Natural Breaks (Jenks)**.
- For Classes, choose **5**.
- Slopehills is now displayed in five class breaks. Because there are so many values of slope in the Slopehills grid, you need to divide these slope values into smaller categories. You can create a new grid that breaks slopehills into five categories (based on the Natural Breaks classification) of 1 through 5, with values of 1 being the flattest slopes and values of 5 being the steepest slopes. You can do this in ArcGIS Pro by using the Slice tool. See **Smartbox 19.5** for more information about using Slice with a raster layer.

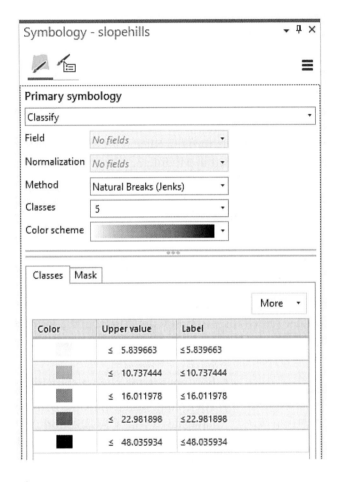

 SMARTBOX 19.5

How does Slice work with raster layers in ArcGIS Pro?

The **slice** process takes grid cells from one layer and assigns new values to them by using a data classification scheme. When you use the Slice tool, the various grid-cell values of a raster are grouped together into a user-defined number of classes. The slice process can be used to group all the unique distance values (potentially thousands of different values) into a smaller number of classes, with each class assigned a new value. In Figure 19.12, the first raster

slice A process that involves grouping several grid-cell values together according to a classification scheme.

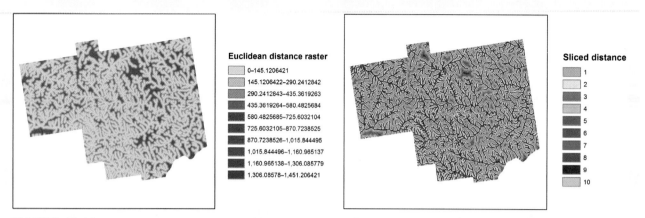

FIGURE 19.12 A Euclidean distance raster that has been sliced into 10 different categories.

depicts each grid cell containing a value for the Euclidean distance away from a set of streams. The second raster shows that same raster after a slice process; every grid cell now has a value of 1 to 10, with 1 as the lowest distance value and 10 as the highest distance value.

The classifications available through Slice are Equal Interval and Natural Breaks (which work as described in Module 3, where you used them for classifying values on choropleth maps) and Equal Area (which is similar to the Quantiles method from Module 3, in that it seeks to have an equal number of cells—not data values but a count of cells—in each category).

- Return to the Geoprocessing pane. The tool you will be using here is Slice. You can either search for it or find it directly from the **Spatial Analyst Tools** toolbox, in the **Reclass** toolset. Using either method, open the **Slice** tool.

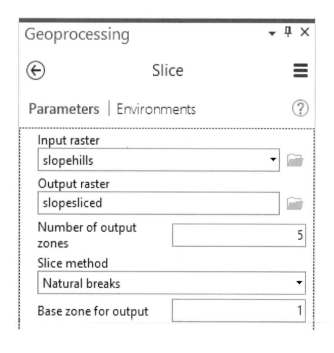

- For Input raster, choose **slopehills**.
- For Output raster, click the **browse** button, navigate to your **C:\GIS\Module19** folder, and name the Output raster **slopesliced**.
- For Number of output zones, choose **5**.
- For Slice method, choose **Natural breaks**.
- Leave the other options alone and run the Slice tool.
- The new grid is added to the Contents pane and shows the slope values placed in one of five new classes, with the lowest slope values represented by cells with a value of 1 and the highest slopes represented by cells with a value of 5. These slope values will be based on the five Natural Breaks categories used to display your slopehills layer; you can change between the display of the slopehills and slopesliced grids to see this.

- You can now create a new shortest path, using this newly recategorized slope as a constraint on the cost-distance calculation. Return to the Geoprocessing pane. The tool you will be using here is Cost Distance. You can either search for it or find it directly from the **Spatial Analyst Tools** toolbox, in the **Distance** toolset. Using either method, open the **Cost Distance** tool.

- For Input raster or feature source data, choose **ranger**.

- For Input cost raster, choose **slopesliced**.

- For Output distance raster, click the **browse** button, navigate to your **C:\GIS\Module19** folder, and name the output raster **slopedist**.

- For Output backlink raster, click the **browse** button, navigate to your **C:\GIS\Module19** folder, and name the output raster **slopelink**.

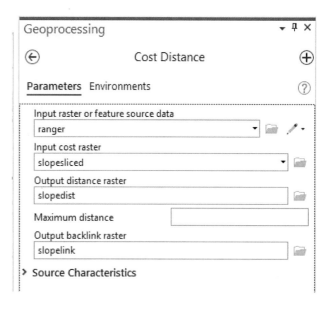

- Leave the other options alone and run the Cost Distance tool to calculate the cost distance based on these parameters.

- Next, you can calculate the shortest path based on the new cost distance. Return to the Geoprocessing pane. The tool you will be using here is again Cost Path As Polyline. You can either search for it or find it directly from the **Spatial Analyst Tools** toolbox, in the **Distance** toolset. Using either method, open the **Cost Path As Polyline** tool.

- For Input raster or feature destination data, choose **hikers**.

- For Input cost distance or euclidean distance raster, choose **slopedist**.

- For Input cost backlink, back direction or flow direction raster, choose **slopelink**.

- For Output polyline features, click the **browse** button, navigate to your **C:\GIS\Module19** folder, go to the **Module19.gdb** geodatabase, and name the output polyline feature **Slopepath**.

- For Path type, choose **Each cell**.

- Leave the other options alone and run the Cost Path As Polyline tool to calculate the new shortest path (a line layer called Slopepath) between the state forest HQ and the hikers, based on these constraints.

- To better visualize the results of this cost-distance calculation, in the Contents pane, right-click on the **Slopepath** layer and choose **Copy**. Then move to the Hocking Hills_3D scene, click on **2D Layers** and then, on the **Map** tab, in the **Clipboard** group, click the **Paste** button. The Slopepath layer is displayed as a 2D layer

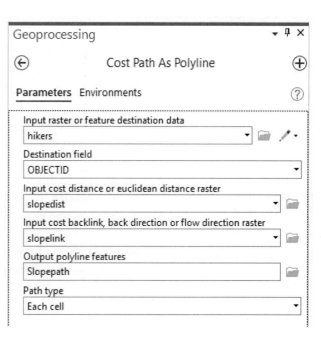

draped over the Topographic basemap. You might want to change the symbology of the Slopepath line to make it thicker and have its color stand out from the basemap and the Euclidpath layer.

- Answer Questions 19.7 and 19.8.

Question 19.7 How did this new cost distance change the shortest path?

Question 19.8 Is this new path one that you would use to rescue the hikers? Why or why not?

Step 19.6 Using More Than One Cost Layer for Cost Distance

- Return to the Hocking Hills map and turn off all layers except the nlcdhills layer. You'll now be creating a third cost-distance surface, but this time you'll use two factors: the steepness of the terrain you already calculated and the type of land cover one would travel across. The assumption being made now is that certain types of land cover "cost" more to travel through than others; for example, it's easier/less costly to traverse developed terrain—such as paved surfaces—than it is to traverse forested terrain or water.

- To create this new cost-distance surface, you'll be creating a new grid and reclassifying the land-cover types into cost values. The higher the cost value, the more difficult (that is, the more costly) it will be to traverse the terrain. In ArcGIS Pro, you can accomplish these goals by using the Reclassify tool. (See **Smartbox 19.6** for more about reclassifying a raster layer.)

SMARTBOX 19.6

How are raster layers reclassified in ArcGIS Pro?

The Slice tool assigns new values to grid cells by placing them in different categories according to a classification scheme. However, you might sometimes want to change the value of cells yourself rather than group them into classes. For instance, if you have a grid with values 1, 2, and 3, but you want to create a grid that assigns each of these cells the value 1, you need a way to tell ArcGIS Pro what new values need to be assigned to these cells. The Reclassify tool allows you do this.

Reclassify An ArcGIS Pro tool that allows you to create a new grid by assigning new grid-cell values to an existing raster.

Reclassify allows you to create a new raster by changing the values assigned to grid cells in another raster. With Reclassify, you specify what new values are assigned to old values. These new values must be numerical; you can also type NODATA for the assigned new value. Reclassify is commonly used to make changes to a grid or update its information. It's also used to change values from one classification to another. For instance, an NLCD raster has all of its developed categories classified as the values 21, 22, 23, and 24. If you are simply examining all built areas, you might want to reclassify the NLCD grid so that each of these four values is assigned the new value 1, while all other NLCD values are assigned the value NODATA. The end result would be a new grid that contains only grid cells representing developed areas, with no other values in the grid.

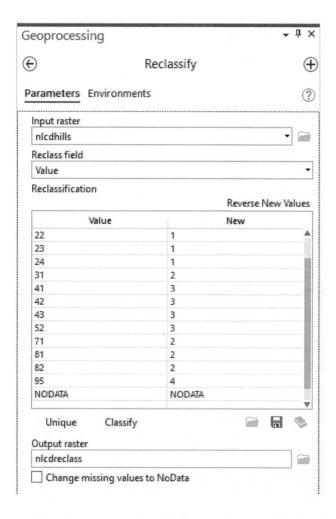

- Return to the Geoprocessing pane. The tool you will be using here is Reclassify. You can either search for it or find it directly from the **Spatial Analyst Tools** toolbox, in the **Reclass** toolset. Using either method, open the **Reclassify** tool.
- For Input raster, choose **nlcdhills**.
- For Reclass field, choose **Value**.
- Click the **Unique** button to separate out the NLCD 2011 values into their own separate numbers to be reclassified.

- Reclassify the NLCD 2011 values as follows:

Old Value	New Value	NLCD 2011 Classification
11	5	Open Water
21	1	Developed, Open Space
22	1	Developed, Low Intensity
23	1	Developed, Medium Intensity
24	1	Developed, High Intensity
31	2	Barren
41	3	Deciduous Forest
42	3	Evergreen Forest
43	3	Mixed Forest
52	3	Shrub Scrub
71	2	Grassland/Herbaceous
81	2	Pasture/Hay
82	2	Cultivated Crops
95	4	Emergent Wetlands

Important Note: Remember that the higher the new value, the more costly it is to traverse that particular land cover.

- For Output raster, click the **browse** button, navigate to your **C:\GIS\Module19** folder, and name the output raster **nlcdreclass**.

- Run the Reclassify tool. A new grid called nlcdreclass is created, and it contains the new cost values.

- Now you'll combine your two cost grids (the reclassified slope and the reclassified land cover) to form a new cost grid. This new grid will take into account both weights. To combine your grids, you have to overlay them. Raster overlay works differently than the polygon overlays (see Module 9). You'll do much more with raster overlay and get into how it works in Module 20. For now, however, you can just follow the next few steps to combine the two layers by simply adding them together. Return to the Geoprocessing pane. The tool you will be using here is Raster Calculator. You can either search for it or find it directly from the **Spatial Analyst Tools** toolbox, in the **Map Algebra** toolset. Using either method, open the **Raster Calculator** tool.

- You need to build a raster calculation to combine the two grids by adding them together. In the Layers and variables list, double-click on **nlcdreclass**. Next, double-click the + button. Finally, double-click on **slopesliced** in the Layers and variables list. Your raster calculation should read **"nlcdreclass" + "slopesliced"**.

- For Output raster, click the **browse** button, navigate to your **C:\GIS\Module19** folder, and name the output raster **slopeandland**.

- Run the Raster Calculator tool. The new slopeandland grid is added to the Contents pane. Slopeandland combines the two rasters by adding the cell values together for cells that would be on top of each other. For instance, a location on the slopesliced grid might have a value of 2. That same location on the nlcdreclass grid might have a value of 3. Thus, because 2 + 3 = 5, the same location on the slopeandland grid would have a value of 5. Answer Questions 19.9 and 19.10.

Step 19.6 Using More Than One Cost Layer for Cost Distance 459

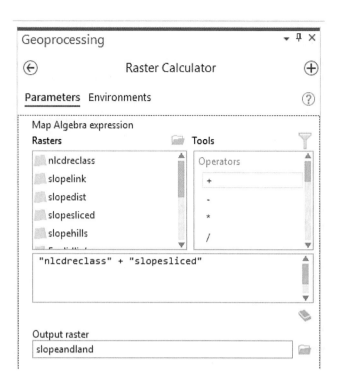

Question 19.9 What (specifically) do values 2 and 8 represent in the new slopeandland grid (beyond simply values added together)?

Question 19.10 Why are there no values of 10 in the slopeandland grid?

- Now you'll compute the cost-weighted distance and the shortest-cost path by using this new slopeandland cost-distance grid. Return to the Geoprocessing pane. The tool you will be using here is again Cost Distance. You can either search for it or find it directly from the **Spatial Analyst Tools** toolbox, in the **Distance** toolset. Using either method, open the **Cost Distance** tool.
- For Input raster or feature source data, choose **ranger**.
- For Input cost raster, choose **slopeandland**.
- For Output distance raster, click the **browse** button, navigate to your **C:\GIS\Module19** folder, and name the output raster **slopelanddist**.
- For Output backlink raster, click the **browse** button, navigate to your **C:\GIS\Module19** folder, and name the output raster **slopelandlink**.
- Leave the other options alone and run the Cost Distance tool to calculate the cost distance based on these parameters.

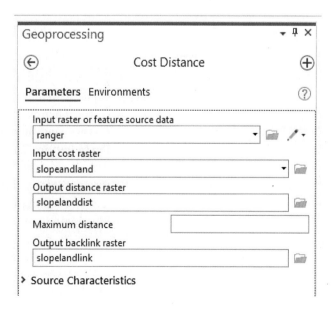

- Next, you'll calculate the shortest path based on this last cost distance. Return to the Geoprocessing pane. The tool you will be using here is again Cost Path As Polyline. You can either search for it or find it directly from the **Spatial Analyst Tools** toolbox, in the **Distance** toolset. Using either method, open the **Cost Path As Polyline** tool.
- For Input raster or feature destination data, choose **hikers**.
- For Input cost distance or euclidean distance raster, choose **slopelanddist**.
- For Input cost backlink, back direction or flow direction raster, choose **slopelandlink**.
- For Output polyline features, click the **browse** button, navigate to your **C:\GIS\Module19** folder, go to the **Module19.gdb** geodatabase, and name the Output polyline feature **Slopelandpath**.
- For Path type, choose **Each cell**.
- Leave the other options alone and run the Cost Path As Polyline tool to calculate the new shortest-cost path (which will be a line layer called Slopelandpath) between the state forest HQ and the hikers, based on these constraints.

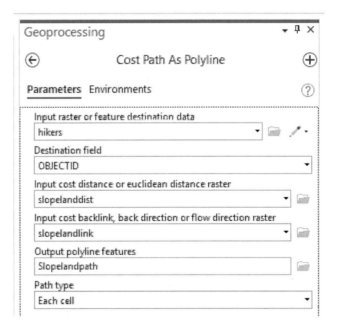

- To better visualize the results of this cost-distance calculation, in the Contents pane, right-click on the **Slopelandpath** layer and choose **Copy**. Then move to the Hocking Hills_3D scene, click on **2D Layers**, and then, on the **Map** tab, in the **Clipboard** group, click the **Paste** button. The Slopelandpath layer is then displayed as a 2D layer draped over the Topographic basemap. You might want to change the symbology of the Slopelandpath line to make it thicker and have its color stand out from the basemap, the Euclidpath layer, and the Slopepath layer.
- Answer Questions 19.11 and 19.12.

> **Question 19.11** How (specifically) did this third cost distance change the shortest path from the one generated using the slope as a constraint on the problem? Why did the path change this way?

> **Question 19.12** Which of these three cost paths would you select to rescue the hikers? Why?

Step 19.7 Applying the Cost-Distance Surfaces

- At this point, you have generated three different "shortest paths" to reach the stranded hikers. You also have three different cost-distance surfaces (Eucliddist, slopedist, and slopelanddist) with corresponding backlink grids (Euclidlink, slopelink, and slopelandlink). Thus, none of these grids needs to be re-created to determine the shortest path from the state forest HQ to another location on the map; only the new cost path to this new point would have to be generated.

- This next scenario assumes that an emergency situation has occurred at a campsite near the main family campground. Back in the Hocking Hills map, turn off all the layers except the DEMhills and NLCDhills layers and add the **campers** layers to the Contents pane.

- Create three new cost-path polylines from the state forest HQ (the rangers layer) to the destination of the campers point as follows:
  - Use Eucliddist and Euclidlink to create a path called **Euclidcamp**.
  - Use slopedist and slopelink to create a path called **slopecamp**.
  - Use slopelanddist and slopelandlink to create a path called **slopelandcamp**.

- When you have these three new cost-path polylines, copy and paste them all into the Hocking Hills_3D scene to examine each of the paths. Answer Question 19.13.

> **Question 19.13** How does each of these three paths to reach the campers differ from the other two? Be specific when referring to terrain features or land-cover features near the campsite that the paths avoid or do not avoid. Also describe which path you would take to reach the campers and why.

Step 19.8 Exporting, Saving, and Sharing Your Results

- Save your Module19 project. You can then present your results in one of two ways:
  - Print a layout (see Module 3) showing only the locations of the ranger, the hikers, and the campers, as well as the six polyline versions of the paths.
  - Share your results as a web scene through ArcGIS Online (see Module 18). If you're sharing your data, to save on space and processing, your web

scene should contain only the locations of the ranger, the hikers, and the campers, as well as the six polyline versions of the paths.

Closing Time

This module explored the concept of representing distance in a GIS as a continuous surface. By creating a surface of Euclidean distance measurements, you can easily determine how far away every location is from a source. Through the use of the cost-distance and least-cost path processes, you can then generate the shortest path across the surface from any location to a source. While the cost-distance tools are very versatile, ArcGIS Pro also provides an additional set of tools for modeling other types of surface movement. (See *Related Concepts for Module 19* for more information.)

In this module you combined two different raster layers to create a single new raster. By combining the two cost layers (the sliced slope layer and the reclassified land-cover layer), you created a new cost raster with elements of both of the original layers. Essentially, you overlaid these two rasters, which is a powerful trick to know when working with raster data. In the next module, you'll do much more with raster overlays in the context of map algebra.

RELATED CONCEPTS FOR MODULE 19

Using Path Distance in ArcGIS Pro

The cost-distance functions in ArcGIS Pro allow for a wide variety of different applications involving movements across a surface. However, ArcGIS Pro contains another set of functions that allow you to apply cost-distance concepts with additional complexity for use in dispersion modeling or more detailed least-cost-path studies. The functions of **path distance** use the same types of accumulative costs, allocation, backlink, and least-cost path algorithms as their cost-distance counterparts but allow you to model many other factors that would influence movement across a surface. When modeling these additional forces, you use the Path Distance tool rather than the Cost Distance tool.

Path distance allows for the inclusion of the actual surface distance being traveled as well as other horizontal-based factors that would influence movement. For instance, a strong wind behind you would allow you to move faster, while a strong headwind would slow your movement. Vertical factors can be included as well. For example, if you were walking uphill on a steeper slope, your progress would be slowed (as you simulated in this module). However, if you were walking downhill on the same steep slope, your movement would speed up. Many types of frictions can be used, depending on the phenomena being modeled with path distance. For instance, smoke plumes, pollution concentrations, and movement of contaminants can all be modeled as continuous surfaces in ArcGIS Pro.

path distance A cost-distance function that allows for a variety of other frictions (such as horizontal or vertical factors) to be included in the analysis.

Key Terms

Euclidean distance (p. 440)
allocation (p. 442)
cost (p. 444)

accumulated cost (p. 444)
least-cost path (p. 449)
backlink (p. 449)

slice (p. 453)
Reclassify (p. 456)
path distance (p. 462)

20

How to Perform Map Algebra in ArcGIS Pro

ArcGIS Pro Skills

In this module, you will learn how to do the following in ArcGIS Pro:
- Use the ArcGIS Pro ModelBuilder environment for setting up and executing workflows.
- Use the Raster Calculator tool to perform map algebra and raster overlay.
- Combine raster layers with various Boolean and map algebra functions.
- Use a relational operator with a raster surface.

Introduction

In the last module, you combined two raster layers (representing slope and land-cover rankings) to form a new, third layer. By adding the two layers together, you performed a raster overlay. In ArcGIS Pro, raster layers can be overlaid through the use of **map algebra**, a means of using Spatial Analyst extension tools and operators to perform actions on rasters. As its name implies, map algebra involves combining datasets through mathematical operations, such as adding or multiplying rasters to create a new raster. You can use map algebra to mathematically combine the values of cells from multiple grids that share the same location.

Suppose, for instance, you want to rank locations based on the difficulty in traversing their terrain, considering factors such as the steepness of the slopes and the type of land cover, and you find the locations that are the most difficult (or easiest) to cross based on these two factors. See Figure 20.1 for an example of this scenario: Two grids are featured—a raster called sloperank (which ranks the difficulty of traversing the slope of the land on a scale from 1 to 5) and a second raster called landrank (which ranks the difficulty of traversing the land cover, also on a scale from 1 to 5). By combining the sloperank and landrank grids, map algebra forms the new output grid. The value for the cell in the upper-left corner in sloperank is 5, and the value for the cell in the upper-left corner in landrank is 3, so when these values are added, the cell in the upper-left corner of output is 8. The output grid represents the overlay of the sloperank and landrank grids, using an addition operation to show the cumulative difficulty of crossing a part of the landscape.

Learning Outcomes

After studying this module, you should be able to:
- Explain what map algebra is.
- Explain how a relational operator is used in map algebra.
- Describe the differences between the AND, OR, NOT, and XOR Boolean operators in map algebra.
- Define how raster overlay operates.
- Explain how weighted overlay operates.

map algebra An operation performed on rasters, such as overlaying two or more grids.

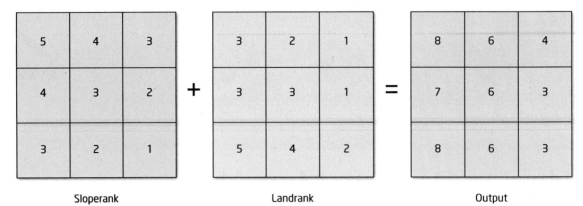

FIGURE 20.1 An example of using an addition operator in map algebra to combine two grids.

Other applications for combining grids are also available. For instance, if you're trying to find an initial set of locations for new housing developments, two criteria would be to build on relatively flat sections of land that are privately owned. See Figure 20.2 for an example of this scenario: There are two grids—a raster called flatareas (which features a value of 1 if the slope of the land is flat enough to build on and a value of 0 if it is not) and a second raster called privpub (which features a value of 1 if the section of land is privately owned and a value of 0 if it is publicly owned). By multiplying these two grids together, map algebra creates a new output grid. As in the previous addition example, the grid cells that correspond with each other (such as the value of 1 in the upper-left corner of the flatareas grid and the value of 0 in the upper-left corner of the privpub grid) are multiplied. The output grid shows that areas that meet both criteria have a value of 1 (because only cells that have a value of 1 in both the flatareas grid and the privpub grid could be multiplied together to get an output value of 1), and areas that meet one or neither criterion have a value of 0. Thus, only those sections of the landscape with an output value of 1 are suitable as the initial areas to build on.

Beyond math operators, there are several other forms of map algebra available (such as relational or Boolean operations) that we'll discuss in this module. In addition, this module will introduce the ArcGIS Pro ModelBuilder environment for laying out the various steps you'll use with map algebra. While all of the analysis in this module can be performed without using ModelBuilder, you'll likely find its visual interface especially helpful in conceptualizing the **workflow**, or the sequence of actions you're performing with various layers and tools.

workflow The sequence of actions performed using GIS layers and tools.

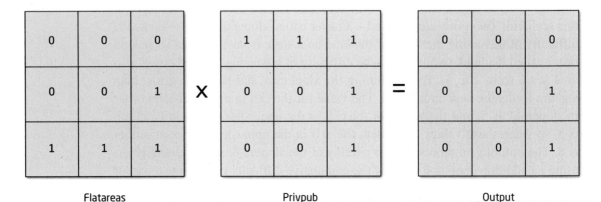

FIGURE 20.2 An example of using a multiplication operator in map algebra for combining two grids.

Module Scenario and Applications

This module puts you in the role of a developer in Columbiana County, Ohio. You are searching for the most suitable locations in the county for building an ecological preserve. The site must meet the following criteria: (1) It must be on a relatively flat area of land, (2) it must contain either forest or wetlands land cover, (3) it must be more than 500 meters from a highway, and (4) it must be within 500 meters of a lake or within 500 meters of a stream. The suitable sites for the ecological preserve will be areas that meet all these criteria; that is, these conditions are mandatory, and your site cannot leave out any of these items.

The following are additional examples of other real-world applications of this module's skills:

- An emergency preparedness center for a coastal city is simulating the effects of the amount of flooding generated by different categories of hurricanes. The center's staff members can use map algebra to examine selected levels of elevation models according to the amount of expected flooding as a starting point for their analysis.

- An environmental planner is examining a multistate distribution of forests and wetlands as part of a larger modeling initiative. She can use map algebra techniques with a large NLCD dataset to extract separate 30-meter layers of different forest and wetland types to use as model inputs.

- An urban planner is examining a statewide distribution of seasonal homes. He can use map algebra techniques in conjunction with layers such as NLCD to determine some of the most suitable places in the state for the locations of seasonal home developments.

Study Area

For this module, you will be working with data from Columbiana County, Ohio.

Data Sources and Localizing This Module

The data for this module focus on features and locations within Columbiana County, Ohio. However, you can easily modify this module to use data from your own local county instead—for instance, if you wanted to perform this module's applications in Beaver County, Pennsylvania. The DEM and NLCD2011 that were downloaded from The National Map would also be available for Beaver County; you could clip the layers to the county boundaries and process them to keep consistent 30-meter cell sizes for this exercise.

The lakes, streams, and UShighways layers were also derived from data downloaded from The National Map. If you were going to use these for Beaver County, Pennsylvania, you would download the Transportation geodatabase for that county and use the Trans_Roadsegment layer as the source for UShighways. Then you would use only road segments with non-null entries in the US_Route field. For lakes

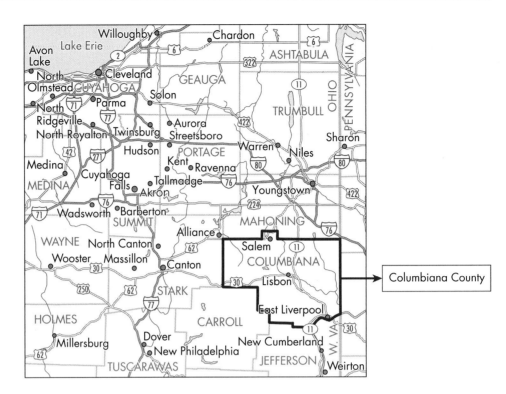

and streams, you would download the Hydrography geodatabase for that county and use the NHD_waterbody layer for lakes and the NHD_Flowline layer for streams.

Step 20.1 Getting Started

- **Important note:** Starting with ArcGIS 2.5, there are many new distance tools and methods available for use. In ArcGIS 2.5, the Euclidean Distance tools used in this module are located in the **Distance** toolset, then within the **Legacy** toolset. For more information on using the new distance tools with this module, see the book's Website at: **https://www.macmillanlearning.com/college/us/product/Discovering-GIS-and-ArcGIS-Pro/p/131923075X.**

- This module's hands-on applications use the data folder called Module20. Your instructor will be able to supply you with this data, or you can download it directly from this book's website at **https://www.macmillanlearning.com/college/us/product/Discovering-GIS-and-ArcGIS-Pro/p/131923075X**. The text in this module assumes that you have this Module20 folder in a computer location referenced as C:\GIS; if you have it somewhere else (for instance, in a flash drive referenced as G:\GISClass), substitute that location and path to the Module20 folder throughout this module.

- Module20 contains a file geodatabase called Colummodeldata that contains the following items:
    - UShighways: a line feature class representing the U.S. highways that run through the county.
    - lakes: a polygon feature class representing the lakes, ponds, and reservoirs in the county.
    - streams: a line feature class representing the various types of rivers and streams in the county.
    - NLCDColum: a raster of 30-meter NLCD 2011 data for Columbiana County.

- DEMcolum: a raster of DEM data for Columbiana County (which has been resampled to 30 meters for purposes of this module).
- Start ArcGIS Pro.
- Sign in with your Esri account username and password.
- Create a new project using the **Map** template. Call this project **Module20** and place it in your **C:\GIS\Module20** folder. Ensure that there is not a checkmark in the box next to **Create a new folder for this project**.
- When ArcGIS Pro opens, change the map's name to **Columbiana**.
- You will be using the ArcGIS Pro Spatial Analyst extension in this module. If you don't have access to the Spatial Analyst extension, see Smartbox 11.1 and Troublebox 11.1 in Module 11 for more information on how to get it set up in ArcGIS Pro.
- Add the **DEMcolum** layer to the Contents pane.
- Add the **NLCDColum** layer to the Contents pane.
- Next, add the **lakes**, **streams**, and **UShighways** feature classes to the Contents pane.
- Use the following geoprocessing environment settings for this module (on the **Analysis** tab, in the **Geoprocessing** group, click the **Environments** button):
  - For Current Workspace, use the **Module20.gdb** geodatabase within your **C:\GIS\Module20** folder.
  - For Scratch Workspace, use the **Module20.gdb** geodatabase within your **C:\GIS\Module20** folder.
  - For Output Coordinate System, choose **DEMcolum**. ArcGIS Pro then updates the coordinate system to that of the **DEMcolum** layer.
  - For Extent, choose **DEMcolum**. ArcGIS Pro then updates the x and y coordinates of the lower-left corner and upper-right corner of the grid extent to those of the DEMcolum layer.
  - For Cell Size, choose **DEMcolum**. ArcGIS Pro then updates the cell size for raster output in this project to 30 meters as that is the cell size of the DEMcolum layer.
- Leave the other settings alone and click **OK** to put these geoprocessing environment settings into place.
- The coordinate system being used in this module is **Albers Conical Equal Area (NAD 83)**, using **meters** as the map units. Check Columbiana Map Properties (on the Coordinate System tab) to verify that this coordinate system is being used.

Step 20.2 Creating a New Model

- You'll be setting up your workflow as a model, and ArcGIS Pro will treat a new model as a tool because it's composed of one or more geoprocessing tools. To create a new, empty model, on the **Analysis** tab, in the **Geoprocessing** group, click the **ModelBuilder** button. A new tabbed view called Model opens, providing a large blank space to work in.

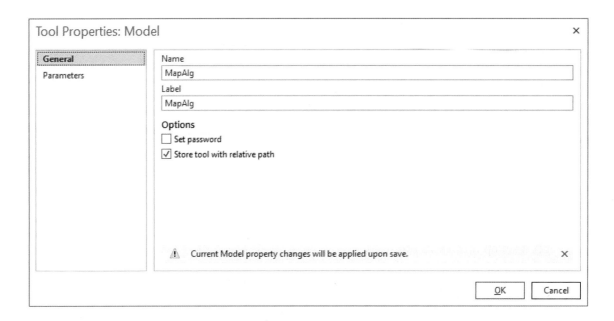

- With the new Model tab chosen, a new tab called ModelBuilder appears on the ribbon. On the **ModelBuilder** tab, in the **Model** group, click the **Properties** button.
- In the Tool Properties: Model dialog box that opens, change the Name field to **MapAlg** and also change the Label field to **MapAlg**. Leave the other settings alone and click **OK** to close the dialog. The name on the tab changes from Model to MapAlg (for "map algebra"), which will be the name of your model.
- On the **ModelBuilder** tab, in the **Model** group, click the **Save** button. ArcGIS Pro writes any changes you've made to the model (and so far, all you've done is change its name) to disk.
- Also, in the Catalog pane, expand the **Module20** toolbox (which was the default toolbox created for the Module20 project). Inside it, you can see an item with the model symbol called MapAlg, which is the MapAlg model you've just created.
  - On the ModelBuilder tab, in the **Insert** group, click the **Tools** button. The Geoprocessing pane opens. In the Geoprocessing pane, click the **Toolboxes** button to display the full set of toolboxes to use. You're now ready to start making a model. For more information about the basics of ModelBuilder, see **Smartbox 20.1**. Refer to **Troublebox 20.1** if you run into later difficulties with saving and reopening a model in ArcGIS Pro.

SMARTBOX 20.1

What is ModelBuilder and how is it used in ArcGIS Pro?

ModelBuilder The interface of ArcGIS Pro used in laying out workflows and creating models.

ModelBuilder is an ArcGIS Pro graphical environment that allows you to build workflows and then run them to create outputs. ModelBuilder is a drag-and-drop environment in which you select data layers from the Contents pane or the Catalog pane (or use tools from the Geoprocessing pane) and place them into

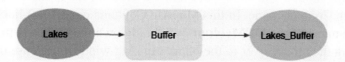

FIGURE 20.3 A simple ModelBuilder workflow showing a layer connected to a tool connected to an output.

a visual layout for designing a workflow. In ModelBuilder, by default, the data layers used in the model are displayed as blue ovals, the tools being used are displayed as yellow rectangles, and the outputs from the tools are displayed as green ovals. Lines and arrows show the connections among these elements. Figure 20.3 shows a workflow that started with a data layer called lakes; the Buffer tool was used on this layer to end up with a resultant layer called lakes_Buffer.

When you place a layer or tool in the ModelBuilder environment, you can move the element to anywhere inside the window. Think of using ModelBuilder the way you would think about designing a flowchart that shows how various layers and tools fit together. ModelBuilder allows you to run the processes and create the various outputs. In this module you will design and run a basic workflow with ModelBuilder. In Module 21, you'll learn many more aspects and features of ModelBuilder.

TROUBLEBOX 20.1

Why can't I reopen my saved model?

You can save a model and then reopen it later, when you want to continue designing or run it. If you have saved and closed a model and want to reopen it to continue to work, navigate to the toolbox containing your model and expand it. (In this module, you will work with the MapAlg model, stored in the Module20 toolbox and located in the C:\GIS\Module20 folder.) You'll see the icon for the model inside the toolbox. Right-click on the model, and a new set of options appear. If you select **Open**, however, a tool opens in the Geoprocessing pane with a statement saying "no parameters," and your design will not all appear.

Don't panic; all your model work is still there. Until you formally assign model parameters to the model (which you will do in Module 21, when you make use of parameters in building a model), the model can still be opened. When you right-click on the model icon, select **Edit**. Your model reopens in a new tab for you to continue to work with it. If you save it and close it again, just use the Edit option to reopen the model.

Step 20.3 First Criterion: Must Be on a Relatively Flat Piece of Land

- The DEMcolum layer contains elevation information, with each cell coded with an elevation value. To get information about relatively flat (versus steep) land, you would first need to calculate the slope of the land. Thus, DEMcolum is the first source grid to use in the model. You should see that the MapAlg tab in the view has its own Contents pane. In this MapAlg Contents pane, left-click once on the name of the **DEMcolum** grid, hold down the mouse button, and drag the mouse over to the blank MapAlg window. Release the mouse button, and the first source is added to the model—a blue oval called DEMcolum.

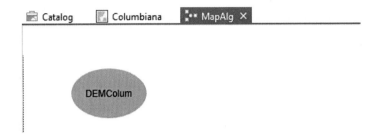

- Next, you need to calculate the slope of the DEM to find places that are relatively flat and places with a relatively steep slope. In the Geoprocessing pane, expand the **Spatial Analyst Tools** toolbox. Then open the **Surface** toolset. Finally, choose the **Slope** tool and drag and drop it from the Geoprocessing pane into the MapAlg window.

- Two new items are added to the MapAlg view: a box labeled **Slope**, and another oval connected to it, labeled **Output raster**.

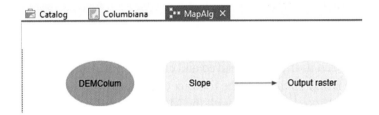

- To complete the workflow, you need to connect DEMcolum to Slope and finally to Output raster. Click on the DEMcolum oval, hold down the mouse button, and drag a line between the DEMcolum oval and the Slope box. You see an arrow appear between the oval and the box (this arrow is referred to as a *connector*). Release the mouse on the Slope box, and a new pop-up menu appears.

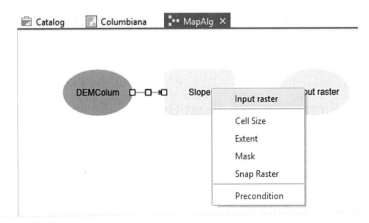

Step 20.3 First Criterion: Must Be on a Relatively Flat Piece of Land   471

- In this new pop-up window, choose **Input raster** to tell ArcGIS Pro that the raster to be used with the Slope tool is the DEMcolum layer.

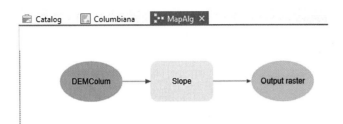

- The Slope box turns yellow, the Output raster oval turns green, and you can see that the basic workflow and connections of the model have been established. Answer Question 20.1.

> **Question 20.1** The model (such as it is) could be run right now. What (specifically) would happen in ArcGIS Pro and what exactly would the model output be (at this stage)?

- To make some further changes, double-click on the **Slope** box.
- A pop-up window appears with the Slope tool (much as with the Geoprocessing pane tools you've been working with). In this pop-up window, use the following settings:
  - For Input raster, **DEMcolum** should already be chosen (from when you drew the connector arrow and chose **Input raster**).
  - For Output raster, click the **browse** button, navigate to your **C:\GIS\Module20** folder, go to the **Module20.gdb** geodatabase, and name the output raster **Sloperaster**.
  - Leave the other settings alone and click **OK**.

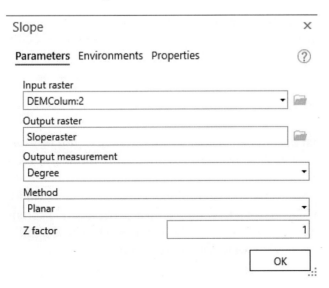

- The green output oval is updated to become Sloperaster instead of the generic default Output raster. You now want to find areas in that Sloperaster grid that have less than a 10-degree slope (to meet the first modeling condition). To do so, you have to query the Sloperaster layer to find values of less than 10. A relational tool is needed to do this. See **Smartbox 20.2** for more about using relational operators in map algebra.

SMARTBOX 20.2

What is a relational operator and how is it used in map algebra?

relational operator One of the six operators (equal, not equal, greater than, greater than or equal to, less than, less than or equal to) used when building a Raster Calculator expression.

A **relational operator** is one of six different connectors used in a Raster Calculator expression (similar to what you used back in Module 2 in a SQL statement): equal to (==), not equal to (!=), greater than (>), greater than or equal to (>=), less than (<), and less than or equal to (<=). In map algebra, each relational operator could be used to build an expression that will create a new grid containing the value 1 for each grid cell for which that expression is true and the value 0 for each grid cell for which that expression is false. For instance, in Figure 20.4 the expression being evaluated is rasterlayer >= 5. Every grid cell in the raster layer contains a value and that value will be compared with 5. An output raster will be created with the same dimensions as the raster layer, but it will contain only values 0 and 1. For each cell in the raster layer greater than or equal to a value of 5, the output raster will contain the value 1. For each cell in the raster layer less than a value of 5, the output raster will contain the value 0.

FIGURE 20.4 An example of a map algebra relational operation creating a new grid.

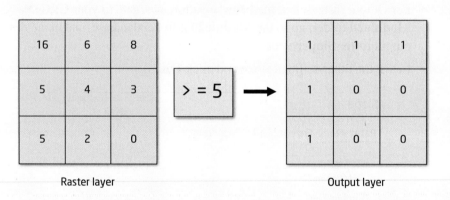

A relational operator can be used to create a new grid of values higher or lower than a number (for instance, to extract only those grid cells of a DEM that are above or below a certain elevation, such as DEMcolum > 300, which would find all grid cells of an elevation above 300 meters). Similarly, a relational operator can be used to extract only grid cells with a certain value (for instance, an expression such as NLCDColum == 11, which would extract only grid cells that represent a water land-cover type).

- Back in the Geoprocessing pane, from the **Spatial Analyst Tools** toolbox, choose the **Map Algebra** toolset and then choose the **Raster Calculator** tool and drag and drop it from the Geoprocessing pane into the ModelBuilder window. Position the Raster Calculator box to the right

Step 20.3 First Criterion: Must Be on a Relatively Flat Piece of Land

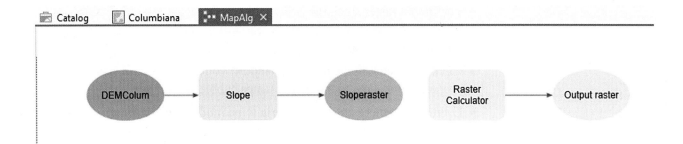

of the Sloperaster output from the Slope tool. **Raster Calculator** is the ArcGIS Pro tool used for performing map algebra.

- Draw a connecting arrow from Sloperaster to the Raster Calculator box.
- In the pop-up menu that appears, choose the option **Map Algebra expression**. Answer Question 20.2.

> **Question 20.2**
> Why did the Raster Calculator tool not turn yellow?

- In the MapAlg window, double-click on the **Raster Calculator** tool (the white box). The Raster Calculator tool dialog box opens, so you can make changes to its parameters.

- Use the Raster Calculator tool to build an expression for Sloperaster grid values less than 10 (because you want slopes of less than 10 degrees). You build Raster Calculator expressions much the way you built a query in Module 2: First double-click the model layer you want to work with (a layer with a blue symbol inside the yellow box, in this case **Sloperaster**) and then choose the operator you want to use (in this case, the < operator for a less-than operation), press the **Spacebar** to separate the relational operator and the value, and finally use the keyboard to type **10** as the value to use.

Raster Calculator An ArcGIS Pro tool used for map algebra operations.

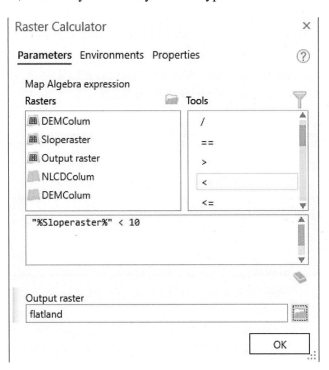

- For Output raster, click the **browse** button, navigate to your **C:\GIS\Module20** folder, go to the **Module20.gdb** geodatabase, and name the output raster **Flatland**.

- Click **OK**. Back in the MapAlg window, you see the Raster Calculator tool properly connected to the output raster bubble and ready to go.

- Even though there are more criteria to be modeled, for now you can save the MapAlg model again. Then to run the model, on the **ModelBuilder** tab, in the **Run** group, click the **Run** button. The two processes (the yellow boxes) run, and you see their boxes turn red as they are running. The final output raster, called **flatland**, is then created.

- Return to the Columbiana map tab view and add the flatland grid to the Contents pane.

    Important Note: If you are adding the flatland grid from the Module20 .gdb geodatabase in the Catalog pane and the geodatabase is empty, right-click on the geodatabase and choose **Refresh**. Doing so should prompt ArcGIS Pro to update the geodatabase with any changes or new layers that have been added to it. Do the same if adding something from the Module20 folder itself if it does not contain the layers you are looking for.

- Examine the flatland grid and answer Question 20.3.

Question 20.3 What do the values 1 and 0 in the flatland grid represent?

Step 20.4 Second Criterion: Must Contain Forests or Wetlands

- The NLCDColum grid contains NLCD classification regarding land-cover information. For the model, you are interested in grid cells representing forested land-cover classes (class 41, class 42, and class 43) and grid cells representing wetlands land-cover classes (class 90 and class 95).

    Important Note: Refer to Module 12 for what, specifically, each of these land-cover classifications represents.

- In the MapAlg tabbed view, drag and drop the **NLCDColum** grid from the Contents pane into the ModelBuilder window.

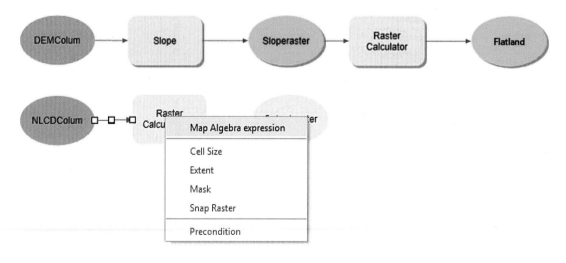

- To run this criterion, in the Geoprocessing pane, choose the **Spatial Analyst Tools** toolbox, then choose the **Map Algebra** toolset, then choose the **Raster Calculator** tool and drag and drop it from the Geoprocessing pane into the MapAlg window.

- Draw a connection between NLCDColum and the Raster Calculator tool. Choose the option **Map Algebra expression** when making the connection.

- The Raster Calculator box is still not yellow, so double-click on it to set its parameters. For more about building more complex expressions using map algebra, see **Smartbox 20.3**.

SMARTBOX 20.3

How are more complex expressions built using map algebra?

Rasters can be overlaid using the basic mathematical operations, but many types of overlays or analysis require you to use more complex map algebra expressions. For instance, if you want to create a new grid that contains all of the input layer's values of 2 and all of the input layer's values of 3, just using a single relational operator or mathematical expression will not work. In essence, you would need ArcGIS Pro to evaluate the input grid with two expressions at the same time (input grid == 2 or input grid == 3) to create an output grid that has a value of 1 where the input grid had a value of either 2 or 3. Several map algebra expressions can be evaluated simultaneously using a **Boolean operator**.

Four Boolean operators are used with the expressions or rasters: AND, OR, XOR, or NOT.

The **AND** Boolean operator is used to find the intersection of two expressions. Where both expressions are true, the output grid has the value 1, and where only one (or neither) of the expressions is true, the output grid has the value 0. For instance, Figure 20.5 contains two rasters: elevation and landcover. Say that you want to find all grid cells with an elevation value less than 300 and also all land-cover values equal to 11. The map algebra expression (elevation < 300) AND (landcover == 11) creates a new grid with values of 1 for cells that meet both criteria and values of 0 for cells that meet one or neither of the criteria. Note that in the Raster Calculator, the AND operation is represented with the **&** symbol.

Boolean operator (map algebra) One of the four operators (AND, OR, XOR, NOT) used when building a Raster Calculator expression.

AND (map algebra) The Boolean operator used with an intersection of two grids.

FIGURE 20.5 An example of using the AND map algebra expression.

310	320	322
305	320	330
295	297	290

Elevation

<300 AND

11	11	22
31	23	22
11	21	11

Landcover

== 11 →

0	0	0
0	0	0
1	0	1

Output

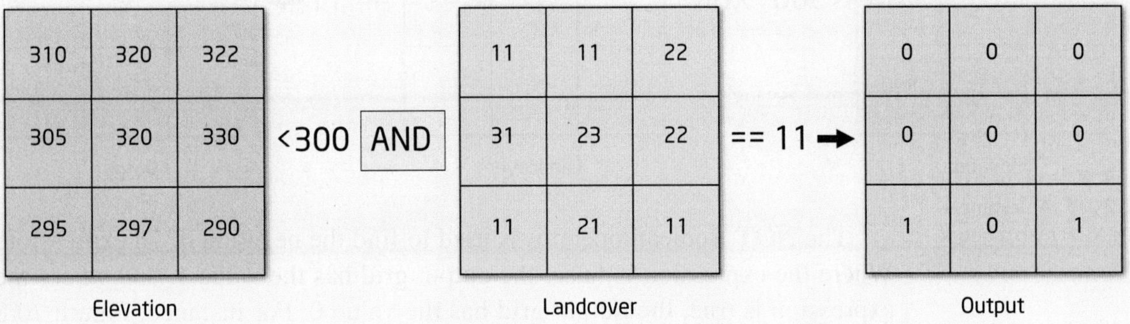

OR (map algebra) The Boolean operator used with a union of two grids.

The **OR** Boolean operator is used to find the union of two expressions. Where one or both of the expressions are true, the output grid has the value 1, and where neither of the expressions is true, the output grid has the value 0. For instance, Figure 20.6 contains two rasters: elevation and landcover. Say that you want to find all grid cells with an elevation value less than 300 as well as all land-cover values equal to 11. The map algebra expression (elevation < 300) OR (landcover == 11) creates a new grid with values of 1 for cells that meet one or both criteria and values of 0 for cells that meet neither of the criteria. Note that, in the Raster Calculator, the OR operation is represented with the | symbol.

FIGURE 20.6 An example of using the OR map algebra expression.

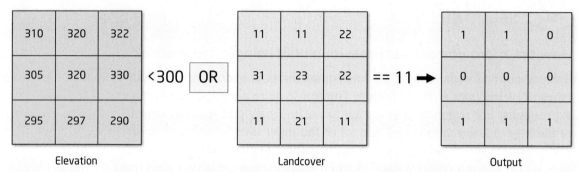

Elevation | Landcover | Output

XOR (map algebra) The Boolean operator used with an "exclusive or" of two grids.

NOT (map algebra) The Boolean operator used with a negation of a grid or expression.

The **XOR** Boolean operator is used to find the "exclusive or" overlay of two expressions. Exclusive or refers to finding everything except what the two expressions have in common with each other. When one of the expressions is true, the output grid has the value 1, and where both (or neither) of the expressions are true, the output grid has the value 0. For instance, Figure 20.7 contains two rasters: elevation and landcover. Say that you want to find all grid cells with an elevation value less than 300 or all land-cover values equal to 11—but not both conditions. The map algebra expression (elevation < 300) XOR (landcover == 11) creates a new grid with values of 1 for cells that meet one criterion and values of 0 for cells that meet neither (or both) of the criteria. Note that in the Raster Calculator, the XOR operation is represented with the ^ symbol.

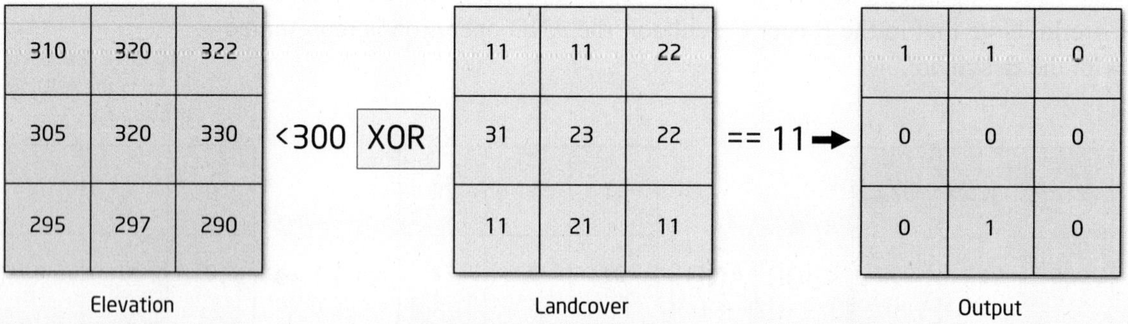

FIGURE 20.7 An example of using the XOR map algebra expression.

The **NOT** Boolean operator is used to find the negation of an expression. Where the expression is false, the output grid has the value 1, and where the expression is true, the output grid has the value 0. For instance, Figure 20.8

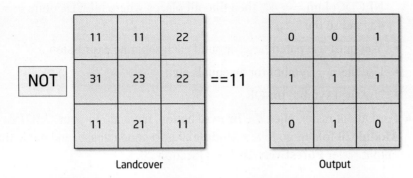

FIGURE 20.8 An example of using the NOT map algebra expression.

contains one raster: landcover. Say that you want to find all grid cells that have a land-cover value not equal to 11. The map algebra expression NOT (landcover == 11) creates a new grid with values of 1 for cells that are any value other than 11 and 0 for cells that have a value of 11. Note that in the Raster Calculator, the NOT operation is represented with the ~ symbol.

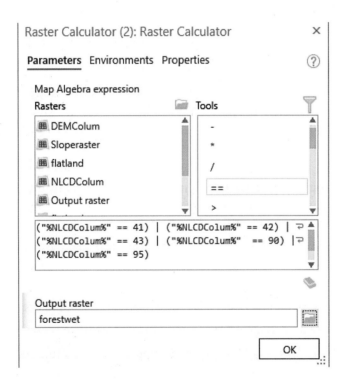

- In the Raster Calculator, build a query to find values from the NLCDColum layer that are equal to 41, 42, 43, 90, or 95. Keep in mind the following when building this expression:
  - Use the model version of NLCDColum (the one with the blue symbol inside the yellow box), not the layer version (the one with only the yellow box next to it).
  - The Raster Calculator requires you to input each expression individually, separated by a Boolean connector. So you would first want to find all places where NLCDColum == 41, then find all places where

NLCDColum == 42, then find all places where NLCDColum == 43, and so on.

- You must use parentheses around each separate expression.
- Use the == symbol for an equals sign.
- Use the | symbol for OR.

• For Output raster, click the **browse** button, navigate to your **C:\GIS\Module20** folder, go to the **Module20.gdb** geodatabase, and name the output raster **Forestwet**. Answer Question 20.4.

> **Question 20.4** What would be the exact output grid result if you used AND instead of OR in each case in this map algebra expression? Why?

• Click **OK** to complete the setup of the tool. Back in the MapAlg window, the new criterion is completed. You should see the arrow stretching from the NLCDColum layer to the now-yellow Raster Calculator tool.

• Save your model.

Step 20.5 Third Criterion: Must Be More Than 500 Meters from a Highway

• Drag and drop the **UShighways** feature class from the Contents pane into the MapAlg window. You might want to rearrange the window, zoom, or expand it to give yourself more room to work (since there are still more criteria to be modeled).

• To compute distance from highways, in the Geoprocessing pane, choose the **Spatial Analyst Tools** toolbox, then choose the **Distance** toolset, then choose the **Euclidean Distance** tool and drag and drop it from the Geoprocessing pane into the MapAlg window.

• Add a connection between the UShighways oval and the Euclidean Distance tool and choose **Input raster or feature source data** when making the connection.

• Double-click on the model's **Euclidean Distance tool** to open it.

• Make sure all parameters (especially the cell size) are set correctly:

- **UShighways** should be set as the input raster or feature source data.
- The default name for the distance calculation should be **EucDist_UShi1**.
- The output cell size should be **30** (Which is the cell size of DEMColum, which should already be set because of your environment settings in Step 20.1).
- Click **OK**. The tool is now properly set up in ModelBuilder. You should see that in addition to your green Output distance raster, a separate output oval labeled Output direction raster and an additional output oval labeled output distance raster are extended from the Euclidean Distance tool. These are both optional outputs from the Euclidean Distance tool that you won't be using in this module, so you can just ignore them or move them out of the way.

- Now that you've calculated distance, you have to find areas greater than 500 meters from a highway, and you need another relational tool to do this. From the Geoprocessing pane, in the **Spatial Analyst Tools** toolbox, choose the **Map Algebra** toolset and then choose the **Raster Calculator** tool and drag and drop it from the Geoprocessing pane into the MapAlg window.

- Add a connection between the Output distance raster and the Raster Calculator tool. Select **Map Algebra expression** when prompted and open the tool itself.

- Build an expression in the Raster Calculator of the Output distance raster grid greater than 500 (because you want to find places more than 500 meters from a highway). To do so, select the layer with which you want to work (in this case, **Output distance raster**), select the operator you want to use (in this case, the > operator for a greater than operation), and then type the value to use (in this case, **500**).

- For Output raster, click the **browse** button, navigate to your **C:\GIS\Module20** folder, go to the Module20.gdb geodatabase, and name the output raster **awayhighs**.

- Click **OK**.

- Save your model.

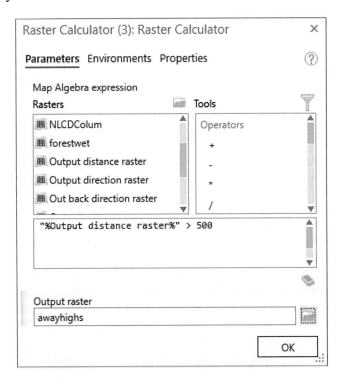

Step 20.6 Fourth Criterion: Must Be Within 500 Meters of a Lake or of a Stream

- Repeat the same process as for the third criterion in Step 20.5 to compute the distance from the Columbiana County streams. This time, use the less than (<) operator to select areas within 500 meters of streams. For Output raster, click the **browse** button, navigate to your **C:\GIS\Module20** folder, go to the **Module20.gdb** geodatabase, and name the output raster **nearstreams**. Also make sure that all grid cell sizes are 30 meters.

- Repeat the same process, computing distance from the Columbiana County lakes. This time, use the less than (<) operator to select areas within 500 meters of lakes. For Output raster, click the **browse** button, navigate to your **C:\GIS\Module20 folder**, go to the **Module20.gdb** geodatabase, and name the output raster **nearlakes**. Also make sure that all grid cell sizes are 30 meters.

- Save your model.

Step 20.7 Completing and Running the Map Algebra Model

- By this point, you should have the results of five sets of outputs:
  - Flatland: places on the landscape with a relatively flat slope
  - Forestwet: land uses that are forests or wetlands
  - Awayhighs: areas that are more than 500 meters from a highway
  - Nearstreams: areas that are within 500 meters of a river
  - Nearlakes: areas that are within 500 meters of a lake

- You now have to combine (or overlay) all five of these grids in order to find the most suitable sites for the ecological preserve, using the criteria laid out at the beginning of this module. You need to use Boolean operators to combine these layers. See **Smartbox 20.4** for more about overlaying multiple grids with Boolean operators.

SMARTBOX 20.4

How are rasters overlaid with Boolean operators using map algebra?

Two or more grids can be overlaid using Boolean operators, in which case the values of one grid are evaluated against the values of a second grid, using a Boolean operator. When a Boolean operator is used in this fashion, nonzero values are considered to be the "true" condition, and zero values are considered to be the "false" condition. For example, Figure 20.9 contains two grids: lowland and water. The lowland grid contains a value of 1 if the cell represents

FIGURE 20.9 Using an AND expression to overlay two grids.

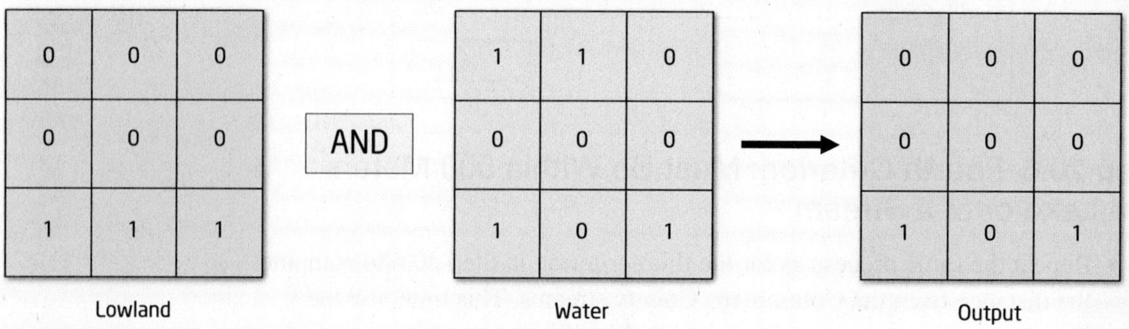

a low elevation and a value of 0 for higher elevation, while the water grid contains a value of 1 if the grid cell is classified as a water land-cover type and a 0 if it is some other land-cover type. You want to overlay the two grids to find only those cells that are both low elevation and classified as water. You can use the expression (lowland AND water) to produce an output grid with a value of 1 for those cells that have a nonzero value in the lowland grid and a nonzero value in the water grid.

Multiple Boolean operators may be used in a single map algebra expression to overlay more than two grids. Figure 20.10 contains three grids: lowland, water, and nearroads. The lowland grid contains a value of 1 if the cell represents a low elevation and a value of 0 for higher elevation. The water grid contains a value of 1 if the grid cell is classified as a water land-cover type and a 0 if it is some other land-cover type. The nearroads grid contains a value of 1 if the grid cell is close to a road and a value of 0 if the grid cell is far from a road. You want to overlay the three grids to find only those cells that are both low elevation and classified as water, in addition to those cells that are near a road. You can use the expression ((lowland AND water) OR nearroads) to produce an output grid with a value of 1 for those cells that have either a 1 in both the lowland grid and the water grid, or a 1 in the nearroads grid.

FIGURE 20.10 Using multiple Boolean operators to overlay three grids.

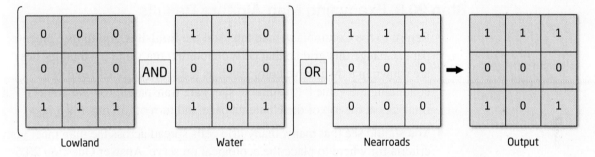

Important Note: Think very carefully about your overall goal in this project and how to combine layers using the Boolean operators. Also keep in mind how parentheses are used in an expression as well as how the order of operations of mathematics is evaluated. Before you do anything else in ModelBuilder, answer Question 20.5.

> Write down the operation needed to find the suitable sites using the five final output grids and appropriate logical operators. For example, you would write something like *flatland XOR (awayhighs OR nearstreams)* — or whatever set of grids and operators find the suitable sites.
>
> **Question 20.5**

- In the Geoprocessing pane, from the **Spatial Analyst Tools** toolbox choose the **Map Algebra** toolset and then choose the **Raster Calculator** tool and drag and drop it from the Geoprocessing pane into the MapAlg window. Position this tool to the right side of all of the five criteria outputs.

- Make a connection from each of the five criteria outputs to the Raster Calculator tool, using the **Map Algebra expression** for each one.

- Next, open the Raster Calculator tool and construct the map algebra expression to properly combine these five grids, using Boolean operators (AND, OR, NOT, XOR) and parentheses.

- For Output raster, click the **browse** button, navigate to your **C:\GIS\Module20** folder, go to the Module20.gdb geodatabase, and name the output raster **finalsites**. Click **OK** to close the Raster Calculator tool. You should see that all five of your criteria outputs now have black arrows connecting them to the tool.

- On the **ModelBuilder** tab, in the **Run** group, click the **Validate** button. ArcGIS Pro checks to verify that everything in the model (elements, connections, and so on) is okay, based on the inputs you've indicated. Finally, on the **ModelBuilder** tab, in the **Run** group, click the **Run** button. ModelBuilder runs each tool and creates the output from each of them. You see each tool turn red as it is run. When the model finishes running, you can close the report that ArcGIS Pro gives you.

Step 20.8 Examining Map Algebra Results

- Return to the Columbiana map tab. Add the **finalsites** grid to its Contents pane. Change this grid so that cells containing the value 1 (suitable sites) are shown in a distinctive color (such as red, blue, or black) and cells containing the value 0 (unsuitable sites) are transparent (or "no color"). Put finalsites at the top of the Contents pane and turn off all other layers.

- You should see that many, many grid cells spread across the map meet your criteria for where to place the ecological preserve. Answer Question 20.6.

Question 20.6 How many grid cells are considered suitable sites for the ecological preserve? In real-world values, how many square meters are considered suitable sites?

- Insert a new map into the Module20 project and change its name to **Most Suitable**. In the Contents pane of the Columbiana map, choose the **finalsites** layer and then, on the **Map** tab, in the **Clipboard** group, click the **Copy** button. Switch to the **Most Suitable** map and then, on the **Map** tab, in the **Clipboard** group, click the **Paste** button. The Most Suitable map now contains a duplicate of the finalsites layer. Copy and paste the **NLCDColum** layer into the Most Suitable map and arrange the layers so that the finalsites layer sits on top of the NLCDColum layer. If necessary, use the layer properties to change the colors of the finalsites layer and the land-cover classifications of the NLCDColum layer to make analysis easier on you (that is, to make the suitable sites really stand out from the land uses).

- You will now assess the locations of the ecological preserve in relation to the other land-cover features. Examine the finalsites layer in relation to the NLCDcolum layer and answer Question 20.7. When answering this question, keep in mind the kinds of land cover you would want to have adjacent to an ecological preserve, what types would be the least desirable, and the sizes of the suitable sites in terms of contiguous grid cells. Zoom the Most Suitable map in closely on the site that corresponds with your answer.

> **Question 20.7** Choose one site that you (as the developer in this scenario) consider to be the most suitable for development of an ecological preserve in terms of its size and, most importantly, in relation to other land uses. Explain (in detail) why you would choose this area the most suitable location.

- Insert a new map into the Module20 project and change its name to **Least Suitable**. As you did before, copy and paste the **finalsites** and **NLCDColum** layers into the Least Suitable map, placing finalsites on top of NLCDColum, change their colors as necessary, and answer Question 20.8. Zoom the Least Suitable map in closely on the site that corresponds with your answer.

> **Question 20.8** Choose one site that you (as the developer in this scenario) consider to be the least suitable for development of an ecological preserve in terms of its size and, most importantly, in relation to other land uses. Explain (in detail) why you would choose this area the least suitable location.

Step 20.9 Saving and Printing Your Results

- Save your model.
- Save your Module20 project.
- Compose and print a layout (see Module 3) that consists of the following elements:
  - Three map frames: The first map frame should show the most suitable site (as described in your answer to Question 20.7), the second map frame should show the least suitable site (as described in your answer to Question 20.8), and the third map frame should show only the finalsites layer on the basemap at the extent of the entire county. Each map frame should have its own scale bar. The most suitable and least suitable sites should be large and very clear.
  - One legend to clearly refer to all three map frames: This legend should contain the land-cover types from NLCDColum as well as a designation of the final sites. The land-cover classifications in the legend should be properly labeled (that is, make sure your legend says things like "woody wetlands" or "open water" instead of things like "95" or "11").
  - Finishing touches: Add a title, a north arrow, text with your name, the date, and the data source for the layout as a whole.

- Finally, using the finalsites layer for the entire county, draw a circle around the site that corresponds with the most suitable and least suitable sites and draw a thick line from them to each of the map frames that contains the zoomed-in views of these sites. To add shapes, lines, and other graphical elements to the layout, on the **Insert** tab, in the **Graphics** group, select the appropriate tools.

 Closing Time

This module examined how to use mathematical, relational, and Boolean operators on rasters with map algebra for a variety of different applications. Map algebra allows you to work with grids in several ways, including querying rasters and performing raster overlay. However, many more options are available when you are overlaying rasters. For instance, this module assumed that all five of the criteria were of equal importance when combining them to find potential sites for the ecological preserve. If one of the criteria were more important than the others (if, for example, the presence of forests or wetlands were twice as important as the other factors), you would have to give more weight to that layer than to the others. See *Related Concepts for Module 20* for more information about performing weighted overlays with rasters in ArcGIS Pro.

In this module, you worked with ModelBuilder to visualize the workflow involved in performing actions on the various layers and overlaying them using the Raster Calculator. ModelBuilder is an excellent tool for setting up a workflow to track how layers and tools connect with one another. In this module, you in essence created a simple "model" to identify possible ecological preserve sites. However, much more can be done with ModelBuilder; as the name implies, it's also used for creating GIS models, which you will do in the next module. In addition, you'll create a more detailed type of model for ranking and assessing the suitability of sites rather than just for visual examination and your own decision making.

RELATED CONCEPTS FOR MODULE 20

Using Weighted Raster Overlay

When combining layers, at times you'll find that some layers are more important, or should carry more weight, than other layers. For instance, say that a historical preservationist is attempting to determine which sections of the landscape are under the greatest development pressures, based on a variety of conditions (such as proximity to existing urban areas or the slope of the land). Based on key informant interviews and other research, she determines that the proximity to water bodies is 80 percent of the overall importance, while the slope of the land is merely 20 percent of the overall importance in selecting a site. Thus, she can't simply add the two grids together. Rather, she needs to give a greater weight to the proximity-to-water-bodies grid and then combine the two grids. This process of assigning relative weights to layers during overlay is referred to as creating a **weighted overlay**.

weighted overlay An overlay that is a combination of two or more grids, with each grid assigned a relative importance.

Figure 20.11 shows an example of a weighted overlay. In this example, two sample grids (proxwater and slopeland) simulate two factors. The first has values that represent the distance, in meters, to a body of water, and the second shows the degree of slope of the land. However, you can see that the values for the proxwater and the slopeland grids are completely different: A value of 15 in the slopeland grid indicates a 15-degree slope, while a value of 15 in the proxwater grid indicates 15 meters from a body of water. Because the numerical value 15 means two completely different things in the two grids, you can't simply add these two values together.

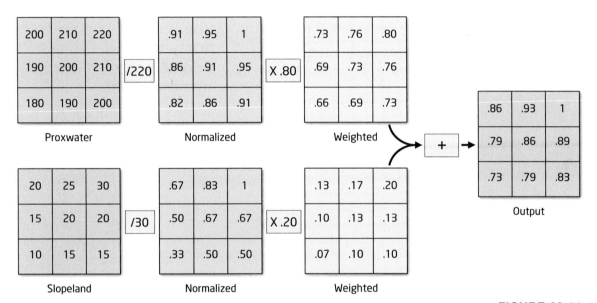

FIGURE 20.11 The weighted overlay process, showing how two grids are normalized, weighted, and then combined together using map algebra techniques.

The value 220 also has different meanings in different grids. In proxwater, this number indicates a location 220 meters from a body of water, while in slopeland, this number means a 220-degree slope (which is impossible). Before you can combine the rasters, you must first bring these raster values to a level at which they mean the same thing, such as expressing them as percentages. Altering all values so that they are on the same level results in values that have been **normalized**. (See Module 3 for more about normalizing numerical values.) A simple way to normalize raster values is to divide each grid cell value by the highest overall value in the raster (using a map algebra division operation). After normalization, the values in the two grids mean the same thing.

Now say that, based on research, you know that the proxwater grid should make up 80 percent of the analytical weight, and the slopeland grid should consist of 20 percent of the weight. The process of weighting each grid involves multiplying the (normalized) values by the assigned weight. In this case, all the normalized proxwater grid values will be multiplied by 0.80, and the normalized slopeland grid values will be multiplied by 0.20 (using a map algebra multiplication operation). Proper selection of weights is often an involved process, sometimes informed by expert opinions on the subject at hand or crafted through complicated mathematical or computerized processes.

normalize To alter count data values so that they represent the same level of data (for example, as a percentage).

Once each grid has been weighted, you can add together the grids (with a map algebra addition operation) to create a ranking of the site locations. The final output can be interpreted as "The higher the grid value, the more appealing the location for urban development, based on the concept that sites farthest from water are much more desirable than sites with a higher degree of slope of the land." In ArcGIS Pro, the Weighted Overlay tool helps with these tasks.

Key Terms

map algebra (p. 463)
workflow (p. 464)
ModelBuilder (p. 468)
relational operator (p. 472)
Raster Calculator (p. 473)
Boolean operator (p. 475)
AND (p. 475)
OR (p. 476)
XOR (p. 476)
NOT (p. 476)
weighted overlay (p. 484)
normalize (p. 485)

21

How to Build a Model in ArcGIS Pro

ArcGIS Pro Skills

In this module, you will learn how to do the following in ArcGIS Pro:
- Construct, validate, and run a model by using ModelBuilder.
- Add model parameters to a model.
- Use multiple tools in a model.
- Use NODATA in reclassification.
- Use the Slice tool to create rankings.
- Display rasters based on attributes other than Value.
- Create and evaluate a suitability index.
- Create a geoprocessing package from your model for sharing online.

Learning Outcomes

After studying this module, you should be able to:
- Define what a model is in GIS.
- Explain the process of cartographic modeling.
- Explain what a suitability index is.
- Describe the basic workflow for creating a site suitability model.

Introduction

When you're working with a GIS project, it is likely to involve the use of multiple types of datasets, tools, and methods. For example, a GIS analysis for determining which sections of coastal and inland areas are most likely to be converted to an urban land use through development would need to examine a wide variety of factors. Proximity to the shoreline, proximity to existing urban areas, and the designation of private versus public lands are just a few of the factors that would be required in this kind of analysis. With GIS, you would use a variety of tools and different spatial datasets to design and then combine all these factors. You would then analyze the results to find which areas of land are most likely to be developed.

A **GIS model** examines all these factors in an attempt to describe or explain a process or to predict results. Numerous types of GIS models can be found in the literature for a variety of different applications. You can use ModelBuilder in ArcGIS Pro to design and share your own models. In ArcGIS Pro, a **model** is a workflow using different layers and tools that can also be used as its own separate tool. In Module 20, you used ModelBuilder to design a simple model to implement a workflow of various map algebra concepts. You can use that same graphical design interface to create a standalone tool that can be executed to combine multiple tools and datasets, which you can then share with others.

A model consists of a set of **variables**, or different types of factors, that are combined as part of the overall analysis. For instance, in the earlier example, "proximity

GIS model A representation of the factors used for explaining the processes that underlie an event or for predicting results.

model An ArcGIS Pro tool that combines data layers, geoprocessing tools, and outputs in a workflow.

variables Different factors that are combined into an overall analysis.

FIGURE 21.1 One part of a sample model created using ModelBuilder.

to the shoreline" would be a variable; so would "proximity to existing urban areas." In ModelBuilder, such variables can be created through the use of input layers and tools. To determine "proximity to the shoreline" in ModelBuilder, you would start with a layer representing the shoreline and use the Euclidean Distance tool to create a new raster in which each grid cell has a value of that cell's proximity to the features of the shoreline layer (similar to what you did in Module 19).

Keep in mind that this is just one simple operation, and models often contain many types of processes used together. See Figure 21.1 for an example. In this case, you must first use the Project Raster tool to project an initial layer called NLCDColum to the coordinate system used by the other layers. ArcGIS Pro generates a new projected raster called Output Raster Dataset. Next, you use the Raster Calculator to query the new projected layer to find all the grid cells with value equal to 11 (that is, open water). All these tools and layers can be chained together in ModelBuilder. As each tool is used, a new layer is generated. This concept of applying a process to a layer and generating a new layer (which can then have another process applied to it, if necessary, to generate a third layer) is called **cartographic modeling**. (Note that the modeling you're doing in this module involves raster layers, but models can be designed using both vector and raster layers.)

cartographic modeling The process of applying a process or an action to a layer and creating a new layer as an output.

Module Scenario and Applications

In this module, you will be taking the role of a planner in Columbiana County, Ohio. You are searching for the most suitable site in the county for relocating a specific animal species. The relocation area has to meet the following criteria: (1) It must be in a forested area of land, and (2) it must be at a low elevation. In addition, the relocation area ideally should be (3) as close as possible to a water area and (4) as far away as possible from an urban or developed area. Finally, (5) the relocation area must be on a section of land that is at least 25 acres. The relocation area has to meet the first two criteria, and end up being the best option based on the third, fourth, and fifth criteria.

The following are additional examples of other real-world applications of this module's skills:

- A historic preservation specialist is trying to determine which sections of unprotected Civil War battlefield land are under the greatest development pressures in order to best direct preservation efforts or funds. She can build and run a GIS model to help locate the sites under the greatest pressures.

- An urban planner is attempting to determine how many residents of a city live on the 100-year and 500-year floodplains. He can use multiple layers in a GIS model to establish the floodplains and then compute the residential populations in these zones.

- An environmental planner is trying to find the most suitable sites in a state for designation as a protected animal habitat based on a variety of different factors. He can build a GIS model to combine these factors to aid in determining which sites are most suited for the habitat.

 ## Study Area

- For this module, you will be working with data from Columbiana County, Ohio.

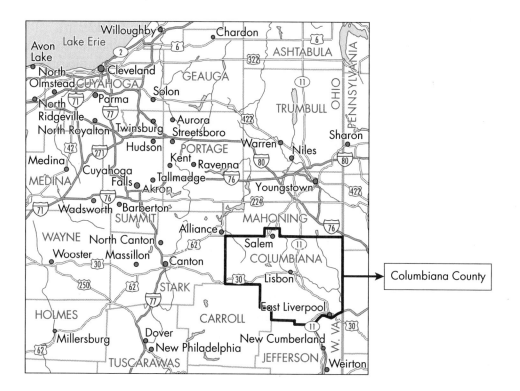

 ## Data Sources and Localizing This Module

The data for this module focus on features and locations within Columbiana County, Ohio. However, you can easily modify this module to use data from your own local county instead—for instance, if you wanted to perform this module's applications in Marlboro County, South Carolina. The DEM and NLCD2011 that were downloaded from The National Map are also available for Marlboro County; you could clip the layers to the county boundaries and process them to keep consistent 30-meter cell sizes for this exercise.

The Maskcolum grid was created by performing map algebra on the NLCD grid to extract all grid cells with value greater than 1 and then reclassifying all the 0 values as NODATA. The same could be done to create a mask grid for Marlboro County, using its NLCD layer.

Step 21.1 Getting Started

- **Important note:** Starting with ArcGIS 2.5, there are many new distance tools and methods available for use. In ArcGIS 2.5, the Euclidean Distance tools used in this module are located in the **Distance** toolset, then within the **Legacy** toolset. For more information on using the new distance tools with this module, see the book's Website at: **https://www.macmillanlearning.com/college/us/product/Discovering-GIS-and-ArcGIS-Pro/p/131923075X**.

- This module's hands-on applications use the data folder called Module21. Your instructor will be able to supply you with this data, or you can download it directly from this book's website at **https://www.macmillanlearning.com/college/us/product/Discovering-GIS-and-ArcGIS-Pro/p/131923075X**. The text in this module assumes that you have this Module21 folder in a computer location referenced as C:\GIS; if you have it somewhere else (for instance, in a flash drive referenced as G:\GISClass), substitute that location and path to the Module21 folder throughout this module.

- Module21 contains a file geodatabase called Columsitedata that contains the following items:
  - NLCDcolum: a raster of 30-meter NLCD 2011 data for Columbiana County.
  - DEMcolum: a raster of DEM data for Columbiana County (which has been resampled to 30 meters for purposes of this module).
  - Maskcolum: a 30-meter raster that conforms to the boundaries of Columbiana County.

- Start ArcGIS Pro.

- Sign in with your Esri account username and password.

- Create a new project using the **Map** template. Call this project **Module21** and place it in your **C:\GIS\Module21** folder. Ensure that there is not a checkmark in the box next to **Create a new folder for this project**.

- When ArcGIS Pro opens, change the map's name to **Columbiana**.

- You will be using the ArcGIS Pro Spatial Analyst extension in this module. If you don't have access to the Spatial Analyst extension, see Smartbox 11.1 and Troublebox 11.1 in Module 11 for more information on how to get it set up in ArcGIS Pro.

- Add the **DEMcolum** layer to the Contents pane.

- Add the **NLCDColum** layer to the Contents pane.

- Add the **Maskcolum** layer to the Contents pane.

- Use the following geoprocessing environment settings for this module (on the **Analysis** tab, in the **Geoprocessing** group, click the **Environments** button):
  - For Current Workspace, use the **Module21.gdb** geodatabase within your **C:\GIS\Module21** folder.
  - For Scratch Workspace, use the **Module21.gdb** geodatabase within your **C:\GIS\Module21** folder.
  - For Output Coordinate System, choose **DEMcolum**. ArcGIS Pro then updates the coordinate system to that of the DEMcolum layer.

- For Extent, choose **DEMcolum**. ArcGIS Pro then updates the x and y coordinates of the lower-left corner and upper-right corner of the grid extent to those of the DEMcolum layer.
- For Cell Size, choose **DEMcolum**. ArcGIS Pro then updates cell size for raster output in this project to 30 meters, as that is the cell size of the DEMcolum layer.
- For Mask, choose **Maskcolum**.
- Leave the other settings alone and click **OK** to put these geoprocessing environment settings into place.
- The coordinate system being used in this module is **Albers Conical Equal Area (NAD 83)**, using meters as the map units. Check Columbiana Map Properties (under the Coordinate System tab) to verify that this coordinate system is being used.
- Turn off the Maskcolum layer.
- To start with a new empty model, on the **Analysis** tab, in the **Geoprocessing** group, click the **ModelBuilder** button. A new tabbed view called Model opens and provides a large blank space to work in.
- With the new Model tab chosen, a new tab called ModelBuilder appears on the ribbon. On this **ModelBuilder** tab, in the **Model** group, click the **Properties** button. Change the Name and Label fields of the model to be **SiteSuit**. (See Module 20 for more about how to do this.)
- On the **ModelBuilder** tab, in the **Model** group, click the **Save** button.
- In the Catalog pane, expand the **Module21 toolbox** (which was the default toolbox created for the Module21 project). Inside it you can see an item with the model symbol called SiteSuit, which is the empty model you've just created.
- On the **ModelBuilder** tab, in the **Insert** group, click the **Tools** button. The Geoprocessing pane opens. In the Geoprocessing pane, click the **Toolboxes** button to display the full set of toolboxes to use. As you saw in Module 20, ModelBuilder provides a drag-and-drop environment. Remember that each item you use in ModelBuilder is a model element. For further information about model elements, see **Smartbox 21.1**.

SMARTBOX 21.1

What are model elements and how are they represented in ModelBuilder?

In ModelBuilder, each of the items you work with is considered a **model element** (Figure 21.2). There are three main items in ModelBuilder that are model elements:

- **Variables:** In ModelBuilder, a variable is either a data layer or a value. For instance, an NLCD layer could be added to ModelBuilder as a data variable, while an input for a query threshold could be used as a value variable. By default, data variables appear as dark blue ovals, and value variables appear as lighter blue ovals. The output from a tool can also be considered a variable and is referred to as *derived data* (and, by default, appears as a green oval).

model element Each of the items (Variables, Tools, and Connectors) that can be placed into the ModelBuilder interface.

variable (ModelBuilder) The specific term in ModelBuilder for a data layer or value input (which, by default, appears as a blue oval in ModelBuilder).

FIGURE 21.2 Various model elements inside the ModelBuilder environment.

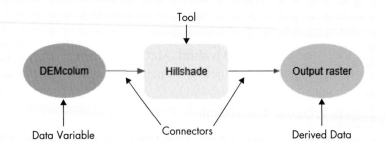

tool (ModelBuilder) The specific term in ModelBuilder for one of the geoprocessing tools, models, or scripts (which, by default, appears as a yellow rectangle).

connector (ModelBuilder) The specific term in ModelBuilder for one of the lines that link Variables and Tools (which, by default, appear as one of four types of lines).

- **Tools:** Tools are the various items from the Geoprocessing pane that are used to perform actions on variables. For instance, the Project tool and the Slice tool from the Geoprocessing pane can be added as tool model elements in ModelBuilder. Python scripts and completed models with parameters can also be added as tool model elements. By default, tools appear as yellow rectangles.

- **Connectors:** Connectors are the lines that link variables and tools. For instance, to project a layer called roads, a connector is drawn between the data variable roads and the Project tool. By default, a solid black line joins data to a tool, a dashed black line indicates a connector based on a geoprocessing environment setting, a dotted black line indicates some sort of precondition that must be met before the tool will run, and dashed blue lines are feedback connectors that loop the output of a tool back into the tool again.

- Drag and drop the **NLCDcolum** grid from the SiteSuit model's Contents pane into the ModelBuilder window. Also drag and drop the **DEMcolum** grid from the SiteSuit model's Contents pane into the ModelBuilder window. These are the two model elements from which you will build everything else in the model.

Step 21.2 Setting Model Parameters

Before you begin building the model, you need to set up the model's parameters. (See **Smartbox 21.2** for more about model parameters.)

SMARTBOX 21.2

What are model parameters and how are they used in ArcGIS Pro?

One advantage of creating a model in ArcGIS Pro is the flexibility to use (or reuse) it with different inputs. For instance, the model you're creating in this module relies on two different inputs: the NLCD layer and the DEM layer. Everything else you'll be working with is created or derived from these two layers. In theory, after completing this module, you could take an

NLCD and a DEM layer for another county and plug them into the model and examine the output for animal-relocation areas for that county. You wouldn't have to remake the model; all those processes would remain the same, and only the two inputs would need to change. The ability to set up differing inputs (whether data variables or value variables) makes these two layers the **model parameters**, which are the only items that the user has to specify when running the model.

> **model parameters** The user-defined inputs or variables of a model.

Using model parameters allows you to rerun a model by using different inputs. For instance, if a model input were a value for population, you could set it as a model parameter and allow the users of the model to type in any population value. In the ModelBuilder dialog, the variables that are used as model parameters have a letter *P* next to them.

For a model to open and run as an independent item, the parameters must be set. For instance, in Module 20, when you tried to reopen your model by selecting **Open**, ArcGIS Pro reported "no parameters." (See Troublebox 20.1 on page 469 in Module 20 for more information.) If you don't set up the inputs of the model as parameters, you cannot simply open the model (as a standalone tool) and run it. Likewise, the model cannot be shared or run by someone else unless the model parameters are set. Also, the model cannot be used as a tool within another model unless parameters are set.

- In the SiteSuit model, two parameters need to be set: the DEM input and the NLCD input. All the other processes of the model that you'll be working with in the rest of this module (such as using the Raster Calculator to extract the urban areas or using the Distance and Slice tools to determine proximity to urban areas) are derived from these two initial parameters. You need to rename these two items to be more useful and intuitive for model inputs than DEMcolum or NLCDcolum. In the SiteSuit model, right-click on the **NLCDcolum** model element and select **Rename**.

- In the box that opens on top of the model element, type **NLCD input layer** and click the left mouse button. You see that the NLCDcolum element now is labeled NLCD input layer.

- Use the same process to rename the DEMcolum model element as **DEM input layer**.

- Save the SiteSuit model. (See Module 20 for how to save a model.)

- You can now make the NLCD input layer a model parameter. In the SiteSuit model, right-click on the blue bubble for **NLCD input layer element** and choose **Model Parameter**. You see a small *P* appear next to the model element, indicating that it's now a parameter. Do the same with the DEM input layer model element. Both of your layers are now considered model parameters.

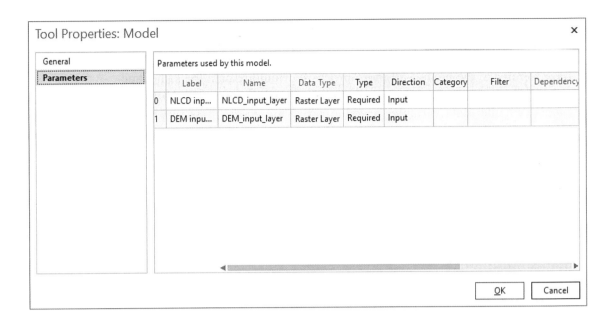

- Next, on the **ModelBuilder** tab, in the **Model** group, click the **Properties** button. In the dialog box that opens, choose the **Parameters** tab. Those two layers are now added as model parameters. Accept the default settings (because both of these layers are required to run the model, Required is already the default state for them as model parameters) and click **OK**.

Step 21.3 First Criterion: Must Be in a Forested Area

- The first criterion for the animal-relocation area is that it must be in a forested area. To extract these areas from the NLCD grid, you have to build an expression using the Raster Calculator. In the Geoprocessing pane, from the **Spatial Analyst Tools** toolbox, choose the **Map Algebra** toolset, then choose the **Raster Calculator** tool and drag and drop it from the Geoprocessing pane into the SiteSuit window.

- Add a connection between the NLCD input layer and the Raster Calculator tool and choose the option **Map Algebra expression** when making the connection.

- Double-click the **Raster Calculator** tool to open it and then use it to build an expression that allows you to select all the grid cells in the model version of the NLCD input layer that represent forested land-cover types (with values 41, 42, or 43). (See Module 20 for how to use the Raster Calculator to build an expression.)

- For Output raster, click the **browse** button, navigate to your **C:\GIS\Module21** folder, go to the **Module21.gdb** geodatabase, and name the Output raster **forested**.

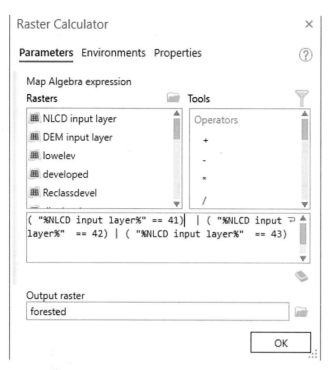

- The end result of this process is a new grid called forested that has all grid cells that met the criteria from the Raster Calculator expression (that is, all NLCD grid cells with values 41, 42, or 43) assigned the value 1 and all cells that did not meet the criteria (that is, all NLCD grid cells with values other than 41, 42, or 43) assigned the value 0. In the SiteSuit model, you can see that the NLCD input layer map element, Raster Calculator tool, and forested derived data output grid are all properly connected.

Step 21.4 Second Criterion: Must Be at a Low Elevation

- The second criterion for the animal-relocation area is that it must be at a low elevation. From examining the values of the DEMcolum grid, you can see that the county's elevations run from 200 meters to 440 meters. To determine lower elevations, you can now build a second Raster Calculator expression to find all the areas in the county that are below 320 meters. In the Geoprocessing pane, from the **Spatial Analyst Tools** toolbox, choose the **Map Algebra** toolset, then choose the **Raster Calculator** tool and drag and drop it from the Geoprocessing pane into the SiteSuit window.

- Add a connection between DEM input layer and the Raster Calculator tool and choose the option **Map Algebra expression** when making the connection.

- Double-click on the **Raster Calculator** tool to open it. Use the Raster Calculator's buttons and menu options to build an expression to find all the values of the model version of the DEM input layer that are less than or equal to (<=) a value of 320.

- For Output raster, click the **browse** button, navigate to your **C:\GIS\Module21** folder, go to the **Module21.gdb** geodatabase, and name the Output raster **lowelev**.

- The end result of this process is a new grid called lowelev that has all grid cells that met the criteria from the Raster Calculator expression (that is, all DEM grid cells that were either less than or equal to the value 320) assigned the value 1 and all cells that did not meet the criteria (that is, all DEM grid cells with value greater than 320) assigned the value 0. In the SiteSuit model, you can see the DEM input layer map element, Raster Calculator tool, and lowelev derived data output grid all properly connected.

Step 21.5 Must Be at Both a Forested Area and at a Low Elevation

- Now, you need to find the areas that are both forested and at a low elevation. Because both of these grids contain only the values 0 (does not meet the criteria) or 1 (meets the criteria), you can simply multiply the two grids together in a map algebra expression. (See Module 20 for further information on map algebra.) The output from this process will be a grid with values of 1 that are both forested and low elevation and values of 0 everywhere else.

- In the Geoprocessing pane, from the **Spatial Analyst Tools** toolbox, choose the **Map Algebra** toolset, then choose the **Raster Calculator** tool and drag and drop it from the Geoprocessing pane into the SiteSuit window. Position this new Raster Calculator tool so that both the forested and lowelev grids can clearly connect to it.

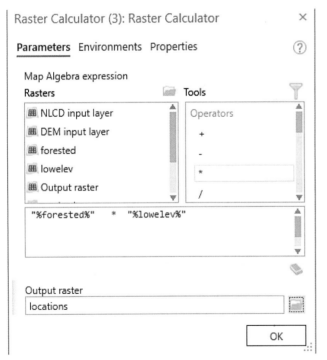

- Draw a connection from forested to this new Raster Calculator tool and choose **Map Algebra expression** when making the connection. Also draw a connection from lowelev to Raster Calculator the same way. When both forested and lowelev are connected to Raster Calculator, double-click on the **Raster Calculator** tool to open it.

- Build a map algebra expression to multiply the forested grid by the lowelev grid.

- For the Output raster, click the **browse** button, navigate to your **C:\GIS\Module21** folder, go to the **Module21.gdb** geodatabase, and name the Output raster **locations**.

- Click **OK**. Answer Question 21.1.

Question 21.1 Why did you use multiplication instead of addition to overlay the two grids? What would be the end result of this expression if you used addition instead of multiplication?

- Back in the SiteSuit model, you can see that your locations grid is set up as the product of the forested and lowelev grids.

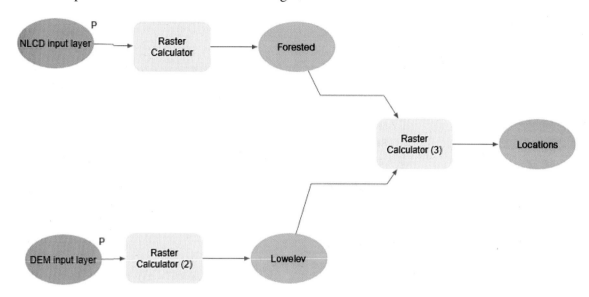

- Run the SiteSuit model. (See Module 20 for how to run a model.)
- In the Columbiana map, add the **locations** layer to the Contents pane and examine it. Your animal-relocation area will have to be built in an area with grid cells equal to 1. The locations grid shows you the places that are potentially viable for the animal-relocation area. However, some of these sites would be more suitable than others, based on the other three criteria you still have to model.

Step 21.6 Third Criterion: Must Be as Far Away as Possible from Developed Areas

- The third criterion for the relocation area is that the location should be as far away from an urban/developed area as possible. To determine the sites that meet this criterion, you first have to extract the grid cells that represent developed areas from the NLCD input layer grid. (See Module 12 for what each value in the NLCD represents.
- In the Geoprocessing pane, from the **Spatial Analyst Tools** toolbox, choose the **Map Algebra** toolset, then choose the **Raster Calculator** tool and drag and drop it from the Geoprocessing pane into the SiteSuit window.
- Add a connection between NLCD input layer and the Raster Calculator tool and choose the option **Map Algebra expression** when making the connection.
- Double-click on the **Raster Calculator** tool to open it. Use the Raster Calculator to build an expression (see Module 20) that allows you to select all the grid cells in the model version of the NLCD input layer that represent developed land-cover types (values 21, 22, 23, or 24).

- For Output raster, click the **browse** button, navigate to your **C:\GIS\Module21** folder, go to the **Module21.gdb** geodatabase, and name the Output raster **developed**.
- Click **OK**.

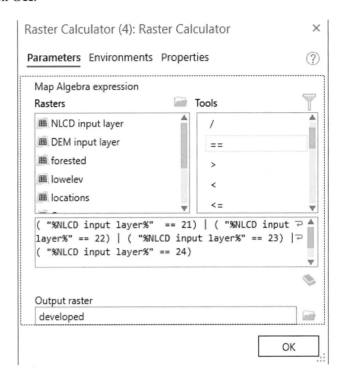

- Run the model point to generate the developed grid. The developed grid will consist of values of 1 for grid cells assigned a developed land-cover type and values of 0 for non-developed areas.
- Save your SiteSuit model.
- The next step is to find the distance away from each developed grid cell. As in Modules 19 and 20, you can do a Euclidean distance calculation to find this information. However, you only want to find the distance from those cells with the value 1, not from all cells in the grid; that is, you don't want to find the distance from cells with the value 0. To do this, you need to reclassify the grid (see Module 19) so that all cells with the value 0 get assigned the value NODATA. In the Geoprocessing pane, from the **Spatial Analyst Tools** toolbox, choose the **Reclass** toolset, then choose the **Reclassify** tool and drag and drop it from the Geoprocessing pane into the SiteSuit window.
- Add a connection between developed and the Reclassify tool and choose the option **Input raster** when making the connection.
- Open the **Reclassify** tool.
- For Input raster, **developed** should already be chosen.
- For Reclass field, choose **Value**.
- Click the **Unique** button (so that you can assign a new value to each old value in the grid).
- For 0 in the Value column, enter **NODATA** in the New column.

Step 21.6 Third Criterion: Must Be as Far Away as Possible from Developed Areas

- For 1 in the Value column, enter **1** in the New column.
- For NODATA in the Value column, leave **NODATA** in the New column.
- For Output raster, click the **browse** button, navigate to your **C:\GIS\Module21** folder, go to the **Module21.gdb** geodatabase, and name the Output raster **reclassdevel**.
- Click **OK** when all settings are correct.

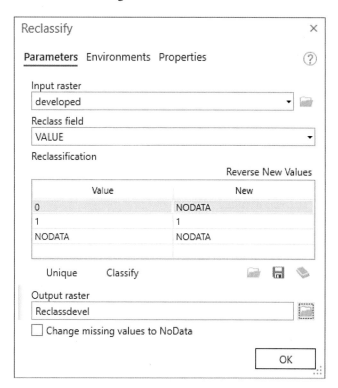

- Run the model again to create a new grid called reclassdevel. Add it to the Contents pane of the Columbiana map. It should consist only of 1 values (and NODATA) that show the locations of developed areas in the county. Now you can calculate distance from each of these developed grid cells. Answer Question 21.2.

> **Question 21.2**
> Why was it necessary to reclassify the grid so that 1 values remained the same but 0 values were assigned to NODATA? (*Hint:* Consider that the next step is calculating distance and think about how the distance calculation will be performed from this source raster.)

- Save your SiteSuit model.
- In the Geoprocessing pane, from the **Spatial Analyst Tools** toolbox, choose the **Distance** toolset, then choose the **Euclidean Distance** tool and drag and drop it from the Geoprocessing pane into the SiteSuit window.
- Add a connection between reclassdevel and the Euclidean Distance tool and choose the option **Input raster or feature source data** when making the connection.
- Open the **Euclidean Distance** tool.

- For Input raster, the **reclassdevel** grid should already be chosen.
- For Output raster, click the **browse** button, navigate to your **C:\GIS\Module21** folder, go to the **Module21.gdb** geodatabase, and name the Output raster **distdevel**.
- Make sure Output cell size is **DEMcolum** (this will be 30 meters).
- Leave the other options alone and click **OK** to calculate the new distance raster.

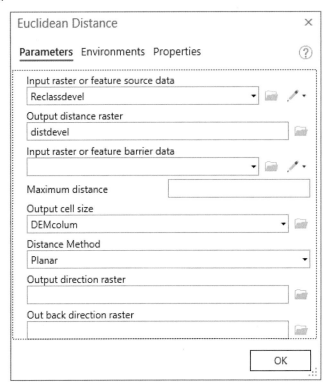

- The new distdevel raster should have a value showing the distance each cell is from a developed cell. However, this distdevel grid has thousands of different values. You need to find which areas are closest to an urban area and which areas are farthest away. The next step is to calculate this by breaking up these thousands of distance values into more manageable groups on a 1-to-10 scale, where the most desirable places (those farthest away) get the value 10, and the least desirable places (those closest to developed areas) get the value 1. In ArcGIS Pro, you can do this by using the Slice tool (see Module 19). In the Geoprocessing pane, from the **Spatial Analyst Tools** toolbox, choose the **Reclass** toolset, then choose the **Slice** tool and drag and drop it from the Geoprocessing pane into the SiteSuit window.
- Add a connection between distdevel and the Slice tool and choose the option **Input raster** when making the connection.
- Open the **Slice** tool.
- For Input raster, **distdevel** should already be chosen for you.
- For Output raster, click the **browse** button, navigate to your **C:\GIS\Module21** folder, go to the **Module21.gdb** geodatabase, and name the Output raster **slicedevel**.

- For Number of output zones, type **10**.
- For Slice method, choose **Equal Interval**.
- For Base zone for output, use **1**.
- Click **OK** when all settings are correct.

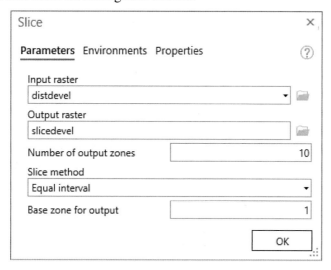

- Run the model and add the **slicedevel** grid to the Columbiana map Contents pane. Each cell will have a value from 1 through 10, with 10 in the cells farthest from a developed area and 1 in the cells closest to a developed area.
- Save the SiteSuit model.

Step 21.7 Fourth Criterion: Must Be as Close as Possible to a Body of Water

- The fourth criterion for the relocation area involves being as close as possible to a body of water. Repeat the previous steps to find these areas as follows:
  - Use the Raster Calculator to extract all cell values of 11 (indicating "Open Water" land cover) from the NLCD input layer variable. Call this new grid **water**.
  - Run the model to create the water grid.
  - Reclassify the water grid so its values of 0 are set to **NODATA**, its values of NODATA are set to **NODATA**, and the other values are set to **1**. Call this new grid **reclasswater**.
  - Calculate the Euclidean distance from the reclasswater grid. Call this new grid **distwater**.
  - Slice the distwater grid into 10 equal-interval output zones. Call this new grid **slicewater**.
- Save the SiteSuit model. Your model is getting more complicated, so you'll probably have to adjust the position of the model elements or resize the model, as you have more to do. On the **ModelBuilder** tab, in the **View** group, click the **Auto Layout** button, and ArcGIS Pro automatically adjusts the position of the model elements for you. You might also want to move things around manually to better suit your own needs.

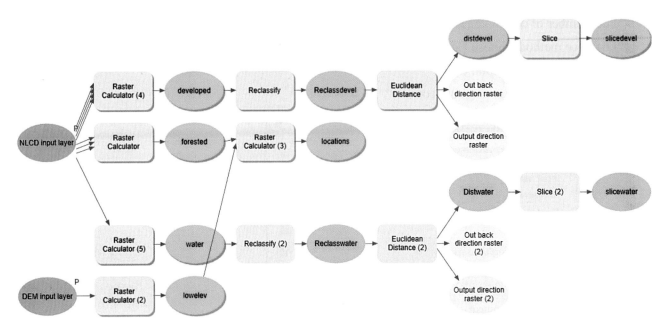

- Run the model to generate the slicewater grid and add it to the Columbiana map Contents pane.
- Notice that the slicewater grid has its highest values (10) as places farthest from a body of water and its lowest values (1) as places closest to a body of water, which is the opposite of what you want. Thus, you need to reclassify the slicewater grid to reorder the values. In the Geoprocessing pane, from the **Spatial Analyst Tools** toolbox, choose the **Reclass** toolset, then choose the **Reclassify** tool and drag and drop it from the Geoprocessing pane into the SiteSuit window.
- Add a connection between slicewater and the Reclassify tool and choose the option **Input raster** when making the connection.
- Open the **Reclassify** tool.
- For Input raster, **slicewater** should already be chosen.
- For Reclass Field, choose **Value**.
- Click the **Unique** button to assign individual new values to each of the old values.
- To give those areas closest to the water bodies higher values and those areas farthest away lower values, reclassify your new values as follows:

Old Value	New Value
1	10
2	9
3	8
4	7
5	6
6	5
7	4
8	3
9	2
10	1

- For Output raster, click the **browse** button, navigate to your **C:\GIS\Module21** folder, go to the **Module21.gdb** geodatabase, and name the Output raster **reslicewater**.
- Click **OK** to close the Reclassify tool.
- Run the model again and add the **reslicewater** grid to the Columbiana map Contents pane. Examine both slicewater and reslicewater and answer Question 21.3.

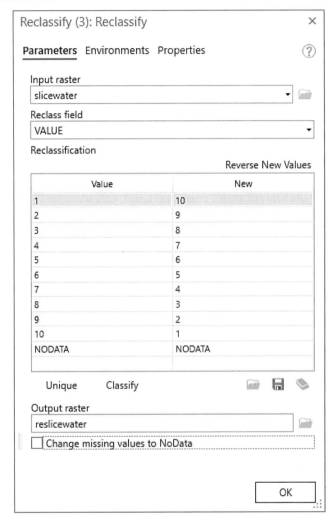

> **Question 21.3**
> How do the slicewater and the reslicewater grids compare with each other? How do they relate to the water grid?

- Save your SiteSuit model.

Step 21.8 Creating a Suitability Index

- You now have the following grids:
  - Slicedevel: This shows the distance rankings from 1 to 10 for proximity to developed areas for the entire county.

- Reslicewater: This shows the rankings from 1 to 10 for proximity to water bodies. You are assuming that values of 10 in each grid are the most desirable areas (as they are the values farthest from developed lands in the slicedevel grid and the closest to water in the reslicewater grid).

- The next step is to overlay these two grids using map algebra (and thus the Raster Calculator) to show the overall rankings from 1 to 20 in terms of suitability. Add a connection between slicedevel and the Raster Calculator tool and choose the option **Map Algebra expression** when making the connection. Also add a connection between reslicewater and the Raster Calculator tool and choose the option **Map Algebra expression** when making this second connection.

  - Open the new **Raster Calculator** tool and use it to build an expression to add the slicedevel grid and the reslicewater grid together.

- For Output raster, click the **browse** button, navigate to your **C:\GIS\Module21** folder, go to the **Module21.gdb** geodatabase, and name the Output raster **ranking**. Click **OK** to close the Raster Calculator.

- Run the model and add this new **ranking** grid to the Columbiana map Contents pane. What you have created here is a suitability index for assessing your suitable sites. (For more information about suitability indexes, see **Smartbox 21.3**.) Examine the ranking grid and answer Questions 21.4, 21.5, and 21.6.

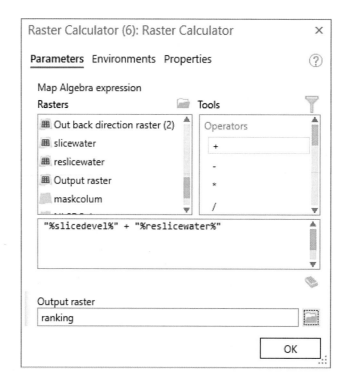

SMARTBOX 21.3

What is a suitability index?

In Module 20, you identified the best locations in a county for an ecological preserve, a type of site-suitability analysis. The results of that analysis were locations that were either suitable (cells with the value 1) or unsuitable (cells with the value 0). Similarly, the locations grid you created in this module shows which locations in the county would be suitable for animal relocation (those that are forested and at a low elevation and that have the value 1) and all other locations in the county, which are unsuitable (that are not forested or wetlands, or at a higher elevation, and have the value 0). The disadvantage of this type of site-suitability analysis is that it tells you only which sites are desirable and which are not. Sometimes, this may be all the information you need, but in some studies, you might want additional information on how suitable a particular site is compared with other suitable sites.

suitability index A ranking of locations according to a set of criteria.

To obtain a ranking of site suitability, you can build a **suitability index**, which can help you develop a wider range of options for suitable sites beyond the basic "suitable or not suitable" binary output. For example, Figure 21.3 shows three sample grids, each one a variable for determining a suitable site for a vacation home: distance to forests, distance to rivers, and distance to roads. In each raster, the cells have values of 1 through 5, with the value representing a ranking of the suitability of that site (1 being the least desirable and 5 being the most desirable). By adding together all these ordinal values (see Module 2), you can obtain a ranking of the relative suitability of each site. For instance, in the suitability index shown in Figure 21.3, grid cells with values 12 or 13 would be the most suitable sites for a vacation home, cells with values 7 or 10 would be the next step down in terms of suitability, and cells with values 3 or 4 would be the least suitable choices.

FIGURE 21.3 An example of overlaying three rasters to create a basic suitability index.

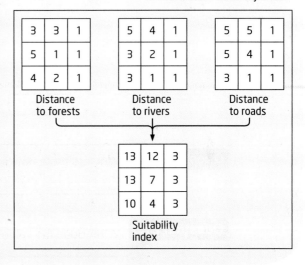

Question 21.4 What do values 2, 3, and 4 represent in the ranking grid? What do values 17, 18, and 19 represent?

Question 21.5 How many grid cells have the value 2? How many grid cells have the value 19?

Question 21.6 As a result of adding together two grids of values of 1 to 10, the ranking grid should have values from 2 to 20. However, there are no values of 20 in the ranking grid. Why?

- You now have the information you need to determine the third and fourth criteria for finding the animal-relocation areas. It's time to combine this ranking of all grid cells in the county with the subset of the chosen locations identified from the first two criteria by overlaying the ranking and locations grids.

- Again, you can perform the overlay of these two layers by using map algebra and the Raster Calculator. Add a connection between **ranking** and the Raster Calculator tool and choose the option **Map Algebra expression** when making the connection. Also add a connection between **locations** and the Raster Calculator tool and choose the option **Map Algebra expression** when making this second connection.

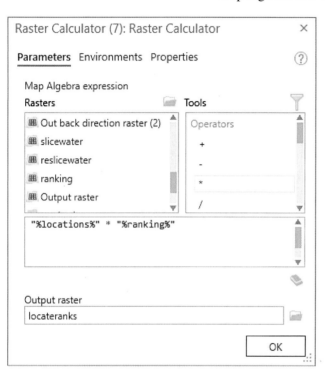

- Open the new **Raster Calculator** tool and use it to build an expression to multiply the ranking grid and the locations grid together.

- For Output raster, click the **browse** button, navigate to your **C:\GIS\Module21** folder, go to the **Module21.gdb** geodatabase, and name the Output raster **locateranks**. Click **OK** to close the Raster Calculator.

- Run the model and add the **locateranks** grid to the Columbiana map Contents pane. Answer Questions 21.7 through 21.10.

Question 21.7 Why are there now no values of 19 in the locateranks grid? (That is, why is the value 18 the highest value in the locateranks grid?)

Question 21.8 What does the value 0 represent in the locateranks grid?

Question 21.9 Create a new grid called locateranks2, calculated by adding together the locations and ranking grid with the Raster Calculator in ModelBuilder. Why is the output of this grid not useful in determining the final sites for the animal-relocation areas? Delete the locateranks2 grid (and any tools and connections for it in the final model) when you're done.

Question 21.10 Create a new grid called locateranks3, calculated by overlaying the locations and ranking grid together with the AND operator in the Raster Calculator in ModelBuilder. Why is the output of this grid only 1s and 0s, compared with the results you obtained from multiplying the two grids? Delete the locateranks3 grid (and any tools and connections for it in the final model) when you're done.

- Save the SiteSuit model.

Step 21.9 Fifth Criterion: Find Large Sections of the Best Possible Land Area

- The last step is to find the best sections of land on the largest possible areas to relocate the animals to (because you need a minimum of 25 acres of land). To do this, you'll have to regiongroup the ranked sites together (see Module 12) to form contiguous sections of land to examine. In the Geoprocessing pane, from the **Spatial Analyst Tools** toolbox, choose the **Generalization** toolset, then choose the **Region Group** tool and drag and drop it from the Geoprocessing pane into the SiteSuit window. Add a connection between **locateranks** and the Region Group tool and choose the option **Input raster** when making the connection.

- Open the **Region Group** tool.
- The input raster should already be **locateranks**.
- For Output raster, click the **browse** button, navigate to your **C:\GIS\Module21** folder, go to the **Module21.gdb** geodatabase, and name the Output raster **locateregions**.
- For Number of neighbors to use, choose **Eight**.
- For Zone grouping method, choose **Within**.
- Be sure to put a checkmark next to **Add link field to output**.
- Click **OK**.
- Run the model.
- Save the SiteSuit model.

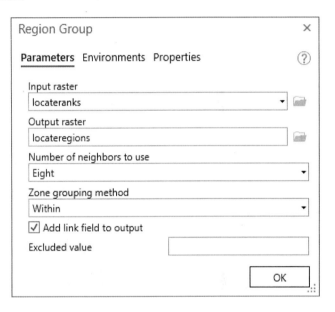

Step 21.10 Validating and Running the Model

- To clean up the SiteSuit model, on the **ModelBuilder** tab, in the **View** group, first click the **Auto Layout** button and then click the **Fit to Window** button. ArcGIS Pro streamlines your model layout and makes the workflow easier to read.

- Next, you can validate and run the complete model. In validating a model, ArcGIS Pro checks to be sure that all of the model's variables and parameters are correct. When a model is validated, all the variables are reset as if they haven't been run while ArcGIS Pro checks all of the model's elements to make sure everything is okay. To validate the model, on the **ModelBuilder** tab, in the **Run** group, click the **Validate** button.

- Now it's time to run the model. In the Catalog pane, return to your **Module21.tbx** toolbox,

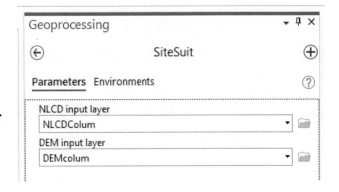

where you should see your SiteSuit model. Right-click on the **SiteSuit** icon and select **Open**. Because the model's parameters have been set, the model opens—not as the graphical ModelBuilder interface but rather as the model itself. The model takes the form of a geoprocessing tool in the Geoprocessing pane that will allow the user to specify two inputs (namely, the two model parameters): a DEM layer and an NLCD layer.

- Make sure that **NLCDColum** is chosen for the NLCD input layer and that **DEMcolum** is chosen for the DEM input layer.

- Run the SiteSuit geoprocessing tool. The text at the bottom of the Geoprocessing pane shows each process of the SiteSuit model as it runs.

- When the SiteSuit tool finishes running, the entire model will have been rerun, and any previously created layers are overwritten. The areas in the locateregions grid have met the first four of the model criteria and have been brought together into large contiguous areas. In the next step, you'll evaluate the results of the model and find which of these large areas best meets the criteria as well as the minimum size requirement.

Step 21.11 Analyzing Site Suitability Model Results

- Add the **locateregions** grid to the Columbiana map's Contents pane. It shows all the regions created by grid cells of similar ranking values that have been merged together; that is, all cells of value 8 that are contiguous are joined to form a region, all cells of value 9 that are contiguous are joined to form another, different region, and so on. With more than 5000 regions, ArcGIS Pro displays this layer as a stretched ramp of colors rather than showing each individual region. To change the symbology of the locationsregions grid to Unique Values, choose this locateregions layer in the Contents pane and then on the ribbon, on the **Appearance** tab, in the **Rendering** group, click the **Symbology** button and choose **Unique Values**.

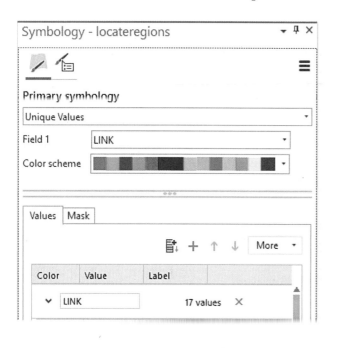

- In the Symbology pane, change Value Field to **LINK**. As noted in Module 12, Link shows you the original values assigned to the grid cells (that is, their original ranking) prior to the regiongrouping process. Change the color scheme to a ramp of colors that shows the lower-ranked grid cells in a more distinctive color than the higher-ranked grid cells.

- Open the attribute table for the locateregions grid and sort the LINK field so that the higher values are sorted at the top of the attribute table. You should have more than 5000 different regions, but only a few of them would be the most desirable places to use as a relocation area for the animals.

- Carefully examine the regions with the highest rankings to determine which one you will select as the relocation area. The values in the attribute table can be read as follows:

- The number under Value indicates the grid-cell identifier for a particular region (that is, the first region created has a Value of 1, the second region created will have a Value of 2, and so on).
- The number under LINK indicates the ranking of that region (the original value from the locateranks grid).
- The number under Count indicates the number of grid cells that comprise that particular region.
- Keep in mind that you're looking at the number of grid cells in each region, and you need a plot of land with at least 25 acres. Before deciding on the best region to relocate the animals, you should calculate how many grid cells are equivalent to 25 acres (and, thus, you can disregard all regions of less than this number). Answer Question 21.11.

Question 21.11

The size of each grid cell is 900 square meters (since they are 30-meter-resolution grid cells). One square meter (m^2) is equal to 0.000247105 acres. How many grid cells represent 25 acres? (Show your work.)

- With the information from Question 21.11, once again examine the regions with the highest rankings in the attribute table. Answer Question 21.12.

Question 21.12

What is the value assigned to the region of land you will select as the relocation site? How many grid cells does this region consist of?

- You can use the Raster Calculator one final time (outside the model environment) to create a new grid that shows your selected area for the animal-relocation land. This grid will have the value 1, and all other regions will have the value 0. In the Geoprocessing pane, from the **Spatial Analyst Tools** toolbox, choose the **Map Algebra** toolset and then choose the **Raster Calculator** tool.
- Build a query to find values from the locateregions grid that are equal to the value of the chosen region. (Note that now you're using the locateregions layer itself, so you should use the one with the yellow shape rather than the model version with the three triangles.).
- Use the == button for an equals sign.
- For Output raster, click the **browse** button, navigate to your **C:\GIS\Module21** folder, go to the **Module21.gdb** geodatabase, and name the Output raster **chosensite**.
- Click **OK**. The chosensite grid consists of 1 values for the region you've selected and 0 values for all other areas.
- Zoom in closely on this area and answer Question 21.13.

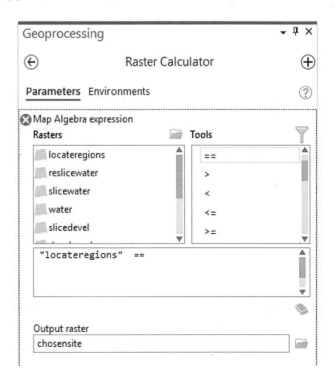

Question 21.13 Why was this one site chosen over other regions that were ranked higher?

Step 21.12 Printing or Sharing Your Results

- Save your Module21 project.
- Turn off all layers except the chosensite layer and change the symbology so that the 1 values are clearly distinguished from the 0 values.
- Compose and print a layout (see Module 3) of the final version of your sites (including all the usual map elements and design for a layout) so that the dimensions of the final chosen site can be clearly seen.
- An advantage of creating a model with parameters is that others can utilize it as well. If you want to share your model, you can accomplish this with a **geoprocessing package** (.gpk) file. A geoprocessing package can take any tool (or a completed and run model with parameters) and save the model or tool (along with its data) as a single file that can be shared (either to disk or to ArcGIS Pro Online). To save your SiteSuit model as a geoprocessing package and share it, do the following:
  - On the **Share** tab, in the **Package** group, choose **Geoprocessing** and then choose the **SiteSuit** model.
  - A new Geoprocessing Package pane opens. It works similarly to a regular Project Package pane, in that you can choose to save the geoprocessing package to disk or upload it to your ArcGIS Online account. You need to provide tags and a summary, and when you click **Analyze**, ArcGIS Pro looks over the SiteSuit tool for any issues that need to be fixed. When you've fixed any outstanding issues (usually related to extra documentation or descriptions), you can click **Package** to create the geoprocessing package.

geoprocessing package An ArcGIS Pro format used for collecting and sharing models.

Closing Time

This module examined how to use ModelBuilder to create a functioning standalone model that can be used as a separate tool or shared via a geoprocessing package. Models are powerful and versatile items and can be as simple or as complex as needed for problem solving. By using model parameters, you can tweak and fine-tune your models for a variety of different inputs and variables. In the next module, you'll continue using ModelBuilder to design a specific type of model that will use several new tools for hydrologic applications.

All the tasks you performed in this module could have been performed using individual tools, such as Slice, Reclassify, and the Raster Calculator, but ModelBuilder provides a more intuitive visual interface for conceptualizing and running workflows. However, there is much more you can do with ModelBuilder to perform more complex processes. See *Related Concepts for Module 21* for more about these advanced options for using ModelBuilder.

RELATED CONCEPTS FOR MODULE 21

Advanced ModelBuilder Options

Although the model you created in this module contains many variables and tools used in different combinations to create an output, the model itself is actually fairly simple. You began with two different inputs, connected them and their derived data with a variety of different tools, and at the end of a large workflow, you had one result to analyze. There may be times when models you want to create in ModelBuilder will be called on to do more complex things. Fortunately, ModelBuilder contains numerous advanced options for variables, tools, and connectors that allow you to handle more complicated workflows and answer more involved analysis questions. See Figure 21.4 for examples of some of these more advanced features.

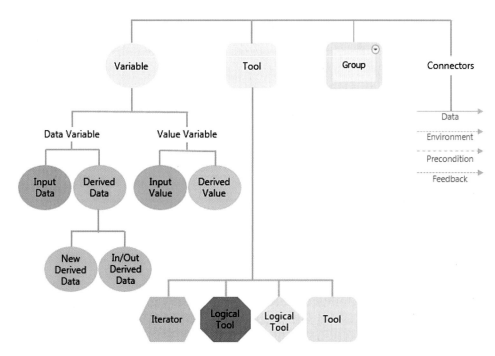

FIGURE 21.4 The various features of ModelBuilder available for variables, tools, and connectors.

When it comes to variables, you can specify not only input data and derived data but also special variables referred to as value variables. If you need to have something such as a specific numerical input or an SQL expression executed by the model or specified by the model user, a value variable will allow you to do that. (See Module 22 for more about using value variables.)

Tools can take other forms in ModelBuilder. One special kind of tool is an **iterator** (by default shown as a six-sided orange box). An iterator allows you to repeat actions several times or perform multiple instances of an action. For example, if you had a geodatabase that contained 20 different DEMs and needed to have the model project each one, you wouldn't want to set up each DEM individually as a data variable and then have 20 copies of the project tool, thus making 20 different derived data layers. Instead, you could use an iterator to repeat the application of the project tool being used on each of the 20 DEM layers with the use of just one set of variables and connections.

Another special type of tool that can be used in ModelBuilder is a **logical tool** (which by default is a four-sided yellow diamond or a six-sided red hexagon).

iterator A ModelBuilder tool that allows you to duplicate and repeat a process multiple times.

logical tool A ModelBuilder tool that allows you to use an if-then-else condition for different actions in a model.

feedback connector A ModelBuilder connector that allows you to take the derived data output from a tool and use it again as an input for the same tool.

With a logical tool, you can have ArcGIS Pro check to see if a specified condition is true or false and take an action accordingly. For instance, if your model requires that a layer have a defined projection before you can take further actions, you would first want ModelBuilder to check whether that projection has been defined. The logical tool would first check whether the projection exists—in which case it would give an indication of "true"—and then proceed with the next step of the model. If the field didn't exist—that is, if it gave an indication of "false"—the model could take an extra series of steps to properly run the Define Projection tool before proceeding with the rest of the model. This is also referred to as an "if-then-else" structure in models: If a condition is true, then do action one, but otherwise ("else")—that is, if the condition is false—do the second action.

In addition, other types of connectors can be used beside the regular data connector you used in this module. A special type of connector (shown with a blue line instead of the default black line) is a **feedback connector**, which allows you to connect derived data outputs back to be used as inputs to the tool.

Key Terms

GIS model (p. 487)
model (p. 487)
variables (p. 487)
cartographic modeling (p. 488)
model element (p. 491)

variable (p. 491)
tool (p. 492)
connector (p. 492)
model parameters (p. 493)
suitability index (p. 505)

geoprocessing package (p. 510)
iterator (p. 511)
logical tool (p. 511)
feedback connector (p. 512)

How to Use Hydrologic Modeling Tools in ArcGIS Pro

ArcGIS Pro Skills

In this module, you will learn how to do the following in ArcGIS Pro:
- Put together a hydrologic workflow as a model with parameters.
- Calculate flow direction.
- Locate and fill sinks to create a "depressionless" elevation surface.
- Calculate flow accumulation.
- Extract stream channels.
- Set a value variable in a model.
- Set up and evaluate stream ordering.
- Extract watersheds based on stream links.
- Account for intermediate data in a model.

Introduction

Numerous types of models can be constructed using the various ArcGIS Pro tools. One set of tools with numerous applications related to modeling water-related issues is the Hydrology toolset, available in the Spatial Analyst toolbox. **Hydrologic modeling** refers to examining factors related to the movement of water across a landscape. As water falls on the terrain, it moves across the land, and an elevation surface is used in GIS to model the effects of the direction and accumulation of the flow of water, as well as to delineate streams and catchment areas. Like the networks discussed in Module 11, a stream is defined by its length (referred to as a *link*) and where it meets another stream (referred to as a *junction*). In ArcGIS Pro, the Hydrology toolset allows you to start with an elevation surface of an area, model the flow and accumulation of water, and extract the area's streams and watershed boundaries.

A **watershed** is an area in which all the water flows to a common location. For example, the Chesapeake Bay watershed covers a large multistate area because the drainage location for the region's water will end up as the Chesapeake Bay. Smaller watersheds can be defined by a river system. For instance, the Mahoning River Watershed covers portions of eight northeastern Ohio counties and is

Learning Outcomes

After studying this module, you should be able to:
- Explain what a watershed is.
- Outline the steps used in extracting a watershed from a raster elevation surface.
- Describe how the D8 method of flow direction operates.
- Explain what a sink is in hydrologic modeling.
- Explain how flow accumulation is computed for a raster surface.
- Explain the difference between the Shreve and Strahler methods of stream ordering.

hydrologic modeling The process of examining the movement of water over a surface.

watershed An area in which all locations flow to a common drainage area.

formed by several streams flowing into other streams and eventually into the river. The boundaries of the watershed (referred to as the *drainage divide*) define the area for which all water will drain to a common location (such as a river, a bay, or a culvert). In this module, you will use the links of the various streams in a county and delineate a watershed area (also called a *sub-basin*) for each stream link (the portion of the stream between junctions; Figure 22.1).

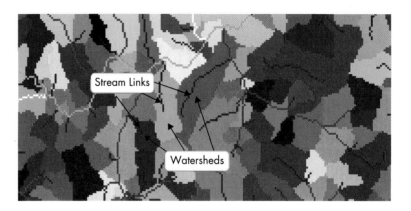

FIGURE 22.1 A portion of a stream network and the watershed area computed for each link of a stream (where each watershed and stream link is a different color).

The source for the various hydrologic processes you'll be working with in this module is a digital elevation model. These hydrologic modeling processes assume that all locations receive equal amounts of rainfall and that the soil and land-cover characteristics don't affect the overland flow of water. Your work in deriving factors like the watersheds or the stream network will be based on the DEM and created as raster surfaces. You'll thus end up with raster surfaces in which you can assess the direction of the flow of water or the accumulation of water at each grid cell location.

Module Scenario and Applications

This module puts you in the role of an environmental scientist examining potentially polluted areas in relation to the stream network of Columbiana County, Ohio. In particular, you want to determine what areas of the landscape will flow or drain into specific streams in the county. To this end, you'll start with a digital elevation model of the county and use it to compute watersheds for each section of a stream in the county and then analyze your results.

The following are additional examples of other real-world applications of this module's skills:

- A county engineer wants to model the amount of water runoff for a study area. She can use the hydrologic modeling tools to determine the direction and accumulation of water flow as layers for her analysis.

- An urban planner is modeling the flood potential of an area where a new subdivision is being planned. She can use the hydrologic modeling tools to examine the watershed area and the amount of flow accumulation for the proposed site.

- An archeologist is working onsite and wants to determine the drainage basin area that would collect rainfall. He can use the watershed tools to generate this area as another layer to add to the GIS data for the site.

 ## Study Area

- For this module, you will be working with data from Columbiana County, Ohio.

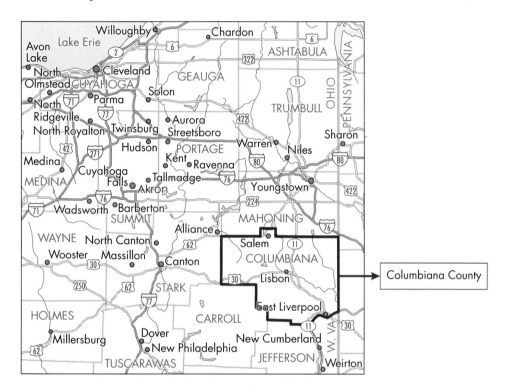

 ## Data Sources and Localizing This Module

The data for this module focus on features and locations within Columbiana County, Ohio. However, you can easily modify this module to use data from your own county instead, such as Harper County, Kansas. The DEM was downloaded from The National Map 3DEP data and processed to align the layer to the county boundaries and keep consistent 30-meter cell sizes for this module; the same DEM data are available for Harper County. The streams data were downloaded from The National Map hydrography data; you can use the NHD_Flowline layer for Harper County for streams.

Step 22.1 Getting Started

- This module's hands-on applications use the data folder called Module22. Your instructor will be able to supply you with this data, or you can download it directly from this book's website at **https://www.macmillanlearning.com/college/us/product/Discovering-GIS-and-ArcGIS-Pro/p/131923075X**. The text in this module assumes that you have this Module22 folder in a computer location referenced as C:\GIS; if you have it somewhere else (for instance, in a flash drive referenced as G:\GISClass), substitute that location and path to the Module22 folder throughout this module.

- Module22 contains a file geodatabase called Columhydrodata that contains the following items:
  - DEMcolum: a raster of 3DEP DEM data for Columbiana County, which has been resampled to 30 meters for purposes of this module.
  - streams: a line feature class representing the various types of rivers and streams in the county.
- Start ArcGIS Pro.
- Sign in with your Esri account username and password.
- Create a new project using the **Map** template. Call this project **Module22** and place it in your **C:\GIS\Module22** folder. Ensure that there is not a checkmark in the box next to **Create a new folder for this project**.
- When ArcGIS Pro opens, change the map's name to **Columbiana**.
- You will be using the ArcGIS Pro Spatial Analyst extension in this module. If you don't have access to the Spatial Analyst extension, see Smartbox 11.1 and Troublebox 11.1 in Module 11 for more information on how to get it set up in ArcGIS Pro.
- Add the **DEMcolum** layer to the Contents pane.
- Use the following geoprocessing environment settings for this module (from the **Analysis** tab, in the **Geoprocessing** group, click the **Environments** button):
  - For Current Workspace, use the **Module22.gdb** geodatabase within your **C:\GIS\Module22** folder.
  - For Scratch Workspace, use the **Module22.gdb** geodatabase within your **C:\GIS\Module22** folder.
  - For Output Coordinate System, choose **DEMcolum**. ArcGIS Pro then updates the coordinate system to that of the DEMcolum layer.
  - For Extent, choose **DEMcolum**. ArcGIS Pro then updates the x and y coordinates of the lower-left corner and upper-right corner of the grid extent to those of the DEMcolum layer.
  - For Cell Size, choose **DEMcolum**. ArcGIS Pro then updates cell size for raster output in this project to 30 meters as that is the cell size of the DEMcolum layer.
- Leave the other settings alone and click **OK** to put these geoprocessing environment settings into place.
- The coordinate system being used in this module is **Albers Conical Equal Area (NAD 83)**, using meters as the map units. Check Columbiana Map Properties (on the Coordinate System tab) to verify that this coordinate system is being used.
- As in Modules 20 and 21, you'll be using ModelBuilder in this module— but now for hydrologic modeling. To start with a new empty model, on the

Analysis tab, in the **Geoprocessing** group, click the **ModelBuilder** button. A new tabbed view called Model opens, providing a large blank space to work in.

- With the new **Model** tab chosen, a new tab called ModelBuilder appears on the ribbon. On this **ModelBuilder** tab, in the **Model** group, click the **Properties** button. Change the Name and Label fields of the model to **Hydromodel**. (See Module 20 for more about how to do this.)

- On the **ModelBuilder** tab, in the **Model** group, click the **Save** button.

- In the Catalog pane, expand the **Module22** toolbox (which was the default toolbox created for the Module22 project). Inside it you can see an item with the model symbol; this is an empty model called Hydromodel that you've just created.

- On the **ModelBuilder** tab, in the **Insert** group, click the **Tools** button. The Geoprocessing pane opens. In the Geoprocessing pane, click the **Toolboxes** button to display the full set of toolboxes available for use.

Step 22.2 Setting the Hydro Model Parameters

- Drag and drop the **DEMcolum** grid from the Hydromodel Contents pane into the ModelBuilder window. This grid will be the model element from which you will build everything else in the model.

- Before you begin building the model, you need to set up the model's parameters. (See Module 21 for more about setting model parameters.) There are two parameters for the hydrologic model that you'll be creating in this module: a data variable and a value variable. In the Hydromodel window, rename DEMcolum as **DEM Input Layer**. (See Module 21 for how to do this.)

- Set the DEM input layer as a required model parameter. (See Module 21 for instructions on how to do this.) The first parameter (the data variable—the DEM Input Layer) is now set. You'll set the value variable later in this module.

Step 22.3 Examining Slopes in Relation to Streams

- The first thing you'll generate for analysis is a slope grid of the DEM input layer. In the Geoprocessing pane, from the **Spatial Analyst Tools** toolbox, choose the **Surface** toolset and then choose the **Slope** tool and drag and drop it from the Geoprocessing pane into the Hydromodel window.

- Add a connection between DEM input layer and the Slope tool and choose the option **Input raster** when making the connection.

- Double-click the **Slope** tool to open it.

- For Output raster, click the **browse** button, navigate to your **C:\GIS\Module22** folder, go to the **Module22.gdb** geodatabase, and name the Output raster **slopegrid**.

- Leave the other parameters alone and click **OK**.

518 CHAPTER 22 How to Use Hydrologic Modeling Tools in ArcGIS Pro

- Run the model to create the slopegrid layer and then add it to the Columbiana map's Contents pane. This slopegrid layer gives you some indication of how water will flow along the terrain surface of Columbiana County. Examine the layer to get a feel for where the stream channels are likely to be found.

- Add the streams feature class to the Contents pane and position it so that you can see the streams layer on top of the slopegrid layer. Answer Question 22.1 and then turn off both the slopegrid and streams layers for now.

> **Question 22.1** Without doing any further hydrologic modeling, and just based on visual inspection, how do you think the streams in the Columbiana County area are distributed with respect to slope?

Step 22.4 Calculating Flow Direction

- The flow direction can be determined by using the Flow Direction tool. In the Geoprocessing pane, from the **Spatial Analyst Tools** toolbox, choose the **Hydrology** toolset and then choose the **Flow Direction** tool and drag and drop it from the Geoprocessing pane into the Hydromodel window.

- Add a connection between DEM input layer and the Flow Direction tool and choose the option **Input surface raster** when making the connection.

- Double-click the **Flow Direction** tool to open it.

- For Output raster, click the **browse** button, navigate to your **C:\GIS\Module22** folder, go to the **Module22.gdb** geodatabase, and name the Output raster **flowdir**.

- For Flow direction type, choose **D8**.

- Leave the other parameters alone and click **OK**.

- Run the model to create the **flowdir** layer. This new grid will show the direction of water flow at each cell along the DEM surface. For more information about flow direction, see **Smartbox 22.1**.

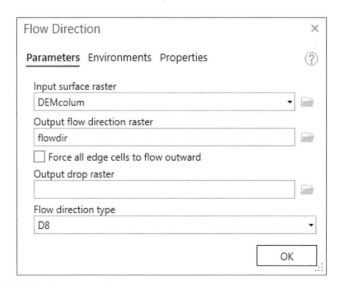

SMARTBOX 22.1

How does flow direction operate in ArcGIS Pro?

Flow direction describes how water will move overland and in which direction it will move. The Flow Direction tool in ArcGIS Pro determines in which of the eight cardinal directions water will move at each cell of the raster elevation surface. The computation is performed in a manner similar to the computation of slope aspect (see Module 15), as it is assumed that water will flow downhill via the steepest slope. The slope is computed from a grid cell to its adjacent grid cell in each of the eight directions. The grid cell with the steepest slope is chosen as the one to which water will flow.

By default, ArcGIS Pro uses the **D8** (deterministic 8) algorithm to calculate flow direction. This algorithm indicates that one (and only one) of the eight directions must be selected for water flow. With D8, water cannot move in multiple directions at once, nor is any randomness involved in which way water will flow. The flow direction is calculated by subtracting the elevation value of each of the eight adjacent cells from the elevation value of the grid cell for which flow direction is being determined. The four lateral directions (N, W, E, S) have this difference divided by the value 1, and the four diagonal directions (NE, NW, SE, SW) have this difference divided by the value 1.414 (because these differences are measured from the center of each cell; see Module 19 for further information). The grid cell with the overall greatest positive value is chosen as the cell to which water should flow. ArcGIS Pro assigns a numerical value to the grid cell to indicate to which of the eight cells the water should flow, based on the direction from the originating cell. These newly assigned values are placed into a new "flow direction grid." The values assigned indicate the direction from which the water will flow (Table 22.1).

flow direction A determination of the direction in which overland water flow will move.

D8 An algorithm for determining flow direction that outputs which of the eight cardinal directions will be the steepest slope for the flow of water.

Assigned Value in Flow Direction Grid	Direction of Flow
1	East
2	Southeast
4	South
8	Southwest
16	West
32	Northwest
64	North
128	Northeast

Table 22.1 The assigned values in the flow direction grid and their corresponding flow directions

Figure 22.2 shows an example of calculating the flow direction for one grid cell—in this case, the center cell in blue. The Elevation Raster contains elevation values at each grid cell (like the DEM). The "Calculation" grid shows the

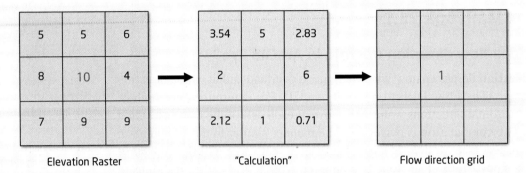

Elevation Raster   "Calculation"   Flow direction grid

FIGURE 22.2 The initial elevation raster, the internal calculation of the steepest slope, and the corresponding output written to the flow direction grid.

values internally calculated by the flow direction process for each cell. These values are determined by subtracting the value of each cell from the center cell and dividing by the appropriate distance value (1 for lateral calculations and 1.414 for diagonal ones). For instance, the adjacent cell to the northeast is calculated to be 2.83 (that is, (10 − 6) / 1.414). The steepest slope is found to be to the adjacent grid cell on the east; the value 6 is the highest calculated value, and this value will be direction of the steepest slope. Thus, at the cell's location in the flow direction grid, the value 1 is written, indicating that water at that cell will flow to the east. A flow direction value is then calculated for each cell in the elevation raster.

- Display the **flowdir** grid as unique values. Note the values computed for flow direction. Answer Question 22.2.

> **Question 22.2** Why are there values other than 1, 2, 4, 8, 16, 32, 64, and 128 in your flow direction grid? (*Hint:* You might want to refer to the ArcGIS Pro Help entry for "Flow Direction" at https://pro.arcgis.com/en/pro-app/tool-reference/spatial-analyst/flow-direction.htm to answer this question.)

Step 22.5 Locating Sinks

- You will now examine the surface for the sinks that exist on it. In the Geoprocessing pane, from the **Spatial Analyst Tools** toolbox, choose the **Hydrology** toolset and then choose the **Sink** tool and drag and drop it from the Geoprocessing pane into the Hydromodel window.
- Add a connection between the flowdir layer and the Sink tool and choose the option **Input D8 flow direction raster** when making the connection.
- Double-click the **Sink** tool to open it.

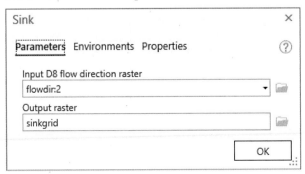

- For Output raster, click the **browse** button, navigate to your **C:\GIS\Module22** folder, go to the **Module22.gdb** geodatabase, and name the Output raster **sinkgrid**.
- Leave the other parameters alone and click **OK**.
- Run the model to create the **sinkgrid** layer. Add sinkgrid to the Columbiana map Contents pane. This layer shows the location of the sinks in the Columbiana County DEM layer. For more information about sinks, see **Smartbox 22.2**.
- Answer Questions 22.3 and 22.4. *Note:* You use the sinkgrid layer's attribute table to answer these two questions. If the layer has no attribute table, see **Troublebox 22.1** for information on how to create an attribute table for a raster layer.

SMARTBOX 22.2

What are sinks in hydrologic modeling?

An elevation surface may have low-elevation values surrounded on all sides by high-elevation values (Figure 22.3). These low-elevation locations, called **sinks**, often result from errors in the elevation surface. Sinks are problematic in hydrologic modeling because they represent grid cells for which flow direction can't be properly computed. The direction of flow is determined by the direction of the steepest slope, but this condition isn't present at a sink (because all eight cells around the sink cell contain higher values, thus indicating that water would have to flow uphill at these spots). When the flow direction algorithm tries to act on a cell with a sink, it can't properly determine which way the water should go, and the resultant flow direction grid ends up with improper values.

sinks Low areas in a digital elevation model surrounded on all sides by high areas.

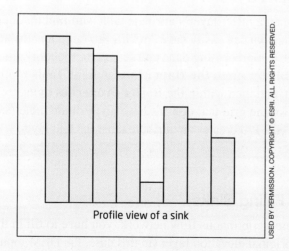

FIGURE 22.3 The appearance of a sink (low-elevation cells among higher-elevation cells).

To accurately determine flow direction (which serves as the basis for many other hydrologic functions), sinks need to be removed from the elevation surface. ArcGIS Pro can "fill" these sink cells by artificially adding elevation values to

depressionless DEM A DEM that has had its sinks "filled."

them so that these cells are no longer surrounded by higher elevations (Figure 22.4). The end result of "filling" the sinks in a DEM is the creation of a "sinkless" DEM, or **depressionless DEM**. Without sinks, this new depressionless elevation surface is used in the remainder of the hydrologic modeling processes.

FIGURE 22.4 An example of "filling" a sink so that water flow across a surface can be determined.

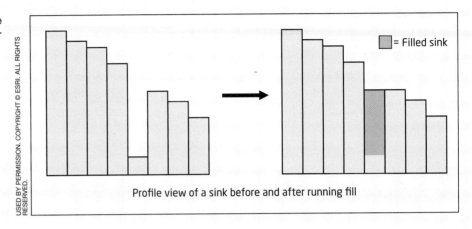

Profile view of a sink before and after running fill

Question 22.3 How many sinks are in the Columbiana County DEM layer?

Question 22.4 What is the total number of cells that make up the sinks?

TROUBLEBOX 22.1

How do I build a raster attribute table for a raster layer?

If you try to open a raster layer's attribute table and find that option grayed out and unavailable, you can easily have ArcGIS Pro build an attribute table for that layer. Open the Geoprocessing pane and search for the tool called **Build Raster Attribute Table** (or go to **the Data Management Tools** toolbox, within the **Raster** toolset, and then within the **Raster Properties** toolset, choose it there). Run this tool for the grid that doesn't have an attribute table (such as the sink-grid), and ArcGIS Pro creates an attribute table for that layer.

Step 22.6 Filling Sinks

- To create a continuous flow network, you have to fill in the sinks back in the original elevation layer (in this case, the DEM input layer). In the Geoprocessing pane, from the **Spatial Analyst Tools** toolbox, choose the **Hydrology** toolset and then choose the **Fill** tool and drag and drop it from the Geoprocessing pane into the Hydromodel window.

- Add a connection between DEM input layer and the Fill tool and choose the option **Input surface raster** when making the connection.
- Double-click the **Sink** tool to open it.

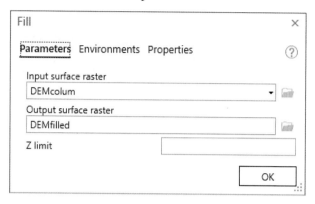

- For Output raster, click the **browse** button, navigate to your **C:\GIS\Module22** folder, go to the **Module22.gdb** geodatabase, and name the Output raster **DEMfilled**.
- Leave the other parameters alone and click **OK**.
- Run the model to create the **DEMfilled** layer, which is a "depressionless" or "sinkless" DEM (in which all of the sinks from the original layer have been filled. Add the DEMfilled layer to the Contents pane.

  Important Note: You will be working with the "sinkless" DEM (DEMfilled) and derivations of it for the remainder of the module.

- Use the ModelBuilder tools to create a new flow direction grid (call it **newflowdir**) from the sinkless DEMfilled layer as follows:
  - In the Geoprocessing pane, from the **Spatial Analyst Tools** toolbox, choose the **Hydrology** toolset and then choose the **Flow Direction** tool and drag and drop it from the Geoprocessing pane into the Hydromodel window.
  - Add a connection between DEM input layer and the Flow Direction tool and choose the option **Input surface raster** when making the connection.
  - Double-click the **Flow Direction** tool to open it.
  - For Output raster, click the **browse** button, navigate to your **C:\GIS\Module22** folder, go to the **Module22.gdb** geodatabase, and name the Output raster **newflowdir**.
  - For Flow direction type, choose **D8**.
  - Leave the other parameters alone and click **OK**.
  - Run the model to create the newflowdir layer.
- Add the **newflowdir** layer to the Columbiana map Contents pane. Note the values that are now computed for flow direction at each cell compared with the previous flow direction grid (which was created from the original DEM input layer, which still contained sinks).

- Save your model. At this point, your Hydromodel dialog should look something like this:

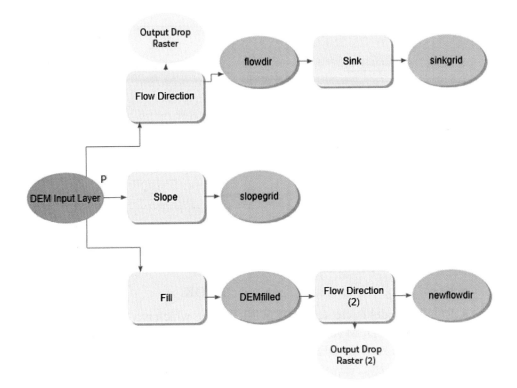

Step 22.7 Calculating Flow Accumulation

- With the direction of flow computed, you can now determine the accumulation of flow for each cell on the elevation surface. The flow direction can be determined by using the Flow Direction tool. In the Geoprocessing pane, from the **Spatial Analyst Tools** toolbox, choose the **Hydrology** toolset and then choose the **Flow Accumulation** tool and drag and drop it from the Geoprocessing pane into the Hydromodel window.

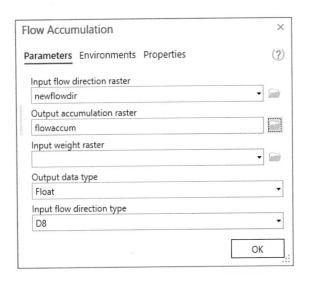

- Add a connection between the newflowdir layer and the Flow Accumulation tool and choose the option **Input flow direction raster** when making the connection.
- Double-click the **Flow Accumulation** tool to open it.
- For Output raster, click the **browse** button, navigate to your **C:\GIS\Module22** folder, go to the **Module22.gdb** geodatabase, and name the Output raster **flowaccum**.
- For Output data type, choose **Float**.
- For Input flow direction type, choose **D8**.
- Leave the other parameters alone and click **OK**.
- Run the model to create the **flowaccum** layer. This new grid shows you how many pixels "flow" into a location. (See **Smartbox 22.3** for more information.)

SMARTBOX 22.3

How does flow accumulation work in ArcGIS Pro?

Once flow direction is computed (and a coded value indicating in which of the eight directions that flow will occur is assigned to the grid), ArcGIS Pro can determine **flow accumulation**, or a count of how many cells have their flow connected to a location cell. The more cells that are connected, the greater the amount of water that flows into those cells. As the amount of connected cells grows, higher values for flow accumulation will be calculated.

Figure 22.5 provides an example of calculating flow accumulation. The flow direction grid is coded with the specific values indicating the direction that water will move from each cell. The "flow network" diagram graphically shows what the flow direction grid represents. For instance, three cells connect to the middle grid cell. Thus, its value for flow accumulation is 3. The grid cell in the bottom-left corner has only one cell that directly connects to it, so its value for flow accumulation is 1. However, the flow accumulation values build up as more and more cells connect to one another and a total cumulative value is calculated. For instance, the grid cell in the bottom center has a flow accumulation value of 8 because there is a cumulative total of 8 cells that flow into it. Three cells flow into the middle cell, and then the middle cell flows into the bottom center, for a total of four. Then each of the bottom corner cells has a total of two cells each that connect to the bottom center. Thus, the total flow accumulation value of the bottom-center cell is 8.

flow accumulation A determination of how many cells will flow into a location.

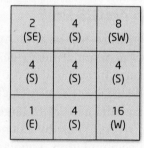
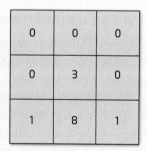

Flow direction   "Flow network"   Flow accumulation

FIGURE 22.5 An example of calculating flow accumulation from a flow direction grid.

- Add the **flowaccum** grid to the Columbiana map Contents pane. The next step will be to display the flowaccum grid visualized with a different classification method. To start, from in the Contents pane, click on **flowaccum** and then, on the **Appearance** tab, click the **Symbology** button and choose **Classify**.

- The Symbology pane opens. For Method, choose **Quantile**, and for Classes, choose **10**. Also change the color ramp to something other than black and white.

- The flowaccum layer is now displayed using quantiles, with 10 different categories. Turn on the **streams** layer and place it over top of the flowaccum grid. Answer Question 22.5, zooming in on some of the streams to see them in relation to the flowaccum grid. Turn off the streams layer when you're finished.

> **Question 22.5** Keeping in mind that flowaccum is showing only the accumulation of overland water flow, how does it compare with the streams layer? What differences are there between the two datasets? What does this say about the channels shown in the streams layer? *Hint:* Select some of the areas on the flowaccum grid that have the highest values and zoom in closely to see those areas in relation to the streams layer.

Step 22.8 Creating a Stream Network with a Threshold Value

- What you will now do is extract from the flowaccum grid the cells that represent the stream network. (See **Smartbox 22.4** for more information about this process.) In the Geoprocessing pane, from the **Spatial Analyst Tools** toolbox, choose the **Conditional** toolset and then choose the **Con** tool and drag and drop it from the Geoprocessing pane into the Hydromodel window.

SMARTBOX 22.4

How can a stream network be extracted in ArcGIS Pro?

Each cell of the flow accumulation grid has a value for how many grid cells connect to that particular cell. As flow accumulation values increase, more and more cells are connected together. In terms of hydrologic modeling, these cells represent areas on the surface where greater amounts of water flow link together and accumulate, and cells with large values of flow accumulation represent streams and rivers. Thus, by examining only cells with a high enough value of flow accumulation, you can determine what cells are representative of streams and model the stream network.

The tricky part of this process is determining what flow accumulation values are high enough to constitute a stream network. This *threshold* value represents the number of cells that connect to constitute a stream channel. Knowledge of the study area and its various factors (such as the climate or the soil conditions) will help in determining the proper threshold value for accurate modeling of streams. For purposes of this module's activities, you're using a sample value of 100 grid cells as a threshold and comparing the results with a National Hydrology Dataset (NHD) layer from The National Map. Figure 22.6 shows an example of identifying areas that have a greater amount of water accumulating at each location and extracting them as streams.

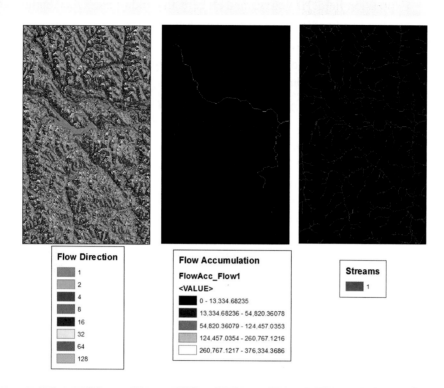

FIGURE 22.6 Flow direction and flow accumulation grids and the resultant stream network of grid cells extracted from them using a threshold value.

- Add a connection between the flowaccum layer and the Con tool and choose the option **Input conditional raster** when making the connection.
- Double-click the **Con** tool to open it.
- In the Expression box, click the **New expression** button to create the SQL expression (see Module 2) to set up the threshold.
- In the box that opens, choose **VALUE** from the first pull-down menu, choose **is Greater than** from the second pull-down menu, and then in the third box type **100**. Then click the green checkmark. The SQL expression in the box now reads "Value is greater than 100." This indicates that for all grid cells in the flowaccum raster that have a value higher than 100, you want to assign a new value, and for all cells that have a value of 100 or less, you want to assign a different value.
- For Input true raster or constant value, type **1**. This value indicates that all cells that meet the condition specified in the Expression field (that is, "Value is greater than 100") will receive a value of 1.
- Leave the Input false raster or constant value (optional) field blank. By leaving this field blank, you instruct ArcGIS Pro to assign the value NODATA to all cells that do not meet the condition specified in the

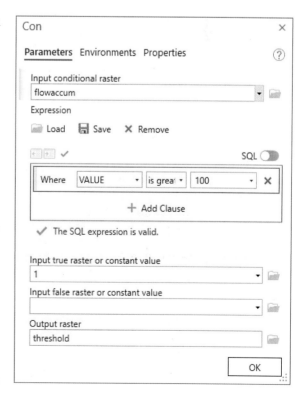

528 CHAPTER 22 How to Use Hydrologic Modeling Tools in ArcGIS Pro

Expression field. (If you wanted all cells that didn't meet the condition to receive a value other than NODATA, you could type that value here.)

- For Output raster, click the **browse** button, navigate to your **C:\GIS\Module22** folder, go to the **Module22.gdb** geodatabase, and name the Output raster **threshold**.
- Leave the other parameters alone and click **OK**.
- The SQL expression that you created ("Value is greater than 100") will also become the second model parameter. As noted in Smartbox 22.4, the value used for the flow accumulation threshold can change, depending on the project. Therefore, you'll want the model's users to be able to define their own threshold values when running the model. To set up this part of the Con tool as its own variable (which you can then set as a parameter), in the Hydromodel window, right-click on the **Con** tool, select **Create Variable**, select **From Parameter**, and finally select **Expression**.

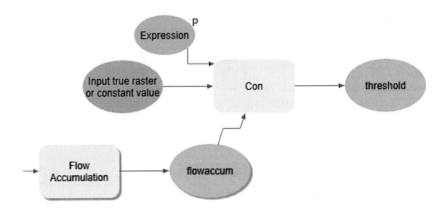

- A new light-blue oval (labeled Expression) appears, indicating that a value variable has been created for the model as an input to the Con tool. Move this variable so that it can be clearly seen as an input to the Con tool and does not cover any other model elements. To set this as a model parameter, right-click on the **Expression** oval and select **Parameter**. The *P* symbol appears next to the oval, indicating that it is now set as a model parameter.

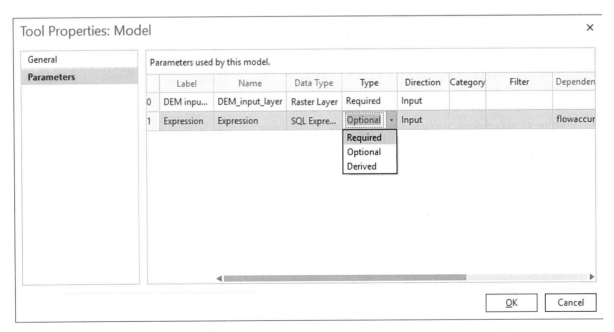

- You now want to set the expression as a required model parameter rather than an optional one (because a threshold level needs to be set for the model to run properly). On the **ModelBuilder** tab, in the **Model** group, click the **Properties** button. In the Tool Properties: Model dialog box that opens, choose the **Parameters** tab.

- You should see that the two parameters are set up: DEM input layer as a required parameter and Expression as an optional parameter. Click next to **Optional** and from the pull-down menu that appears, change its state to **Required**. Click **OK** when both model parameters are set as Required.

- Run the model to create the threshold layer. Add the threshold grid to the Columbiana map's Contents pane. Turn off the other layers, except for threshold and streams, and answer Question 22.6. Turn off both threshold and streams when you're done.

> **Question 22.6**
> How does the threshold layer compare with the streams layer? What differences are there between the two datasets? *Hint:* Choose some of the areas on the threshold grid and zoom in closely to see those areas in relation to the streams layer.

- Save the model.

Step 22.9 Stream Ordering

- Now that you have a set of cells that represent the layout of the stream network, you can examine the stream ordering. (See **Smartbox 22.5** for more about the two methods of stream ordering used in ArcGIS Pro.) In the Geoprocessing pane, from the **Spatial Analyst Tools** toolbox, choose the **Hydrology** toolset and then choose the **Stream Order** tool and drag and drop it from the Geoprocessing pane into the Hydromodel window.

 SMARTBOX 22.5

How does stream ordering work in ArcGIS Pro?

Once you've identified the grid cells that represent the set of streams with which you'll be working, you can assign an ordering to these streams. **Stream order** is a way of ranking streams; lower orders are minor streams with little in the way of cells that flow into them, while higher orders are streams with a larger number of smaller streams that flow together to produce larger streams. In other words, stream ordering is a means of classifying streams by how many other streams flow into them. This ordering is assigned to the links of the streams themselves, and the stream junctions are used to determine where streams join together.

Two different types of stream-ordering techniques are available in ArcGIS Pro. The first of these is the **Shreve method**, which determines the order of a

stream order A computed ranking of streams based on how they flow into one another.

Shreve method A stream-ordering method that allows for the stream-order ranking to be computed by adding the ranks of the two streams flowing into each another.

Strahler method A stream-ordering method that allows for stream order to increase in rank only if two streams of the same order of magnitude flow into each another.

stream by adding the values of the streams that join together. For instance, two first-order streams join to produce a second-order stream. A second-order stream and a first-order stream join to produce a third-order stream. A third-order and a fourth-order stream join to become a seventh-order stream, and so on (Figure 22.7a). The Shreve ordering method is very simple and tends to produce a high range of stream-ordering values as more and more streams flow together.

The second stream-ordering technique is the **Strahler method**, which allows an increase in stream ranking only when two streams of the same order join. For instance, two first-order streams join to form a second-order stream. Two second-order streams join to form a third-order stream. However, when a first-order stream and a second-order stream join, the result is a second-order stream; because the two initial streams are of different rankings, the ranking of the new stream does not increase. The Strahler method tends to produce an overall lower range of values for the ranking of streams. Figure 22.7b shows an example of the Strahler method.

FIGURE 22.7 A stream network ordered using (a) the Shreve method and (b) the Strahler method.

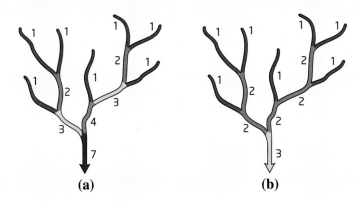

- Add a connection between the newflowdir layer and the Stream Order tool and choose the option **Input flow direction raster** when making the connection.
- Add a connection between the threshold layer and the Stream Order tool and select the option **Input stream raster** when doing so.
- Double-click the **Stream Order** tool to open it.
- For Output raster, click the **browse** button, navigate to your **C:\GIS\Module22** folder, go to the **Module22.gdb** geodatabase, and name the Output raster **strahstream**.
- For Method of stream ordering, choose **Strahler**.
- Click **OK**.
- Run the model to create the Strahler Stream Order layer.
- Use the ModelBuilder tools to create a second, separate stream-order grid called **shrevestream**, but this time using the **Shreve** method of stream ordering.

Again, use **threshold** as your input stream raster and **newflowdir** as your input flow direction raster.

- Run the model and then add both the strahstream and shrevestream grids to the Columbiana map's Contents pane and examine them. Answer Question 22.7.

> **Question 22.7**
> What stream orders of magnitude are represented by both the Strahler and Shreve stream-ordering methods in your model?

Step 22.10 Watershed Delineation

- ArcGIS Pro can delineate a series of watersheds for each one of the streams calculated from the DEM surface (based on the surface flow). First, however, it needs to compute the stream links (the junctions of each stream) as part of the watershed delineation process. In the Geoprocessing pane, from the **Spatial Analyst Tools** toolbox, select the **Hydrology** toolset and then select the **Stream Link** tool and drag and drop it into the Hydromodel window.

- Add a connection between the newflowdir layer and the Stream Link tool and select the option **Input flow direction raster** when doing so.

- Add a connection between the threshold layer and the Stream Link tool and select the option **Input stream raster** when doing so.

- Double-click on the **Stream Link** tool to open it.

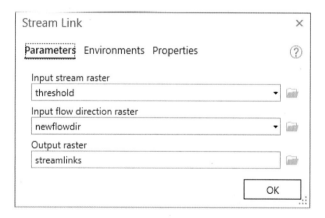

- For Output raster, click the **browse** button, navigate to your **C:\GIS\Module22** folder, go to the **Module22.gdb** geodatabase, and name the Output raster **streamlinks**.

- Click **OK** to close the Stream Link tool.

- In the Geoprocessing pane, from the **Spatial Analyst Tools** toolbox, select the **Hydrology** toolset and then select the **Watershed** tool and drag and drop it into the Hydromodel window.

- Add a connection between the streamlinks layer and the Watershed tool and select the option **Input raster or feature pour point data** when doing so.

- Add a connection between the newflowdir variable and the Watershed tool and select the option **Input D8 flow direction raster** when doing so.
- Double-click on the **Watershed** tool to set its parameters.

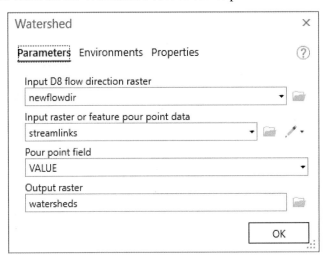

- For Output raster, click the **browse** button, navigate to your **C:\GIS\Module22** folder, go to the **Module22.gdb** geodatabase, and name the Output raster **watersheds**.
- For Pour point field (which is optional), choose **VALUE**.
- Click **OK** to close the Watershed tool.
- Run the model and add the watersheds grid to the Columbiana map's Contents pane. Remember that this watershed grid is based on the stream links extracted from the threshold value of 100 previously set. Answer Question 22.8.

Question 22.8 How many watersheds are extracted?

- Save the model.

Step 22.11 Validating and Running the Model

- In the model, you've designated several output layers to be created along 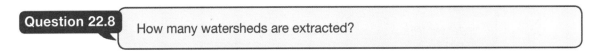 the way, and many of them are used as inputs to tools to eventually create that final watersheds layer. However, ArcGIS Pro usually does not permanently keep these intermediate layers and deletes them after the model is run. When you run a model from the ModelBuilder interface, the intermediate layers are not deleted, but when you run a model as a standalone tool (as you did in Module 21 and as you will do in the next several steps), these intermediate layers are automatically removed. So if you want to keep one of these layers, you have to do a couple extra steps. For instance, you will want to keep the output from the Stream Link tool for further analysis; therefore, you need to have the model create a second, duplicate, set of stream links. To begin, in the Geoprocessing pane, from the **Spatial Analyst Tools** toolbox, select the **Hydrology** toolset and then select the **Stream Link** tool and drag and drop it into the Hydromodel window. You now have a second Stream Link tool in the model.

- Add a connection between the newflowdir layer and the second Stream Link tool and select the option **Input flow direction raster** when doing so.
- Add a connection between the threshold layer and the second Stream Link tool and select the option **Input stream raster** when doing so.
- Double-click on the second **Stream Link tool** to open it.
- For Output raster, click the **browse** button, navigate to your **C:\GIS\Module22** folder, go to the **Module22.gdb** geodatabase, and name the Output raster **streamanalyze**.
- Click **OK** to close the Stream Links tool. You now have a second output from the model called streamanalyze that will be produced along with the watersheds layer and so will not be deleted when the model is run from the Geoprocessing pane.
- Save the model.
- Next, you want to validate the model. On the **ModelBuilder** tab, in the **Run** group, click the **Validate** button.
- Next, you're now going to run the entire model as a separate tool. In the Catalog pane, navigate to the **C:\GIS\Module22** folder and expand the **Module22** toolbox. Right-click on the **Hydromodel** icon and select **Open**. Because the model's parameters have been set, the model opens—not as the graphical ModelBuilder interface but as the model itself. The model takes the form of a tool in the Geoprocessing pane that will allow the user to specify two inputs (namely, the two model parameters): a DEM layer and the expression used to determine the flow accumulation threshold.

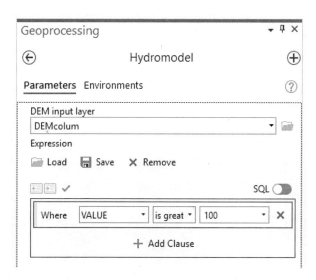

- You can see that DEMcolum and Value is Greater Than 100 are the defaults used when running the model. If you wanted to use a different DEM layer or a new value for the threshold, you can specify it here. Run the tool with these two default parameters. A new dialog box appears, showing each process of the model as it runs. When it completes, close the dialog box.
- At this point, the entire model has been rerun, and any previously created layers have been overwritten. Back in the Catalog pane, right-click on the **Module22.gdb** geodatabase and choose **Refresh**. You see that only

the output variables that were the final outputs of the model (that is, outputs that did not connect to another tool) are available; all the other intermediate output layers have been deleted.

Step 22.12 Analyzing the Results of Hydrologic Modeling

- You can now take a look at the watershed results of the model and how they relate to the streams. As part of this module's scenario, you're examining potentially polluted areas around Columbiana County streams. What you want to do is find the area of the watershed around McCormick Run for use in future analysis related to pollution sources near that stream. Back in the Columbiana map, add the **streamanalyze** and **watersheds** layers to the Contents pane and turn off all other layers except for the **streams** vector layer. Arrange these three layers so that streams is on top and watersheds is on the bottom.

- Change the symbology of the streams and streamanalyze layers so that you can easily distinguish the two layers.

- Change the symbology of the watersheds layer to **Unique Values** by using **Value** for the Value field. In this way, you can assign each of the watersheds its own color to distinguish them from one another.

- Open the attribute table for the streams layer and locate the McCormick Run stream. (You might want to sort the GNIS_Name field and find McCormick Run.) There will be two links. Select the main one (it has the OBJECTID field equal to 3532) and zoom in to that stream. You can see the stream line laid over its 30-meter-resolution raster counterpart in the streamlinks layer, along with the computed watershed for that stream link. Answer Questions 22.9 and 22.10.

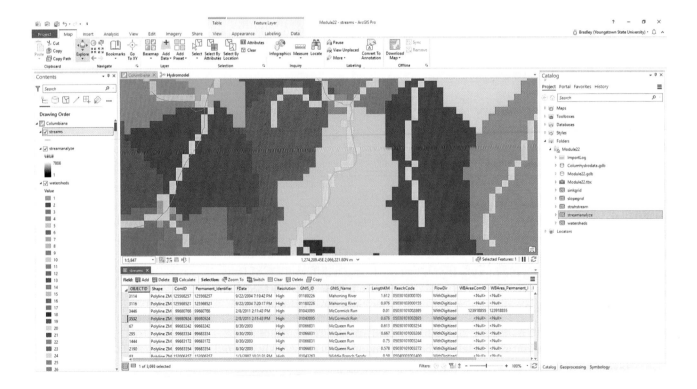

> **Question 22.9** How does the vector version of McCormick Run (from the National Hydrology Dataset layer from The National Map) compare with its 30-meter-resolution counterpart derived from the DEM?

> **Question 22.10** What order of magnitude stream is McCormick Run, according to both the Strahler method and the Shreve method of stream ordering? (*Hint:* Examine your strahstream and shrevestream grids to the map answer this question.)

- Using the Explore tool and the watersheds layer's attribute table, answer Questions 22.11 and 22.12.

> **Question 22.11** Which watershed number (in terms of its OBJECTID value) has been computed for McCormick Run?

> **Question 22.12** How large (in terms of acreage) is the McCormick Run watershed? (*Hint:* The size of each grid cell is 900 square meters—because they are 30-meter-resolution grid cells—and 1 square meter (m^2) is equal to 0.000247105 acres.) Show your work.

Step 22.13 Saving, Printing, or Sharing Your Results

- Save your Module21 project.
- Compose and print a layout (see Module 3) of your final version of the McCormick Run stream, the streamlinks version of the same, and the computed watershed for that stream link (including all of the usual map elements and design for a layout).
- If you want to share your model, you can share it as a geoprocessing package. (See Module 21 for how to create and share a model as a geoprocessing package.)

Closing Time

This module explored how to work with the various hydrology tools available in ArcGIS Pro by starting with a digital elevation model and ending with a stream network and a set of watersheds delineated from the streams. Hydrologic modeling techniques allow for many types of analysis related to the overland flow of water and are commonly utilized by engineers, water resource managers, and environmental scientists. Beyond the Hydrology toolset, Esri has an extended set of hydrologic tools available through the Arc Hydro data model and tools (for links to download this, see **http://downloads.esri.com/archydro/archydro/Setup/Pro/**).

Watershed delineation can be performed for locations other than stream links. For instance, if you have identified specific drainage points throughout the area and want to determine the watershed for those areas, you'll need to use a different technique than the one you used for computing stream links. See *Related Concepts for Module 22* for more information.

RELATED CONCEPTS FOR MODULE 22

Using Pour Points in ArcGIS Pro

In hydrologic modeling, you're not limited to delineating watersheds based on stream links. For instance, if you have a set of culverts throughout a region, you might want to determine the catchment area for each one. In this case, you could use a point feature class of the culvert locations (rather than the stream links) as the input into the Watershed tool. ArcGIS Pro refers to a feature class that represents an outlet for a drainage system as a **pour point**. By using pour points, you can specify the sub-basin for each of these locations (Figure 22.8).

pour point A feature class that represents the outlet location for the drainage of water.

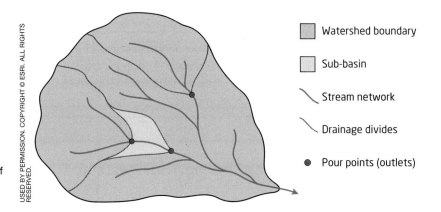

FIGURE 22.8 A sample set of watersheds (sub-basins) created using pour points.

Pour points are locations to which all of the water in the catchment area flows. In the hydrologic modeling process, using pour points allows you to go straight from creating the flow direction grid (from a depressionless DEM) into the watershed delineation process. The Watershed tool has two required inputs: a flow direction raster and a layer representing the sources for the water outlets. In this module, we derived the stream links for use as the outlet source layer, but pour points can be used instead. If you use pour points in this fashion, the Snap Pour Point tool (in the Hydrology toolset) may be useful. It locates the highest-value cell of the flow accumulation grid that is near the point and then moves the point to that location. The location of the point will be in the area of the largest nearby water accumulation.

Key Terms

Hydrologic modeling (p. 513)
watershed (p. 513)
Flow direction (p. 519)
D8 (p. 519)
sinks (p. 521)
depressionless DEM (p. 522)
flow accumulation (p. 525)
stream order (p. 529)
Shreve method (p. 529)
Strahler method (p. 530)
pour point (p. 536)

APPENDIX A

Transitioning from ArcMap to ArcGIS Pro

Esri has been very clear about the fact that while ArcMap (and its related components in ArcGIS Desktop such as ArcCatalog, ArcScene, and ArcGlobe) are not going to be unsupported programs anytime soon, the focus of their desktop GIS efforts is on ArcGIS Pro. As such, transitioning from the older software to ArcGIS Pro is very much in one's best interests to stay current and up-to-date with the applications of Esri's GIS software.

One important factor with ArcGIS Pro is that a user will always be signed in using his or her Esri account in order to work with the program. Whereas ArcMap seemed designed around creating maps and layouts, ArcGIS Pro is built to integrate with the cloud, allowing easier sharing of maps and layers with ArcGIS Online as well as ease of downloading and working with online content. So not only will a user have a license to work with the software served online, but that license will be tied to the user's ArcGIS Online account as well.

If you have done a lot of work in ArcMap and have a variety of map documents filled with layers, all that work is not lost. ArcGIS Pro allows you to import map documents created in ArcMap to become projects. On the **Insert** tab, in the **Project** group, you can click the **Import Map** button to activate this process.

Similarly, the version of Python that works with ArcMap and the other components of ArcGIS Desktop uses Python 2.7, but ArcGIS Pro uses Python 3.x. For instance, some tool functions that were available in the old Python 2.7 version of ArcPy will not work in ArcGIS Pro. If you have access to Python 2.7 scripts, they have to be converted to Python 3 if you want to use them in ArcGIS Pro. A geoprocessing tool called Analyze Tools For Pro can help identify issues with previous Python scripts written for ArcGIS Pro to determine compatibility with ArcGIS Pro. This tool is available in the **Geoprocessing** pane, within the **Data Management** toolbox and inside the **General** toolset. Also, the 2to3 Python utility (see **https://docs.python.org/2/library/2to3.html**) may be of help in converting Python scripts to version 3.

One ArcMap utility that did not make the switch to ArcGIS Pro is Personal Geodatabase (and its files in Access format with the extension .mdb). In fact, ArcGIS Pro's catalog will not recognize Personal Geodatabase files. To use this data in ArcGIS Pro, you have to move your data from a Personal Geodatabase file into a file geodatabase in ArcMap first, and then you can use the file geodatabase in ArcGIS Pro. For more information on migrating from Personal Geodatabase files to a file geodatabases, see **http://desktop.arcgis.com/en/arcmap/latest/manage-data/geodatabases/migrating-to-the-file-geodatabase.htm**.

Unfortunately, making the full transition from ArcMap to ArcGIS Pro isn't the most intuitive process; not only does ArcGIS Pro have a different layout and tools, but ArcGIS Pro also uses different terminology and functions than ArcMap.

However, one of the easiest ways to transition from ArcMap to ArcGIS Pro is to learn the ArcGIS Pro analogues of ArcMap's functionality. Table A-1 lays out common ArcMap concepts, what their ArcGIS Pro counterparts are, and how to access them in ArcGIS Pro.

ArcMap Concept	ArcGIS Pro Concept	How to Access in ArcGIS Pro
ArcCatalog	Catalog View	View tab, Windows group, Catalog View button
Add Data button	Add Data button	Map tab, Layer group, Add Data button
Add Data (ArcGIS Online)	Portal	Map tab, Layer group, Add Data button, choose All Portal
Add Data (basemap)	Basemap	Map tab, Layer group, Basemap button
ArcGlobe	Global Scene	Insert tab, Project group, New Map button, then click New Global Scene
ArcScene	Local Scene	Insert tab, Project group, New Map button, then click New Local Scene
ArcToolbox	Toolboxes	Analysis tab, Geoprocessing group, Tools button, then in Geoprocessing pane click Toolboxes
Attribute Table	Attribute Table	Data tab, Table group, Attribute Table button
Catalog	Catalog pane	View tab, Windows group, Catalog Pane button
Data Frame	Map	Insert tab, Project group, New Map button
Editor Toolbar	Edit tab	Edit tab, Features group, Create button or Modify button
Environment Settings	Environments	Analysis tab, Geoprocessing group, Environments button
Find Tool	Locate	Map tab, Inquiry group, Locate button
Identify Tool	Explore	Map tab, Navigate group, Explore button
Layout	Layout	Insert tab, Project group, New Layout button
Map Document (.mxd)	Project (.aprx)	Project tab, New, Map (also on Quick Access Toolbar)
Map Package (.mpk)	Project Package (.ppkx)	Share tab, Package group, Project button
Measure Tool	Measure	Map tab, Inquiry group, Measure button
ModelBuilder	ModelBuilder	Analysis tab, Geoprocessing group, ModelBuilder button
New Item (Server, Geodatabase, etc.)	Connections	Insert tab, Project group, Connections button
Open (Map Document)	Open (Project)	Project tab, Open (also on Quick Access Toolbar)
Pan	Explore	Map tab, Navigate group, Explore button
Print	Print	Share tab, Print group, either Map button or Layout button
Publish a Service to ArcGIS Online	Share As Web Layer or Share As Web Map	Share tab, Share As group, Web Layer or Web Map button
Python Window	Python Window	View tab, Windows group, Python button
Save (Map Document)	Save (Project)	Project tab, Save (also on Quick Access Toolbar)
Search	Find Tools	Analysis tab, Geoprocessing group, Tools button, then in Geoprocessing pane click Find Tools
Select By Attribute	Select Layer By Attribute	Map tab, Selection group, Select by Attributes button
Select By Location	Select Layer By Location	Map tab, Selection group, Select by Location button
Table of Contents	Contents pane	View tab, Windows group, Contents button
Zoom In and Out	Explore	Map tab, Navigate group, Explore button

Table A.1 ArcMap functions, their ArcGIS Pro versions, and how to access them

Glossary

3D Analyst An ArcGIS Pro extension used for 3D visualization and working with data that have z-values. (p. 352)

3D Elevation Program (3DEP) A digital elevation model of the entire United States, provided by the USGS. (p. 357)

accumulated cost The total cost of traveling from a source to a grid cell. (p. 444)

accuracy How closely a measurement matches up with its real-world counterpart. (p. 168)

active remote sensing A process in which a sensor generates its own energy source and measures the reflection or backscatter of that energy. (p. 394)

address locator ArcGIS Pro functionality that specifies the reference data, geocoding style, and necessary attributes for geocoding. (p. 244)

address locator style The format that will be used for geocoding addresses. (p. 245)

aerial photography The acquisition of imagery of the ground taken from an airborne platform. (p. 303)

Albers Equal Area Conic A projected coordinate system used for east–west data at mid-latitudes. (p. 19)

algorithm A set of steps used in a process. (p. 269)

allocation The determination of which source is closest in distance to each grid cell. (p. 442)

AND (map algebra) The Boolean operator used with an intersection of two grids. (p. 475)

AND (query) The Boolean operator that chooses features that meet both criteria in a query. (p. 51)

animation The capture of movements, actions, or changes to layers in a video format. (p. 387)

annotation A method of adding text to a map that allows for individual pieces of text to be edited separately. (p. 91)

Append A geoprocessing tool that combines one or more layers of data together with an already existing layer. (p. 220)

Arcade A scripting language for writing expressions in ArcGIS. (p. 33)

ArcGIS A software platform developed by Esri. (p. 2)

ArcGIS Desktop The version of ArcGIS that runs on a personal computer. (p. 2)

ArcGIS Living Atlas of the World A set of GIS data and resources set up by Esri for access via ArcGIS Online and ArcGIS Pro. (p. 133)

ArcGIS Online Esri's cloud-based GIS platform, where data and web mapping software can be accessed via the Internet. (p. 92)

ArcGIS Pro A 64-bit standalone GIS program that comes with ArcGIS Desktop. (p. 2)

ArcGIS World Geocoding Service An ArcGIS Online service that allows you to geocode addresses using Esri resources in the cloud. (p. 255)

ArcPad An Esri software product for mobile devices that allows for GPS integration and field data collection. (p. 166)

ArcPy The Python add-in that enables ArcGIS Pro tools and settings to be used in Python scripts. (p. 32)

aspect A determination of the direction the steepest slope is facing at a location. (p. 361)

attachment A function that allows a user to connect non-spatial media, such as documents, files, photos, or charts, to a geospatial feature and make them part of the triggered pop-up. (p. 57)

attribute Non-spatial data that can be associated with a geospatial location. (p. 35)

attribute accuracy How closely the non-spatial features of a dataset match their real-world counterparts. (p. 182)

attribute table A spreadsheet-style form in which the rows consist of individual objects and the columns are the attributes associated with these objects. (p. 34)

backlink The grid that shows the direction to be traveled for the least-cost amount of movement from cell to cell. (p. 449)

band A narrow range of wavelengths being measured by a remote sensing device. (p. 314)

basemap (ArcGIS Pro) A map service available from ArcGIS Online (such as topographic maps or high-resolution imagery) that can be used in ArcGIS Pro. (p. 135)

basemap (ArcGIS Online) An image layer that serves as a backdrop for the other layers used in ArcGIS Online. (p. 106)

batch geocoding The process of matching a group of addresses together at once. (p. 239)

blue Electromagnetic energy with wavelengths between 0.4 and 0.5 micrometers. (p. 315)

Boolean operator (map algebra) One of the four operators (AND, OR, XOR, NOT) used when building a Raster Calculator expression. (p. 475)

Boolean operator (query) One of the connectors used in building a compound query. (p. 51)

539

brightness value A term used synonymously with digital number. (p. 315)

buffer A zone of spatial proximity around a feature or set of features. (p. 226)

cartographic modeling The process of applying a process or an action to a layer and creating a new layer as an output. (p. 488)

cartography The art and science of making maps. (p. 61)

Catalog pane A dockable window component of ArcGIS Pro that you use to manage your data. (p. 9)

catalog view A view used in ArcGIS Pro to manage data. (p. 11)

channel The display of bands of imagery in shades of red, green, or blue. (p. 315)

choropleth map A type of thematic map in which data are displayed according to one of several different classifications. (p. 62)

CityEngine Powerful 3D design software that can quickly create large-scale 3D models (such as models of cities). (p. 413)

Clip A geoprocessing tool that extracts objects from one layer based on the geometry of a second layer. (p. 225)

Clip function An ArcGIS Pro raster function used for creating subsets of raster layers. (p. 300)

closest facility A tool used to determine which locations on a network are nearest (via network distances) to a different set of locations. (p. 279)

cloud A computing structure in which data, content, and resources are all stored at another location and served to the user via the Internet. (p. 93)

COLLADA Collaborative Design Activity. An XML-based file type (with the extension.dae) that works as an interchange format from one kind of 3D modeling file type to another. (p. 424)

Collector for ArcGIS A free Esri app that allows you to collect location-based data in the field with a smartphone or tablet and upload the data to ArcGIS Online. (p. 166)

color composite An image formed by placing a band of imagery into each of the three channels (red, green, and blue) to view a color image instead of a grayscale one. (p. 316)

color ramp A range of colors that are applied to the thematic data of a map. (p. 72)

color-infrared An image arranged by placing the near-infrared band in the red channel, the red band in the green channel, and the green band in the blue channel. (p. 317)

completeness A measure of the wholeness of a dataset. (p. 185)

compound query A query that contains more than one operator. (p. 51)

connectivity The linkages between edges and junctions of a network. (p. 266)

connector (ModelBuilder) The specific term in ModelBuilder for one of the lines that link Variables and Tools (which, by default, appear as one of four types of lines). (p. 492)

Content Standard for Digital Geospatial Metadata (CSDGM) A set format created by the FGDC for what items a metadata file should contain. (p. 141)

Contents pane A component of ArcGIS Pro that shows all items or layers that are being used in a project. (p. 8)

continuous field view A conceptualization of the world in which all items vary across Earth's surface as constant fields, and values are available at all locations along the field. (p. 284)

contour An imaginary line drawn on a map that connects points of common elevation. (p. 379)

contour interval The vertical distance between contour lines. (p. 379)

control points Point locations where the coordinates are known that are used in aligning an unreferenced image to a source. (p. 326)

cost The number of units it takes to move from one grid cell to another. (p. 442)

credits A system used by Esri to control the amount of content that can be served by an organization in the ArcGIS Online subscription model. (p. 94)

crowdsourcing Leveraging the knowledge and resources of multiple individual users for a larger project. (p. 136)

D8 An algorithm for determining flow direction that outputs which of the eight cardinal directions will be the steepest slope for the flow of water. (p. 519)

dashboard A type of web app that can combine maps, data, and analytics together. (p. 110)

data classification Various methods used for grouping together (and displaying) values on a choropleth map. (p. 67)

data interoperability A concept that allows GIS data of many different file types to be imported into ArcGIS Pro or Esri file formats to be converted to other file types. (p. 140)

datum A reference surface of Earth. (p. 17)

Delaunay triangle A triangle that can have a circle drawn through its three points that does not pass through any additional points. (p. 381)

Delaunay triangulation The process of connecting sets of points to build the triangles and edges of a TIN. (p. 381)

depressionless DEM A DEM that has had its sinks "filled." (p. 522)

deterministic An interpolation method that does not take spatial variation into account. (p. 346)

digital elevation model (DEM) A representation of a terrain surface created by measuring elevations at evenly spaced sampling points. (p. 350)

Digital Line Graphs (DLGs) The digitized features from USGS topographic maps. (p. 138)

digital number The energy measured at a single pixel according to a predetermined scale. (p. 315)

Digital Raster Graphics (DRGs) Scanned and georeferenced versions of USGS topographic maps. (p. 138, p. 391)

digital surface model (DSM) A representation of the surface and features generated from a set of lidar points. (p. 403)

digital terrain model (DTM) A representation of a terrain surface calculated by measuring elevation values at a series of locations. (p. 349)

digitizing The process of creating digital GIS data from a secondary source. (p. 143)

Dijkstra's algorithm A mathematical process that determines the shortest path along a network from an origin node to the other nodes on the network. (p. 269)

discrete object view A conceptualization of the world in which all items can be represented by objects. (p. 12)

Dissolve A geoprocessing tool that combines polygons with similar attributes. (p. 223)

DOQQ Digital Orthophoto Quarter Quad; a product showing orthoimagery covering 3.75 degrees of latitude by 3.75 degrees of longitude. (p. 311)

draping A process in which GIS layers are given z-values to match the corresponding heights in a digital terrain model. (p. 355)

drive-time polygon Another term used for a service area. (p. 277)

edges The lines that connect the points of the triangles of a TIN. (p. 382)

edges The links of a network. (p. 261)

elevation surface The source of z-values used for a location from a digital terrain model surface. (p. 354)

environment settings Geoprocessing options that control activities in a map document, such as the extent of outputs, the coordinate system of outputs, or the cell size of outputs. (p. 218)

Equal Interval A data classification method that involves selecting class break levels by dividing the total span of values (from highest to lowest) by the number of desired classes. (p. 67)

Erase A type of GIS overlay that retains all the features from the first layer except for what they have in common with the second layer. (p. 231)

error of commission An error in which extra items that should not be part of a dataset are added to it. (p. 185)

error of omission An error in which items that should be part of a dataset are left out of it. (p. 185)

Esri Environmental Systems Research Institute—the market leader in GIS software and the developer of ArcGIS. (p. 2)

ETM+ The Enhanced Thematic Mapper instrument onboard Landsat 7. (p. 319)

Euclidean distance The straight-line distance between two points. (p. 440)

extension An add-on set of functions for ArcGIS Pro. (p. 6, 265)

Extract By Mask An ArcGIS Pro tool that performs extraction using a polygon boundary. (p. 301)

extraction The process of creating a subset raster from a larger raster. (p. 301)

extrusion A GIS technique used to give an object height. (p. 416)

facility A location from which service areas are created. (p. 277)

false color composite An image arranged by not placing the red band in the red channel, the green band in the green channel, and the blue band in the blue channel. (p. 316)

feature class A GIS layer stored inside a geodatabase that can contain multiple items of the same data type. (p. 149)

feature dataset A grouping of related feature classes stored within a geodatabase. (p. 149)

feature layer A hosted web layer that allows a user to share GIS data layers that can also be displayed, queried, or edited. (p. 100)

Federal Geographic Data Committee (FGDC) A committee that is responsible for setting GIS metadata content standards. (p. 141)

feedback connector A ModelBuilder connector that allows you to take the derived data output from a tool and use it again as an input for the same tool. (p. 512)

field A column of an attribute table. (p. 34)

file geodatabase A single folder that can hold an essentially unlimited amount of data, including multiple datasets, and has the file extension .gdb. (p. 12, 149)

first return The initial reflection of the lidar laser pulse, which usually indicates the height of an object. (p. 403)

flow accumulation A determination of how many cells will flow into a location. (p. 525)

flow direction A determination of the direction in which overland water flow will move. (p. 519)

functional surface A continuous representation of values that can have a single value assigned to each location. (p. 350)

geocoding The process of using the numbers and letters of an address to plot a point at the appropriate location. (p. 239)

geocoding service An online utility that allows for geocoding one or more addresses. (p. 259)

geodatabase An ArcGIS Pro structure for a single item that contains multiple datasets, each as its own feature class. (p. 149)

geodatabase topology A set of rules applied to the feature classes of a feature dataset in order to remove errors. (p. 188)

geographic coordinate system (GCS) A set of global latitude and longitude measurements used as a reference system for finding locations. (p. 18)

geographic information systems (GIS) Hardware and software that allow for computer-based analysis, manipulation, visualization, and retrieval of location-related data. (p. 2)

geographic scale The real-world size or extent of an area. (p. 78)

GeoPDF A format for maps to be used as PDFs that can contain geographic information and multiple layers. (p. 392)

geoprocessing The process of taking an action on a dataset that results in a new dataset being created. (p. 215)

geoprocessing package An ArcGIS Pro format used for collecting and sharing models. (p. 510)

Geoprocessing pane A window in ArcGIS Pro where tools are run. (p. 21)

geoprocessing tools Specific functions in ArcGIS Pro that are used to analyze and manipulate GIS data. (p. 21)

georeferencing A process whereby spatial referencing is given to data. (p. 326)

geospatial or spatial Referring to items that are tied to a specific real-world location. (p. 2)

geospatial technologies A term that encompasses many types of methods and techniques for the collection, analysis, modeling, and visualization of geospatial data. (p. 2)

geostatistical An interpolation method that accounts for spatial variation. (p. 347)

Geostatistical Analyst An ArcGIS Pro extension that contains several functions related to examining and working with spatial statistics. (p. 348)

geotag A process whereby non-spatial media are linked to geospatial features. (p. 57)

GIS model A representation of the factors used for explaining the processes that underlie an event or for predicting results. (p. 487)

global interpolation The process of using all known values in a dataset when approximating an unknown value. (p. 329)

Global Navigation Satellite System (GNSS) An overall term for the technologies that use signals from satellites for finding locations on Earth's surface. (p. 166)

Global Positioning System (GPS) A technology that uses signals broadcast from satellites for position determination on Earth. (p. 165)

GPX The standard format for data collected by a GPS receiver. (p. 165)

graduated colors The use of various hues in representing ranges of values on a map. (p. 72)

graduated symbols The use of different sized symbols to convey thematic information on a map. (p. 72)

green Electromagnetic energy with wavelengths between 0.5 and 0.6 micrometers. (p. 315)

grid cell A single square unit of a raster. (p. 284)

Ground The source of z-values used to represent the base heights of the digital terrain model at the surface level. (p. 354)

hard breaklines Lines used in a TIN to enforce surface discontinuities. (p. 382)

heads-up digitizing Using a map or imagery as a backdrop in the digitizing process. (p. 143)

hillshade A shaded relief map of the terrain created by modeling the position of an illumination source (such as the Sun) in relation to the terrain. (p. 359)

histogram stretch A technique used to artificially enhance the visual brightness and colors of pixels from a remotely sensed image. (p. 317)

hosted A term denoting that the layers are stored on a server in the cloud. (p. 99)

hot spot analysis A technique used to determine spatial clusters of high values and low values. (p. 214)

hull The polygon boundary around the area of a TIN. (p. 382)

hydrologic modeling The process of examining the movement of water over a surface. (p. 513)

hyperspectral Sensing hundreds of bands of energy at once. (p. 314)

Identity A type of GIS overlay that retains all the features from the first layer along with the features it has in common with the second layer. (p. 230)

impedance A value that represents how many units (of time or distance) are used in moving across a network edge. Also referred to as *transit cost*. (p. 267)

incident The locations for which the closest facility function tries to determine the nearest facility. (p. 279)

index contour The main contour lines on a map. (p. 379)

intensity The strength of the reflected return of the laser pulse from an object. (p. 403)

intermediate contour The contour lines drawn in between index contours at the contour interval. (p. 379)

Intersect A type of GIS overlay that retains the features that are common to both layers. (p. 229)

interval data A type of numerical data in which the difference between numbers is significant but there is no fixed non-arbitrary zero point associated with the data. (p. 42)

inverse distance weighted (IDW) An interpolation process that involves assigning a higher weight to the values of known points closer to the location being interpolated and lower weights to the values of known points that are farther away. (p. 335)

ISO 191xx An international metadata standard adopted in the United States as the North American Profile (NAP) of ISO 191 *xx*. (p. 141)

iterator A ModelBuilder tool that allows you to duplicate and repeat a process multiple times. (p. 511)

join A method of linking two or more tables together. (p. 43)

junctions The nodes of a network (or the places where edges come together). (p. 261)

key The field that two tables have to have in common with each other in order for the tables to be joined. (p. 43)

keyframe A snapshot of a frame used in an animation. (p. 387)

KML Keyhole Markup Language, an XML-based file format that can hold georeferenced information and that is the file format used by Google Earth. (p. 434)

KMZ A compressed version of a KML file. (p. 434)

Kriging A local geostatistical interpolation method. (p. 347)

Lambert Conformal Conic A projected coordinate system used for east–west data. (p. 19)

Landsat A long-running U.S. remote sensing program that had its first satellite launched in 1972 and continues today. (p. 319)

Landsat scene A single image obtained by a Landsat satellite sensor. (p. 320)

landscape The horizontal orientation of a layout. (p. 75)

large-scale map A map with a larger value for its representative fraction. Such maps usually show a smaller geographic area. (p. 78)

LAS dataset An ArcGIS Pro file structure used for storing, viewing, and analyzing one or more LAS files. (p. 398)

LAS file A standard file format for holding lidar data. (p. 397)

last return The final reflection of the lidar laser pulse, which usually indicates the elevation height of the ground. (p. 403)

layer package A single file that contains one data layer. (p. 31)

layout The ArcGIS Pro interface used for composing a printed map. (p. 73)

LAZ file A compressed file format for holding lidar data. (p. 397)

least-cost path The set of grid cells that represents the lowest value for accumulated cost to move from a source to a destination. (p. 449)

legend A graphical device used on a map to explain what the various map symbols and colors represent. (p. 84)

lidar A method for measuring elevation values using laser pulses sent to the ground from an aircraft. (p. 394)

line A one-dimensional vector object. (p. 12)

line of sight (LOS) The visibility in a straight line between two points. (p. 367)

lineage The data sources and processes used for creating a dataset. (p. 169)

linear interpolation A method used in geocoding to place an address location among a range of addresses. (p. 250)

line-on-polygon overlay A type of GIS overlay that results in a new line layer where the lines have the attributes of the polygons within which they are. (p. 234)

local interpolation The process of using a subset of known values when approximating an unknown value. (p. 329)

logical consistency A measure of whether the same rules were used throughout a dataset. (p. 185)

logical tool A ModelBuilder tool that allows you to use an if-then-else condition for different actions in a model. (p. 511)

many-to-one join A join in which many records in an attribute table are linked to one record from another table. (p. 44)

map A visual representation of geospatial data that conveys some sort of message to its reader. (p. 61)

map (view) The area used for displaying GIS data in two dimensions. (p. 13)

map algebra An operation performed on rasters, such as overlaying two or more grids. (p. 463)

map elements The various items used in composing a layout. (p. 73)

map frame The map element that allows for the display of the contents of a map on a layout. (p. 74)

map package A single file that contains the contents of one map in ArcGIS Pro. (p. 31)

map scale A metric used to determine the relationship between measurements made on a map and their real-world equivalents. (p. 78)

Maplex Label Engine Functionality in ArcGIS Pro that allows for a greater degree of flexibility and options for labeling features. (p. 91)

mass points The selection of points used in creating a TIN. (p. 381)

Merge A geoprocessing tool that combines two or more layers of data together into a single layer. (p. 220)

metadata Descriptive information about data. (p. 141)

middle infrared (MIR) A term used synonymously with shortwave infrared. (p. 315)

model An ArcGIS Pro tool that combines data layers, geoprocessing tools, and outputs in a workflow. (p. 487)

model element Each of the items (Variables, Tools, and Connectors) that can be placed into the ModelBuilder interface. (p. 491)

model parameters The user-defined inputs or variables of a model. (p. 493)

ModelBuilder The interface of ArcGIS Pro used in laying out workflows and creating models. (p. 468)

mosaic A raster created by joining several smaller rasters. (p. 301)

mosaic dataset An ArcGIS Pro file structure designed for managing one or more rasters or collections of raster data. (p. 302)

MSS The Multispectral Scanner onboard Landsats 1 through 5. (p. 319)

multipatch An ArcGIS Pro data format that allows for fully 3D objects and structures to be represented. (p. 418)

multipoint A geodatabase feature class that can hold large amounts of point data. (p. 410)

multispectral Sensing several bands of energy at once. (p. 314)

NAD27 The North American Datum of 1927. (p. 17)

NAD83 The North American Datum of 1983. (p. 17)

nadir The location under the sensor or camera in remote sensing. (p. 304)

NAIP National Agriculture Imagery Program; a program maintained by the USDA. (p. 311)

NAPGD2022 The North American-Pacific Geopotential Datum of 2022, a new vertical datum set to be adopted for U.S. geospatial data in 2022. (p. 350)

National Geospatial Program (NGP) A U.S. federal initiative for managing and distributing geospatial resources. (p. 122)

National Land Cover Dataset (NLCD) A 30-meter raster dataset of land cover for the entire United States. (p. 291)

NATRF2022 The North American Terrestrial Reference Frame of 2022. (p. 18)

Natural Breaks A data classification method that involves selecting class break levels by searching for spaces in the data values. (p. 67)

NAVD88 The North American Vertical Datum of 1988, a commonly used vertical datum for U.S. digital terrain models. (p. 350)

near infrared (NIR) Electromagnetic energy with wavelengths between 0.7 and 1.3 micrometers. (p. 315)

network A series of junctions and edges connected together for modeling concepts such as streets and routes. (p. 261)

Network Analyst An ArcGIS Pro extension used with transportation networks. (p. 265)

network dataset An ArcGIS Pro structure that is a series of connected junctions and edges, commonly used for transportation-related problems. (p. 266)

NoData A raster data cell that contains a null value. (p. 290)

nominal data A type of data that is a unique identifier of some kind. If the data are numerical, the differences between numbers are not significant. (p. 41)

non-spatial data Descriptive information that does not have location-based qualities. (p. 34)

normalize To alter count data values so that they are at the same level of representing the data (such as using them as percentages). (p. 65, 485)

north arrow A graphical device used to show the orientation of the map. (p. 83)

NOT (map algebra) The Boolean operator used with a negation of a grid or expression. (p. 476)

NOT (query) A Boolean operator used to negate a function or query. (p. 51)

oblique image A remotely sensed image acquired at an angle. (p. 304)

off-nadir viewing The capability of a sensor to observe areas other than the ground directly underneath it. (p. 304)

offset A value that is used in ArcGIS Pro to raise an object above the ground surface. (p. 427)

OLI The Operational Land Imager instrument onboard Landsat 8. (p. 319)

one-to-one join A join in which a single record is linked to another single record. (p. 43)

OpenStreetMap An open-source collaborative mapping project whose data are also available as a basemap in ArcGIS Pro. (p. 136)

OR (map algebra) The Boolean operator used with a union of two grids. (p. 476)

OR (query) The Boolean operator that chooses features that meet one or the other (or both) of the criteria in a query. (p. 51)

ordinal data A type of data that refers solely to a ranking of some kind. (p. 41)

Ordinary Kriging A Kriging method that assumes an unknown mean and a lack of trend in the data. (p. 347)

organization The ArcGIS Online version utilized by a group, such as a business or a school. (p. 94)

orthoimage (orthophoto) A spatially referenced aerial photo with uniform scale. (p. 307)

overlay The combination of two or more layers in GIS. (p. 215)

overshoot A digitizing error in which a vertex of a line goes beyond its target location. (p. 179)

package A single file that contains work or data used in ArcGIS Pro. (p. 29)

panchromatic Black-and-white imagery acquired by sensing the entire visible portion of the spectrum at once. (p. 314)

pane A movable and dockable window that contains items or commands used in ArcGIS Pro. (p. 8)

pan-sharpening The technique of fusing a higher-resolution panchromatic band with lower-resolution multispectral bands to improve the clarity and detail seen in an image. (p. 324)

parsing Breaking an address into its component parts. (p. 248)

path distance A cost-distance function that allows for a variety of other frictions (such as horizontal or vertical factors) to be included in the analysis. (p. 462)

pattern The arrangement of objects in an image, used as an element of image interpretation. (p. 309)

perspective view A view of GIS data at an oblique angle that causes it to take on a "three-dimensional" appearance. (p. 354)

point A zero-dimensional vector object. (p. 12)

point cloud The processed elevation points of a lidar dataset. (p. 395)

point digitizing A digitizing method in which a vertex is placed with each click of a mouse. (p. 157)

point-in-polygon overlay A type of GIS overlay that results in a new point layer where the points have the attributes of the polygons within which they are. (p. 233)

polygon A two-dimensional vector object. (p. 12)

portal A central repository of available online data. (p. 132)

portrait The vertical orientation of a layout. (p. 75)

positional accuracy How closely the spatial features of a dataset match their real-world locations. (p. 176)

pour point A feature class that represents the outlet location for the drainage of water. (p. 536)

precision How consistent or exact a measurement is. (p. 168)

project (.aprx) A file that contains information about where all the data layers used in a session are located, as well as their appearance and settings. (p. 6)

project package A single file that contains all the tools, scripts, tasks, and data layers used in a project, as well as their appearance. (p. 29)

projected coordinate system (PCS) A set of measurements made on a flat grid system, initially derived from a GCS. (p. 18)

projection A process that changes a layer from one coordinate system into another. (p. 20)

pseudo-3D The perspective view of a digital terrain model, which is often a 2.5D model rather than a 3D model. (p. 354)

publish To place data or content onto a cloud server. (p. 99)

Python An object-oriented, open-source programming language that is used for developing scripts for ArcGIS Pro. (p. 32)

Quantiles A data classification method that involves placing an equal number of values in each class. (p. 68)

query The conditions used to retrieve data from a database or table. (p. 47)

Quick Access Toolbar The section of ArcGIS Pro that is available to perform basic functions such as opening or saving projects. (p. 9)

radiometric resolution A sensor's ability to determine fine differences in a band of energy measurements. (p. 315)

Raster Calculator An ArcGIS Pro tool used for map algebra operations. (p. 473)

raster data model A model that represents geospatial data using a series of equally spaced and sized grid cells. (p. 284)

ratio data A type of numerical data in which the difference between numbers is significant, and there exists a fixed, non-arbitrary zero point associated with the data. (p. 42)

Reclassify An ArcGIS Pro tool that allows you to create a new grid by assigning new grid-cell values to an existing raster. (p. 456)

record A row of an attribute table. (p. 34)

red Electromagnetic energy with wavelengths between 0.6 and 0.7 micrometers. (p. 315)

reference data The base street layer used as a source for geocoding. (p. 239)

reference map A map that shows the location of features rather than thematic information. (p. 61)

region A set of contiguous grid cells that have the same value. (p. 294)

regularized A Spline method that results in an overall smoother surface but that may also result in values that are beyond the range of the known point values. (p. 342)

relate A join operation that establishes a connection between tables but does not append fields from one table to another table. (p. 44)

relational operator (query) One of six connectors used to build a query. (p. 48)

relational operator (map algebra) One of the six operators (equal, not equal, greater than, greater than or equal to, less than, less than or equal to) used when building a Raster Calculator expression. (p. 472)

remote sensing The process of collecting information related to the reflected or emitted electromagnetic energy from a target by a device a considerable distance away from the target onboard an aircraft or spacecraft platform. (p. 303)

representative fraction (RF) A value indicating how many units of measurement on the map are the equivalent of a number of units of measurement in the real world. (p. 78)

return The reflection of energy from a lidar laser pulse. (p. 403)

ribbon The section of ArcGIS Pro that contains tabs, groups, and buttons for various functions. (p. 9)

RMSE The root mean square error, an error measure used in determining the accuracy of the overall transformation of unreferenced data. (p. 326)

SaaS (Software as a Service) A cloud structure wherein the software being used is stored on a server at another location and accessed on demand via the Internet. (p. 93)

satellite imagery Digital images of Earth acquired by sensors onboard orbiting spaceborne platforms. (p. 303)

scale bar A graphical device used to represent scale on a map. (p. 81)

scene The ArcGIS Pro environment used for working with 3D data and visualization. (p. 352)

script A short section of computer code. (p. 32)

segment (streets) A single part of a line that corresponds to one portion of a street. (p. 239)

segment (digitizing) One part of a line or polygon object created through digitizing. (p. 157)

Select Layer By Location An ArcGIS Pro tool used to select features from one or more layers based on their spatial relationship to another layer. (p. 206)

selection The process of choosing certain records or features and setting them aside from the remainder of the records or features. (p. 46)

semivariogram A model of the variance between points and their distance apart. (p. 347)

service area A polygon boundary created around sections of a network to determine which areas of the network are within a certain distance of a location. (p. 277)

service A format for data that can be accessed as web content. (p. 104)

shadow The dark shapes in an image caused by a light source shining on an object, used as an element of image interpretation. (p. 309)

shape The distinctive form of an object, used as an element of image interpretation. (p. 310)

shapefile A series of files (with extensions including .shp, .shx, and .dbf) that make up one vector data layer. (p. 150)

share To distribute data, content, maps, or applications across the Internet. (p. 99)

shortwave infrared (SWIR) Electromagnetic energy with wavelengths between 1.3 and 3.0 micrometers. (p. 315)

Shreve method A stream-ordering method that allows for the stream-order ranking to be computed by adding the ranks of the two streams flowing into each another. (p. 529)

simple query A query that contains only one operator. (p. 48)

sinks Low areas in a digital elevation model surrounded on all sides by high areas. (p. 521)

site and association Information related to the locations of objects and their related attributes in an image, used as elements of image interpretation. (p. 309)

size The physical dimensions (length, width, and area on the ground) of objects, used as an element of image interpretation. (p. 309)

SketchUp 3D design software owned and distributed by Trimble. (p. 434)

slice A process that involves grouping several grid-cell values together according to a classification scheme. (p. 453)

slope A measurement of the rate of elevation change at a location found by dividing the vertical height (the rise) by the horizontal length (the run). (p. 360)

small-scale map A map with a lower value for its representative fraction. Such maps usually show a large geographic area. (p. 78)

snapping tolerance How closely a vertex must be placed within a feature to activate snapping. (p. 179)

snapping Linking two vertices together to become a single vertex. (p. 178)

soft breaklines Enforced lines in a TIN that do not represent discontinuities. (p. 382)

source layer (spatial analysis) The layer that serves as the starting point for a spatial analysis operation. (p. 191)

spatial analysis The process of examining the characteristics or features of spatial data or how features spatially relate to one another. (p. 190)

Spatial Analyst An extension that enables several raster functions and tools in ArcGIS Pro. (p. 287)

spatial autocorrelation A measure of the degree of clustering of objects and their data values. (p. 213)

Spatial Data Transfer Standard (SDTS) A neutral file format used for the distribution of GIS data. (p. 140)

spatial interpolation The process of determining an unknown value at a location based on known values at other locations. (p. 328)

spatial join A join in which attributes from one layer are appended to another based on the spatial relationship of the layers. (p. 200)

spatial query A query that selects records or objects from a layer based on their spatial relationships with other layers rather than their attributes. (p. 206)

spatial resolution The size of the area on the ground represented by 1 pixel's worth of energy measurement. (p. 312)

SPCS The State Plane Coordinate System projected coordinate system, which divides states into zones with measurements made in feet or meters. (p. 18)

SPCS2022 The State Plane Coordinate System of 2022, a revised and updated version of SPCS. (p. 18)

Spline An interpolation method that involves using a mathematical process to fit a surface exactly through known points. (p. 342)

SQL The Structured Query Language—a formal language for building queries. (p. 48)

Standard Deviation A data classification method that involves computing break values by using the mean of the data values and the average distance of a value from the mean. (p. 69)

step A single action (either a command or geoprocessing tool) that is executed when running a task. (p. 237)

stops Destinations to visit on a network. (p. 268)

story map A web application designed to convey a specific theme or concept to the user using a special set of web templates. (p. 116)

Strahler method A stream-ordering method that allows for stream order to increase in rank only if two streams of the same order of magnitude flow into each another. (p. 530)

stream order A computed ranking of streams based on how they flow into one another. (p. 529)

street centerline A file containing line segments representing roads. (p. 244)

subset A raster created by removing a set of grid cells from a larger raster. (p. 300)

suitability index A ranking of locations according to a set of criteria. (p. 505)

Summarize tool A tool to calculate a statistic (such as the sum) of the objects or their attributes within an area. (p. 198)

surface A raster that contains a value for some phenomena at all locations. (p. 328)

Survey 123 A free Esri app that allows you to collect form based data along with location data using a smartphone or tablet and upload the data to ArcGIS Online. (p. 166)

Symmetrical Difference A type of GIS overlay that retains all the features from both layers except for the features that they have in common. (p. 231)

target layer (spatial analysis) The layer that contains the features that will be examined in the end result of a spatial analysis operation. (p. 191)

task A preset interactive series of actions used to establish the steps, settings, and tools in a workflow. (p. 236)

Task Designer The tools used to edit a task item, task group, task, or step. (p. 237)

task group A collection of tasks inside a task item. (p. 236)

task item An overall container that can hold many tasks and can also be exported to a task file and imported to another project. (p. 236)

template The ArcGIS Pro project function that establishes what will be available when a new project is created. (p. 6)

temporal accuracy The time period and currentness of a dataset. (p. 172)

tension A Spline method that results in a less smooth surface than a regularized Spline but that contains values closer to the range of the known point values. (p. 342)

terrain dataset An ArcGIS Pro file structure that can hold large amounts of elevation data and render them quickly at a variety of scales and resolutions. (p. 410)

texture Repeated shadings or colors in an image, used as an element of image interpretation. (p. 310)

The National Map An online resource that distributes U.S. geospatial data. (p. 122)

thematic map A map that displays a particular theme or feature. (p. 62)

thermal infrared (TIR) Electromagnetic energy with wavelengths between 3.0 and 14.0 micrometers. (p. 315)

three-dimensional (3D) A model of the terrain that allows for multiple z-values to be assigned to each x/y coordinate. (p. 354)

TIGER/Line Topologically Integrated Geographic Encoded Referencing boundary and road network GIS data created by the U.S. Census Bureau. (p. 139, p. 244)

tile layer A hosted, cached web layer that sets up GIS data as a series of image tiles that can be displayed but not queried or edited. (p. 100)

TIN Triangulated irregular network, a terrain model formed from non-overlapping triangles that allows for unevenly spaced elevation points. (p. 375)

TIRS The Thermal Infrared Sensor onboard Landsat 8. (p. 319)

TM The Thematic Mapper instrument onboard Landsats 4 and 5. (p. 319)

tone The grayscale levels (from black to white) or range of intensity of a particular color discerned as a characteristic of particular features present in an image, used as an element of image interpretation. (p. 310)

tool (ModelBuilder) The specific term in ModelBuilder for one of the geoprocessing tools, models, or scripts (which, by default, appears as a yellow rectangle). (p. 492)

topographic map A map created by the USGS to show landscape and terrain as well as the location of features on the land. (p. 62, 391)

topology How objects relate or connect to one another, independent of their coordinates. (p. 188)

trace A problem-solving technique used with a utility network. (p. 283)

transformation A process in which data are altered from unreferenced to having spatial reference. (p. 326)

Transverse Mercator A projected coordinate system used for north–south data. (p. 18)

trend A very general pattern or overriding process that affects measurements. (p. 329)

true color composite An image arranged by placing the red band in the red channel, the green band in the green channel, and the blue band in the blue channel. (p. 316)

TSP Traveling Salesman Problem, a mathematical process that involves determining the optimal configuration for rearranging a series of stops on a network. (p. 273)

turns Information used by ArcGIS Pro to determine information about valid flows along a network. (p. 266)

two-and-a-half dimensional (2.5D) A model of the terrain that allows for a single z-value to be assigned to each x/y coordinate. (p. 354)

UAS Unmanned aircraft system; drone aircraft flown by pilot on the ground using remote control. (p. 303)

undershoot A digitizing error in which a vertex of a line falls short of its target location. (p. 178)

Union A type of GIS overlay that retains all the features from both layers. (p. 231)

Universal Kriging A Kriging method that assumes a trend within the data (and that removes the trend in order to perform interpolation). (p. 348)

Update A type of GIS overlay that retains all features from the second layer as well as those features from the first layer that are not in common with the second. (p. 231)

US Topo A digital topographic map series created by the USGS to allow multiple layers of data to be used on a map in GeoPDF file format. (p. 391)

utility network An ArcGIS Pro structure of a series of connected junctions and edges, commonly used for utility-related problems. (p. 282)

UTM The Universal Transverse Mercator projected coordinate system, which divides the world into 60 reference zones with measurements in meters. (p. 18)

variable (ModelBuilder) The specific term in ModelBuilder for a data layer or value input (which, by default, appears as a blue oval in ModelBuilder). (p. 491)

variables Different factors that are combined into an overall analysis. (p. 487)

vector data model A model that represents geospatial data with a series of vector objects. (p. 12)

vector objects Points, lines, and polygons that are used to model real-world phenomena using the vector data model. (p. 12)

vector tile layer A hosted web layer that displays using the vector representation of the data. (p. 100)

vertex A defining point placed at the start or end of a segment during digitizing. (p. 157)

vertical datum A baseline used in measuring elevation values (above or below this level). (p. 350)

vertical exaggeration A 3D visualization technique that alters the vertical scale but keeps the horizontal scale the same. (p. 386)

vertical image A remotely sensed image in which the camera is looking down at the landscape. (p. 304)

view The window serving as the central work area in ArcGIS Pro. (p. 8)

view dome A volumetric version of a viewshed, which shows what can or can't be seen in a spherical area around an observer point. (p. 373)

viewshed The visibility within 360 degrees around a location. (p. 367)

visibility analysis Techniques used in GIS to determine what areas can be seen and what areas cannot be seen from a particular vantage point. (p. 367)

visible light Electromagnetic energy with wavelengths between 0.4 and 0.7 micrometers. (p. 314)

visual hierarchy The way features are displayed on a map to emphasize their level of prominence. (p. 88)

visual image interpretation The process of examining information to identify objects in an aerial (or other remotely sensed) image. (p. 309)

volunteered geographic information (VGI) User-generated geospatial data. (p. 136)

watershed An area in which all locations flow to a common drainage area. (p. 513)

web app A lightweight, standalone website that allows the user to customize its appearance and tools. (p. 109)

web layer A format used when publishing a map of one or more layers to ArcGIS Online. (p. 99)

web map An interactive online representation of GIS data that can be accessed via a web browser. (p. 103)

Web Mercator auxiliary sphere The projected coordinate system used by ArcGIS Online. (p. 99)

web scene An interactive version of an ArcGIS Pro scene that can be viewed and used through a web browser. (p. 419)

WebGL Web Graphics Library, a web standard for viewing 3D graphics. (p. 419)

weighted overlay An overlay that is a combination of two or more grids, with each grid assigned a relative importance. (p. 484)

WGS84 Web Mercator A projected coordinate system often used with web maps. (p. 19)

WGS84 The World Geodetic System datum of 1984. (p. 18)

workflow A sequence of actions performed using GIS layers and tools. (p. 236, 464)

XOR (map algebra) The Boolean operator used with an "exclusive or" of two grids. (p. 476)

Zonal Statistics A tool for calculating statistical values based on the raster grid-cell values that fall within a boundary. (p. 340)

zone One of several different values assigned to a raster grid cell. (p. 290)

z-value The elevation assigned to an x/y coordinate. (p. 349)

Index

Note: Page numbers with f indicate figures; those with t indicate tables.

A

accumulated cost, 444–446, 445f, 446f, 449
accuracy
 attribute, 182
 defined, 168, 169f
 of geocoded results, 254, 254f
 positional, 176–177
 temporal, 172–173
Activated Map Frame, 78–79
active remote sensing, 394. *See also* lidar; remotely sensed imagery
active sensors, 313
Add Clause button, 53
address locators
 in ArcGIS Pro, 244–245, 259
 defined, 244
address locator styles, 249–250
address matching, 245. *See also* geocoding
adjacency, 188
aerial photography, 303–304. *See also* remotely sensed imagery
AFF. *See* American FactFinder
Airbus Defence and Space, 324
airports. *See also* spatial analysis
 McCarran International, 148, 148f
 National Map data, 195
 precipitation data, 331. *See also* spatial interpolation
Albers Equal Area Conic, 19
Algebra. *See* Map Algebra
algorithms
 cost distance, 443–448, 445f, 446f, 447f
 D8, 519
 defined, 269
 Dijkstra's algorithm, 269–271, 269f, 270f, 271f
 least-cost path, 449–450
 Traveling Salesman Problem, 273
allocation, 442–443, 442f
altitude, 359, 359f
American FactFinder
 described, 60
 Ohio counties attribute data, 37, 64, 97
American Society of Photogrammetry and Remote Sensing (ASPRS), 401, 401t
analysis. *See also* network analysis; spatial analysis
 hot spot, 214, 214f
 3D visualization, 371–372

using queries, 55–56
visibility, 367–371, 373–374, 374f
Analyzing Patterns toolset, 213
AND (Map Algebra), 475, 475f
AND operator, 51
animations
 in ArcGIS Pro, 387–390
 creating, 387–390
 defined, 387
annotation
 defined, 91
 maps, 90–91
Append, in ArcGIS Pro, 220
append operation, 220
.aprx (Project files), 6–7
Arcade, 32–33
ArcCatalog, 2, 537, 538t
ArcGIS, 2. *See also* ArcMap; *other ArcGIS entries*
 coordinate systems in, 17–19
 datums in, 17–19
ArcGIS Advanced, 2
ArcGIS Basic, 2
ArcGIS Desktop, 2. *See also* ArcMap
ArcGIS Living Atlas of the World, 133
ArcGIS Online. *See also* Web maps; Web Scenes; *specific topics*
 account setup, 98
 building Web maps in, 103–104
 data from, for use in ArcMap, 99–100
 data from, for use in ArcMap Pro, 132–134
 feature layer, 100
 free version, 94
 interface, 93f
 introduction, 92–93
 organization version, 94
 portal, 132–133
 publishing to, 5, 99
 sharing map from, 99, 103
 tile layer, 100
 vector tile layer, 100
 web maps, built in, 103–104
 Web Scenes, 421–423, 422f
 World Geocoding Service, 259–260, 259f
ArcGIS Pro. *See also specific topics*
 adding Microsoft Excel tables to, 151
 adding online data to, 132–134
 animations, 387–390
 attachments in, 57–58
 attributes in, 59–60
 attribute tables in, 34–35, 35f, 39–40, 40f
 catalog view, 11

changing appearance of, 25–29, 427
changing projections in, 20–23
closing map (view) in, 14
compound queries in, 51
contour lines used in, 379–380, 380f
coordinate systems in, 17–19
creating animation in, 387–390
data accessibility, 10–11
datums in, 17–19
defined, 2
defining projections in, 23–25
distance calculations, 439–447
and Esri, 5–6
extensions in, 6
extrusions, 416–417, 416f
geocoding, 239–260
geoprocessing, 21, 215–238
geospatial data, 1–33, 143–167, 412–434
getting started, 5–13
hydrology tools, 513, 518
LAS files in, 397–400
layout, 8f
lidar data in, 400–406
Map Algebra, 463–486
map (view) in, 13, 14f
ModelBuilder, 468–469, 484
multipatches, 417–418
navigating a scene in, 356–357
network analysis, 261–283
non-spatial data, 34–35, 41, 43
offsets in, 426–428
projections in, 17–19
projects in, 6–7
publishing from, 97, 99, 421
Python, 9, 32–33, 537
raster tools, 301
records selection in, 46–47
remotely sensed imagery, 303–326
saving work as package in, 29–31
scene in, 354–356
setting up layers, 97–98
simple query in, 48–50
spatial analysis, 190–214
spatial interpolation, 328–348
symbol options in, 28, 28f
tables joined in, 43–44, 44f
templates in, 6–7
terrain datasets in, 410, 411f
3D and pseudo-3D visualizations, 354–355, 412–413
3D editing, 423–426
3D visualization operation and, 354–356, 356f

548

TINs and Web Scenes, 375–376, 391
toolboxes, 9, 21
units of measurement, 377–378
using U.S. Census Data in, 59, 59f, 60
vector data referenced in, 12–13
ArcGIS Pro skills, 350
   attribute tables, 34
   digital elevation models, 349
   distance calculations, 435
   editing data, 168
   geocoding, 239
   geoprocessing, 215
   geospatial data, 1, 143, 412
   hydrologic modelling, 513
   lidar data, 394
   Map Algebra, 463
   maps, 61
   models, 487
   network analysis, 261
   online data, 119
   raster data, 284
   remotely sensed imagery, 303
   spatial analysis, 190
   spatial interpolation, 328
   3D visualization, 375
   TINs, 375
   Web maps, 92
ArcGIS Standard, 2
ArcGIS World Geocoding Service, 255, 256, 259–260
ArcGlobe, 2, 537, 538t. *See also* geospatial data
Arc Hydro, 535
ArcMap, 2, 266, 537 538, 538t. *See also* maps; *specific topics*
ArcMap Pro
   metadata in, 141–142
   using ArcGIS Online data in, 132–134
ArcPad, 166
ArcPy, 32, 537
ArcScene, 2, 537, 538t. *See also* digital elevation
Ashtabula County, 217f. *See also* geoprocessing
aspect
   DEMs, 360–363, 361f
   TINs, 383–385
ASPRS, 401, 401t
attachments, 56–59
   in ArcGIS Pro, 57–58
   defined, 57
attribute accuracy, 182
attributes
   from American FactFinder, 37
   in ArcGIS Pro, 59–60
   configuring pop-ups, 56–59
   defined, 35
   information in ArcGIS Pro, 38, 38f
   interactively obtaining information, 38–39
   selection by, 47–53
   types, 41–42, 41f
   updating, 182
attribute tables
   analysis, using queries, 55–56
   in ArcGIS Pro, 34–35, 35f, 39–40, 40f
   ArcGIS Pro skills, 34

closing time, 59
compound queries, 50–53
data sources, 37
defined, 34
examining, 39–42
exporting selection results, 54
functions of, 38–39, 38f
getting started, 38
interactively obtaining information, 38–39
interface, 39–40, 40f
introduction, 34–35, 35f
joining, 43–46, 44f
on pop-up window, 56–59
raster data, 293, 293t
real-world applications, 36
related concepts, 59–60, 59f
scenario, 35–36
selecting records, 46–47
simple queries, 47–50
study area, 36, 36f
working with attachments, 56–59
Autocomplete Freehand polygon digitizing, 159
Autocomplete Polygon option digitizing, 158
autocorrelation, spatial, 213, 213f
azimuth, 359, 359f

B

backlink grid, 449–450, 450f
bands of energy, 314
bare-Earth, 404
barriers, 275–276
base heights, 416
Basemap button, 80, 106
basemaps
   from ArcGIS Online, 135–137
   defined, 106, 135
   for digitizing, 147
   types of, 106–107
BatchGeo, 259
batch geocoding, 239
bathymetric lidar, 395
Bing Maps, 99
biosolids, 1–2, 2f
blue, 315
bookmarking
   scenes, 419, 422, 422f
Boolean operators, 51, 475–477
Boolean operators (Map Algebra), 475–477, 475f, 476f, 477f, 480–481, 480f, 481f
boundaries (polygon), dissolving, 223–224
breaklines, 382, 382f
   hard, 382, 382f
   soft, 382, 382f
brightness value, 315
buffers, 226–228, 227f
Bureau of Labor Statistics, unemployment data, 65

C

CAD (computer-aided design), 150
Calculate Field, 204
Calculator, Raster. *See* Raster Calculator
campus (YSU), 145f, 146f, 170f, 187f, 414f. *See also* editing data; geospatial data creation

Cartographic Boundary files, 37, 60
cartographic modeling, 488
cartography, 61. *See also* maps
Catalog pane
   accessing folders and data through, 11
   ArcGIS Pro, 8f
   defined, 9
catalog view, 11
Census Bureau, U.S., 4, 37, 59–60, 59f, 139–140. *See also* American FactFinder; TIGER/Line
channels, 315–316
checklists
   maps, 89
   Web maps, 115–116
choropleth maps
   in ArcGIS Pro, 65, 67–69
   color choice, 72–73
   data classification, 67–71
   defined, 62
   normalized data, 65–66
   setting up, 64–67
Circle, Rectangle, and Ellipse options digitizing, 158
CityEngine, 413, 413f
Cleveland, Ohio, 305. *See also* remotely sensed imagery
Clip function, 300, 301f
clipping, described, 224–226, 225f
Clip tool, 225, 225f
closest facilities, 279–282, 279f
cloud, 93, 94f, 99. *See also* ArcGIS Online
cluster analysis, 213
clusters, 213–214
COLLADA, 424–426
Collector for ArcGIS, 166, 166f
color choice, 72–73
color composite, 316, 316f
color infrared imagery, 313–318
color ramps, 72
Columbiana County, 36f, 217f, 466f, 489f, 515f. *See also* geoprocessing; Map Algebra; models
Compass button, 422, 422f
completeness
   data quality, 185
   errors of commission, 185
   errors of omission, 185
compound queries, 50–53
   in ArcGIS Pro, 51
   defined, 51
connectivity
   network, 266–267
   topology, 188
Connectors (ModelBuilder), 492, 492f
consistency, logical, 185
construction feature, 275
containment, 188
Contents pane
   in active map, 14f
   ArcGIS Pro, 8f
   defined, 8
   displaying data layers in, 16, 16f
Content Standard for Digital Geospatial Metadata (CSDGM), 141–142

continuous field view, 284
contour interval, 379
contours, 379–380
control point, 326
coordinate systems. *See* geographic coordinate systems (GCS); projected coordinate systems (PCS)
Coordinate Systems tab, 20
cost distance. *See also* distance calculations
    algorithm, 443–448, 445f, 446f, 447f
    applying surfaces to create new paths, 461
    further calculations, 452–456
    with multiple cost layers, 456–461
cost paths, 449–451, 455
costs
    accumulated, 444–446, 445f, 446f, 448
    defined, 444
    transit. *See* impedance
counties, 36f, 96f, 194f, 217f, 330f, 336, 337f. *See also* attribute tables; geoprocessing; maps; spatial analysis; spatial interpolation; Web maps; *specific counties*
Count of Selected Records, 40, 40f
county boundaries data, 63–64
Create Features box, 156
Create Utility Network tool, 283
creating
    animation, 387–390
    buffer around features, 226–228
    feature classes, 152–155
    features, 183
    geospatial data. *See* geospatial data creation
    initial layout, 73–75
    models, 467–469
    network dataset, 265–267
    routes. *See under* routes
    stream network, 526–529
    suitability index, 503–506
    TINs, 380–383
    Web maps, 108–112
    Web Scene, 418–421
credits, 94
crowdsourcing, 136
CSDGM, 141–142
cutting
    features, 183–184
    polygons, 177–178
Cuyahoga County, 336, 337f

D

D8 algorithm, 519
Dark Gray Canvas basemaps, 107
Darwin, Charles, 117
dashboard, 110
data about data (metadata), 141–142
data classification, 67–71, 71f
data editing. *See* editing data
data formats, 150. *See also specific formats and extensions, e.g. .mdb*
data interoperability, 140, 150
data layers. *See* layers
Data Management Tools, 21, 24, 57, 522

data quality. *See also* accuracy; editing data
    completeness, 185
    introduction, 168–169
    lineage and, 169
    logical consistency, 185
    precision, 168, 169f
data sources. *See* American FactFinder; National Map; TIGER/Line; *under specific topics*
datums
    in ArcGIS, 17–19
    defined, 17
    NAD27, 17
    NAD83, 17
    NATRF2022, 18
    vertical, 350
    WGS84, 18
.dbf (shapefiles), 150
Defined Interval method, 70
degrees (GCS), 18
Delaunay triangles and triangulation, 381–382, 382f
deleting features, 183
DEMs. *See* digital elevation models (DEMs)
depressionless DEMs, 522. *See also* hydrologic modeling
derived data, 491
deselecting features, 223
deterministic interpolation, 346
digital elevation models (DEMs)
    ArcGIS Pro skills, 349
    bare-Earth, 404
    closing time, 373
    data sources, 351–352
    defined, 350
    depressionless, 522. *See also* hydrologic modeling
    and derivations, 358–363
    getting started, 352–353
    introduction, 350, 350f
    line-of-sight, 367
    profile graphs, 364–367
    related concepts, 373–374, 374f
    saving and printing results, 373
    scenario, 350–351
    slope, 360–363, 360f
    study area, 351
    3DEP layers, 357–358
    3D visualization layers in a scene, 363–364
    3D visualization results analysis, 371–372, 372f
    view domes, 373
    viewsheds, 367
    visibility analysis, 367–371, 373–374, 374f
Digital Line Graphs (DLGs), 138
digital numbers, 315, 316f
Digital Orthophoto Quad (DOQ), 311
Digital Orthophoto Quarter Quad (DOQQ), 311
Digital Raster Graphics (DRGs), 138, 391
Digital Surface Models (DSMs), 403
digital terrain models (DTMs), 349. *See also* digital elevation models (DEMs); triangulated irregular networks (TINs)

digitizing. *See also* heads-up digitizing
    basemap for, 147
    defined, 143
    finishing up, 163
    measuring length of objects, 163–164
    options for, 157–159
    process of, 157–159
    scale's importance, 148, 148f
Dijkstra's algorithm, 269–271, 269f, 270f, 271f
discrete object view, 12
displaying
    and visualization of 3D layers in a scene, 363–364
Display Units, 15, 147, 243, 267
dissolving polygon boundaries, 223–224
distance calculations. *See also* cost distance
    in ArcGIS Pro, 439–447
    ArcGIS Pro skills, 435
    closing time, 462
    data sources, 437–438
    Euclidean distance, 439–443, 440f, 441f
    exporting, saving, and sharing results, 461–462
    getting started, 438–439
    introduction, 435–436, 436f
    path distance, 462
    related concepts, 462
    scenario, 436–437
    shortest paths over raster surfaces, 443–451
    visualizing cost-path results, 451–452
DLGs, 138
DOQ, 311
DOQQ, 311
drainage divide, 514, 536f
draping, 355
DRGs, 138, 391
DSMs, 403
DTMs, 349. *See also* digital elevation models (DEMs); triangulated irregular networks (TINs)
Dual Ranges style, 245, 246, 249
dynamic text, 74

E

EarthExplorer, 138, 139f, 140, 305, 320, 396
ecological preserve model. *See* Map Algebra
edges
    Delaunay triangles, 382
    network, 261
Edge Snapping, 179, 179f, 180–181, 181f
editing data
    ArcGIS Pro skills, 168
    closing time, 187–188
    creating features, 183
    cutting polygons, 177–178
    data sources, 170
    deleting features, 183
    errors, 185–187
    getting started, 170–173
    introduction, 168–169
    moving features, 183–184
    printing or sharing results, 187

Index **551**

real-world applications, 169
related concepts, 188–189, 189f
reshaping feature, 173–176
scenario, 169
snapping, 178–181, 179f
splitting lines, 177–178
study area, 170
topological, 188
updating attributes, 182
Editor toolbar, 538t
elements, model, 491–492, 492f
elevation. *See* digital elevation models (DEMs); lidar; triangulated irregular networks (TINs)
elevation surface, 346, 354–355, 521–522
End Snapping, 179, 179f
Enhanced Thematic Mapper (ETM+), 319, 320t
Environmental Protection Agency, 1
Environmental Systems Research Institute (Esri)
    account setup, 98
    and ArcGIS Pro, 5–6
    CityEngine, 413, 413f
    defined, 2
    described, 5
    Landsat Explorer tool, 304, 304f
    need for, 5–6
    USA Counties dataset, 195, 217
environment settings, 218–219
Equal Interval method, 67–68, 69f
Erase overlay, 230f, 231
errors
    of commission, 185
    locating and editing, 185–187
    of omission, 185
    RMSE, 326
Esri. *See* Environmental Systems Research Institute (Esri)
ETM+, 319, 320t
Euclidean distance, 439–443, 440f, 441f
Explore tool, 362, 363
exporting
    distance calculations, 461–462
    scenes, 409–410
    selection results, 54
expression language, 32
extensions
    in ArcGIS Pro, 6
    defined, 6, 265
    Geostatistical Analyst, 347f, 348, 348f
    Network Analyst, 265
    Spatial Analyst, 287
Extract By Mask, 301, 301f
extracting data layers, 221–223
extraction, 301, 301f
extrusions, 416–417, 426–427

F

facilities, 277–282, 279f
FactFinder. *See* American FactFinder
false color composite, 316–317, 316f
Farm Service Agency, 311
feature classes
    buffers around, 226–228, 227f

clipping, 224–226, 225f
creating, 152–155
defined, 149, 150f
multipoint, 410
feature datasets, 149–150, 150f
feature layer
    ArcGIS Online, 100
    defined, 100
features
    buffers around, 226–228, 227f
    creating, 183
    reshaping, 173–176
    spatial autocorrelation of, 213
    working with, 183–184
Federal Geographic Data Committee (FGDC), 141–142
feedback connector, 512
FGDC, 141–142
fields, 34, 39–42
Field tools, 40, 40f
file geodatabases, 9–10, 12, 149
filling sinks, 522–524
FIPS codes, 4, 37, 63, 96
first return, 403–404, 404f
flow accumulation, 524–526, 525f
flow direction, 518–520, 519t, 520f
fonts for map text, 87
Freehand option digitizing, 158
functional surface, defined, 350

G

Garmin satellite navigation system, 136
.gcpk (locator packages), 258
GCS. *See* geographic coordinate system (GCS)
.gdb (geodatabases), 149
geocoding
    Address Locators, 244–247
    ArcGIS Pro skills, 239
    batch, 239
    closing time, 259
    data sources, 241–242
    defined, 239
    examining results, 253–256
    getting started, 242–243
    introduction, 239–240, 241f
    process, 247–250
    real-world applications, 240–241
    related concepts, 259–260, 259f
    rematching addresses, 250–253
    scenario, 240–241
    sharing or printing, 258–259
    spatial analysis of geocoded points, 256–258
    study area, 241, 241f
geocoding services, 259–260, 259f
geodatabases, 11, 148–150
geodatabase topology, 188–189, 189f
Geodesic measurement, 196
geographic coordinate system (GCS), 18
geographic information systems (GIS), 2. *See also specific topics*
    data examining, 17–19
    vector data, 11
geographic scale, 78
Geometrical Interval method, 71

GeoPDF, 392
geoprocessing
    in ArcGIS Pro, 21, 238–239, 353
    ArcGIS Pro skills, 215
    clipping feature classes, 224–226, 225f
    closing time, 235–236
    creating buffers, 226–228
    data source, 217
    defined, 215
    dissolving polygon boundaries, 223–224
    getting started, 217–219
    introduction, 215–216
    merging layers, 219–221
    point-in-polygon overlays, 23f, 233–235
    polygon overlays, 228–233, 229f
    printing or sharing, 235
    real-world applications, 216
    related concepts, 236–238, 237f, 238f
    scenario, 216
    selecting and extracting layers, 221–223
    study area, 216, 217f
    tasks and, 236–237
    workflow and, 236
geoprocessing packages, 30, 510
Geoprocessing pane, 21, 24, 492, 537
geoprocessing tools, 21
georeferencing
    defined, 326
    images, 326
geospatial data
    adding, 15–17
    analyzing. *See* spatial analysis
    applications, 3
    creating. *See* geospatial data creation
    data sources, 4–5, 122–123
    defined, 2
    examining, 17–23
    getting started, 5–13
    scenario, 3
    study area, 4f
    in 3D. *See* geospatial data in 3D
    VGI, 136–137
    working with map, 13–15
geospatial data creation
    closing time, 164–165
    converting x/y coordinates to point feature class, 150–152
    creating feature classes, 152–155
    data sources, 144–145
    finishing up digitizing, 163
    getting started, 145–147
    heads-up digitizing of lines, 155–161
    heads-up digitizing of polygons, 161–163
    introduction, 143–144
    measuring length of digitized objects, 163–165
    printing or sharing, 164–165
    related concepts, 165–167, 166f
    scenario, 144
    study area, 144
    using GNSS and mobile GIS for, 165–167
    working with geodatabases, 148–150

552 Index

geospatial data in 3D
    closing time, 433
    data sources, 414
    editing, 423–426
    extrusions, 416–417
    getting started, 415
    introduction, 412–413
    KML files, 434
    KMZ files, 434
    multipatches, 417–418
    offsets, 426–428
    related concepts, 434, 434f
    saving and sharing results, 433
    scenario, 413–414
    SketchUp, 434, 434f
    study area, 414f
    3D objects, 429–433
    Web Scene, 418–423
geospatial technologies, 2, 187, 239
Geostatistical Analyst, 348
geostatistical interpolation methods, 346–348, 347f, 348f
geotag, 57
GIS models, 487. *See also* models
GIS overlays. *See* overlays
Global Account, Esri. *See* ArcGIS Online
global interpolation, 329
Global Navigation Satellite Systems (GNSS), 165–167
Global Positioning System (GPS), 165–167
Global Visualization Viewer, 320
GloVis, 305, 320
GNSS, 165–167
Google Earth, 303
Google Maps, 136, 259, 303
Go to a Specific Record, 39–42
.gpk (geoprocessing packages), 510
GPS, 165–167
GPX, 165–166
graduated colors, 72–73, 73f
graduated symbols, 73, 73f
graphics, 484
grayscale (black and white) imagery, 307, 316
green, 315
grid cells, 284
gridded format, 331, 334
Ground, 354

H

hard breaklines, 382
heads-up digitizing
    of lines, 155–161
    of polygons, 161–163
High/Low Clustering, 213
high-resolution satellites, 324–325, 324t, 325t
hillshades, 359–360, 359f
histogram stretch, 317
HMS Beagle, 117
Hocking County, 351f, 355f, 437f. *See also* digital elevation models (DEMs)
hosted layers, 99
hot spot analysis, 214, 214f
hulls, 382
hydrography data, 125–127
hydrologic modeling
    analyzing results, 534–535
    ArcGIS Pro skills, 513
    Arc Hydro, 535
    closing time, 535–536
    data sources, 515
    defined, 513
    filled sinks, 522–524
    flow accumulation, 524–526, 525f
    flow direction calculation, 518–520, 519t, 520f
    getting started, 515–517
    introduction, 513–514
    pour points, 536, 536f
    related concepts, 536, 536f
    saving, printing, or sharing results, 535
    scenario, 514–515
    setting parameters, 517
    sinks, 520–524, 521f, 522f
    slopes and streams, 517–518
    stream networks, 526–529, 527f
    stream ordering, 529–531, 530f
    study area, 515f
    validating and running model, 532–534
    watershed delineation, 531–532
hyperspectral sensor, 314

I

Identify tool, 538t
Identity overlay, 230, 230f
IDW, 334–339
Imagery basemaps, 106
Imagery with Labels basemap, 107, 135
impedance
    closest facilities and, 279, 279f
    defined, 267
    distance calculations and, 444
    shortest distance and, 269–271
import arcpy, 32
incidents, 279
index contours, 379, 380f
Initial view button, 422, 422f
intensity, 403
intermediate contours, 379, 380f
interpolation
    deterministic, 346
    global, 329
    linear, 250, 250f, 254, 254f
    local, 329, 329f
    spatial. *See* spatial interpolation
Intersection Snapping, 179f, 180
Intersect overlay, 229–230
interval data, 41f, 42
inverse distance weighted (IDW), 334–339
ISO 191 xx, 141–142
iterator, 511

J

Jenks Optimization method, 67, 68f
joins
    attribute tables, 43–46, 44f
    defined, 43
    spatial, 200–205, 201f, 202f, 203f
junctions, 261

K

keyframes, 387
Keyhole Markup Language (KML), 150, 434
keys, 43–46, 44f
KML (Keyhole Markup Language), 150, 434
KML files, 434, 434f
KMZ files, 434, 434f
Kriging, 347–348, 347f, 348f

L

labels, maps, 90–91
Lambert Conformal Conic, 19, 20f
Landfire, 138, 139f
Landsat, 318–324. *See also* remotely sensed imagery
Landsat 8 scene, 320
Landsat Explorer tool, 304, 304f, 305
Landsat scene, 320
Landscape orientation, 75
large-scale maps, 78
LAS classification values, 401–403, 401t
LAS datasets, 396. *See also* lidar
    in ArcGIS Pro, 397–400
    creating, 398
    defined, 398
laser. *See* lidar
LAS files, 410. *See also* lidar
    in ArcGIS Pro, 397–400
    classification system, 401–403
    defined, 397
last return, 403–404, 404f
latitude, 18
layer package, 31
layers. *See also specific topics*
    from ArcGIS Online
    changing, to different projection, 20–23
    changing the appearance of, 25–29, 25f–28f
    coordinate systems in, 20–21
    defining projections for, 23–25
    displaying, in the Contents pane, 16, 16f
    exporting to new, 54
    extracting data, 221–223
    merging, 219–221
    publishing, 102
    roads/streets, 243–244
    setting up in ArcGIS Pro, 97–98
layouts, 73–75. *See also* maps
    in ArcGIS Pro, 8f, 73–88
    map frame inserting into, 76–77
    map frame working in, 77–78
    placing inset map into layout, 79–81
LAZ files, 397
least-cost path, 449–450
legend, 84–86
length of digitized objects, 163–165
libraries, 242–243. *See also* geocoding; network analysis
Libremap Project, 138
lidar
    closing time, 410
    data sources, 396–397
    defined, 394
    examining data as a surface, 406–408, 407f
    examining data in 3D scene, 408–409, 409f
    examining data in ArcGIS Pro, 400–406, 405f
    exporting and saving results, 409–410

Index 553

getting started, 397
introduction, 394–395, 395f
LAS files, 397–400, 398f
measurements, 403–406
reclassified points, 403, 405f
related concepts, 410, 411f
restoring measurements to points, 408
scenario, 395–396
study area, 396f
terrain datasets, 410, 411f
lidar data
  in ArcGIS Pro, 400–406
  LAS file, 397
  as surface, 406–407
  in 3D Scene, 408–409
lidar points
  reclassified, 406
  restoring measurements to, 408
Light Gray Canvas basemap, 107
lineage, 169
linear interpolation, 250, 250f, 254, 254f
line-of-sight (LOS), 367
line-on-polygon overlays, 234
Line option digitizing, 157
lines. *See also* vector objects
  defined, 12
  heads-up digitizing of, 155–161
  splitting, 177–178
  vector object, 12, 12f
List by drawing order, 16, 16f, 17
List by editing, 16, 16f
List by labeling, 16, 16f
List by perspective imagery, 16, 16f
List by selection, 16, 16f
List by snapping, 16, 16f
List by source, 16, 16f
local interpolation, 329, 329f
locator packages, 258–259
logical consistency, 185
logical tool, 511–512
longitude, 18, 150

M

Magellan satellite navigation system, 136
Mahoning County. *See also* geocoding; geoprocessing; geospatial data; networks; Ohio counties; raster data
  geodatabase, 12, 13
  GIS data from, 138
  high schools' address data, 264
  libraries' address data, 241–242
  map, 36f, 217f, 241f, 263f, 286f
  road network, 262
MahoningCounty.gdb file, 10
Manual Interval method, 70
many-to-one joins, 44
map (view)
  adding data, 15–17
  ArcGIS Pro, 13, 14f
  closing, 14
  Contents pane in, 14f
  defined, 13
  in project (.aprx), 13–15
Map Algebra
  ArcGIS Pro skills, 463
  "away from highway" criterion, 478–479

Boolean operators, 480–481, 480f, 481f
"close to lake or river" criterion, 479–480
closing time, 484
completing and running model, 480–482
"contains forests or wetlands" criterion, 474–478
creating model, 467–469
data sources, 465–466
defined, 463
examining results, 482–483
getting started, 466–467
introduction, 463–464, 464f
"of a lake or of a stream" criterion, 479–480
related concepts, 484–486, 485f
relational operators, 472
"relatively flat land" criterion, 469–474
saving and printing results, 483–484
scenario, 465
study area, 465, 466f
weighted overlays, 484–486, 485f
map documents, 538. *See also* Web maps
map elements, 73–74, 82
map frame, 74
  inserting into layout, 76–77
  working in layout, 77–78
Maplex Label Engine, 91
map package, 31
map projections, 561–562, 562f
MapQuest, 136
maps. *See also* basemaps; choropleth maps; Web maps
  annotation, 90–91
  from ArcGIS Online, 132–133
  ArcGIS Pro skills, 61
  checklist, 89
  closing time, 90
  color choice, 72–73
  data sources, 63–64
  defined, 61
  design strategies, 88–89
  editing elements, 88–89
  evaluation, 88–89
  exporting, 89–90
  frame in layout, 76–79
  getting started, 64
  labels, 90–91
  large-scale, 78
  layout view, 73–75
  legend, 84–86
  north arrow, 83–84
  placing inset map into layout, 79–81
  printing, 89–90
  publishing, 99–100
  reference, 61–62
  refining elements, 88–89
  related concepts, 90–91
  scale, 78–79
  scenario, 62–63
  setting up, 64–67
  sharing, 99–103
  small-scale, 78
  text, 86–88
  thematic, 62
  topographic, 62, 138, 391, 392

types, 61–62
  of Wellington, New Zealand, 62f
map scale, 78
mass points, 381, 382f
McCarran International Airport, 148, 148f
.mdb (personal geodatabases), 537
measurements. *See also* distance calculations
  spatial analysis, 196–198
  units of, 377–378
Measure tool, 196
  snap to a point, 197
merging layers, 219–221
metadata, 141–142
Michigan Department of Natural Resources, 438
Microsoft Excel tables, 151
middle infrared (MIR), 315
Midpoint Snapping, 179f, 180
minus button, 422, 422f
MIR, 315
mobile map package, 31
ModelBuilder
  advance options, 511–512, 511f
  animal relocation model. *See* models
  defined, 468–469, 469f
  ecological preserve model. *See* Map Algebra
  model elements in, 491–492, 492f
models. *See also* digital elevation models (DEMs); hydrologic modeling
  analyzing results, 508–510
  "at low elevation" criterion, 495–496
  "away from developed areas" criterion, 497–501
  "close to water" criterion, 501–503
  closing time, 510
  combining "in forested area" and "low elevation" criteria, 496–497
  creating, 467–469
  data sources, 489
  defined, 487
  ecological preserve model. *See* Map Algebra
  elements, 491–492, 492f
  getting started, 490–492
  "in forested area" criterion, 494–495
  introduction, 487–488, 488f
  "large section" criterion, 507
  ModelBuilder, 511–512, 511f
  parameters, 492–494
  printing or sharing results, 510
  Python and, 32–33, 32f
  related concepts, 511–512, 511f
  re-opening, 469
  scenario, 488–489
  study area, 489f
  suitability index, 503–506, 505f
  validating and running, 507–508
Modify scene button, 422, 422f
Moran's I, 213
mosaic dataset, 302
mosaics
  defined, 301
  of raster data, 301–302, 301f
moving
  features, 183–184

.mpk (map packages), 538
MRLC, 286, 291, 351, 437
MSS, 319, 320t
multipatches, 417–418
  defined, 418
  used in ArcGIS Pro, 418
multipoint feature classes, 410
Multi-Resolution Land Characteristics Consortium (MRLC), 286, 291, 351, 437
multispectral imagery
  working with, 313–318
Multispectral Scanner (MSS), 319, 320t
multispectral sensor, 314
.mxd (map documents), 5. *See also* Web maps

N

NAD27, 17, 18
NAD83, 17
nadir, 304, 305
NAIP imagery, 305, 311–313, 323f
NAPGD2022, 350
National Agriculture Imagery Program (NAIP) imagery, 305, 311–313, 323f. *See also* triangulated irregular networks (TINs)
National Geographic basemap, 107
National Geospatial Program (NGP), 122–123
National Land Cover Database (NLCD), 286, 291, 292t, 351, 437. *See also* distance calculations; raster data
The National Map, 396
  described, 122, 123f
  downloading from, 37, 122–126
  examining data from, 128–132
  unpacking data from, 126–127
  Website, 4, 122
National Weather Service, 334
  FIPS codes, 4, 37, 63, 96, 242
National Weather Service Advanced Hydrologic Prediction Service, 331
NATRF2022, 18
Natural Breaks, 67, 68f
NAVD88, 350
Navstar GPS, 166
near infrared (NIR), 315
network analysis. *See also* routes; stops
  ArcGIS Pro skills, 261
  closest facilities, 279–282, 279f
  closing time, 282
  creating network datasets, 265–267
  data sources, 263–264
  getting started, 265–266
  introduction, 261–262
  printing or sharing results, 282
  real-world applications, 262–263
  related concepts, 282–283, 283f
  saving results, 282
  scenario, 262–263
  service areas, 277–278, 277f
  study area, 263
Network Analyst, 265. *See also* network analysis
network datasets, 265–267

networks. *See also* network analysis
  defined, 261
  TINs. *See* triangulated irregular networks (TINs)
  utility, 282–283
NGP, 122–123
NIR, 315
NLCD, 286, 291, 292t, 351–352, 437. *See also* distance calculations; raster data
NOAA
  FIPS codes, 37
NoData, 290, 290f
node snapping, 178–181, 179f
nominal data, 41, 41f
non-spatial data. *See also* attributes; attribute tables; records
  attribute accuracy, 182
  defined, 34
  fields, 34, 39–42
normalized data, 65–66, 485–486, 485f
North America Albers Equal Area Conic (NAD 83), 332, 335
North American Datum of 1927, 17
North American Datum of 1983, 17
north arrow, 83–84
NOT (Map Algebra), 476–477, 477f
NOT operator, 51

O

oblique imagery, 304
Oceans basemap, 107
off-nadir viewing, 304, 305
offsets, 426–428
Ohio counties, 36f, 96f, 194f, 217f, 330f. *See also* attribute tables; geoprocessing; maps; spatial analysis; spatial interpolation; Web maps; *specific counties*
Ohio Department of Natural Resources, 437
OLI, 319, 320t
one-to-one joins, 43–44
online GIS data
  from ArcGIS Online, 132–134
  closing time, 140–141
  data sources, 120–121
  getting started, 121–122
  introduction, 119
  other sources, 138–140
  real-world applications, 120
  related concepts, 141–142
  scenario, 120
  study area, 120, 121f
  VGI, 136–137
OpenStreetMap, 136, 137f
OpenStreetMap basemap, 107
Operational Land Imager (OLI), 319, 320t
Operations Dashboard, 110, 110f
OR (Map Algebra), 476, 476f
ordinal data, 41–42, 41f
Ordinary Kriging, 347–348
organization, ArcGIS Online, 94
orientation, 75
OR operator, 51
orthoimagery and orthoimages (orthophotos)
  data source, 305

  defined, 306–307
  draping on TINs, 385–387
  in grayscale, 307–310
  Landsat, 318–325
  multispectral, 313–318, 314f, 316f, 318f
  NAIP, 311–313
  working with, 306–310
orthorectification, 307
overlays. *See also* Map Algebra
  defined, 215, 216f
  point-in-polygon, 233–235, 233f
  polygons, 228–233, 229f, 230f
overshoot, 178, 178f

P

package
  defined, 29
  geoprocessing, 30
  layer, 31
  map, 31
  project, 29–30
  saving work as, ArcGIS Pro, 29–31
Package Project pane, 30, 30f
page setup, 75
panchromatic sensor, 314
pane
  closing, 9
  defined, 8
pan-sharpening, 324
parameters, 492–494, 517
parsing, 249, 250t
Pasda, 138
path distance, 462
pattern, visual image interpretation, 309
patterns analysis, 213
PCS. *See* projected coordinate system (PCS)
PDF files
  GeoPDF, 392
  saving as, 31
Pennsylvania Spatial Data Access (Pasda), 138
PerSeas attributes, 64, 65f, 66–67
personal geodatabases, 537
perspective view, 354
Planar measurement, 196
plus button, 422, 422f
Point at End of Line option digitizing, 159
point cloud, 395. *See also* lidar
point digitizing, 157, 159
point feature class, from x/y coordinates, 150–152
point-in-polygon overlays, 233–235, 233f
Point option digitizing, 159
points. *See also* vector objects
  defined, 12
  geocoded, spatial analysis of, 256–258
  from geocoding process, 240
  vector object, 12, 12f
Points Along a Line option digitizing, 159
Point Snapping, 179, 179f
polygon overlays, 228–233, 229f
polygons. *See also* vector objects
  cutting, 177–178
  defined, 12

digitizing, 158–159, 158f
dissolving boundaries, 223–224
extrusions, 416–417
heads-up digitizing of, 161–163
Thiessen, 382, 382f
vector object, 12, 12f
polylines, 12
pop-ups
   attribute, 56–59
   Web map, 103
Portage County, 36f. *See also* Ohio counties
portal, 132–133
Portrait orientation, 75
positional accuracy, 176–177
pour points, 536, 536f
precipitation, 328–329, 329f, 331, 345. *See also* spatial interpolation
precision, 168, 169f
presenting
   results of 3D visuaization analysis, 371–372, 372f
preserve model, ecological. *See* maps
principal point, 306
printing
   editing data, 187
   geocoding, 258–259
   geoprocessing results, 235
   geospatial data creation, 164–165
   hydrologic model, 535
   Map Algebra results, 483–484
   models, 510
   network analysis results, 282
   raster data, 300
   remotely sensed imagery, 325
   spatial analysis, 212
   spatial interpolation results, 346
profile graphs, 364–367
project (.aprx)
   in ArcGIS Pro, 6–7
   defined, 6
   map (view) in, 13–15
projected coordinate system (PCS), 18–19
   Calculate Geometry and, 167
projections
   in ArcGIS, 17–19
   changing layers to different, 20–23
   defined, 20
   defining in ArcGIS Pro, 23–25
project package, 29–30
Project Pane. *See* Package Project pane
pseudo 3D, 417
pseudo-3D visualization, 413
   defined, 354
publishing
   defined, 99
   Web apps, 112–114
   web layers, 102
   Web maps, 99–100
pyramids, 307
Pythagorean theorem, 440, 440f
Python
   in ArcGIS Pro, 32–33
   defined, 32
   models and, 32–33, 32f

Q

Quantiles method, 68–69, 70f
queries
   analysis, using, 55–56
   compound, 50–53
   defined, 47
   simple, 47–50
Quick Access Toolbar
   ArcGIS Pro, 9f
   defined, 9

R

Radial Lines option digitizing, 158
radiometric resolution, 315
Raster Calculator. *See also* Map Algebra; models
   Boolean operators, 475–477
   for cost distance, 458–459
   defined, 473
   relational operators, 472
raster data, 13. *See also specific topics*
   ArcGIS Pro skills, 284
   attribute tables, 293, 293t
   basics, 289
   closing time, 300
   converting vector data to, 296–299, 296f, 297f, 299f
   data sources, 286
   environment settings and, 287–288
   georeferencing, 326
   getting started, 286–288
   introduction, 284–285, 285f
   mosaics of, 301–302, 301f
   printing or sharing, 300
   real-world applications, 285
   regions of, 294–295, 294f
   related concepts, 300–302
   resolution and, 299, 299f
   scenario, 285
   study area, 285, 286f
   subsets of, 300–301, 301f
   vector data *vs.*, 296–297, 297f
   zones of, 290–293, 290f
raster data model, 284
raster surfaces, shortest paths over, 443–451
ratio data, 41f, 42
real-world applications
   animal relocation model. *See* models
   ArcGIS Pro presentation, 3
   attribute tables, 36
   DEMs, 350–351
   distance calculations, 436–437
   ecological preserve model. *See* Map Algebra
   editing data, 169
   geocoding, 240–241
   geoprocessing, 216
   geospatial data creation, 144
   geospatial data in 3D, 413–414
   hydrologic modeling, 514–515
   lidar, 395–396
   Map Algebra, 465
   maps of, 62–63
   models, 488–489
   network analysis, 262–263
   online GIS data, 120
   raster data, 285
   remotely sensed imagery, 305
   spatial analysis, 193–194
   spatial interpolation, 329–330
   TINs, 376–377
   Web apps, 95
   Web maps, 95
Reclassify tool, 456–458
records. *See also* attribute tables
   defined, 34
   examining, 39–42
   selecting, 46–47. *See also* queries
red, 315
Redo icon, 9f
reference data
   accuracy of geocoded results and, 254
   defined, 239
   streets layer used as, 243–244
reference maps, 61–62
regions of raster data, 294–295, 294f
regularized Spline method, 342
relate operation, 44
relational operators, 48, 472, 472f
relief displacement, 306
relocation of animals model. *See* models
rematching addresses, 250–253
remotely sensed imagery. *See also* multispectral imagery; satellite imagery
   aerial photography, 303
   ArcGIS Pro skills, 303
   closing time, 325–326
   color infrared imagery, 313–318
   data sources, 305
   georeferencing, 326
   getting started, 306
   introduction, 303–305
   Landsat, 318–324
   multispectral imagery, 313–318
   NAIP imagery, 311–313
   orthoimagery and orthoimages, 306–310
   printing or sharing, 325
   process, 314f
   real-world applications, 305
   related concepts, 326
   scenario, 305
   study area, 305
   visual image interpretation and, 309–310
   working with Landsat photography, 318–324
remote sensing
   defined, 303
   process, 313–317, 314f, 316f
re-opening models, 469
representative fraction (RF), 78
reshaping features, 173–176
resolution
   DEM, 350, 350f
   high-resolution imagery, 135
   radiometric, 315
   raster data and, 299, 299f
   spatial, 312, 312f
   TIN image, 376
returns, 403, 404f. *See also* lidar

556 Index

RF, 78
ribbon
    ArcGIS Pro, 9f
    basic layout of, 9f
    defined, 9
Right-Angle Line option digitizing, 157
Right-Angle Polygon option digitizing, 158
RMSE, 326
roads layer, for geocoding, 243–244
Rock & Roll Hall of Fame, 307–308, 308f
root mean square error (RMSE), 326
Rotate button, 422, 422f
rotating features, 183–184
routes
    calculating shortest, 269–271, 269f, 270f, 271f
    creating, with barriers on network, 275–276
    creating by rearranging stops, 272–275
    creating by visiting stops in order, 267–272

S

SaaS (Software as a Service) application, 93
sampling, 334
satellite imagery
    defined, 303
    high-resolution sources, 324, 324t
    Landsat, 300
satellites
    high-resolution capabilities, 324–325, 324t, 325t
saving
    animation, 390
    layout, 461–462, 535
    network analysis results, 282
    3D view, 373
    work as package in ArcGIS Pro, 29–31
scale
    aerial photos, 307
    digitizing and, 148, 148f
    maps, 78–79
    representation on layouts, 78–79
scale bars, 81–83
scenarios. *See* real-world applications
scene
    in ArcGIS Pro, 354–356
    defined, 352
    examining lidar data in 3D, 408–408, 409f
    visualization of 3D layers in a, 363–364
schools. *See* geoprocessing
script. *See also* Python
    and ArcGIS Pro, 32–33
    defined, 32
scripting. *See* Python
SDTS, 140
seasonal attributes, 64, 65f, 66–67
seasonal home data. *See* maps; Web maps
segments. *See also* vertices
    defined, 157, 157f, 239
Select By Attributes, 205
Select By Location
    main discussion, 206–208, 206f, 207f
    spatial analysis using, 205, 208–210

selecting
    by attributes, 205
    data layers, 221–223
    defined, 46
    records, 46–47
Selection Tools for attribute tables, 40, 40f
Select Layer By Location
    buffers *vs.*, 226–227
semivariograms, 347, 347f
service areas, 277–278, 277f
services, 104
sewage sludge, 1
shadow, visual image interpretation, 309–310
shape, visual image interpretation, 310
shapefiles, 12, 150
sharing
    Address Locators, 258
    distance calculations, 461–462
    editing data, 187
    geocoding, 258–259
    geoprocessing results, 235
    geospatial data creation, 164–165
    hydrologic model, 535
    models, 510
    network analysis results, 282
    raster data, 300
    remotely sensed imagery, 325
    spatial analysis, 212
    spatial interpolation results, 346
    3D scene, 410
    3D visualizations, 433
    Web apps, 115
    Web maps, 99–103
shortest paths over raster surfaces, 443–451
shortest routes
    affect of reordering stops on, 273–274, 273f
    ArcMap Pro calculation, 269–271
shortest straight-line distance (Euclidean), 439–443, 440f, 441f
shortwave infrared (SWIR), 315
Show All Records, 40, 40f
Show Selected Records, 40, 40f
.shp (shapefiles), 150
Shreve method, 529, 530f
.shx (shapefiles), 150
simple queries, 47–50
    in ArcGIS Pro, 47–50
    defined, 48
sinks, 520–524, 521f, 522f
site and association, visual image interpretation, 309
site suitability models
    animal relocation model. *See* models
    ecological preserve model. *See* Map Algebra
size, visual image interpretation, 309
sketching, 157. *See also* digitizing
SketchUp, 434, 434f
slice, 453–454, 454f
slope and slope layers
    DEMs, 360–363, 360f
    streams and, 517–518
    TINs, 383–385

Slopehills, 453, 454
small-scale maps, 78
Snapping, 364
snapping tolerance, 178–179
Snapping toolbar, 197, 197f
snapshots, 387
soft breaklines, 382
Software as a Service (SaaS) application, 93
sorting, 46
source, 205. *See also* Select By Location
source layer, 191, 191t–193t
spatial analysis
    ArcGIS Pro skills, 190
    basic measurements, 196–198
    closing time, 212–213
    data sources, 194–195
    of geocoded points, 256–258
    getting started, 195
    introduction, 190–193
    printing or sharing, 212
    real-world application, 193–194
    related concepts, 213–214, 214f
    scenario, 193–194
    Select By Location, 205, 208–210
    spatial joins, 200–205, 201f, 202f, 203f, 210–212
    spatial queries, 206, 210–212
    study area, 194, 194f
    for summarizing features, 198–200
    types of relationship rules, 191t–193t
Spatial Analyst, 287. *See also* spatial analysis
spatial autocorrelation, 213, 213f
spatial data, 2. *See also* geospatial data
Spatial Data Transfer Standard (SDTS), 140
spatial interpolation
    ArcGIS Pro skills, 328
    closing time, 346
    data sources, 331
    defined, 328
    evaluating interpolated surfaces, 344–345
    examining data points, 332–334
    geostatistical methods, 346–348, 347f, 348f
    getting started, 331–332
    with IDW, 334–339
    introduction, 328–329
    printing or sharing, 346
    real-world applications, 329–330
    related concepts, 346–348, 347f, 348f
    scenario, 329–330
    with Spline, 341–344
    study area, 330
    Zonal tools, 340–341, 340f
spatial joins
    in ArcGIS Pro, 200–205, 201f, 202f, 203f
    for spatial analysis, 210–212
spatial queries
    defined, 206
    spatial analysis using, 210–212
spatial resolution, 312, 312f
spatial statistics, 213

SPCS, 18
SPCS2022, 18
splines, 341–344, 342f
Split Line option digitizing, 157
splitting lines, 177–178
SQL, 48
Standard Deviation method, 69–71, 71f
Stark County, 36f. *See also* Ohio counties
State Plane Coordinate System (SPCS), 18
State Plane Coordinate System of 2022 (SPCS2022), 18
steps, 237, 237f
stops
    defined, 268
    rearranging, to create route, 272–275
    visiting in order, to create route, 267–272
story map, 116–117, 117f
Strahler method, 530, 530f
straight-line distance (Euclidean), 439–443, 440f, 441f
stream order, 529
streams
    networks, 526–529, 527f
    ordering, 529–531, 530f
    slopes and, 517–518
street centerline files, 244
Streets basemaps, 107
streets layer, for geocoding, 243–244
Structured Query Language (SQL), 48
structures layer, from National Map, 37
study areas. *See under specific topics*
sub-basins, 514
subsets
    defined, 300
    of raster data, 300–301, 301f
suitability index, 503–506, 505f
Summarize tools, 198–199
Summarize Within tool, 198–199
surface analysis of TINs, 383–385
surfaces, 328. *See also* spatial interpolation
    examining lidar data as, 406–408, 407f
swipe tool, 345
SWIR, 315
.sxd (scene documents), 381. *See also* Web Scenes
symbology
    of GIS layers, 25–28
    3D, in ArcGIS Pro, 355
Symbology pane, 25–28
Symmetrical Difference overlay, 230f, 231

T

tables. *See* attribute tables
Table tabs, 40, 40f
Tangent Snapping, 179f, 180
target, 205. *See also* Select By Location
target layer, 191, 191t–193t
Task Designer, 237
task group, 236, 236f
task item, 236, 236f
tasks, 236–237
temperature data types, 42
templates
    in ArcGIS Pro, 6–7
    on Create Features box, 156
    defined, 6

temporal accuracy, 172–173
tension Spline method, 342
TerraGo, 392
terrain datasets, 410, 411f
Terrain with Labels basemap, 107
text
    dynamic, 74
    maps, 86–88
texture, visual image interpretation, 310
Thematic Mapper (TM), 319, 320t
thematic maps, 62
thermal infrared (TIR), 315
Thermal Infrared Sensor (TIRS), 319, 320t
thermal sensors, 313
Thiessen polygons, 382, 382f
3D, 354. *See also* geospatial data in 3D
    examining lidar data in, 408–409, 409f
    sharing scene, 410
    viewing TINs in, 385–387
    visibility analysis, 367–371, 373–374, 374f
    visualization of layers in a scene, 363–364
    visualization operation, 354–356, 356f
    visualization results analysis, 371–372, 372f
3D Analyst, 352. *See also* digital elevation models (DEMs)
3D Elevation Program (3DEP), 357–358, 358f
3D objects, 429–433
3D visibility analysis
    interactive, 373–374, 374f
3D visualization
    COLLADA format, 424–426
    of layers in a scene, 363–364
    operating with, 354–356, 356f
    results analysis, 371–372, 372f
three-dimensional (3D), 354
    examining lidar data in, 408–409, 409f
    sharing scene, 410
    viewing TINs in, 385–387
    visibility analysis, 367–371, 373–374, 374f
    visualization of layers in a scene, 363–364
    visualization operation, 354–356, 356f
    visualization results analysis, 371–372, 372f
threshold value, creating stream networks with, 526–529, 527f
TIGER/Line, 139–140
    described, 37, 60, 244
    Ohio counties dataset, 37, 63, 96
    streets data, 244
    water bodies' data, 4
tile layer
    ArcGIS Online, 100
    defined, 100
tiles, 100
TIR, 315
TIRS, 319, 320t
TM, 319, 320t
tone, visual image interpretation, 310

toolboxes
    ArcGIS, 21
    ArcGIS Pro, 9
Tools (ModelBuilder), 492, 492f
Topographic basemaps, 107
topographic maps, 138, 391, 392
topology, 188–189, 189f
Trace option digitizing, 158–159
traces, 283
transformation, 326
transparency levels, TINs, 383
Transverse Mercator, 18
Traveling Salesman Problem (TSP), 273
trends, 329. *See also* spatial interpolation
triangulated irregular networks (TINs)
    ArcGIS Pro skills, 375
    closing time, 391
    creating, 380–383
    data sources, 377–378
    defined, 375, 376f
    examining contour lines, 379–380
    getting started, 378
    introduction, 375–376
    related concepts, 391–392, 392f
    saving and sharing results, 390
    scenario, 376–377
    study area, 377f
    surface analysis, 383–385
    terrain datasets, 410, 411f
    3D view, 385–387
true color composite, 316, 316f
Trumbull County, 36f, 217f. *See also* geoprocessing; Ohio counties
TSP, 273
turns, 266
2-Point Line option digitizing, 158
two-and-a-half-dimensional (2.5D), 354, 413, 417
typographic maps, 62

U

UAS, 303
undershoot, 178, 178f
unemployment rates, 65
Union overlay, 230f, 231
United States Department of Agriculture, 311
United States Geological Survey (USGS), 396
    EarthExplorer, 138, 139f, 140, 320
    National Map basemap, 107
    topographic maps, 391
    US Topos, 391, 392f
units of measurement, in ArcGIS Pro, 377–378
Universal Kriging, 348
Universal Transverse Mercator (UTM), 18, 24
unmanned aircraft systems (UAS), 303
unpacking National Map data, 126–127
Update overlay, 230f, 231
US Addresses - Zip 5-Digit, 245
US Address-Street Name style, 245
USA Topo Maps basemap, 107, 391, 392f
USDA, 311
USGS National Map basemap, 107

U.S. Interagency Elevation Inventory, 396
US Topos, 391–392, 392f
utility networks, 282–283
UTM (Universal Transverse Mercator), 18, 24

V

Van Rossum, Guido, 32
variables, 487–488
Variables (ModelBuilder), 491, 492f, 511–512
vector data
    converting to raster data, 296–298
    defined, 11
    raster data *vs.*, 296–297, 297f
    types of, 12–13
vector data model
    defined, 12
    file geodatabase, 12
    shapefile, 12
vector objects, 12, 12f. *See also* lines; points; polygons
    spatial joins, 200–205, 201f, 202f, 203f
vector tile layer
    ArcGIS Online, 100
    defined, 100
Vertex Snapping, 179, 179f, 181f
vertical datums, 350
vertical exaggeration, 386–387, 386f
vertical images, 304
vertices
    in ArcGIS Pro, 152
    defined, 157
    editing location of, 160–161
    reshaping features, 173–176
    snapping, 178–181, 179f
    topology rules and, 189
VGI, 136–137
videos. *See* animations
view
    ArcGIS Pro, 8f
    ArcGIS Pro layout, 82
    catalog, 11
    closing, 9
    defined, 8
    discrete object, 12
view domes, 373
viewing
    off-nadir, 304, 305
    TIN and imagery in 3D, 385–386, 385–387
    Web apps, 115
viewsheds
    creating interactive, 374f
    defined, 367
visibility analysis, 367–371
    interactive, 373–374, 374f
visible light, 314

visual hierarchy, 88-89
visual image interpretation
    defined, 309
    pattern, 309
    remotely sensed imagery and, 309–310
    shadow, 309–310
    shape, 310
    site and association, 309
    size, 309
    texture, 310
    tone, 310
volunteered geographic information (VGI), 136–137
Voronoi regions, 382

W

watersheds
    defined, 513, 514f
    delineation, 531–532, 536
    pour points, 536, 536f
Weather Underground, 331, 345
Web AppBuilder, 110
Web apps
    built in ArcGIS Online, 109–110
    closing time, 116
    configuring for online publishing, 112–114
    created using *Operations Dashboard*, 110f
    creating, 108–112
    data sources, 96–97
    defined, 109
    getting started, 97
    introduction, 92–93
    related concepts, 116–117
    scenario, 95
    sharing, 115
    study area, 95–96
    viewing, 115
WebGL, 419
web layer, 99
Web maps. *See also* ArcGIS Online
    ArcGIS Online, 103–106
    ArcGIS Pro skills, 92
    closing time, 116
    configuring from published layers, 103–108
    data sources, 96–97
    defined, 103–104
    getting started, 97
    introduction, 92–93
    publishing, 99–100
    related concepts, 116–117
    scenario, 95
    sharing, 99–103
    study area, 95–96
Web Mercator auxiliary sphere, 99
Web Scenes
    in ArcGIS Online, 421–423, 422f

    in ArcGIS Pro, 412, 419
    creating, 418–421
    defined, 419
    publishing, 421
Websites
    ArcGIS Online, 421
    Census Bureau, 4, 37, 59, 60, 64, 97, 140
    Landfire, 138, 139f
    Mahoning County Public Library, 241–242, 263
    MRLC, 291, 351, 437
    National Map, 37, 122
    TerraGo, 392
    USGS EarthExplorer, 138, 139f, 396
    Wunderground (Weather Underground), 345
weighted overlays, 484–486, 485f
Wellington, New Zealand map of, 62f
WGS84, 18
WGS84 Web Mercator, 19
Wikimapia, 136
workflows, 236, 464
workspaces, 218
World Geocoding Service, 259–260, 259f
World Geodetic System, 18
World Imagery Clarity, 147
Wunderground, 331, 345

X

XOR (Map Algebra), 476, 476f
x/y coordinates, conversion to point feature class, 150–152

Y

Youngstown, Ohio, 120, 121f, 136f, 377f, 396f. *See also* lidar
Youngstown State University, 145f, 146f, 170f, 187f, 414f. *See also* editing data; geospatial data creation; geospatial data in 3D

Z

Zip 5-Digit style, 245
Zonal Statistics, 340, 340f
Zonal tools, 340–341, 340f
zones
    buffers, 226–228, 227f
    of raster data, 290–293, 290f
zoom
    icons on ArcScene Tools toolbar, 422
    icons on Explore toolbar, 38
    icons on Layout toolbar, 73, 73f
z-values. *See also* digital elevation models (DEMs); triangulated irregular networks (TINs)
    defined, 349
    vertical exaggeration, 386